Information-Theoretic Radar Signal Processing

IEEE Press
445 Hoes Lane
Piscataway, NJ 08854

Information-Theoretic Radar Signal Processing

Edited by

Yujie Gu
Aptiv, USA

Yimin D. Zhang
Temple University, USA

IEEE Press
WILEY

Published by John Wiley & Sons, Inc., Hoboken, New Jersey.
Published simultaneously in Canada.

Library of Congress Cataloging-in-Publication Data Applied for:

Hardback ISBN: 9781394216925

Cover Design: Wiley
Cover Image: © ThinkNeo/Getty Images

Set in 9.5/12.5pt STIXTwoText by Straive, Chennai, India

To my devoted family for their endless love, unwavering support, and constant encouragement

—Yujie Gu

To my loving family whose unwavering support and boundless love have illuminated my entire academic career

—Yimin D. Zhang

Contents

About the Editors

Yujie Gu received his PhD degree from Zhejiang University, Hangzhou, China, in 2008. Currently, he works at Aptiv Advanced Engineering Center as a Senior Radar Scientist focusing on research and development of advanced automotive radar systems. His research experience was in statistical and array signal processing, including robust adaptive beamforming, information-theoretic compressive sensing, spatial spectrum estimation, and performance bound analysis, with applications to radar, wireless communications, and radio astronomy imaging. Dr. Gu is a senior member of the Institute of Electrical and Electronics Engineers (IEEE). He is an elected member of the IEEE Sensor Array and Multichannel (SAM) Technical Committee and the IEEE Signal Processing Theory and Methods (SPTM) Technical Committee. Dr. Gu is an Associate Editor for *IEEE Transactions on Signal Processing*, a Subject Editor-in-Chief for IET *Electronics Letters*, and an Editor for Elsevier *Signal Processing*. He was a Special Sessions Co-Chair of the 2020 IEEE Sensor Array and Multichannel Signal Processing Workshop. Dr. Gu was the recipient of the 2019 IET Communications Premium Award. He is the corresponding author of a paper that received the 2021 IEEE Signal Processing Society Young Author Best Paper Award.

Yimin D. Zhang received his PhD degree from the University of Tsukuba, Tsukuba, Japan, in 1988. He is currently an Associate Professor with the Department of Electrical and Computer Engineering, Temple University, Philadelphia, PA, USA. His research interests include array signal processing, compressive sensing, machine learning, information theory, convex optimization, and time-frequency analysis with applications to radar, wireless communications, and satellite navigation. Dr. Zhang is a Senior Area Editor for *IEEE Transactions on Signal Processing* and an Editor for *Signal Processing*. He was an Associate Editor for *IEEE Transactions on Signal Processing*, *IEEE Transactions on Aerospace and Electronic Systems*, *IEEE Signal Processing Letters*, and *Journal of the Franklin*

Institute. He was a Technical Co-Chair of the 2018 IEEE Sensor Array and Multichannel Signal Processing Workshop. He was the recipient of the 2016 IET Radar, Sonar & Navigation Premium Award, the 2017 IEEE Aerospace and Electronic Systems Society Harry Rowe Mimno Award, the 2019 IET Communications Premium Award, and the 2021 EURASIP Best Paper Award for Signal Processing. He coauthored two papers that, respectively, received the 2018 and 2021 IEEE Signal Processing Society Young Author Best Paper Awards. Dr. Zhang is an IEEE Signal Processing Society Distinguished Lecturer. He is a Fellow of the Institute of Electric and Electronics Engineers (IEEE) and of International Society for Optical Engineering (SPIE).

List of Contributors

Pia Addabbo
Università degli Studi Giustino Fortunato
Benevento
Italy

Kristine Bell
Metron, Inc.
Reston, VA
USA

Daniel W. Bliss
Center for Wireless Information Systems and Computational Architectures (WISCA)
Arizona State University
Tempe, AZ
USA

Nianxia Cao
Advanced Safety and User Experience
Aptiv
Agoura Hills, CA
USA

Badong Chen
Institute of Artificial Intelligence and Robotics
Xi'an Jiaotong University
Xi'an
China

Lujuan Dang
Institute of Artificial Intelligence and Robotics
Xi'an Jiaotong University
Xi'an
China

Fuwang Dong
Department of Electronic and Electrical Engineering
Southern University of Science and Technology
Shenzhen
China

Alejandro C. Frery
School of Mathematics and Statistics
Victoria University of Wellington
Wellington
New Zealand

Gaetano Giunta
Department of Industrial, Electronic and Mechanical Engineering
University of Roma Tre
Rome
Italy

Hana Godrich
Department of Electrical and Computer Engineering
Rutgers University
Piscataway, NJ
USA

Nathan A. Goodman
Advanced Radar Research Center
School of Electrical and Computer Engineering
University of Oklahoma
Norman, OK
USA

Yujie Gu
Advanced Safety & User Experience
Aptiv
Agoura Hills, CA
USA

Sora Haley
College of Engineering and Computer Science
Syracuse University
Syracuse, NY
USA

Lei Huang
State Key Laboratory of Radio Frequency Heterogeneous Integration
Shenzhen University
Shenzhen
China

Zacharie Idriss
Radar Division
U.S. Naval Research Laboratory
Washington, DC
USA

Chris Kreucher
KBR Wyle Services
LLC
Ann Arbor, MI
USA

Bin Liao
Guangdong Key Laboratory of Intelligent Information Processing
College of Electronics and Information Engineering
Shenzhen University
Shenzhen
China

Fan Liu
Department of Electronic and Electrical Engineering
Southern University of Science and Technology
Shenzhen
China

Engin Masazade
Department of Electrical and Electronics Engineering
Faculty of Engineering
Marmara University
Istanbul
Türkiye

Ram M. Narayanan
School of Electrical and Computer Engineering
The Pennsylvania State University
University Park, PA
USA

Abraão D. C. Nascimento
Departamento de Estatística
Universidade Federal de Pernambuco
Recife
Brazil

Arye Nehorai
Department of Electrical and Systems Engineering
Washington University in St. Louis
St. Louis, MO
USA

Danilo Orlando
Department of Information Engineering
University of Pisa
Pisa
Italy

Bryan Paul
General Dynamics Mission Systems
Fairfax, VA
USA

Siyuan Peng
School of Information Engineering
Guangdong University of Technology
Guangzhou
China

Athina P. Petropulu
Department of Electrical and Computer Engineering
Rutgers University
Piscataway, NJ
USA

H. Vincent Poor
Department of Electrical and Computer Engineering
Princeton University
Princeton, NJ
USA

Jose C. Principe
Department of Electrical and Computer Engineering
University of Florida
Gainesville, FL
USA

Raghu G. Raj
Radar Division
U.S. Naval Research Laboratory
Washington, DC
USA

Muralidhar Rangaswamy
Air Force Research Laboratory
Dayton, OH
USA

Zhiguo Shi
College of Information Science and Electronic Engineering
Zhejiang University
Hangzhou, Zhejiang
China

Hing Cheung So
Department of Electrical Engineering
City University of Hong Kong
Hong Kong
China

Petre Stoica
Department of Information
Technology
Uppsala University
Uppsala
Sweden

Bo Tang
College of Electronic Engineering
National University of Defense
Technology
Hefei
China

Jun Tang
Department of Electronic Engineering
Tsinghua University
Beijing
China

Pramod K. Varshney
College of Engineering and Computer
Science
Syracuse University
Syracuse, NY
USA

Peng Xiao
School of Marine Engineering and
Technology
Sun Yat-sen University
Zhuhai
China

Yifeng Xiong
School of Information and
Communication Engineering
Beijing University of Posts and
Telecommunications
Beijing
China

Qianhui You
Guangdong Key Laboratory of
Intelligent Information Processing
College of Electronics and Information
Engineering
Shenzhen University
Shenzhen
China

Yimin D. Zhang
Department of Electrical and
Computer Engineering
Temple University
Philadelphia, PA
USA

Zongyu Zhang
College of Information Science and
Electronic Engineering
Zhejiang University
Hangzhou, Zhejiang
China

Preface

The roots of information theory can be traced back 100 years to the early works of Fisher [1], Nyquist [2], Hartley [3], and others. In [1], Fisher laid down a cornerstone by defining statistical information as the reciprocal of the variance of a statistical sample. This fundamental concept was later independently derived by Cramér and Rao, who utilized it to establish the lower bound on parameter estimation variance [4, 5]. Therefore, information theory has been naturally closely related to signal processing from the beginning.

After World War II, both disciplines of information theory and signal processing witnessed remarkable advancements. In 1948, Shannon published his groundbreaking article, "A mathematical theory of communication" [6], which served as a lodestar for the burgeoning discipline of information theory. In the same year, the Institute of Radio Engineers (IRE), the predecessor of the Institute of Electrical and Electronics Engineers (IEEE), inaugurated its first society, the Signal Processing Society. This historical coincidence further symbolized the close relationship between information theory and signal processing.

Information theory originally emerged to explore the communication of messages. Its core principles include quantification of information, source coding, and channel coding. Shannon's contributions fostered a vibrant community of scholars, leading to the establishment of the IEEE Information Theory Society in 1951. His legacy continues to influence the fields of communications, information theory, and signal processing.

The earlier works of Woodward and Davies in the early 1950s [7, 8] marked the beginning of information-theoretic signal processing for radar applications. However, unlike communication systems, which aim to extract information from received signals in a cooperative framework, radar systems operate in noncooperative environments, seeking information about unknown targets, such as their range, velocity, and angle. This inherent disparity in understanding of information has resulted in a notable lag in the development of information-theoretic processing for radar systems compared with the communication counterparts.

Information theory experienced a resurgence within the radar signal processing community following a period of stagnation spanning over three decades. A key turning point was the publication of Bell's PhD dissertation in 1988 [9], wherein mutual information was used for the first time as an optimization criterion to design radar waveforms that maximize the target information extracted from the received measurements. Since then, information theory has witnessed a renaissance in radar signal processing, especially in the era of multiple-input multiple-output (MIMO) radar. In 2005, Guo et al. [10] provided further insights into the theoretical connection between mutual information in information theory and minimum mean square error (MMSE) in estimation theory. Nowadays, information-theoretic criteria, including Fisher information, Shannon entropy, mutual information, and Kullback–Leibler divergence (also known as relative entropy), have become the foundation of adaptive radar systems and found great success in solving various radar signal processing problems. This book provides a comprehensive introduction to information-theoretic criteria, methods, and applications in radar signal processing with a focus on the latest theoretical and practical advances in selected important topics.

This book is intended not only for radar signal processing researchers, but also for postgraduate and PhD students, researchers, and engineers working on a broad spectrum of signal processing and its applications, including but not limited to radar systems. The readers are assumed to have some background in linear algebra, probability and statistics, compressive sensing, detection and estimation, control and optimization, information theory, as well as radar systems.

The book contains fourteen self-contained chapters, each focusing on a specific topic. These chapters, contributed by leading experts from academia, industry, and government, can be broadly categorized into six areas: target detection (Chapters 1–3), parameter estimation (Chapters 4 and 5), radar imaging (Chapters 6 and 7), target tracking (Chapters 8–10), resource management (Chapters 11 and 12), and performance bound analysis (Chapters 13 and 14).

- *Target detection* is the primary mission of many radar systems. In Chapter 1, the authors (Bo Tang, Jun Tang, and Petre Stoica) employ the relative entropy metric to design MIMO radar waveforms under various practical constraints in order to enhance the detection performance of radar systems. In Chapter 2, the authors (Pia Addabbo, Danilo Orlando, and Gaetano Giunta) utilize the Kullback–Leibler divergence criterion to develop adaptive detection architectures for solving multiple hypothesis testing problems in radar applications. In Chapter 3, the authors (Lei Huang and Hing Cheung So) develop a linear shrinkage minimum description length (MDL) criterion and two shrinkage coefficient-based detectors for source enumeration in a computationally efficient way.
- *Parameter estimation* is one of the fundamental functions of radar systems. In Chapter 4, the authors (Yujie Gu, Nathan A. Goodman, and Yimin D. Zhang)

adopt the maximum mutual information criterion to optimize the compressive sensing kernel over the Stiefel manifold to achieve enhanced time delay estimation performance. In Chapter 5, the authors (Bin Liao, Qianhui You, and Peng Xiao) exploit the normalized $\ell 1$ Shannon entropy function to develop an entropy-enhanced one-bit compressive sensing algorithm for direction-of-arrival (DOA) estimation.

- *Radar imaging* becomes increasingly important in modern high-resolution radar systems, where man-made radar targets often extend over multiple range bins when illuminated by wideband radar signals. In Chapter 6, the authors (Zacharie Idriss, Raghu G. Raj, and Ram M. Narayanan) optimize radar waveforms for multistatic radar imaging by maximizing the mutual information between the received signal and the scene of interest, where natural scenes are sparsely represented by wavelet coefficients. In Chapter 7, the authors (Alejandro C. Frery and Abraão D. C. Nascimento) present information-theoretic results in conjunction with statistical inference for the processing and comprehension of synthetic aperture radar (SAR) and polarimetric SAR (PolSAR) imagery.
- *Target tracking* is another important task in radar applications to locate moving targets over time. In Chapter 8, the authors (Nianxia Cao, Pramod K. Varshney, Engin Masazade, and Sora Haley) propose an information-theoretic sensor selection algorithm for target tracking in sensor networks that strikes a balance between the computational complexity and the tracking performance. In Chapter 9, the authors (Siyuan Peng, Lujuan Dang, Badong Chen, and Jose C. Principe) introduce the minimum error entropy (MEE) criterion for the derivation of adaptive filters that exhibit robustness against non-Gaussian noise. In Chapter 10, the authors (Bryan Paul and Daniel W. Bliss) utilize the estimation information rate as a metric to dynamically manage radar systems, resulting in a target scheduling algorithm and a resource allocation scheme for the generalized radar tracking Bayesian filter.
- *Resource management* helps to improve cost efficiency of radar systems, including reducing system size, enhancing spectrum utilization, and minimizing power consumption. In Chapter 11, the authors (Hana Godrich, Athina Petropulu, and H. Vincent Poor) propose two power allocation strategies for target localization in distributed multiple-radar architectures that, respectively, prioritize estimation accuracy and power consumption. In Chapter 12, the authors (Kristine Bell, Chris Kreucher, and Muralidhar Rangaswamy) develop a fully adaptive radar resource management scheme that emulates the perception–action cycle of cognition and uses the mutual information criterion to design future radar waveforms for the optimized estimation of the state of a surveillance region.
- *Performance bound analysis* provides an efficient way to understand the performance of radar systems and insightful guidelines for radar system design

and algorithm development. In Chapter 13, the authors (Yifeng Xiong, Fuwang Dong, and Fan Liu) derive the information-theoretic performance boundaries that govern the tradeoff between the sensing accuracy and communication rate in integrated sensing and communication (ISAC) systems. In Chapter 14, the authors (Zongyu Zhang, Zhiguo Shi, and Arye Nehorai) devise a global tight Ziv–Zakai bound as a linear combination of the *a priori* bound and the Cramér–Rao bound for evaluating the performance of multisource DOA estimation.

The organization of this book allows readers to gain a comprehensive and deep understanding of key radar signal processing topics through the lens of information theory, and to learn how information-theoretic criteria and methods can be applied to design and implement efficient and effective radar systems. The independence of each chapter also provides the flexibility for instructors to select a subset of chapters for university courses or short seminars.

References

1 Fisher, R.A. (1925). Theory of statistical estimation. *Mathematical Proceedings of the Cambridge Philosophical Society* 22(5): 700–725.

2 Nyquist, H. (1924). Certain factors affecting telegraph speed. *Bell System Technical Journal* 3(2): 324–346.

3 Hartley, R.V.L. (1928). Transmission of information. *Bell System Technical Journal* 7(3): 535–563.

4 Rao, C.R. (1945). Information and the accuracy attainable in the estimation of statistical parameters. *Bulletin of the Calcutta Mathematical Society* 37: 81–91.

5 Cramér, H. (1946). *Mathematical Methods of Statistics*. Princeton, NJ: Princeton University Press.

6 Shannon, C.E. (1948). A mathematical theory of communication. *Bell System Technical Journal* 27: 379–423 and 623–656.

7 Woodward, P.M. and Davies, I.L. (1951). A theory of radar information. *Philosophical Magazine* 41: 1101–1117.

8 Woodward, P.M. (1953). *Probability and Information Theory with Application to Radar*, 2e. London: Pergamon.

9 Bell, M.R. (1988). Information theory and radar: mutual information and the design and analysis of radar waveforms and systems. PhD dissertation. Pasadena, CA: California Institute of Technology.

10 Guo, D., Shamai, S.(S.), and Verdú, S. (2005). Mutual information and minimum mean-square error in Gaussian channels. *IEEE Transactions on Information Theory* 51(4): 1261–1282.

1

Information-Theoretic Waveform Design for MIMO Radar Target Detection

Bo Tang[1], Jun Tang[2], and Petre Stoica[3]

[1] *College of Electronic Engineering, National University of Defense Technology, Hefei, China*
[2] *Department of Electronic Engineering, Tsinghua University, Beijing, China*
[3] *Department of Information Technology, Uppsala University, Uppsala, Sweden*

1.1 Introduction

Information theory is a fundamental tool to study how to communicate reliably in noisy channels [1]. The analysis of the ultimate data compression rate and the highest data transmission rate in communications is a main focus of information theory [2]. After nearly one century of research, many elegant results about the data compression and transmission rate have been developed in information theory. These studies have achieved tremendous success in many areas, including but not limited to communication system design, computer science, statistics, and stock market investment.

An early attempt at using information theory as a guideline for designing radar systems dates back to Woodward's work in the 1950s [3–5]. These cited references pointed out that a well-established and easy-to-understand concept in the radar community, viz., the signal-to-interference-plus-noise ratio (SINR), did not measure information except for the Gaussian case and certain special elliptical contoured distributions [6]. Thus, an "information function" was defined, which was related to the posterior distributions of the observations in the radar receiver. In addition, a posterior receive filter was proposed based on the information function. Besides the receiver design based on information-theoretic approaches, there have been even more attempts to design radar transmit waveforms based on information theory. In [7], the author considered the design of radar waveforms to maximize the mutual information between the target response and the received signals. By connecting the mutual information to the rate-distortion function, the study showed that the mutual information maximization resulted in the

Information-Theoretic Radar Signal Processing, First Edition. Edited by Yujie Gu and Yimin D. Zhang.

reduction of the parameter estimation uncertainty. Therefore, it is meaningful to maximize the mutual information through waveform design. The results in [7] showed that the optimal waveforms maximizing the mutual information had a "water-filling" interpretation. In [8, 9], the authors investigated the design of clutter-resistant radar waveforms. The results therein also showed that the optimal waveforms admitted a "water-filling"-like solution. Note that only the spectral distribution of the radar waveform was derived in [7, 9]. In real-world radar systems, we need to synthesize radar waveforms under practical constraints. In [10, 11], several iterative algorithms, which leveraged fast Fourier transform, were proposed to synthesize waveforms with arbitrary spectral shape under some practical constraints, including the constant-modulus constraint, the low peak-to-average-power ratio (PAPR) constraint, and the similarity constraint. Different from the studies in [7, 9], which focused on the design of continuous-time waveforms, the authors of [12] considered the design of discrete-time waveforms to maximize the mutual information. Moreover, they showed the connection between the waveform maximizing mutual information and that minimizing minimum mean square error (MMSE).

Intuitively, the mutual information between the target response and the received signals stands for the amount of useful information that can be extracted from the received signals. Therefore, the waveforms designed based on mutual information maximization are beneficial for target parameter estimation as well as classification. However, the waveforms maximizing the mutual information might not have a satisfactory detection performance. To improve the target detection performance via designing waveforms, information-theoretic approaches have also been investigated [13–15]. Indeed, Stein's lemma [2] states that, for a fixed probability of false alarm, the probability of miss detection asymptotically decays with an exponential rate equal to the relative entropy (aka Kullback–Leibler divergence) between the distributions of the observations under the two hypotheses (viz., the target is present/absent). Therefore, maximizing the relative entropy results in minimizing the probability of miss detection. In [13], the author studied the problem of detecting point-like targets in colored noise. He proved that the performance of Neyman–Pearson detector is determined by the relative entropy. Moreover, the optimal waveform based on relative entropy maximization was derived. Interestingly, the optimal waveforms also had a "water-filling" interpretation. In [15], the authors considered the detection of weak extended targets based on an information-theoretic approach. Different from the results in [13], the authors showed that the optimal waveforms achieving the maximum relative entropy place all the transmit energy into the frequency bin with the highest signal-to-noise-ratio (SNR) (a similar observation is also presented in [16]).

Note that the above works focus on information-theoretic methods for designing single-input single-output (SISO) radar systems. Unlike SISO radar systems, multiple-input multiple-output (MIMO) radar employs a multi-antenna array for transmitting diverse waveforms and receiving returned signals, showing significantly improved target detection and parameter estimation performance [17–19]. This improvement is partly attributable to the waveform diversity provided by MIMO radar. Therefore, waveform design for MIMO radar is an important topic, and there are many studies devoted to designing MIMO radar waveforms based on information theory [20–24]. In [20], the authors considered waveform design based on maximizing the mutual information for MIMO radar in the white noise case. It is shown that the optimal waveforms that maximize the mutual information also achieve the lowest MMSE. In [21], the authors used an assumption different from that in [20]. By assuming that the target response is uncorrelated (i.e., the target covariance matrix is a scaled identity matrix) and the receiver noise is colored, it is proved that the optimal waveforms maximizing the mutual information are identical to those achieving the Chernoff bound.

In this chapter, we consider information-theoretic approaches for MIMO radar waveform design. Our purpose is to boost the detection performance of MIMO radar in the presence of external disturbance. The external disturbance can be due to hostile jamming (which is independent of the transmit waveforms) or clutter (which is dependent on the waveforms). Considering that the detection probability associated with the detector has a complicated expression, we resort to the relative entropy between the distributions of the observations under the two hypotheses as the waveform design metric. We present an algebraic solution to the waveform design problem for a distributed MIMO radar (namely, the radar antennas are widely separated) in the absence of clutter, as well as a computational approach to design MIMO radar waveforms for more general cases. The performance of the proposed solutions is demonstrated through numerical simulations.

The remaining parts of this chapter are structured as follows. Section 1.2 introduces the signal model and formulates the waveform design problem. Section 1.3 investigates the design of waveforms for a distributed MIMO radar in the absence of clutter. Section 1.4 presents an iterative algorithm to design waveforms for MIMO radar (which can be either distributed or colocated) in the range-spread clutter case. Section 1.5 provides numerical examples to assess the performance of the synthesized MIMO radar waveforms. Finally, Section 1.6 concludes this chapter.

Notations: See Table 1.1.

Table 1.1 List of notations.

Symbol	Meaning
$\mathbb{R}, \mathbb{C}$	Domain of the real and complex numbers
$\mathbb{R}^{M\times N}(\mathbb{C}^{M\times N})$	$M \times N$ real (complex)-valued matrix
$\boldsymbol{X}, \boldsymbol{x}$, and x	Matrix, vector, and scalar
$\boldsymbol{X}(:, m)$	The mth column of $\boldsymbol{X}$
$\boldsymbol{I}$	Identity matrix
$\boldsymbol{0}$	Matrix of zeros
$\mathrm{Diag}(\boldsymbol{x})$	The diagonal matrix formed by $\boldsymbol{x}$
$\mathrm{BlkDiag}(\boldsymbol{A}; \boldsymbol{B})$	The block diagonal matrix formed by $\boldsymbol{A}$ and $\boldsymbol{B}$
$(\cdot)^*, (\cdot)^\top, (\cdot)^\dagger$	Conjugate, transpose, and conjugate transpose
$(\cdot)^{1/2}$	Square root of a positive semi-definite matrix
$\det(\cdot)$	Determinant of a square matrix
$\mathrm{tr}(\cdot)$	Trace of a square matrix
$\lvert\cdot\rvert$	Magnitude
$\lVert\cdot\rVert_2$	Euclidean norm
$\lVert\cdot\rVert_\infty$	Infinity norm
$\lVert\cdot\rVert_F$	Frobenius norm
$\mathbb{E}\{x\}$	Expectation of the random variable x
$\mathrm{vec}(\cdot)$	Vectorization
$\otimes$	Kronecker product
$\arg(\cdot), \mathrm{Re}(\cdot), \mathrm{Im}(\cdot)$	The argument, real and imaginary part of a complex number
$\mathcal{CN}(\boldsymbol{m}, \boldsymbol{R})$	Complex Gaussian distribution with mean $\boldsymbol{m}$ and covariance matrix $\boldsymbol{R}$
$\boldsymbol{A} \succ \boldsymbol{0}\ (\boldsymbol{A} \succeq \boldsymbol{0})$	$\boldsymbol{A}$ is positive definite (semi-definite)
$\lambda_{\min}(\boldsymbol{X})$	The smallest eigenvalue of $\boldsymbol{X}$
$\lambda_{\max}(\boldsymbol{X})$	The largest eigenvalue of $\boldsymbol{X}$
$(x)^+$	$\max(x, 0)$

1.2 Signal Model and Problem Formulation

1.2.1 Signal Model

Consider a MIMO radar with N_T transmit antennas and N_R receive antennas. Let $\boldsymbol{S} = [\boldsymbol{s}_1, \dots, \boldsymbol{s}_{N_T}] \in \mathbb{C}^{L\times N_T}$ denote the transmit waveform matrix, where $\boldsymbol{s}_m = [s_m(1), s_m(2), \dots, s_m(L)]^\top \in \mathbb{C}^L$ is the (discrete-time) transmit waveform of the mth transmitter, $m = 1, \ldots, N_T$, and L is the code length. By assuming that the Doppler shift is negligible, the complex baseband signal of the target return received by the radar system can be written as [25–28]:

$$SH, \tag{1.1}$$

where $\boldsymbol{H} \in \mathbb{C}^{N_T \times N_R}$ denotes the target response matrix, whose (m, n)th element, denoted as h_{mn}, represents the response from the mth transmit antenna to the nth receive antenna. The expression and the statistical properties of $\boldsymbol{H}$ are determined by the array configuration. For a colocated MIMO radar in the presence of D targets, the target response matrix can be described as $\boldsymbol{H} = \sum_{d=1}^{D} a_{t,d} \boldsymbol{a}_T(\theta_{t,d}) \boldsymbol{a}_R^{\mathrm{T}}(\theta_{t,d})$, where $a_{t,d} \in \mathbb{C}$ is the amplitude of the dth target signal, $\theta_{t,d}$ denotes the associated direction-of-arrival (DOA), and $\boldsymbol{a}_T(\theta) \in \mathbb{C}^{N_T}$ and $\boldsymbol{a}_R(\theta) \in \mathbb{C}^{N_R}$ are the transmit and the receive array steering vectors at θ. For a distributed MIMO radar, it is usually assumed (see, e.g., [23–25, 29]) that the columns of $\boldsymbol{H}$ are independent and identically distributed, such that the covariance matrix of $\boldsymbol{h} = \mathrm{vec}(\boldsymbol{H})$ is given by $\boldsymbol{I} \otimes \bar{\boldsymbol{R}}_h$, where $\bar{\boldsymbol{R}}_h$ is the covariance matrix of each column of $\boldsymbol{H}$.

In radar systems, the target return is usually contaminated by clutter as well as disturbances from inside and outside the receiver. The clutter refers to unwanted returns, e.g., from ground, sea, and clouds, and it might be in the same range cell as the target or occupy the neighboring cells, as illustrated in Figure 1.1. By taking the clutter and disturbance into consideration, the received signals of a MIMO radar can be modeled by:

$$\boldsymbol{Y} = \boldsymbol{S}\boldsymbol{H} + \sum_{p=-P}^{P} \boldsymbol{J}_p \boldsymbol{S} \boldsymbol{C}_p + \boldsymbol{N}, \tag{1.2}$$

where $\boldsymbol{J}_p \in \mathbb{R}^{L \times L}$ is the shift matrix, whose (i, j)th element equals 1 if $i - j + p = 0$, and equals 0 otherwise. $\boldsymbol{C}_p \in \mathbb{C}^{N_T \times N_R}$ denotes the response matrix of the clutter

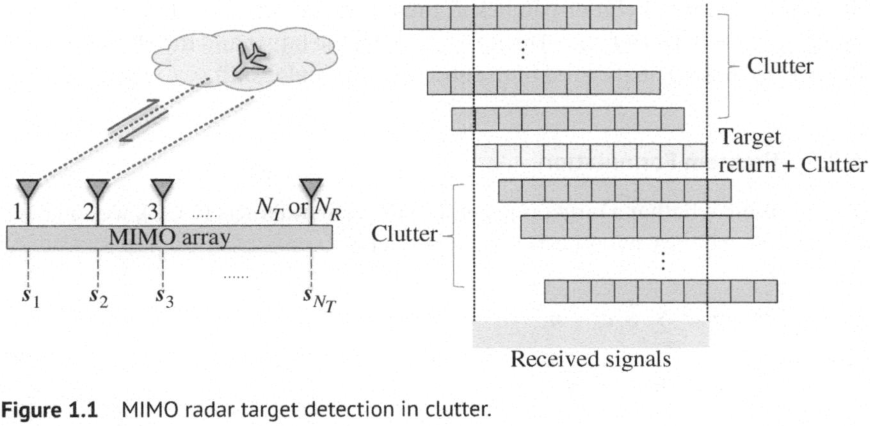

Figure 1.1 MIMO radar target detection in clutter.

in the pth range cell (note that the number of clutter range cells adjacent to the target range cell is $N_c = 2P + 1$), and $\boldsymbol{N} \in \mathbb{C}^{L\times N_R}$ is the (signal-independent) interference (including thermal noise inside the receiver and disturbances outside the receiver).

To facilitate the algebraic derivations in the sequel, we define $\boldsymbol{y} = \text{vec}(\boldsymbol{Y}) \in \mathbb{C}^{LN_R}$, which can be written as

$$\boldsymbol{y} = \widetilde{\boldsymbol{S}}\boldsymbol{h} + \sum_{p=-P}^{P} \widetilde{\boldsymbol{S}}_p\boldsymbol{c}_p + \boldsymbol{n}, \tag{1.3}$$

where $\widetilde{\boldsymbol{S}} = \boldsymbol{I}_{N_R} \otimes \boldsymbol{S}$, $\widetilde{\boldsymbol{S}}_p = \boldsymbol{I}_{N_R} \otimes \boldsymbol{S}_p$, $\boldsymbol{S}_p = \boldsymbol{J}_p\boldsymbol{S}$, $\boldsymbol{c}_p = \text{vec}(\boldsymbol{C}_p)$, and $\boldsymbol{n} = \text{vec}(\boldsymbol{N})$.

Remark 1.1 If the Doppler shift of the signals is taken into consideration, the signal model will be more complicated. Take a colocated MIMO radar as a case study. Denote the Doppler frequency of the dth target by $f_{t,d}$. Then the target return is given by

$$\boldsymbol{Y}_t = \sum_{d=1}^{D} a_{t,d} \,\text{Diag}(\boldsymbol{f}_{t,d})\boldsymbol{S}\boldsymbol{a}_T(\theta_{t,d})\boldsymbol{a}_R^\top(\theta_{t,d}), \tag{1.4}$$

where $\boldsymbol{f}_{t,d} = [1, \exp(j2\pi f_{t,d}T_s), \dots, \exp(j2\pi(L-1)f_{t,d}T_s)]^\top \in \mathbb{C}^L$, and T_s is the sampling interval. Similarly, the clutter can be modeled by

$$\boldsymbol{Y}_c = \sum_{p=-P}^{P} \sum_{m_p=1}^{M_p} a_{c,p,m_p}\boldsymbol{J}_p \,\text{Diag}(\boldsymbol{f}_{c,p,m_p})\boldsymbol{S}\boldsymbol{a}_T(\theta_{c,p,m_p})\boldsymbol{a}_R^\top(\theta_{c,p,m_p}), \tag{1.5}$$

where $\boldsymbol{f}_{c,p,m_p} = [1, \exp(j2\pi f_{c,p,m_p}T_s), \dots, \exp(j2\pi(L-1)f_{c,p,m_p}T_s)]^\top \in \mathbb{C}^L$, a_{c,p,m_p}, f_{c,p,m_p}, and θ_{c,p,m_p} are, respectively, the amplitude, the Doppler frequency, and the DOA associated with the m_pth clutter patch in the pth range cell, and M_p is the number of clutter patches in the pth range cell. Note that if $f_{t,d}T_s \approx 0$, and $f_{c,p,m_p}T_s \approx 0$, $\text{Diag}(\boldsymbol{f}_{t,d}) \approx \text{Diag}(\boldsymbol{f}_{c,p,m_p}) \approx \boldsymbol{I}_L$, the target and the clutter models degrade to those that are valid in the absence of Doppler shift (see, e.g., (1.2)).

1.2.2 Problem Formulation

To determine whether a target is present in the cell under test (CUT), we consider the following binary hypothesis testing problem:

$$\begin{cases} \mathcal{H}_0\colon \boldsymbol{y} = \sum_{p=-P}^{P} \widetilde{\boldsymbol{S}}_p\boldsymbol{c}_p + \boldsymbol{n}, \\ \mathcal{H}_1\colon \boldsymbol{y} = \widetilde{\boldsymbol{S}}\boldsymbol{h} + \sum_{p=-P}^{P} \widetilde{\boldsymbol{S}}_p\boldsymbol{c}_p + \boldsymbol{n}. \end{cases} \tag{1.6}$$

Assume that $\boldsymbol{h} \sim \mathcal{CN}(\boldsymbol{0}, \boldsymbol{R}_h)$, $\boldsymbol{n} \sim \mathcal{CN}(\boldsymbol{0}, \boldsymbol{R}_n)$, and $\{\boldsymbol{c}_p \sim \mathcal{CN}(\boldsymbol{0}, \boldsymbol{R}_{c,p})\}_{p=-P}^{P}$. Moreover, the target response vector (i.e., $\boldsymbol{h}$) and the clutter response vectors (i.e., $\boldsymbol{c}_p, p = -P, \ldots, P$) are independently distributed. Given these assumptions, the probability density functions (PDFs) of $\boldsymbol{y}$ under the two hypotheses are, respectively, given by

$$P_0(\boldsymbol{y}) = \frac{1}{\pi^{LN_R} \det(\boldsymbol{R}_0)} \exp(-\boldsymbol{y}^\dagger \boldsymbol{R}_0^{-1} \boldsymbol{y})$$

and

$$P_1(\boldsymbol{y}) = \frac{1}{\pi^{LN_R} \det(\boldsymbol{R}_1)} \exp(-\boldsymbol{y}^\dagger \boldsymbol{R}_1^{-1} \boldsymbol{y}),$$

where $\boldsymbol{R}_0 = \sum_{p=-P}^{P} \widetilde{\boldsymbol{S}}_p \boldsymbol{R}_{c,p} \widetilde{\boldsymbol{S}}_p^\dagger + \boldsymbol{R}_n$ and $\boldsymbol{R}_1 = \boldsymbol{R}_0 + \widetilde{\boldsymbol{S}} \boldsymbol{R}_h \widetilde{\boldsymbol{S}}^\dagger$.

Using the Neyman–Pearson criterion [30], one can verify that the optimal detector for (1.6) is given by

$$\boldsymbol{y}^\dagger (\boldsymbol{R}_0^{-1} - \boldsymbol{R}_1^{-1}) \boldsymbol{y} \underset{\mathcal{H}_0}{\overset{\mathcal{H}_1}{\gtrless}} T_h, \tag{1.7}$$

where T_h is the detection threshold. Moreover, the test statistic under either hypothesis is a weighted sum of chi-squared random variables (which follows from Lemma 8 in [25]). However, the probability of detection has a very complicated expression. Consequently, an information-theoretic approach is employed herein to design the waveforms, namely, the waveforms are designed to maximize the relative entropy between $P_0(\boldsymbol{y})$ and $P_1(\boldsymbol{y})$.

The relative entropy between $P_0(\boldsymbol{y})$ and $P_1(\boldsymbol{y})$ is given by

$$\begin{aligned} D(P_0 \| P_1) &= \int P_0(\boldsymbol{y}) \log \left(\frac{P_0(\boldsymbol{y})}{P_1(\boldsymbol{y})} \right) d\boldsymbol{y} \\ &= \log \det(\boldsymbol{R}_0^{-1} \boldsymbol{R}_1) + \operatorname{tr}(\boldsymbol{R}_1^{-1} \boldsymbol{R}_0) - LN_R. \end{aligned} \tag{1.8}$$

Thus, the waveform design problem based on relative entropy maximization can be stated as:

$$\begin{aligned} &\max_{\boldsymbol{S}} f(\boldsymbol{S}) = \log \det(\boldsymbol{R}_0^{-1} \boldsymbol{R}_1) + \operatorname{tr}(\boldsymbol{R}_1^{-1} \boldsymbol{R}_0) \\ &\text{s.t.}\ \boldsymbol{S} \in \mathcal{S}, \end{aligned} \tag{1.9}$$

where $\mathcal{S}$ represents the constraint set of the transmit waveforms. The following constraints are considered in this chapter.

1. *Energy constraint*: Considering that the energy of the transmit waveforms in practical radar systems is limited, we impose the following constraint:

   $$\operatorname{tr}(\boldsymbol{S}\boldsymbol{S}^\dagger) = e_t, \tag{1.10}$$

   where e_t is the transmit energy.

2. *Low PAPR constraint*: To facilitate the radio frequency amplifier to operate in the saturated mode and reduce unnecessary distortions, low PAPR waveforms are preferable. To make the transmit energy across the transmit antennas uniformly distributed and control the waveform PAPR, we enforce the constraint below

$$\boldsymbol{s}_m^\dagger \boldsymbol{s}_m = e_t/N_T,\ \mathrm{PAPR}(\boldsymbol{s}_m) \leq \rho,\ m = 1, \dots, N_T, \tag{1.11}$$

where $1 \leq \rho \leq L$, and $\mathrm{PAPR}(\boldsymbol{s}_m) = \frac{\max_l |s_m(l)|^2}{\frac{1}{L}\sum_{l=1}^{L} |s_m(l)|^2}$, $m = 1, \dots, N_T$. Particularly, if $\rho = 1$, the PAPR constraint becomes the constant-modulus constraint:

$$|s_m(l)| = a_s,\ m = 1, \dots, N_T,\ l = 1, \dots, L, \tag{1.12}$$

where $a_s = \sqrt{e_t/(LN_T)}$.

3. *Similarity constraint*: To achieve desired properties like some existing waveforms, a similarity constraint can be enforced on the waveforms:

$$\mathrm{tr}(\boldsymbol{S}\boldsymbol{S}^\dagger) = e_t,\ \|\boldsymbol{S} - \boldsymbol{S}_0\|_F^2 \leq \epsilon, \tag{1.13}$$

where $\boldsymbol{S}_0 \in \mathbb{C}^{L\times N_T}$ stands for the matrix of reference waveforms with certain favorable properties, $\mathrm{tr}(\boldsymbol{S}_0\boldsymbol{S}_0^\dagger) = e_t$, $\epsilon \in [0, 2e_t]$ is the user-specified parameter determining the similarity between the designed waveforms and the reference waveforms.

4. *Constant-modulus and similarity (CMS) constraints*: To ensure that the designed waveforms have a constant modulus and exhibit certain desirable properties, we consider the following constraint:

$$|s(l)| = a_s,\ \|\boldsymbol{s} - \boldsymbol{s}_0\|_\infty \leq \epsilon_\infty,\ l = 1, \dots, LN_T, \tag{1.14}$$

where $s(l)$ represents the lth element of $\boldsymbol{s} = \mathrm{vec}(\boldsymbol{S})$, $\boldsymbol{s}_0 = \mathrm{vec}(\boldsymbol{S}_0)$, $\boldsymbol{S}_0$ denotes a reference waveform matrix (the modulus of each element of $\boldsymbol{S}_0$ is equal to a_s), and $\epsilon_\infty \in [0, 2a_s]$ controls the similarity between $\boldsymbol{s}$ and $\boldsymbol{s}_0$.

In the formulation of the waveform design problem (1.9), we have assumed that the prior knowledge on $\boldsymbol{R}_h$, $\boldsymbol{R}_n$, and $\{\boldsymbol{R}_{c,p}\}_{p=-P}^{P}$ are available. To justify the assumption, we highlight that $\boldsymbol{R}_h$ and $\{\boldsymbol{R}_{c,p}\}_{p=-P}^{P}$ can be estimated from the received signals in previous scans of the MIMO radar (see, e.g., [31–33]). Moreover, the knowledge on the statistical property of the external disturbance can be gained by configuring the radar in a passive mode.

1.3 Optimal Waveforms for Distributed MIMO Radar in the Absence of Clutter

In this section, we consider a simple case, in which the receive antennas of the MIMO radar are widely separated (i.e., a distributed MIMO radar), and the clutter in the received signals is ignored. Due to the widely separated receive antennas, the spatial correlation at the receiver end is negligible. Moreover, similar to [23, 25, 29], we assume that $\boldsymbol{R}_h = \boldsymbol{I} \otimes \bar{\boldsymbol{R}}_h$ and $\boldsymbol{R}_n = \boldsymbol{I} \otimes \bar{\boldsymbol{R}}_n$, i.e., the columns of $\boldsymbol{H}$ and $\boldsymbol{N}$ are independent and identically distributed, where $\bar{\boldsymbol{R}}_n$ stands for the temporal correlation matrix of the interference. Based on this assumption, we have that $\boldsymbol{R}_0 = \boldsymbol{I} \otimes \bar{\boldsymbol{R}}_n$ and $\boldsymbol{R}_1 = \boldsymbol{I} \otimes (\bar{\boldsymbol{R}}_n + \boldsymbol{S}\bar{\boldsymbol{R}}_h\boldsymbol{S}^\dagger)$. Thus, $f(\boldsymbol{S})$ is given by

$$f(\boldsymbol{S}) = N_R \left(\log\det \left(\boldsymbol{I} + \bar{\boldsymbol{R}}_n^{-1}\boldsymbol{S}\bar{\boldsymbol{R}}_h\boldsymbol{S}^\dagger \right) + \mathrm{tr}\left[(\boldsymbol{I} + \bar{\boldsymbol{R}}_n^{-1}\boldsymbol{S}\bar{\boldsymbol{R}}_h\boldsymbol{S}^\dagger)^{-1} \right] \right). \tag{1.15}$$

Therefore, the energy constrained waveform design problem based on relative entropy maximization can be formulated as

$$\begin{aligned} &\max_{\boldsymbol{S}} \ \log\det \left(\boldsymbol{I} + \bar{\boldsymbol{R}}_n^{-1}\boldsymbol{S}\bar{\boldsymbol{R}}_h\boldsymbol{S}^\dagger \right) + \mathrm{tr}\left[\left(\boldsymbol{I} + \bar{\boldsymbol{R}}_n^{-1}\boldsymbol{S}\bar{\boldsymbol{R}}_h\boldsymbol{S}^\dagger \right)^{-1} \right] \\ &\text{s.t.}\ \mathrm{tr}(\boldsymbol{S}\boldsymbol{S}^\dagger) = e_t, \end{aligned} \tag{1.16}$$

where we have ignored the constant terms.

In [25], it is proved that the optimal solution to (1.16) is given by

$$\boldsymbol{S}_{\mathrm{opt}} = \boldsymbol{V}_n \left[\boldsymbol{0}_{N_T\times(L-N_T)}\, \boldsymbol{\Sigma}_1^{1/2}\boldsymbol{\Sigma}_h^{-1/2} \right]^{\mathrm{T}} \boldsymbol{V}_h^\dagger, \tag{1.17}$$

where $\boldsymbol{V}_n$ is a unitary matrix whose columns are the eigenvectors corresponding to the eigenvalues of $\bar{\boldsymbol{R}}_n$ in decreasing order, $\boldsymbol{V}_h$ is a unitary matrix whose columns are the eigenvectors corresponding to the eigenvalues of $\bar{\boldsymbol{R}}_h$ in increasing order, $\boldsymbol{\Sigma}_h = \mathrm{Diag}([\sigma_{h,1}, \dots, \sigma_{h,N_T}])$, $\sigma_{h,1} \le \cdots \le \sigma_{h,N_T}$ are the eigenvalues of $\bar{\boldsymbol{R}}_h$, $\boldsymbol{\Sigma}_1 = \mathrm{Diag}([\sigma_{1,N_T}, \dots, \sigma_{1,1}])$, $\sigma_{1,m}$ $(m = 1, \dots, N_T)$ is given by

$$\sigma_{1,m} = \frac{-p_m - \sqrt{p_m^2 - 4\varpi^2\sigma_{n,L+1-m}^2}}{2\varpi},$$

$p_m = 2\varpi\sigma_{n,L+1-m} + \sigma_{h,N_T+1-m}$, ϖ is the solution to the equation $\sum_{m=1}^{N_T} (\sigma_{1,m})^+ \sigma_{h,N_T+1-m}^{-1} = e_t$ (which can be obtained, for instance, using Newton's method), and $\sigma_{n,1} \ge \cdots \ge \sigma_{n,L}$ are the eigenvalues of $\bar{\boldsymbol{R}}_n$. According to (1.17), the maximum relative entropy that can be achieved by an MIMO radar is given by

$$D_{\mathrm{opt}} = \sum_{m=1}^{N_T} N_R \left(\log\left(1 + \frac{\sigma_{1,m}}{\sigma_{n,L+1-m}} \right) + \left(1 + \frac{\sigma_{1,m}}{\sigma_{n,L+1-m}} \right)^{-1} - N_T \right). \tag{1.18}$$

We can see from (1.17) that the left and right singular vectors of the optimal waveforms should be aligned with the eigenvectors of $\bar{\boldsymbol{R}}_n$ and $\bar{\boldsymbol{R}}_h$, respectively. Moreover, the eigenvector corresponding to the mth smallest eigenvalue of $\bar{\boldsymbol{R}}_n$ should be paired with the eigenvector corresponding to the mth largest eigenvalue of $\bar{\boldsymbol{R}}_h$. This pairing is referred to as "mode alignment" in the area of compressive sensing [34]. We highlight that the relevance of "mode alignment" is also revealed in waveform design for MIMO radar in spectrally crowded environments [28, 35].

In [25], it is also proved that the maximum relative entropy that can be achieved by a MIMO radar transmitting orthogonal waveforms (i.e., the waveform matrix $\boldsymbol{S}$ is a semi-unitary matrix that satisfies $\boldsymbol{S}^{\dagger}\boldsymbol{S} = e_t/N_T \cdot \boldsymbol{I}$) is given by

$$D_{\text{orth}} = \sum_{m=1}^{N_T} N_R \left(\log\left(1 + \frac{e_t}{N_T}\frac{\sigma_{h,m}}{\sigma_{n,L+1-m}}\right) + \left(1 + \frac{e_t}{N_T}\frac{\sigma_{h,m}}{\sigma_{n,L+1-m}}\right)^{-1} - N_T \right), \tag{1.19}$$

and the orthogonal waveform matrix achieving the maximum relative entropy is given by

$$\boldsymbol{S}_{\text{orth}} = \sqrt{\frac{e_t}{N_T}} \boldsymbol{V}_n \left[\boldsymbol{0}_{N_T \times (L-N_T)}\, \boldsymbol{I}_{N_T}\right]^{\mathsf{T}} \boldsymbol{V}_h^{\dagger}. \tag{1.20}$$

We proceed to give a simple case to showcase the performance of the optimal waveforms maximizing the relative entropy. In this example, the distributed MIMO radar has $N_T = 4$ transmit antennas and $N_R = 4$ receive antennas. The code length is $L = 5$. The eigenvalues of $\bar{\boldsymbol{R}}_h$ and $\bar{\boldsymbol{R}}_n$ are $\{0.5, 1, 2, 5\}$ and $\{10, 8, 4, 3, 2\}$, respectively. The eigenvectors of $\bar{\boldsymbol{R}}_h$ and $\bar{\boldsymbol{R}}_n$ are randomly generated. First, we comment on the optimal power allocation of the transmit waveforms. Figure 1.2 shows the allocated power on each mode of the optimal waveform matrix for $e_t = 10$ dB and $e_t = 25$ dB, respectively. We can observe that for the mode that corresponds to a larger $\sigma_{h,m}/\sigma_{n,L+1-m}$ (i.e., the target response is stronger and the interference is weaker), more power is allocated. Moreover, in the case of $e_t = 10$ dB, only two modes are active, implying a higher coherency between the transmit waveforms. If we increase the transmit energy to $e_t = 25$ dB, all the modes are activated, meaning a higher waveform diversity.

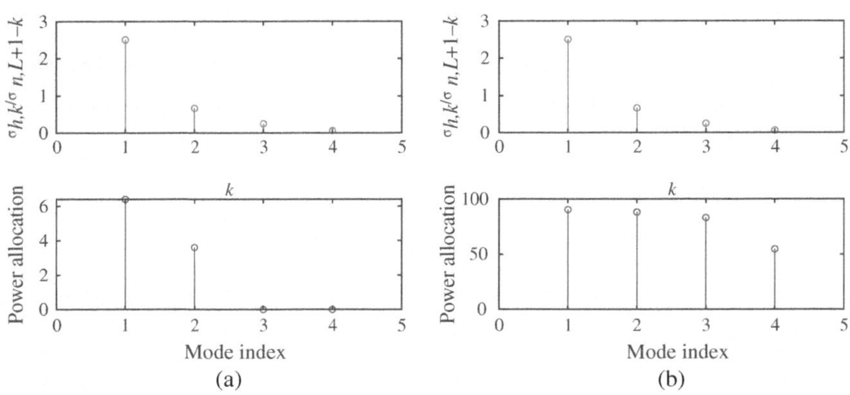

Figure 1.2 Optimal power allocation of the transmit waveforms. (a) $e_t = 10$ dB. (b) $e_t = 25$ dB.

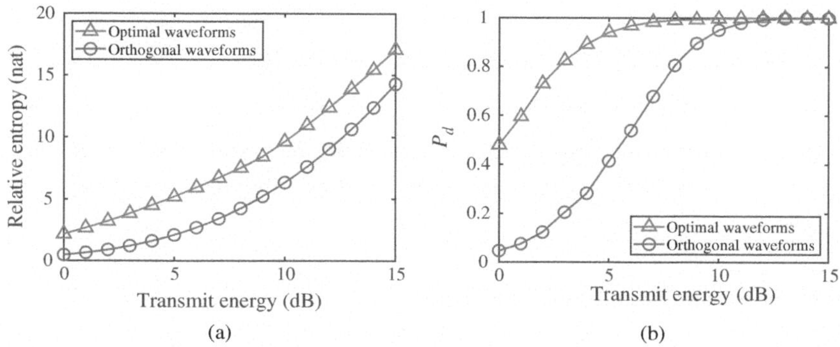

Figure 1.3 Performance comparison between the optimal waveforms and the optimal orthogonal waveforms. (a) Relative entropy versus transmit energy. (b) P_d versus transmit energy.

Now, we compare the performance of the optimal waveforms in (1.17) with that of the optimal orthogonal waveforms in (1.20). Figure 1.3(a) shows the relative entropy of the optimal waveforms and the orthogonal waveforms versus the transmit energy. We can see that, for a given transmit energy, the optimal waveforms achieve a larger relative entropy than the orthogonal waveforms. Figure 1.3(b) compares the detection probability (P_d) of the waveforms corresponding to Figure 1.3(a). In the analysis, the optimal Neyman–Pearson detector (see (1.7)) is used. In addition, the probability of false alarm is $P_{fa} = 10^{-3}$, and 10^5 independent Monte Carlo trials are conducted to determine the detection threshold and the detection probability, respectively. The results in Figure 1.3(b) are consistent with those in Figure 1.3(a), demonstrating that the performance of the optimal waveforms is visibly superior to that of the optimal orthogonal waveforms. This is attributed to the fact that the optimal waveforms have a larger number of degrees of freedom and thus achieve additional waveform diversity.

1.4 MM-Based Waveform Design in the Presence of Range-Spread Clutter

In this section, we design waveforms to boost the detection performance of MIMO radar in the presence of range-spread clutter. To this end, we consider the optimization problem (1.9). In general, this problem is nonconvex. In the following, we derive a minorization–maximization (MM) algorithm [36] to tackle this nonconvex problem. To this end, we split the objective function in (1.9) into three parts:

$$\begin{aligned} f(\boldsymbol{S}) &= \log\det(\boldsymbol{R}_n^{-1}\boldsymbol{R}_1) - \log\det(\boldsymbol{R}_n^{-1}\boldsymbol{R}_0) + \mathrm{tr}(\boldsymbol{R}_1^{-1}\boldsymbol{R}_0) \\ &= f_1(\boldsymbol{S}) + f_2(\boldsymbol{S}) + f_3(\boldsymbol{S}), \end{aligned} \tag{1.21}$$

where $f_1(\boldsymbol{S}) = \log\det(\boldsymbol{R}_n^{-1}\boldsymbol{R}_1)$, $f_2(\boldsymbol{S}) = -\log\det(\boldsymbol{R}_n^{-1}\boldsymbol{R}_0)$, and $f_3(\boldsymbol{S}) = \mathrm{tr}(\boldsymbol{R}_1^{-1}\boldsymbol{R}_0)$. Next, we derive a quadratic surrogate function (i.e., a minorizer) for each $f_r(\boldsymbol{S})$, $r = 1, 2, 3$.

1.4.1 Surrogate Function Construction

Define $\widetilde{\boldsymbol{X}} = [\widetilde{\boldsymbol{S}}_{-P}, \dots, \widetilde{\boldsymbol{S}}_P] \in \mathbb{C}^{LN_R \times N_T N_R N_c}$, $\boldsymbol{R}_C = \mathrm{BlkDiag}(\boldsymbol{R}_{c,-P}; \dots; \boldsymbol{R}_{c,P}) \in \mathbb{C}^{N_T N_R N_c \times N_T N_R N_c}$, and $\boldsymbol{R}_{TC} = \mathrm{BlkDiag}(\boldsymbol{R}_{hc,-P}; \dots; \boldsymbol{R}_{hc,P}) \in \mathbb{C}^{N_T N_R N_c \times N_T N_R N_c}$, where $\boldsymbol{R}_{hc,p} = \boldsymbol{R}_{c,p}$ if $p \neq 0$, and $\boldsymbol{R}_{hc,0} = \boldsymbol{R}_{c,0} + \boldsymbol{R}_h$. Then $\boldsymbol{R}_0 = \widetilde{\boldsymbol{X}}\boldsymbol{R}_C\widetilde{\boldsymbol{X}}^\dagger + \boldsymbol{R}_n$ and $\boldsymbol{R}_1 = \widetilde{\boldsymbol{X}}\boldsymbol{R}_{TC}\widetilde{\boldsymbol{X}}^\dagger + \boldsymbol{R}_n$. Following the derivations in [27, 37], we can show that

$$f_1(\boldsymbol{S}) = \log\det(\boldsymbol{R}_n^{-1}\boldsymbol{R}_1) = \log\det(\boldsymbol{K}_1\boldsymbol{T}^{-1}\boldsymbol{K}_1^\top), \tag{1.22}$$

where $\boldsymbol{K}_1 = [\boldsymbol{I}_{N_T N_R N_c}, \boldsymbol{0}_{N_T N_R N_c \times LN_R}]$, and

$$\boldsymbol{T} = \begin{bmatrix} \boldsymbol{I}_{N_T N_R N_c} & \boldsymbol{R}_{TC}^{\frac{1}{2}}\widetilde{\boldsymbol{X}}^\dagger \\ \widetilde{\boldsymbol{X}}\boldsymbol{R}_{TC}^{\frac{1}{2}} & \boldsymbol{R}_n + \widetilde{\boldsymbol{X}}\boldsymbol{R}_{TC}\widetilde{\boldsymbol{X}}^\dagger \end{bmatrix}.$$

It follows from Lemma 1 in [27] that $\log\det(\boldsymbol{K}_1\boldsymbol{T}^{-1}\boldsymbol{K}_1^\top)$ is convex with respect to $\boldsymbol{T}$. By using a property of convex functions [38], we have

$$\log\det(\boldsymbol{K}_1\boldsymbol{T}^{-1}\boldsymbol{K}_1^\top) \geq \log\det(\boldsymbol{K}_1\boldsymbol{T}_k^{-1}\boldsymbol{K}_1^\top) + \mathrm{tr}(\boldsymbol{G}_{T_k}(\boldsymbol{T} - \boldsymbol{T}_k)),$$

where $\boldsymbol{G}_{T_k} = -\boldsymbol{T}_k^{-1}\boldsymbol{K}_1^\top(\boldsymbol{K}_1\boldsymbol{T}_k^{-1}\boldsymbol{K}_1^\top)^{-1}\boldsymbol{K}_1\boldsymbol{T}_k^{-1}$ is the gradient of $\log\det(\boldsymbol{K}_1\boldsymbol{T}^{-1}\boldsymbol{K}_1^\top)$ at $\boldsymbol{T}_k$ (which can be shown using results from [39]). Let us partition $\boldsymbol{G}_{T_k}$ as:

$$\boldsymbol{G}_{T_k} = \begin{bmatrix} \boldsymbol{G}_{T_k}^{11} & \boldsymbol{G}_{T_k}^{12} \\ \boldsymbol{G}_{T_k}^{21} & \boldsymbol{G}_{T_k}^{22} \end{bmatrix},$$

where $\boldsymbol{G}_{T_k}^{11} \in \mathbb{C}^{N_T N_R N_c \times N_T N_R N_c}$, $\boldsymbol{G}_{T_k}^{12} = (\boldsymbol{G}_{T_k}^{21})^\dagger \in \mathbb{C}^{N_T N_R N_c \times LN_R}$, and $\boldsymbol{G}_{T_k}^{22} \in \mathbb{C}^{LN_R \times LN_R}$. One can verify that $\boldsymbol{G}_{T_k}^{21} = \boldsymbol{R}_n^{-1}\widetilde{\boldsymbol{X}}_k\boldsymbol{R}_{TC}^{\frac{1}{2}}$, $\boldsymbol{G}_{T_k}^{22} = -\boldsymbol{G}_{T_k}^{21}(\boldsymbol{K}_1\boldsymbol{T}_k^{-1}\boldsymbol{K}_1^\top)^{-1}\boldsymbol{G}_{T_k}^{12}$, and $\boldsymbol{G}_{T_k}^{22} = \boldsymbol{R}_{1,k}^{-1} - \boldsymbol{R}_n^{-1}$ [37]. As a result,

$$\mathrm{tr}(\boldsymbol{G}_{T_k}\boldsymbol{T}) = \mathrm{tr}(\boldsymbol{G}_{T_k}^{11}) + \mathrm{tr}(\boldsymbol{G}_{T_k}^{22}\boldsymbol{R}_n) + 2\mathrm{Re}(\mathrm{tr}(\boldsymbol{G}_{T_k}^{12}\widetilde{\boldsymbol{X}}\boldsymbol{R}_{TC}^{\frac{1}{2}})) + \mathrm{tr}(\boldsymbol{G}_{T_k}^{22}\widetilde{\boldsymbol{X}}\boldsymbol{R}_{TC}\widetilde{\boldsymbol{X}}^\dagger).$$

Define $c_1 = \log\det(\boldsymbol{K}_1\boldsymbol{T}_k^{-1}\boldsymbol{K}_1^\top) + \mathrm{tr}(\boldsymbol{G}_{T_k}^{11}) + \mathrm{tr}(\boldsymbol{G}_{T_k}^{22}\boldsymbol{R}_n) - \mathrm{tr}(\boldsymbol{G}_{T_k}\boldsymbol{T}_k)$. Then a surrogate function of $f_1(\boldsymbol{S})$ is given by

$$g_1(\boldsymbol{S}; \boldsymbol{S}_k) = 2\mathrm{Re}\left(\mathrm{tr}(\widetilde{\boldsymbol{X}}\boldsymbol{Z}_k^\dagger)\right) + \mathrm{tr}\left(\boldsymbol{G}_{T_k}^{22}\widetilde{\boldsymbol{X}}\boldsymbol{R}_{TC}\widetilde{\boldsymbol{X}}^\dagger\right) + c_1. \tag{1.23}$$

Next, we rewrite $f_2(\boldsymbol{S})$ as

$$f_2(\boldsymbol{S}) = -\log\det(\boldsymbol{R}_n^{-1}\boldsymbol{R}_0) = -\log\det(\boldsymbol{R}_0) + \log\det(\boldsymbol{R}_n). \tag{1.24}$$

Using the fact that $-\log\det(\boldsymbol{A})$ is convex with respect to $\boldsymbol{A} \succeq \boldsymbol{0}$ [38], we can obtain

$$-\log\det(\boldsymbol{R}_0) \geq -\log\det(\boldsymbol{R}_{0,k}) - \operatorname{tr}\left(\boldsymbol{R}_{0,k}^{-1}(\boldsymbol{R}_0 - \boldsymbol{R}_{0,k})\right).$$

Define $c_2 = \log\det(\boldsymbol{R}_n) + LN_R - \log\det(\boldsymbol{R}_{0,k}) - \operatorname{tr}(\boldsymbol{R}_{0,k}^{-1}\boldsymbol{R}_n)$. Then one can verify that a surrogate function of $f_2(\boldsymbol{S})$ can be written as

$$g_2(\boldsymbol{S}; \boldsymbol{S}_k) = -\operatorname{tr}\left(\boldsymbol{R}_{0,k}^{-1}\widetilde{\boldsymbol{X}}\boldsymbol{R}_C\widetilde{\boldsymbol{X}}^\dagger\right) + c_2. \tag{1.25}$$

Finally, we note that $f_3(\boldsymbol{S}) = \operatorname{tr}(\boldsymbol{R}_1^{-1}\boldsymbol{R}_0) = -\operatorname{tr}(\widetilde{\boldsymbol{S}}^\dagger\boldsymbol{R}_1^{-1}\widetilde{\boldsymbol{S}}\boldsymbol{R}_h) + LN_R$. Moreover, by using the matrix inversion lemma, we have $\widetilde{\boldsymbol{S}}^\dagger\boldsymbol{R}_1^{-1}\widetilde{\boldsymbol{S}} = (\boldsymbol{K}_3\boldsymbol{M}^{-1}\boldsymbol{K}_3^\top)^{-1}$, where $\boldsymbol{K}_3 = [\boldsymbol{I}_{N_TN_R}, \boldsymbol{0}_{N_TN_R\times N_TN_RN_c}]$,

$$\boldsymbol{M} = \begin{bmatrix} \boldsymbol{M}^{11} & \boldsymbol{M}^{12} \\ \boldsymbol{M}^{21} & \boldsymbol{M}^{22} \end{bmatrix},$$

$\boldsymbol{M}^{11} = \widetilde{\boldsymbol{S}}^\dagger\boldsymbol{R}_n^{-1}\widetilde{\boldsymbol{S}}$, $\boldsymbol{M}^{12} = (\boldsymbol{M}^{21})^\dagger = \widetilde{\boldsymbol{S}}^\dagger\boldsymbol{R}_n^{-1}\widetilde{\boldsymbol{X}}\boldsymbol{R}_{TC}^{\frac{1}{2}}$, and $\boldsymbol{M}^{22} = \boldsymbol{I} + \boldsymbol{R}_{TC}^{\frac{1}{2}}\widetilde{\boldsymbol{X}}^\dagger\boldsymbol{R}_n^{-1}\widetilde{\boldsymbol{X}}\boldsymbol{R}_{TC}^{\frac{1}{2}}$. Thus,

$$f_3(\boldsymbol{S}) = -\operatorname{tr}\left(\boldsymbol{R}_h\left(\boldsymbol{K}_3\boldsymbol{M}^{-1}\boldsymbol{K}_3^\top\right)^{-1}\right) + LN_R. \tag{1.26}$$

Note that $-\operatorname{tr}(\boldsymbol{R}_h(\boldsymbol{K}_3\boldsymbol{M}^{-1}\boldsymbol{K}_3^\top)^{-1})$ is convex with respect to $\boldsymbol{M}$ [27]. Hence, we have

$$-\operatorname{tr}(\boldsymbol{R}_h(\boldsymbol{K}_3\boldsymbol{M}^{-1}\boldsymbol{K}_3^\top)^{-1}) \geq -\operatorname{tr}(\boldsymbol{R}_h(\boldsymbol{K}_3\boldsymbol{M}_k^{-1}\boldsymbol{K}_3^\top)^{-1}) + \operatorname{tr}(\boldsymbol{G}_{M_k}(\boldsymbol{M} - \boldsymbol{M}_k)),$$

where $\boldsymbol{G}_{M_k} = -\boldsymbol{M}_k^{-1}\boldsymbol{K}_3\boldsymbol{A}(\boldsymbol{S}_k)\boldsymbol{R}_h\boldsymbol{A}(\boldsymbol{S}_k)\boldsymbol{K}_3^\top\boldsymbol{M}_k^{-1}$ is the gradient of $-\operatorname{tr}(\boldsymbol{R}_h(\boldsymbol{K}_3\boldsymbol{M}^{-1}\boldsymbol{K}_3^\top)^{-1})$ at $\boldsymbol{M}_k$, and $\boldsymbol{A}(\boldsymbol{S}) = \widetilde{\boldsymbol{S}}^\dagger\boldsymbol{R}_1^{-1}\widetilde{\boldsymbol{S}}$. Let us partition $\boldsymbol{G}_{M_k}$ as

$$\boldsymbol{G}_{M_k} = \begin{bmatrix} \boldsymbol{G}_{M_k}^{11} & \boldsymbol{G}_{M_k}^{12} \\ \boldsymbol{G}_{M_k}^{21} & \boldsymbol{G}_{M_k}^{22} \end{bmatrix},$$

where $\boldsymbol{G}_{M_k}^{11} \in \mathbb{C}^{N_TN_R\times N_TN_R}$, $\boldsymbol{G}_{M_k}^{12} = (\boldsymbol{G}_{M_k}^{21})^\dagger \in \mathbb{C}^{N_TN_R\times N_TN_RN_c}$, and $\boldsymbol{G}_{M_k}^{22} \in \mathbb{C}^{N_TN_RN_c\times N_TN_RN_c}$. It can be checked that $\boldsymbol{G}_{M_k}^{11} = -\boldsymbol{R}_h$, $\boldsymbol{G}_{M_k}^{12} = \boldsymbol{R}_h\boldsymbol{M}_k^{12}(\boldsymbol{M}_k^{22})^{-1}$, and $\boldsymbol{G}_{M_k}^{22} = -(\boldsymbol{M}_k^{22})^{-1}\boldsymbol{M}_k^{21}\boldsymbol{R}_h\boldsymbol{M}_k^{12}(\boldsymbol{M}_k^{22})^{-1}$ [37]. Therefore, we have

$$\operatorname{tr}(\boldsymbol{G}_{M_k}\boldsymbol{M}) = \operatorname{tr}(\boldsymbol{G}_{M_k}^{11}\boldsymbol{M}^{11}) + \operatorname{tr}(\boldsymbol{G}_{M_k}^{22}\boldsymbol{M}^{22}) + 2\operatorname{Re}(\operatorname{tr}(\boldsymbol{G}_{M_k}^{12}\boldsymbol{M}^{21})).$$

Define $\hat{c}_3 = LN_R - \operatorname{tr}(\boldsymbol{R}_h(\boldsymbol{K}_3\boldsymbol{M}_k^{-1}\boldsymbol{K}_3^\top)^{-1}) - \operatorname{tr}(\boldsymbol{G}_{M_k}\boldsymbol{M}_k)$. Then $f_3(\boldsymbol{S})$ is minorized by

$$g_3(\boldsymbol{S}; \boldsymbol{S}_k) = \operatorname{tr}(\boldsymbol{G}_{M_k}^{11}\boldsymbol{M}^{11}) + \operatorname{tr}(\boldsymbol{G}_{M_k}^{22}\boldsymbol{M}^{22}) + 2\operatorname{Re}(\operatorname{tr}(\boldsymbol{G}_{M_k}^{12}\boldsymbol{M}^{21})) + \hat{c}_3. \tag{1.27}$$

Based on the above results, it can be inferred that a surrogate function of $f(\boldsymbol{S})$ is given by

$$g(\boldsymbol{S}; \boldsymbol{S}_k) = g_1(\boldsymbol{S}; \boldsymbol{S}_k) + g_2(\boldsymbol{S}; \boldsymbol{S}_k) + g_3(\boldsymbol{S}; \boldsymbol{S}_k). \tag{1.28}$$

Moreover, we can verify that $g(\boldsymbol{S};\boldsymbol{S}_k) \leq f(\boldsymbol{S})$ and $g(\boldsymbol{S}_k;\boldsymbol{S}_k) = f(\boldsymbol{S}_k)$. Thus, $f(\boldsymbol{S})$ is minorized by $g(\boldsymbol{S};\boldsymbol{S}_k)$.

To show that $g(\boldsymbol{S};\boldsymbol{S}_k)$ can be written as a quadratic function of $\boldsymbol{s}$, we let $\boldsymbol{Z}_k = [\boldsymbol{Z}_{k,-P},\dots,\boldsymbol{Z}_{k,P}]$, where $\boldsymbol{Z}_{k,p} \in \mathbb{C}^{LN_R\times N_TN_R}$, and let $\boldsymbol{F}_k = \boldsymbol{G}_{M_k}^{12}\boldsymbol{R}_{TC}^{\frac{1}{2}} = [\boldsymbol{F}_{k,-P},\dots,\boldsymbol{F}_{k,P}]$, where $\boldsymbol{F}_{k,p} \in \mathbb{C}^{N_TN_R\times N_TN_R}$, $p=-P,\dots,P$. Partition $\boldsymbol{W}_k = \boldsymbol{R}_{TC}^{\frac{1}{2}}\boldsymbol{G}_{M_k}^{22}\boldsymbol{R}_{TC}^{\frac{1}{2}}$ as

$$\begin{bmatrix} \boldsymbol{W}_{11,k} & \boldsymbol{W}_{12,k} & \cdots & \boldsymbol{W}_{1N_c,k} \\ \vdots & \vdots & \ddots & \vdots \\ \boldsymbol{W}_{N_c1,k} & \cdots & \cdots & \boldsymbol{W}_{N_cN_c,k} \end{bmatrix},$$

where $\boldsymbol{W}_{\bar{p}\bar{q},k} \in \mathbb{C}^{N_TN_R\times N_TN_R}$, $\bar{p}=1,\dots,N_c$, $\bar{q}=1,\dots,N_c$. Then according to Proposition 1 in [37],

$$g(\boldsymbol{S};\boldsymbol{S}_k) = 2\mathrm{Re}(\boldsymbol{b}_k^\dagger\boldsymbol{s}) + \boldsymbol{s}^\dagger\boldsymbol{B}_k\boldsymbol{s} + c_0, \tag{1.29}$$

where $\boldsymbol{b}_k = \sum_{p=-P}^{P}\widetilde{\boldsymbol{E}}_p^{\mathsf{T}}\boldsymbol{z}_{k,p}$, $\boldsymbol{z}_{k,p} = \mathrm{vec}(\boldsymbol{Z}_{k,p})$, $\widetilde{\boldsymbol{E}}_p = [\widetilde{\boldsymbol{E}}_{1,p}^{\mathsf{T}},\dots,\widetilde{\boldsymbol{E}}_{N_TN_R,p}^{\mathsf{T}}]^{\mathsf{T}} \in \mathbb{R}^{N_TN_R^2L\times LN_T}$, $\widetilde{\boldsymbol{E}}_{r,p} = \boldsymbol{E}_r^{\mathsf{T}}\otimes\boldsymbol{J}_p$, $\mathrm{vec}(\boldsymbol{E}_r) = \boldsymbol{e}_r$ with $\boldsymbol{e}_r \in \mathbb{R}^{N_TN_R}$ being the rth column of $\boldsymbol{I}_{N_TN_R}$ ($r=1,\dots,N_TN_R$), $\boldsymbol{B}_k = \boldsymbol{B}_{4,k} + \boldsymbol{B}_{32,k} + \boldsymbol{B}_{33,k} + \boldsymbol{B}_{33,k}^\dagger$, $\boldsymbol{B}_{4,k} = \sum_{p=-P}^{P}\widetilde{\boldsymbol{E}}_p^{\mathsf{T}}(\boldsymbol{R}_{c,p}^*\otimes(\boldsymbol{G}_{T_k}^{22} - \boldsymbol{R}_{0,k}^{-1}))\widetilde{\boldsymbol{E}}_p + \widetilde{\boldsymbol{E}}_0^{\mathsf{T}}(\boldsymbol{R}_h^*\otimes(\boldsymbol{G}_{T_k}^{22} - \boldsymbol{R}_n^{-1}))\widetilde{\boldsymbol{E}}_0$, $\boldsymbol{B}_{32,k} = \sum_{q=-P}^{P}\sum_{p=-P}^{P}\widetilde{\boldsymbol{E}}_q^{\mathsf{T}}(\boldsymbol{W}_{\bar{q}\bar{p},k}^*\otimes\boldsymbol{R}_n^{-1})\widetilde{\boldsymbol{E}}_p$, $\boldsymbol{B}_{33,k} = \sum_{p=-P}^{P}\widetilde{\boldsymbol{E}}_0^{\mathsf{T}}(\boldsymbol{H}_{k,p}^*\otimes\boldsymbol{R}_n^{-1})\widetilde{\boldsymbol{E}}_p$, $\bar{p} = p+P+1$, $\bar{q} = q+P+1$, $c_0 = c_1+c_2+c_3$, and $c_3 = \mathrm{tr}(\boldsymbol{G}_{M_k}^{22}) + \hat{c}_3$.

Therefore, we can formulate the minorized problem based on (1.29) at the $(k+1)$th iteration as the following quadratic programming problem:

$$\begin{aligned} \max_{\boldsymbol{S}} \quad & 2\mathrm{Re}(\boldsymbol{b}_k^\dagger\boldsymbol{s}) + \boldsymbol{s}^\dagger\boldsymbol{B}_k\boldsymbol{s} \\ \text{s.t.} \quad & \boldsymbol{S}\in\mathcal{S}. \end{aligned} \tag{1.30}$$

1.4.2 Efficient Solutions to the Quadratic Programming Problem (1.30)

• Energy Constraint

We can formulate the quadratic programming problem under the energy constraint as

$$\begin{aligned} &\max_{\boldsymbol{s}}\ 2\mathrm{Re}(\boldsymbol{s}^\dagger\boldsymbol{b}_k) + \boldsymbol{s}^\dagger\boldsymbol{B}_k\boldsymbol{s} \\ &\text{s.t. } \boldsymbol{s}^\dagger\boldsymbol{s} = e_t, \end{aligned} \tag{1.31}$$

where we have used the fact that $\mathrm{tr}(\boldsymbol{S}\boldsymbol{S}^\dagger) = \boldsymbol{s}^\dagger\boldsymbol{s}$. The optimization problem (1.31) can be solved by the Lagrange multiplier method (see, e.g., [12, 27, 37] for details), and the solution is given by

$$\boldsymbol{s}^{(k+1)} = \left(\varrho^{(k+1)}\boldsymbol{I} - \boldsymbol{B}_k\right)^{-1}\boldsymbol{b}_k, \tag{1.32}$$

where $\varrho^{(k+1)} \geq \lambda_{\max}(\boldsymbol{B}_k)$ is the optimal Lagrange multiplier corresponding to the energy constraint, which can be obtained via solving the equation $\boldsymbol{b}_k^{\dagger}(\varrho \boldsymbol{I} - \boldsymbol{B}_k)^{-2}\boldsymbol{b}_k = e_t$.

- **PAPR Constraint**

We formulate the quadratic programming problem under the PAPR constraint as

$$\begin{aligned}&\max_{\boldsymbol{s}}\ 2\mathrm{Re}(\boldsymbol{s}^{\dagger}\boldsymbol{b}_k) + \boldsymbol{s}^{\dagger}\boldsymbol{B}_k\boldsymbol{s}\\&\text{s.t. } \boldsymbol{s}_m^{\dagger}\boldsymbol{s}_m = e_t/N_T, \mathrm{PAPR}(\boldsymbol{s}_m) \leq \rho, m = 1, \ldots, N_T.\end{aligned} \tag{1.33}$$

The above quadratic programming problem can be tackled by MM. To this purpose, note that $(\boldsymbol{s} - \boldsymbol{s}^{(k,q)})^{\dagger}(\boldsymbol{B}_k - \lambda_{\min}(\boldsymbol{B}_k)\boldsymbol{I})(\boldsymbol{s} - \boldsymbol{s}^{(k,q)}) \geq 0$, where $\boldsymbol{s}^{(k,q)}$ denotes the solution obtained by the proposed algorithm at the (k, q)th iteration. Here, the index k denotes the outer loop and the index q denotes the inner loop. Using the above inequality and some algebraic manipulations, we can show that $\boldsymbol{s}^{\dagger}\boldsymbol{B}_k\boldsymbol{s}$ is minorized by $2\mathrm{Re}(\boldsymbol{s}^{\dagger}(\boldsymbol{B}_k - \lambda_{\min}(\boldsymbol{B}_k)\boldsymbol{I})\boldsymbol{s}^{(k,q)}) - (\boldsymbol{s}^{(k,q)})^{\dagger}\boldsymbol{B}_k\boldsymbol{s}^{(k,q)} + 2\lambda_{\min}(\boldsymbol{B}_k)e_t$. Thus, a linear surrogate function of $g(\boldsymbol{S}; \boldsymbol{S}_k)$ is given by

$$h(\boldsymbol{s}; \boldsymbol{s}^{(k,q)}) = 2\mathrm{Re}(\boldsymbol{s}^{\dagger}\boldsymbol{d}_{k,q}) + c_4, \tag{1.34}$$

where

$$\boldsymbol{d}_{k,q} = \boldsymbol{b}_k + (\boldsymbol{B}_k - \lambda_{\min}(\boldsymbol{B}_k)\boldsymbol{I})\boldsymbol{s}^{(k,q)}, \tag{1.35}$$

and $c_4 = c_0 + 2\lambda_{\min}(\boldsymbol{B}_k)e_t - (\boldsymbol{s}^{(k,q)})^{\dagger}\boldsymbol{B}_k\boldsymbol{s}^{(k,q)}$. As a consequence, the following problem will be solved at the $(k, q+1)$th iteration:

$$\begin{aligned}&\max_{\boldsymbol{s}}\ \mathrm{Re}(\boldsymbol{s}^{\dagger}\boldsymbol{d}_{k,q})\\&\text{s.t. } \boldsymbol{s}_m^{\dagger}\boldsymbol{s}_m = e_t/N_T, \mathrm{PAPR}(\boldsymbol{s}_m) \leq \rho, m = 1, \ldots, N_T.\end{aligned} \tag{1.36}$$

Let $\boldsymbol{d}_{k,q} = [\boldsymbol{u}_1^{\mathrm{T}}, \ldots, \boldsymbol{u}_{N_T}^{\mathrm{T}}]^{\mathrm{T}}$, where $\boldsymbol{u}_m \in \mathbb{C}^L$, $m = 1, \ldots, N_T$. Note that $\boldsymbol{s}^{\dagger}\boldsymbol{d}_{k,q} = \sum_{m=1}^{N_T} \boldsymbol{s}_m^{\dagger}\boldsymbol{u}_m$. Thus, the optimization problem (1.36) can be split into N_T independent subproblems, the mth of which is given by

$$\begin{aligned}&\max_{\boldsymbol{s}_m}\ \mathrm{Re}(\boldsymbol{s}_m^{\dagger}\boldsymbol{u}_m)\\&\text{s.t. } \boldsymbol{s}_m^{\dagger}\boldsymbol{s}_m = e_t/N_T, \mathrm{PAPR}(\boldsymbol{s}_m) \leq \rho.\end{aligned} \tag{1.37}$$

Algorithm 2 in [40] can be used to solve the above problem. Particularly, if $\rho = 1$, the solution to the optimization problem (1.37) is: $s_m^{(k,q+1)}(l) = a_s \exp(\jmath \arg(u_m(l)))$, where $s_m^{(k,q+1)}(l)$ and $u_m(l)$ are the lth element of $\boldsymbol{s}_m^{(k,q+1)}$ and $\boldsymbol{u}_m$.

• Similarity Constraint

We formulate the quadratic programming problem under the similarity constraint as

$$\max_{\boldsymbol{s}} \ 2\mathrm{Re}(\boldsymbol{s}^{\dagger}\boldsymbol{b}_k) + \boldsymbol{s}^{\dagger}\boldsymbol{B}_k\boldsymbol{s}$$
$$\text{s.t.} \ \boldsymbol{s}^{\dagger}\boldsymbol{s} = e_t, \|\boldsymbol{s} - \boldsymbol{s}_0\|_2^2 \leq \epsilon. \tag{1.38}$$

Similarly, we use the linear function in (1.34) as a surrogate of the objective in (1.38) and the surrogate problem at the $(k, q+1)$th iteration is given by:

$$\max_{\boldsymbol{s}} \ \mathrm{Re}(\boldsymbol{s}^{\dagger}\boldsymbol{d}_{k,q})$$
$$\text{s.t.} \ \boldsymbol{s}^{\dagger}\boldsymbol{s} = e_t, \|\boldsymbol{s} - \boldsymbol{s}_0\|_2^2 \leq \epsilon. \tag{1.39}$$

The solution to this optimization problem is given by [26] (see also in [11, 27, 41] for similar solutions)

$$\boldsymbol{s}^{(k,q+1)} = \frac{\mu_2\boldsymbol{s}_0 + \boldsymbol{d}_{k,q}}{\mu_1}, \tag{1.40}$$

where $\mu_1 = e_t^{-\frac{1}{2}}\|\mu_2\boldsymbol{s}_0 + \boldsymbol{d}_{k,q}\|_2$, and μ_2 can be obtained by solving (details can be found in [[37], Appendix B]):

$$\max_{\mu_2} \ -2e_t^{\frac{1}{2}}\|\mu_2\boldsymbol{s}_0 + \boldsymbol{d}_{k,q}\|_2 + \mu_2(2e_t - \epsilon)$$
$$\text{s.t.} \ \mu_2 \geq 0.$$

• CMS Constraints

Note that the constraint in (1.14) is tantamount to

$$|s(l)| = a_s, \varphi_0^-(l) \leq \arg(s(l)) \leq \varphi_0^+(l), l = 1, \ldots, LN_T,$$

where $\varphi_0^-(l) = \varphi_0(l) - \delta_c$, $\varphi_0^+(l) = \varphi_0(l) + \delta_c$, $\varphi_0(l) = \arg(s_0(l))$, $s_0(l)$ denotes the lth element of $\boldsymbol{s}_0$, and $\delta_c = \arccos(1 - (\epsilon_\infty/a_s)^2/2)$. Thus, the quadratic programming problem under the CMS constraints is stated as:

$$\max_{\boldsymbol{s}} \ 2\mathrm{Re}(\boldsymbol{s}^{\dagger}\boldsymbol{b}_k) + \boldsymbol{s}^{\dagger}\boldsymbol{B}_k\boldsymbol{s}$$
$$\text{s.t.} \ |s(l)| = a_s, \varphi_0^-(l) \leq \arg(s(l)) \leq \varphi_0^+(l), l = 1, \ldots, LN_T. \tag{1.41}$$

By using the minorizer derived in (1.34) again, the surrogate problem at the $(k, q+1)$th iteration is given by:

$$\max_{\boldsymbol{s}} \ \mathrm{Re}(\boldsymbol{s}^{\dagger}\boldsymbol{d}_{k,q})$$
$$\text{s.t.} \ |s(l)| = a_s, \varphi_0^-(l) \leq \arg(s(l)) \leq \varphi_0^+(l), l = 1, \ldots, LN_T. \tag{1.42}$$

The solution to (1.42) can be written as

$$s^{(k,q+1)}(l) = a_s \exp(\jmath\varphi_{s,k}(l)), \tag{1.43}$$

where

$$\varphi_{s,k}(l) = \begin{cases} \varphi_d(l), & \text{if } \varphi_0^-(l) \le \varphi_d(l) \le \varphi_0^+(l), \\ \varphi_0^-(l), & \text{if } \cos(\varphi_d^+(l)) < \cos(\varphi_d^-(l)), \\ \varphi_0^+(l), & \text{if } \cos(\varphi_d^-(l)) < \cos(\varphi_d^+(l)), \end{cases}$$

$\varphi_d(l) = \arg(d_{k,q}(l))$, $\varphi_d^-(l) = \varphi_d(l) - \varphi_0^-(l)$, and $\varphi_d^+(l) = \varphi_d(l) + \varphi_0^+(l)$.

1.4.3 Algorithm Summary and Properties

The proposed MIMO radar waveform design algorithm is summarized in Algorithm 1.1, where $f(\boldsymbol{s}) \triangleq f(\boldsymbol{S})$, $g(\boldsymbol{s}; \boldsymbol{s}^{(k)}) \triangleq g(\boldsymbol{S}; \boldsymbol{S}_k)$ ($\boldsymbol{s}^{(k)} = \text{vec}(\boldsymbol{S}_k)$), and $\varepsilon_1 > 0$ and $\varepsilon_2 > 0$ are predefined small values. The squared iterative methods (SQUAREM) [42] can be used to speed up the convergence of Algorithm 1.1.

Algorithm 1.1: Waveform design algorithm for MIMO radar target detection in range-spread clutter.

Input: $\{\boldsymbol{R}_{\text{hc},p}\}_{p=-P}^{P}$, $\boldsymbol{R}_\text{n}$, e_t.
Output: $\boldsymbol{s}^\star$.
Initialize: $k = 0$, $\boldsymbol{s}^{(0)}$.
repeat
  Compute $\boldsymbol{B}_k$ and $\boldsymbol{b}_k$;
  `// a) Energy constraint`
  Update $\boldsymbol{s}^{(k+1)}$ using (1.32);
  $q = 0$; $\boldsymbol{s}^{(k,q)} = \boldsymbol{s}^{(k)}$;
  repeat
    Compute $\boldsymbol{d}_{k,q}$ using (1.35);
    `// b) PAPR constraint`
    Update $\boldsymbol{s}^{(k,q+1)}$ by solving (1.36);
    `// c) Similarity constraint`
    Update $\boldsymbol{s}^{(k,q+1)}$ by (1.40);
    `// d) CMS constraints`
    Update $\boldsymbol{s}^{(k,q+1)}$ by (1.43);
    $q = q + 1$;
  until $|g(\boldsymbol{s}^{(k,q+1)}; \boldsymbol{s}^{(k)}) - g(\boldsymbol{s}^{(k,q)}; \boldsymbol{s}^{(k)})| / g(\boldsymbol{s}^{(k,q+1)}; \boldsymbol{s}^{(k)}) < \varepsilon_2$;
  $\boldsymbol{s}^{(k+1)} = \boldsymbol{s}^{(k,q)}$;
  $k = k + 1$;
until $|f(\boldsymbol{s}^{(k+1)}) - f(\boldsymbol{s}^{(k)})| / f(\boldsymbol{s}^{(k)}) < \varepsilon_1$;
$\boldsymbol{s}^\star = \boldsymbol{s}^{(k)}$.

(for more details, see [27]). Note that for simplicity the correlations between the clutter across different range cells have been ignored. Nevertheless, it is not difficult to extend the proposed waveform design algorithm to deal with correlated clutter.

Next we analyze the convergence property of the proposed algorithm. To demonstrate that the objective values rise progressively with each iteration, we first prove that $g(\boldsymbol{s}^{(k+1)}; \boldsymbol{s}^{(k)}) \geq g(\boldsymbol{s}^{(k)}; \boldsymbol{s}^{(k)})$. This is evident for the energy constraint because $\boldsymbol{s}^{(k+1)}$ is the optimal solution for given $\boldsymbol{s}^{(k)}$. For the other constraints, it can be checked that

$$\begin{aligned} g(\boldsymbol{s}^{(k,q+1)}; \boldsymbol{s}^{(k)}) &\geq h(\boldsymbol{s}^{(k,q+1)}; \boldsymbol{s}^{(k,q)}) \\ &\geq h(\boldsymbol{s}^{(k,q)}; \boldsymbol{s}^{(k,q)}) \\ &= g(\boldsymbol{s}^{(k,q)}; \boldsymbol{s}^{(k)}), \end{aligned} \tag{1.44}$$

where the first line and the third line are results of the properties of the constructed surrogate function, and the second line holds since $\boldsymbol{s}^{(k,q+1)}$ is the optimal solution when $\boldsymbol{s}^{(k,q)}$ is given. Thus, we can verify that

$$g(\boldsymbol{s}^{(k)}; \boldsymbol{s}^{(k)}) = g(\boldsymbol{s}^{(k,0)}; \boldsymbol{s}^{(k)}) \leq g(\boldsymbol{s}^{(k,N_{\text{iter}})}; \boldsymbol{s}^{(k)}) = g(\boldsymbol{s}^{(k+1)}; \boldsymbol{s}^{(k)}), \tag{1.45}$$

where N_{iter} denotes the required number of iterations for the problem (1.30) to achieve convergence. Note that $g(\boldsymbol{s}; \boldsymbol{s}^{(k)})$ is a minorizer of $f(\boldsymbol{s})$. As a consequence, $f(\boldsymbol{s}^{(k)}) = g(\boldsymbol{s}^{(k)}; \boldsymbol{s}^{(k)})$ and $f(\boldsymbol{s}^{(k+1)}) \geq g(\boldsymbol{s}^{(k+1)}; \boldsymbol{s}^{(k)})$. Combining with the result in (1.45), we have $f(\boldsymbol{s}^{(k+1)}) \geq f(\boldsymbol{s}^{(k)})$.

Remark 1.2 It is worth remarking that the MM algorithm derived in this section is suitable for a target covariance matrix with an arbitrary rank. Note that for a point-like target, which has a rank-one covariance matrix, the maximization of relative entropy is equivalent to the maximization of SINR. To prove this claim, we note that if $\boldsymbol{R}_h$ is of rank one, it can be factorized as $\boldsymbol{R}_h = \boldsymbol{h}_0\boldsymbol{h}_0^\dagger$. Then, $\log\det(\boldsymbol{R}_0^{-1}\boldsymbol{R}_1) = \log\det(\boldsymbol{I} + \boldsymbol{R}_0^{-1}\widetilde{\boldsymbol{S}}\boldsymbol{h}_0\boldsymbol{h}_0^\dagger\widetilde{\boldsymbol{S}}^\dagger) = \log(1 + \text{SINR})$, where $\text{SINR} = \boldsymbol{h}_0^\dagger\widetilde{\boldsymbol{S}}^\dagger\boldsymbol{R}_0^{-1}\widetilde{\boldsymbol{S}}\boldsymbol{h}_0$. In addition, $\text{tr}(\boldsymbol{R}_1^{-1}\boldsymbol{R}_0) = \text{tr}\,(\boldsymbol{I} + \boldsymbol{R}_0^{-1}\widetilde{\boldsymbol{S}}\boldsymbol{h}_0\boldsymbol{h}_0^\dagger\widetilde{\boldsymbol{S}}^\dagger)^{-1} = LN_R + (1 + \text{SINR})^{-1} - 1$. As a result, the relative entropy corresponding to a point-like target is $D(P_0||P_1) = \log(1 + \text{SINR}) + (1 + \text{SINR})^{-1} - 1$. Note that $\log(1 + x) + (1 + x)^{-1}$ is a monotonically increasing function of $x \geq 0$ [25]. Therefore, for a point-like target, maximization of SINR is equivalent to maximization of relative entropy. In [[12], Lemma 1], it was proved that $\text{tr}(\boldsymbol{A}\boldsymbol{B}^{-1}\boldsymbol{A}^\dagger)$ is jointly convex with respect to $\boldsymbol{A}$ and $\boldsymbol{B}$ for $\boldsymbol{B} \succeq \boldsymbol{0}$. Thus, a minorizer of $\text{SINR} = \boldsymbol{h}_0^\dagger\widetilde{\boldsymbol{S}}^\dagger\boldsymbol{R}_0^{-1}\widetilde{\boldsymbol{S}}\boldsymbol{h}_0$ can be found by using the following inequality:

$$\boldsymbol{h}_0^\dagger\widetilde{\boldsymbol{S}}^\dagger\boldsymbol{R}_0^{-1}\widetilde{\boldsymbol{S}}\boldsymbol{h}_0 \geq 2\text{Re}\left(\boldsymbol{h}_0^\dagger\widetilde{\boldsymbol{S}}_k^\dagger\boldsymbol{R}_{0,k}^{-1}\widetilde{\boldsymbol{S}}\boldsymbol{h}_0\right) - \text{tr}\left(\boldsymbol{R}_{0,k}^{-1}\widetilde{\boldsymbol{S}}_k\boldsymbol{h}_0\boldsymbol{h}_0^\dagger\widetilde{\boldsymbol{S}}_k^\dagger\boldsymbol{R}_0\right). \tag{1.46}$$

As a result, we can use MM to tackle the SINR maximization problem. Moreover, it is not difficult to show that the minorizer on the right-hand side of (1.46) is a quadratic function of $\boldsymbol{s}$. Thus, we need to solve a quadratic programming problem at each iteration of the MM algorithm, similar to that in (1.30).

1.5 Performance Assessment

In this section, we assess the performance of the synthesized waveforms for a colocated MIMO radar in the presence of range-spread clutter. The MIMO radar has $N_T = 4$ transmit antennas and $N_R = 4$ receive antennas. Both the antenna arrays on the transmit and receive end are uniform and linear, with inter-element spacings of $d_T = 2\lambda$ and $d_R = \lambda/2$, respectively, where λ denotes the wavelength. The transmit waveforms have a code length of $L = 16$. The target and clutter distribution over the angle-range region is illustrated in Figure 1.4. Specifically, we consider two targets with directions of 10° and 11°, respectively. We denote the range cell where the targets are present as cell 0. Both targets have an average power of $\sigma_{t,k}^2 = \mathbb{E}(|a_{t,k}|^2) = 1$. The clutter is either in the same range cell as the targets (i.e., cell 0) or in the nearest cells (i.e., cell -1 or 1). The clutter distributions in each range cell are presented in Table 1.2. All the clutter patches have the same average power, i.e., $\sigma_{c,p,j}^2 = \mathbb{E}(|a_{c,p,j}|^2) = 100$. The disturbance $\boldsymbol{n}$ follows an AR(1) process with the correlation coefficient equal to 0.4. Quasi-orthogonal waveforms, which are randomly generated and have constant modulus, are used to initialize the proposed waveform design algorithm. For the proposed iterative algorithm, we set $\varepsilon_1 = 10^{-3}$ and $\varepsilon_2 = 10^{-10}$ to terminate the outer and the inner loop.

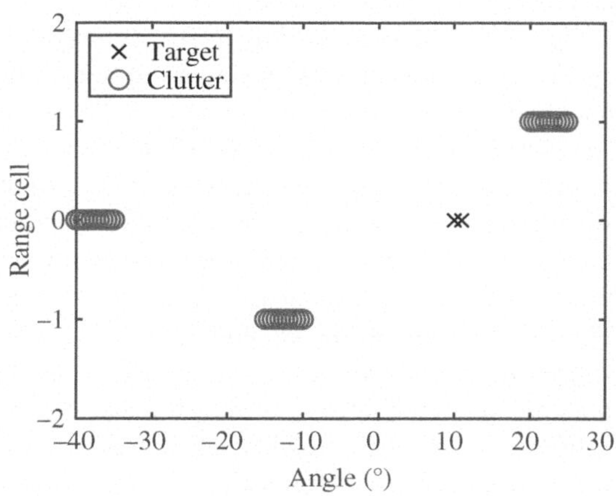

Figure 1.4 Target and clutter spatial distributions.

Table 1.2 Clutter distributions.

Range cell number	Maximum angle	Minimum angle	Grid step
−1	−15°	−10°	0.5°
0	−40°	−35°	0.5°
1	20°	25°	0.5°

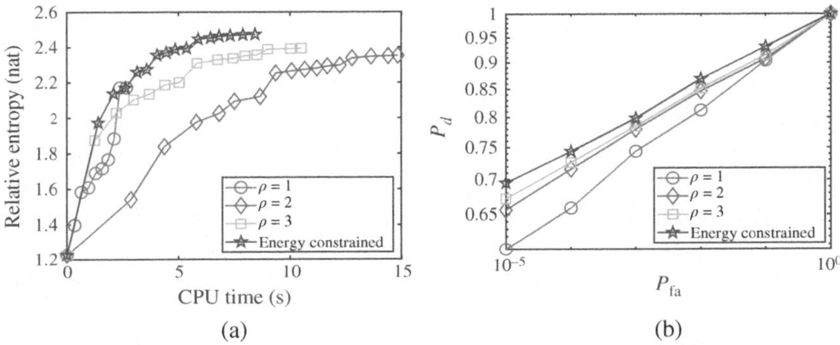

Figure 1.5 (a) Relative entropy of the waveforms synthesized under the PAPR constraint versus the iteration time. (b) P_d versus P_{fa} for the waveforms in (a). $e_t = 1$.

In the detection performance analysis of the synthesized waveforms versus the transmit energy, the probability of false alarm is fixed to be $P_{fa} = 10^{-3}$. Moreover, $100/P_{fa}$ Monte Carlo trials are performed to determine the detection threshold and calculate the detection probability, respectively. Finally, all the computations are conducted on a standard laptop with Intel Core i7-8550U CPU and 16 GB RAM.

Figure 1.5(a) analyzes the relative entropy convergence of the waveforms synthesized under the PAPR constraint, where the transmit waveforms have an energy of $e_t = 1$. The maximum allowed PAPR values (i.e., ρ) of the three sets of waveforms are 1, 2, and 3, respectively. To demonstrate the performance of the synthesized waveforms, we include the relative entropy of the energy-constrained waveforms (i.e., $\rho = L$) as a benchmark. It is seen that the proposed algorithm increases the objective value of the synthesized waveforms monotonically until convergence. Moreover, a larger PAPR value raises the relative entropy. Figure 1.5(b) shows the detection probability of the waveforms in Figure 1.5(a) versus the probability of false alarm. The results show that a larger PAPR value not only raises the relative entropy but also enhances detection performance.

In Figure 1.6(a), the relative entropy of the waveforms synthesized under the constant-modulus constraint (i.e., $\rho = 1$) is plotted for different transmit energy,

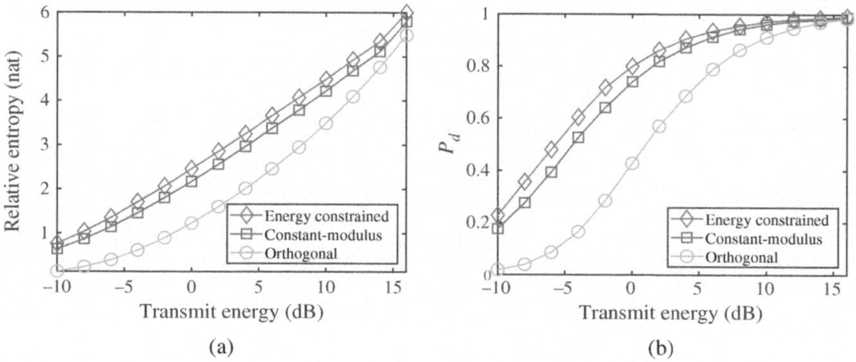

Figure 1.6 Performance assessment of the constant-modulus waveforms synthesized by the proposed algorithm. (a) Relative entropy versus transmit energy. (b) P_d versus transmit energy.

where the transmit energy of the waveforms is varied from −10 to 16 dB. Note that, as expected, the relative entropy rises as the transmit energy increases. Moreover, compared with the orthogonal waveforms, the performance of the waveforms synthesized under the constant-modulus constraint is closer to that of the energy-constrained waveforms. The detection performance associated with the waveforms in Figure 1.6(a) is shown in Figure 1.6(b). Observe that the result in Figure 1.6(b) is consistent with that in Figure 1.6(a), showing that the synthesized waveforms with a larger relative entropy achieve improved target detection performance.

Next, the performance of the waveforms synthesized under the similarity constraint is evaluated. Figure 1.7(a) plots the relative entropy convergence of the waveforms synthesized under the similarity constraint, where $e_t = 1$, and four different similarity parameters (i.e., ϵ) are considered. We use orthogonal waveforms as a reference as well as an initial point of the proposed algorithm. Figure 1.7(b) analyzes the detection probability of the waveforms in Figure 1.7(a) versus the probability of false alarm. Once again, the convergence of the proposed algorithm can be observed for all the four cases. For a smaller similarity parameter, the proposed algorithm converge slower. Because the feasibility region increases with the similarity parameter, the performance of the synthesized waveforms boosts as the similarity parameter increases.

Figure 1.8(a) shows the relative entropy of the waveforms synthesized under the similarity constraint, where the similarity parameter is equal to $0.5e_t$, and the transmit energy ranges from −10 to 16 dB. The detection probability of the synthesized waveforms in Figure 1.8(a) is analyzed in Figure 1.8(b). We can see that

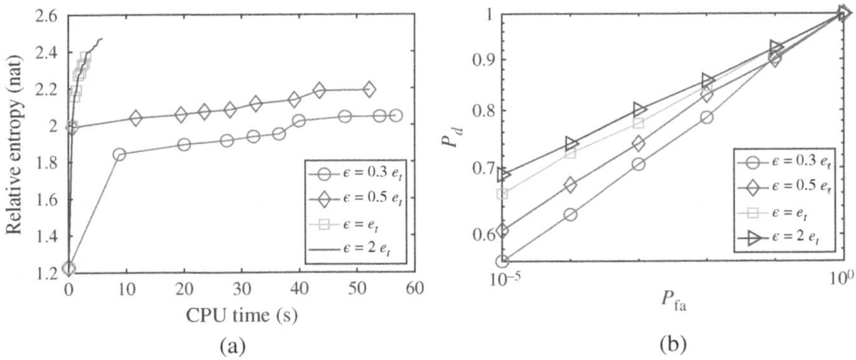

Figure 1.7 (a) Relative entropy of the waveforms synthesized under the similarity constraint. (b) P_d versus P_{fa} for the waveforms in (a). $e_t = 1$.

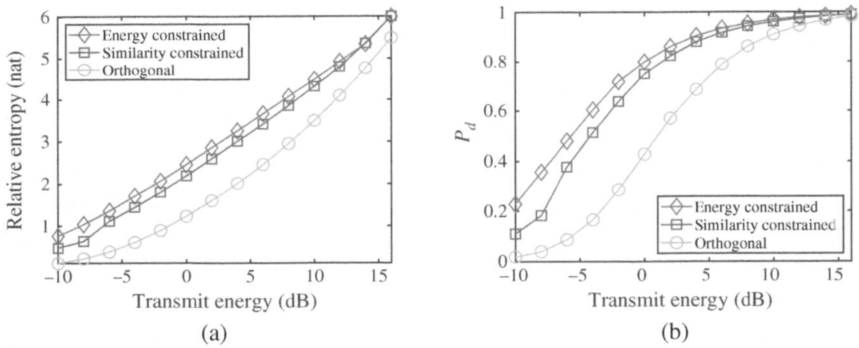

Figure 1.8 Performance assessment of the waveforms synthesized under the similarity constraint. (a) Relative entropy versus transmit energy. (b) P_d versus transmit energy. $\epsilon = 0.5e_t$.

the synthesized waveforms achieve a better performance as the transmit energy increases and demonstrate significant superiority over the orthogonal waveforms.

Finally, the performance of the waveforms synthesized under the CMS constraints is assessed. Figure 1.9(a) shows the relative entropy of the waveforms synthesized under the CMS constraints for $e_t = 1$ and four different values of ϵ_∞. We use orthogonal waveforms as a reference as well as an initial point. Figure 1.9(b) displays the detection probability of the waveforms in Figure 1.9(a) versus the probability of false alarm. It can be seen that the convergence of the proposed algorithm is ensured in all four cases. We also observe that the performance of the synthesized waveforms for the case of $\epsilon_\infty = a_s$ is very close to that for the case of $\epsilon_\infty = 2a_s$. Figure 1.10 shows the performance of the waveforms synthesized under

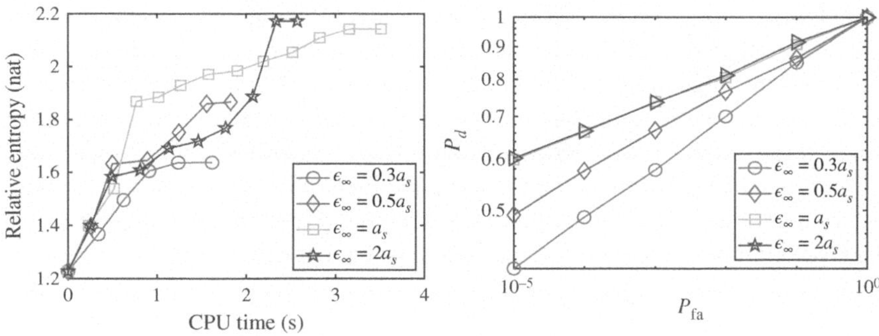

(a) (b)

Figure 1.9 (a) Relative entropy of the waveforms synthesized under the CMS constraints. (b) P_d versus P_{fa} for the waveforms in (a). $e_t = 1$.

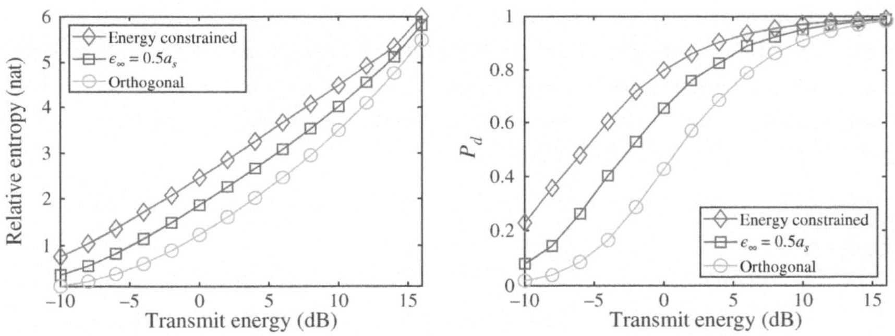

(a) (b)

Figure 1.10 Performance assessment of the waveforms synthesized under the CMS constraints. (a) Relative entropy versus transmit energy. (b) P_d versus transmit energy. $\epsilon_\infty = 0.5a_s$.

the CMS constraints, where $\epsilon_\infty = 0.5a_s$, and the transmit energy ranges from −10 to 16 dB. It is evident that the waveforms synthesized under the CMS constraints outperform the orthogonal waveforms.

1.6 Conclusion

To improve the detection performance of MIMO radar in the presence of disturbance, we have addressed the problem of designing waveforms based on maximizing the relative entropy. We showed that for the case of a distributed MIMO radar in the absence of clutter, the optimal waveform matrix can be obtained in closed form. Moreover, the left singular vectors of the optimal waveform matrix should be aligned with the eigenvectors of the interference covariance matrix, and the right singular vectors should be aligned with the eigenvectors of the target covariance matrix. For the more general case of a MIMO radar operating in range-spread clutter, the proposed MM algorithm can synthesize waveforms under several practical constraints, and it has guaranteed local convergence. Numerical results showed that the waveforms synthesized by the proposed algorithm can achieve better detection performance than orthogonal waveforms.

Acknowledgments

The work of Bo Tang was supported in part by the National Natural Science Foundation of China under Grant 62171450, the Anhui Provincial Natural Science Foundation under Grant 2108085J30, and the Research Plan of National University of Defense Technology under Grant 23-ZZCX-JDZ-42. The work of Petre Stoica was supported in part by the Swedish Research Council (VR Grants 2021-05022, 2017-04610, and 2016-06079).

References

1 Shannon, C.E. (1948). A mathematical theory of communication. *Bell System Technical Journal* 27(3): 379–423.
2 Cover, T.M. and Thomas, J.A. (2012). *Elements of Information Theory*. Wiley.
3 Woodward, P. (1951). Information theory and the design of radar receivers. *Proceedings of the IRE* 39(12): 1521–1524.
4 Woodward, P. (1953). Theory of radar information. *Transactions of the IRE Professional Group on Information Theory* 1(1): 108–113.
5 Woodward, P. (1953). *Probability and Information Theory, with Applications to Radar*. London: Pergamon.
6 Richmond, C.D. (1996). PDF's, confidence regions, and relevant statistics for a class of sample covariance-based array processors. *IEEE Transactions on Signal Processing* 44(7): 1779–1793.
7 Bell, M.R. (1993). Information theory and radar waveform design. *IEEE Transactions on Information Theory* 39(5): 1578–1597.
8 Romero, R. and Goodman, N. (2009). Waveform design in signal-dependent interference and application to target recognition with multiple transmissions. *IET Radar, Sonar and Navigation* 3(4): 328–340.

9 Romero, R.A., Bae, J., and Goodman, N.A. (2011). Theory and application of SNR and mutual information matched illumination waveforms. *IEEE Transactions on Aerospace and Electronic Systems* 47(2): 912–927.

10 Rowe, W., Stoica, P., and Li, J. (2014). Spectrally constrained waveform design. *IEEE Signal Processing Magazine* 31(3): 157–162.

11 Tang, B. and Liang, J. (2019). Efficient algorithms for synthesizing probing waveforms with desired spectral shapes. *IEEE Transactions on Aerospace and Electronic Systems* 55(3): 1174–1189.

12 Tang, B., Liu, J., Wang, H., and Hu, Y. (2021). Constrained radar waveform design for range profiling. *IEEE Transactions on Signal Processing* 69: 1924–1937.

13 Kay, S. (2007). Optimal signal design for detection of Gaussian point targets in stationary Gaussian clutter/reverberation. *IEEE Journal of Selected Topics in Signal Processing* 1(1): 31–41.

14 Kay, S. (2009). Waveform design for multistatic radar detection. *IEEE Transactions on Aerospace and Electronic Systems* 45(3): 1153–1166.

15 Zhu, Z., Kay, S., and Raghavan, R.S. (2017). Information-theoretic optimal radar waveform design. *IEEE Signal Processing Letters* 24(3): 274–278.

16 Garren, D., Osborn, M., Odom, A. et al. (2001). Enhanced target detection and identification via optimised radar transmission pulse shape. *IEE Proceedings: Radar, Sonar and Navigation* 148(3): 130–138.

17 Li, J. and Stoica, P. (2009). *MIMO Radar Signal Processing*. Hoboken, NJ: Wiley-IEEE Press.

18 Li, J. and Stoica, P. (2007). MIMO radar with colocated antennas. *IEEE Signal Processing Magazine* 24(5): 106–114.

19 Haimovich, A.M., Blum, R.S., and Cimini, L.J. (2008). MIMO radar with widely separated antennas. *IEEE Signal Processing Magazine* 25(1): 116–129.

20 Yang, Y. and Blum, R.S. (2007). MIMO radar waveform design based on mutual information and minimum mean-square error estimation. *IEEE Transactions on Aerospace and Electronic Systems* 43(1): 330–343.

21 De Maio, A. and Lops, M. (2007). Design principles of MIMO radar detectors. *IEEE Transactions on Aerospace and Electronic Systems* 43(3): 886–898.

22 Aubry, A., Lops, M., Tulino, A.M., and Venturino, L. (2010). On MIMO detection under non-Gaussian target scattering. *IEEE Transactions on Information Theory* 56(11): 5822–5838.

23 Grossi, E. and Lops, M. (2012). Space-time code design for MIMO detection based on Kullback-Leibler divergence. *IEEE Transactions on Information Theory* 58(6): 3989–4004.

24 Naghsh, M.M., Modarres-Hashemi, M., Kerahroodi, M.A., and Alian, E.H.M. (2017). An information theoretic approach to robust constrained code design for MIMO radars. *IEEE Transactions on Signal Processing* 65(14): 3647–3661.

25 Tang, B., Tang, J., and Peng, Y. (2010). MIMO radar waveform design in colored noise based on information theory. *IEEE Transactions on Signal Processing* 58(9): 4684–4697.

26 Tang, B., Naghsh, M.M., and Tang, J. (2015). Relative entropy-based waveform design for MIMO radar detection in the presence of clutter and interference. *IEEE Transactions on Signal Processing* 63(14): 3783–3796.

27 Tang, B., Zhang, Y., and Tang, J. (2018). An efficient minorization maximization approach for MIMO radar waveform optimization via relative entropy. *IEEE Transactions on Signal Processing* 66(2): 400–411.

28 Tang, B. and Li, J. (2019). Spectrally constrained MIMO radar waveform design based on mutual information. *IEEE Transactions on Signal Processing* 67(3): 821–834.

29 Grossi, E., Lops, M., and Venturino, L. (2011). Robust waveform design for MIMO radars. *IEEE Transactions on Signal Processing* 59(7): 3262–3271.

30 Kay, S. (1998). *Fundamentals of Statistical Signal Processing-Volume II: Detection Theory*. New Jersey: Prentice Hall.

31 Huleihel, W., Tabrikian, J., and Shavit, R. (2013). Optimal adaptive waveform design for cognitive MIMO radar. *IEEE Transactions on Signal Processing* 61(20): 5075–5089.

32 Herbert, S., Hopgood, J.R., and Mulgrew, B. (2018). MMSE adaptive waveform design for active sensing with applications to MIMO radar. *IEEE Transactions on Signal Processing* 66(5): 1361–1373.

33 Gurbuz, S.Z., Griffiths, H.D., Charlish, A. et al. (2019). An overview of cognitive radar: past, present, and future. *IEEE Aerospace and Electronic Systems Magazine* 34(12): 6–18.

34 Carson, W.R., Chen, M., Rodrigues, M.R.D. et al. (2012). Communications-inspired projection design with application to compressive sensing. *SIAM Journal on Imaging Sciences* 5(4): 1185–1212.

35 Tang, B., Zhang, N., Zhang, S., and Huang, Z. (2019). MMSE-based waveform design for the distributed MIMO radar in spectrally crowded environments. *The Journal of Engineering*. 2019(20): 6603–6607.

36 Hunter, D.R. and Lange, K. (2004). A tutorial on MM algorithms. *American Statistician* 58(1): 30–37.

37 Tang, B. and Stoica, P. (2021). Information-theoretic waveform design for MIMO radar detection in range-spread clutter. *Signal Processing* 182: 107961.

38 Boyd, S. and Vandenberghe, L. (2004). *Convex Optimization*. Cambridge: Cambridge University Press.

39 Hjorungnes, A. and Gesbert, D. (2007). Complex-valued matrix differentiation: techniques and key results. *IEEE Transactions on Signal Processing* 55(6): 2740–2746.

40 Tropp, J.A., Dhillon, I.S., Heath, R.W., and Strohmer, T. (2005). Designing structured tight frames via an alternating projection method. *IEEE Transactions on Information Theory* 51(1): 188–209.

41 Tang, B., Li, J., and Liang, J. (2018). Alternating direction method of multipliers for radar waveform design in spectrally crowded environments. *Signal Processing* 142: 398–402.

42 Varadhan, R. and Roland, C. (2008). Simple and globally convergent methods for accelerating the convergence of any EM algorithm. *Scandinavian Journal of Statistics* 35(2): 335–353.

2

Multiple Alternative Hypotheses in Adaptive Radar Detection: An Information-Theoretic Approach

Pia Addabbo[1], Danilo Orlando[2], and Gaetano Giunta[3]

[1] *Università degli Studi Giustino Fortunato, Benevento, Italy*
[2] *Department of Information Engineering, University of Pisa, Pisa, Italy*
[3] *Department of Industrial, Electronic and Mechanical Engineering, University of Roma Tre, Rome, Italy*

2.1 Introduction

Radar technology has witnessed significant advancements in recent years owing to the fast progress in computational power of electronic devices and, hence, in signal processing. These advancements have paved the way for the development of sophisticated radar systems with improved detection and tracking capabilities. In particular, search radars, which are primarily used for target detection in the presence of thermal noise, clutter, and intentional interference, have benefited greatly from these advancements. These operating scenarios set numerous challenges for modern systems. For instance, multiple composite alternative hypotheses often arise when the number of signal parameters (hereafter also indicated as model order) is unknown. Multiple nested hypotheses make very challenging the estimation of unknown parameters. This is due to the fact that the likelihood function monotonically increases with the size of the parameter space and, consequently, the application of the generalized likelihood ratio test (GLRT) and the maximum likelihood (ML) approach may fail [1, 2].

Adaptive detection of targets embedded in unknown interference represent a very difficult task in many radar applications. The situation where the radar window contains range-spread targets with an unknown size, or, an unknown number of multiple point-like targets, represents classical examples of practical interest. Range-spread targets consist of objects of interest characterized by a physical size that exceeds the radar range resolution. This implies scattering centers distributed over consecutive range bins and the adaptive detection becomes more challenging. In fact, the target position within the radar window is not *a priori* known as well as

Information-Theoretic Radar Signal Processing, First Edition. Edited by Yujie Gu and Yimin D. Zhang.

the number of target-contaminated consecutive range bins. Differently from [3], the size of the extended target is supposed unknown.

Similarly, multiple point-like targets occupy nonconsecutive range bins, making it difficult to detect them individually [3, 4]. In such cases, the traditional approach does not take advantage of the available energy since it only processes one range bin at time. Conversely, jointly processing the entire window, to increase the signal-to-interference-plus-noise ratio (SINR), would lead to better detection performance. As a drawback, a multiple alternative hypothesis test must be solved and this task can yield difficult mathematics with a significant computational load. This problem can be formulated with only one null hypothesis comprising thermal noise and clutter against multiple alternative hypotheses associated with the different target sizes and positions within the radar window.

Another challenging situation, in the radar context, is when the system is under the electronic attacks [5]. To enhance radar performance and address the challenges posed by intentional interference, adaptive detection architectures with electronic counter-countermeasure (ECCM) capabilities have been devised in the past [6]. ECCM strategies aim to protect radar systems from electronic attacks, such as masking and deception techniques employed by adversaries. Coherent jammers inject false information into the radar processor by transmitting low-duty cycle signals, whereas noncoherent jammers, also known as noise-like jammers (NLJs), attempt to mask targets by generating nondeceptive interference that blends into the radar receiver's thermal noise [7]. Many adaptive radar detection strategies have been proposed in the past to handle this problem. A suitable estimate of the interference covariance matrix, obtained exploiting a radar returns' set (training samples) from the range bins around the cell under test, is used in most cases. Two sets of training samples are exploited in [8]: the first set is obtained from the radar reference window surrounding the cell under test, and the second training set is collected in a manner that eliminates the clutter component and is affected only by thermal noise and possible jamming signals. Hence, a two-step estimation procedure is proposed using the second dataset to estimate the thermal noise and NLJ components, followed by the estimation of the clutter component using the conventional dataset. The number of NLJs impinging on the radar system, which is generally unknown, can be estimated using the model order selection (MOS) rules, such as the Bayesian information criterion (BIC), the Akaike information criterion (AIC), and the generalized information criterion (GIC) [9].

In this chapter, one-stage adaptive detection architectures, for solving the above mentioned problem, are designed through a systematic framework relying on the Kullback–Leibler information criterion (KLIC), as introduced in [10, 11]. This can help when one out of several alternative hypotheses may be in force against only one null hypothesis. The classical binary tests, widely used in radar as well as in other application, are naturally extended by this type of problem, which introduces an extra degree of uncertainty regarding the number of unknown parameters (e.g., model order). Final KLIC-based architectures consist of the sum between the compressed log-likelihood ratio (CLR) based upon available data and a penalty term accounting for the amount of unknown parameters. It is interesting to note that such penalty term arises from suitable approximations of the Kullback–Leibler divergence (KLD) between the true and a candidate probability density function (PDF), and, under specific constraints, the proposed decision schemes can share the constant false alarm rate (CFAR) property by virtue of the invariance principle. The detection architectures are viewed as the result of a suitable regularization of the log-likelihood. In the chapter, we also provide some numerical examples that have a great practical appeal and highlight the effectiveness of the proposed theoretically founded design framework.

The remainder of this chapter is organized as follows: Section 2.2 provides a formal statement of the considered radar detection problems in terms of multiple hypothesis tests. Section 2.3 describes the proposed design framework and includes a discussion related to the CFAR property of the proposed architectures in the context of the invariance principle. Section 2.4 provides some numerical examples related to the considered radar applications, whereas concluding remarks are contained in Section 2.5.

2.1.1 Notation

Boldface lowercase and uppercase letters are used for indicating vectors and matrices, respectively. $\boldsymbol{x} \sim \mathcal{CN}_N(\boldsymbol{\mu}, \boldsymbol{\Sigma})$ is used when $\boldsymbol{x}$ is a complex circular N-dimensional normal vector with mean $\boldsymbol{\mu}$ and covariance matrix $\boldsymbol{\Sigma} > \boldsymbol{0}$, whereas $\boldsymbol{x} \sim f(\boldsymbol{x}; \boldsymbol{\theta})$ means that f is the PDF of $\boldsymbol{x}$ with parameter vector $\boldsymbol{\theta}$. The other notations used in the chapter are listed in Table 2.1.

Table 2.1 List of notations.

Notation	Description	Notation	Description
$\det(\cdot)$	Determinant	$\mathbb{R}$	Set of real numbers
$\mathrm{Tr}(\cdot)$	Trace	$\mathbb{R}^{N\times M}$	Euclidean space of
$(\cdot)^{\mathrm{T}}$	Transpose		$(N \times M)$ real matrices
$(\cdot)^{\dagger}$	Conjugate transpose	$\mathbb{C}$	Set of complex numbers
$(\cdot)^{-1}$	Inverse	$\mathbb{C}^{N\times M}$	Euclidean space of
$\boldsymbol{I}$	Identity matrix		$(N \times M)$ complex matrices
η	Generic detection threshold		

2.2 Radar Detection Problems with Multiple Alternative Hypotheses

As stated in Section 2.1, typical radar applications naturally lead to multiple alternative hypotheses. In this section, we formalize the previously mentioned examples that are the adaptive radar detection of

- range-spread targets in thermal noise and clutter;
- multiple point-like targets against thermal noise and clutter;
- multiple NLJs embedded in thermal noise only.

In all the cases, the number of target- or NLJ-contaminated range bins is unknown and only an upper bound on it is considered known.[1]

2.2.1 Range-Spread Targets

Let us consider a radar system that uses N spatial, temporal, or space-time channels to sense its surroundings. After baseband downconversion and filtering matched to the transmitted pulse waveform, the incoming signal is suitably sampled. The signal's samples are organized into N-dimensional complex vectors representing the range bins. It is worth noticing that in the context of space-time adaptive processing (STAP), N denotes the number of space-time channels. In contrast, when a single pulse is transmitted by a radar system through a sensor array or a pulse train with a single antenna, N denotes the number of the elements of the spatial array or the transmitted pulses, respectively [12].

Let us denote by $\boldsymbol{Z}_L = [\boldsymbol{z}_1, \dots, \boldsymbol{z}_L] \in \mathbb{C}^{N\times L}$ a matrix whose columns are the vectors belonging to the window under test (WUT), and by $\boldsymbol{Z}_{K_0} = [\boldsymbol{z}_1, \dots, \boldsymbol{z}_{K_0}] \in \mathbb{C}^{N\times K_0}$ the secondary data matrix,[2] that, is gathered by the system close to the WUT, under the homogeneous environment hypothesis, and it is employed to attain adaptivity with regard to the interference covariance matrix [12, 14].

As depicted in Figure 2.1, the set of integers indexing the range bins of the radar window is $\Omega_T = \{1, \dots, L\}$, and, the secondary data are indexed by

1 The number of contaminated range bins can be set according to the operating and/or computational requirements.

2 Notice that the constraint $K_0 \geq N$ might not be met in real-world situations where several constraints limit the number of homogeneous training samples, leading to the so-called sample-starved scenarios [13].

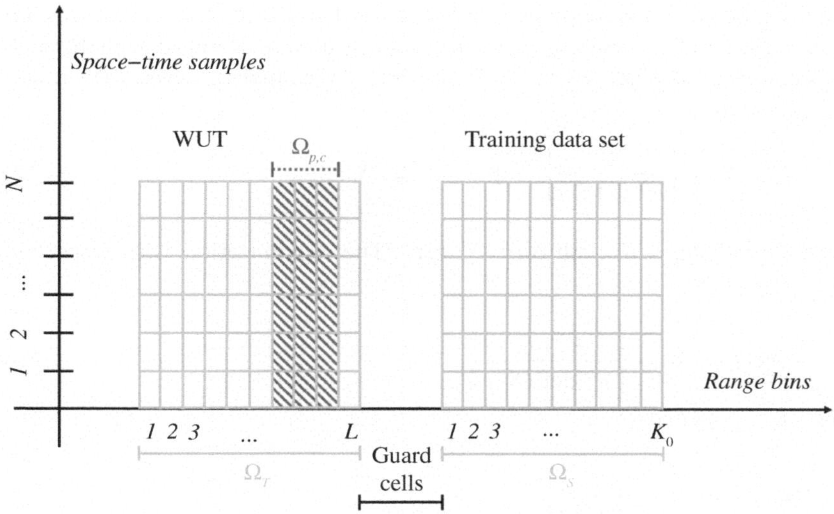

Figure 2.1 Window under test and training data set for range-spread targets.

$\Omega_S = \{L+1, \dots, K\}$, with $K = L + K_0$ and $K_0 \geq N$. The associated multiple hypothesis test is

$$\begin{cases} H_0 : & \begin{cases} \boldsymbol{z}_l = \boldsymbol{n}_l, & l \in \Omega_T, \\ \boldsymbol{z}_k = \boldsymbol{n}_k, & k \in \Omega_S, \end{cases} \\ H_{1,m} : & \begin{cases} \boldsymbol{z}_l = \alpha_l \boldsymbol{v}(\theta_T) + \boldsymbol{n}_l, & l \in \Omega_{p,c} \subseteq \Omega_T, \\ \boldsymbol{z}_l = \boldsymbol{n}_l, & l \in \Omega_T \backslash \Omega_{p,c}, \\ \boldsymbol{z}_k = \boldsymbol{n}_k, & k \in \Omega_S, \end{cases} \end{cases} \tag{2.1}$$

where:

- $m = (p-1)M + c$, with $p \in \Omega_T$ and $1 \leq c \leq M$ indicating the position of the first range bin and the number of consecutive range bins occupied by the extended target, respectively;
- M is a known upper bound on the target size;
- $\Omega_{p,c} = \{p, p+1, \dots, p+c-1\} \subseteq \Omega_T$ is the set of consecutive range bins containing useful signal components;
- $\boldsymbol{v}(\cdot)$ is the nominal steering vector of the target and θ_T is the nominal angle of arrival (AOA) of the target;
- $\alpha_l \in \mathbb{C}$, $l \in \Omega_{p,c}$, accounts for target and channel responses;
- $\boldsymbol{n}_l, \boldsymbol{n}_k \sim \mathcal{CN}_N(\boldsymbol{0}, \boldsymbol{\Sigma})$, $k \in \Omega_S$ and $l \in \Omega_T$, are statistically independent.

In the following, it is supposed that m is known and used to find an expression for the CLR for $H_{1,m}$ versus H_0 that will be shortly used in Section 2.3. To this end, let us denote by $\boldsymbol{Z} = [\boldsymbol{Z}_L, \boldsymbol{Z}_{K_0}] \in \mathbb{C}^{N\times K}$ the overall data matrix, whose PDF, under H_0, is given by

$$f_0(\boldsymbol{Z}; \boldsymbol{\Sigma}) = \left[\frac{1}{\pi^N \det(\boldsymbol{\Sigma})}\right]^K \exp\left\{-\mathrm{Tr}\left[\boldsymbol{\Sigma}^{-1}\boldsymbol{S}_0\right]\right\}, \tag{2.2}$$

where $\boldsymbol{S}_0 = \sum_{l=1}^{L} \boldsymbol{z}_l\boldsymbol{z}_l^\dagger + \boldsymbol{S}$, with $\boldsymbol{S} = \sum_{k=L+1}^{K} \boldsymbol{z}_k\boldsymbol{z}_k^\dagger$. The maximization with respect to $\boldsymbol{\Sigma}$ leads to

$$\max_{\boldsymbol{\Sigma}} f_0(\boldsymbol{Z}; \boldsymbol{\Sigma}) = \left[\frac{K^N e^{-N}}{\pi^N \det(\boldsymbol{S}_0)}\right]^K. \tag{2.3}$$

Under $H_{1,m}$, the PDF is

$$\begin{aligned} f_1(\boldsymbol{Z}; \boldsymbol{\alpha}_{p,c}, \boldsymbol{\Sigma}) &= \left[\frac{1}{\pi^N \det(\boldsymbol{\Sigma})}\right]^K \\ &\times \exp\left\{-\mathrm{Tr}\left[\boldsymbol{\Sigma}^{-1}\left(\sum_{l\in\Omega_{p,c}} (\boldsymbol{z}_l - \alpha_l \boldsymbol{v}(\theta_T))(\boldsymbol{z}_l - \alpha_l \boldsymbol{v}(\theta_T))^\dagger + \boldsymbol{S}_1\right)\right]\right\}, \end{aligned} \tag{2.4}$$

where $\boldsymbol{S}_1 = \sum_{l\in\Omega_T\setminus\Omega_{p,c}} \boldsymbol{z}_l\boldsymbol{z}_l^\dagger + \boldsymbol{S}$ and $\boldsymbol{\alpha}_{p,c} = [\alpha_p, \dots, \alpha_{p+c-1}]^{\mathrm{T}}$. Thus, the maximization with respect to $\boldsymbol{\Sigma}$ leads to

$$\max_{\boldsymbol{\Sigma}} f_1(\boldsymbol{Z}; \boldsymbol{\alpha}_{p,c}, \boldsymbol{\Sigma}) = \left[\frac{K^N e^{-N}}{\pi^N \det\left(\sum_{l\in\Omega_{p,c}} (\boldsymbol{z}_l - \alpha_l \boldsymbol{v}(\theta_T))(\boldsymbol{z}_l - \alpha_l \boldsymbol{v}(\theta_T))^\dagger + \boldsymbol{S}_1\right)}\right]^K, \tag{2.5}$$

while the maximization with respect to $\boldsymbol{\alpha}_{p,c}$ is tantamount to

$$\begin{aligned} &\min_{\boldsymbol{\alpha}_{p,c}} \det\left(\sum_{l\in\Omega_{p,c}} (\boldsymbol{z}_l - \alpha_l \boldsymbol{v}(\theta_T))(\boldsymbol{z}_l - \alpha_l \boldsymbol{v}(\theta_T))^\dagger + \boldsymbol{S}_1\right) \\ &\Rightarrow \min_{\boldsymbol{\alpha}_{p,c}} \det\left((\boldsymbol{Z}_{p,c} - \boldsymbol{v}(\theta_T)\boldsymbol{\alpha}_{p,c}^{\mathrm{T}})^\dagger \boldsymbol{S}_1^{-1}(\boldsymbol{Z}_{p,c} - \boldsymbol{v}(\theta_T)\boldsymbol{\alpha}_{p,c}^{\mathrm{T}}) + \boldsymbol{I}\right), \end{aligned} \tag{2.6}$$

where $\boldsymbol{Z}_{p,c} = [\boldsymbol{z}_p, \dots, \boldsymbol{z}_{p+c-1}]$. By completing a square form within the argument of the determinant, it is not difficult to show that the maximizer with respect to $\boldsymbol{\alpha}_{p,c}$ is

$$\widehat{\boldsymbol{\alpha}}_{p,c}^{\mathrm{T}} = \frac{\boldsymbol{v}^\dagger(\theta_T)\boldsymbol{S}_1^{-1}\boldsymbol{Z}_{p,c}}{\boldsymbol{v}^\dagger(\theta_T)\boldsymbol{S}_1^{-1}\boldsymbol{v}(\theta_T)}. \tag{2.7}$$

Thus, neglecting the constant terms, the CLR for $H_{1,m}$ versus H_0 can be written as [15]

$$\widehat{\Lambda}_1(\boldsymbol{Z}; m) = K \left[\log\det\left(\boldsymbol{S}_0\right) - \log\det\left(\sum_{l\in\Omega_{p,c}} \tilde{\boldsymbol{z}}_l \tilde{\boldsymbol{z}}_l^\dagger + \boldsymbol{S}_1 \right) \right], \tag{2.8}$$

where $\tilde{\boldsymbol{z}}_l = \boldsymbol{z}_l - \frac{\boldsymbol{v}^\dagger(\theta_T)\boldsymbol{S}_1^{-1}\boldsymbol{z}_l}{\boldsymbol{v}^\dagger(\theta_T)\boldsymbol{S}_1^{-1}\boldsymbol{v}(\theta_T)}\boldsymbol{v}(\theta_T)$ and its dependence on m is through $\Omega_{p,c}$.

2.2.2 Multiple Point-Like Targets

The detection problem for multiple point-like targets can also be expressed as in (2.1), with the exception that the targets' contaminated range bins (and, consequently, their corresponding indices) are no longer consecutive. An illustrative description of this scenario is depicted in Figure 2.2 and the constraint $K_0 \geq N$ still holds. It is noted that, for the case at hand, we have to take into account $\binom{L}{m}$ combinations related to the placements of the targets within the radar window (of size L) for each plausible number of targets ($m = 1, \dots, M$). Therefore, from a computational standpoint, maximizing over the complete set of hypotheses might be prohibitive. To limit the computational burden, following the rationale of [16], it is possible to focus only on the first selected m range indices from the vectors $\boldsymbol{z}_1, \boldsymbol{z}_2, \dots, \boldsymbol{z}_L$ ordered according to $|\boldsymbol{v}^\dagger(\theta_T)\boldsymbol{S}^{-1}\boldsymbol{z}_{h_1}|^2 > |\boldsymbol{v}^\dagger(\theta_T)\boldsymbol{S}^{-1}\boldsymbol{z}_{h_2}|^2 > \dots > |\boldsymbol{v}^\dagger(\theta_T)\boldsymbol{S}^{-1}\boldsymbol{z}_{h_L}|^2$, with $\{h_1, h_2, \dots, h_L\}$ the new ordered indexes. With this in mind,

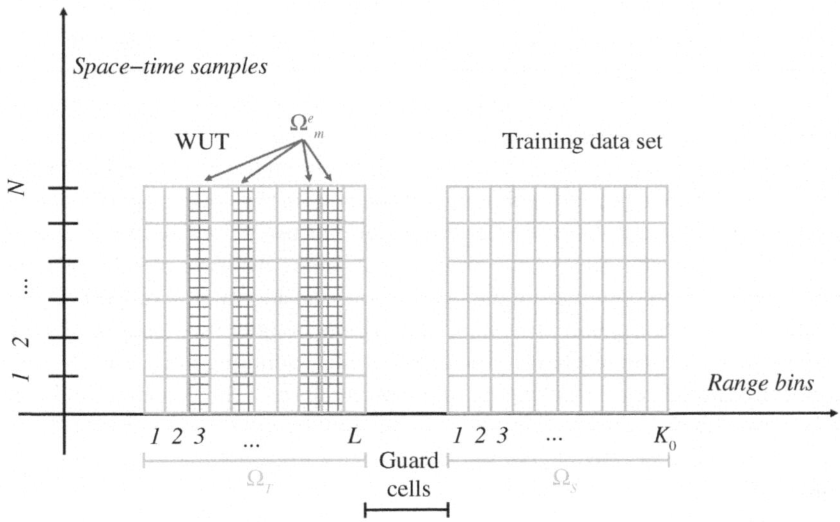

Figure 2.2 Window under test and training data set for multiple point-like targets.

the detection problem becomes

$$\begin{cases} H_0 : & \begin{cases} z_l = n_l, & l \in \Omega_T, \\ z_k = n_k, & k \in \Omega_S, \end{cases} \\ H_{1,m} : & \begin{cases} z_l = \alpha_l v(\theta_T) + n_l, & l \in \Omega_m^e \subseteq \Omega_T, \\ z_l = n_l, & l \in \Omega_T \backslash \Omega_m^e, \\ z_k = n_k, & k \in \Omega_S, \end{cases} \end{cases} \tag{2.9}$$

$m = 1, \dots, M \leq L$, with all the terms as defined in (2.1) except for Ω_m^e which is formed by using the first m selected range indices. Following the same line of reasoning as before, the mth CLR can be written as

$$\hat{\Lambda}_1(Z; m) = K \left\{ \log\det\left(S_0\right) - \log\det\left[\sum_{l \in \Omega_m^e} \tilde{z}_l \tilde{z}_l^\dagger + S_1\right]\right\}, \tag{2.10}$$

where $S_1 = \sum_{l \in \Omega_T \backslash \Omega_m^e} z_l z_l^\dagger + S$, $\tilde{z}_l$, S_0 and S are defined as above.

2.2.3 Multiple Noise-Like Jammers

Here, the radar system listens to the environment, searching for an electronic attack from NLJs, whose number is not *a priori* known [8, 17, 18]. Generally speaking, the interference covariance matrix is composed by different interference sources' contributions, i.e., the thermal noise, the clutter, and the jammer components. Conventional interference covariance matrix estimation procedures exploit training samples collected in the proximity of the cell under test, but, here, the radar system is supposed to be able to acquire data free of clutter components and affected only by the remaining interference sources. As shown in Figure 2.3, it is possible to record clutter-free data before transmitting the pulse waveform for a radar system that transmits one pulse (subfigure on the top), or, it is possible to record clutter-free data (see subfigure on the bottom) by collecting data before transmitting the next pulse and after the maximum usable range, in the case of radar systems transmitting a train of coherent pulse. Another possible way to remove clutter components from data consists in considering the range-Doppler domain and in selecting bins corresponding to high Doppler frequency due to the fact that clutter is concentrated around zero Doppler. Denoting by z_k, $k = 1, \dots, K$, the N-dimensional vectors belonging to the set under test, the detection problem can be formulated as a hypothesis test with one null hypothesis (absence of NLJs) against multiple alternative hypotheses, each one corresponding to an unknown number of NLJs:

$$\begin{cases} H_0 : & z_k \sim \mathcal{CN}_N(\mathbf{0}, \sigma_n^2 I), & k = 1, \dots, K, \\ H_{1,m} : & z_k \sim \mathcal{CN}_N(\mathbf{0}, \sigma_n^2 I + \Sigma_J(m)), & k = 1, \dots, K, \end{cases} \tag{2.11}$$

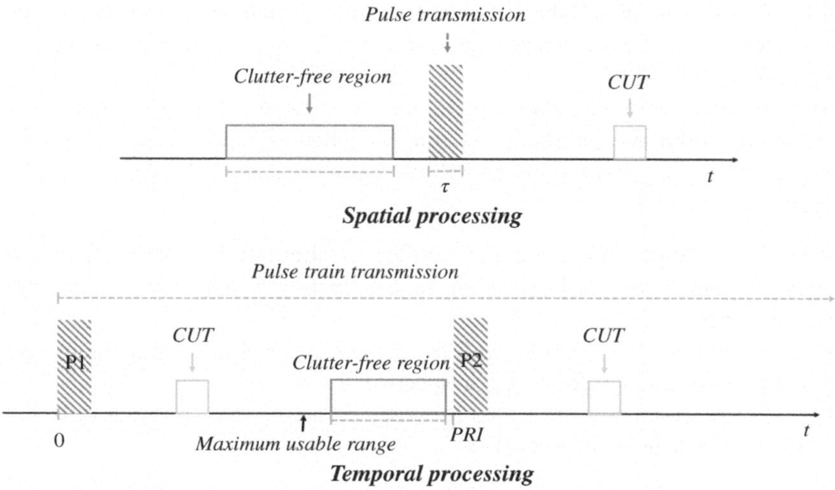

Figure 2.3 Clutter-free region for interference covariance matrix estimation in presence of NLJs.

$m = 1, \dots, M < N$, where $\boldsymbol{\Sigma}_J(m) \in \mathbb{C}^{N\times N}$ is the unknown positive semi-definite matrix (the jammer component) with m denoting the rank that is related to the actual number of directions from which interfering signals hit the system, and $\sigma_n^2 \boldsymbol{I}$ is the noise component with unknown constant $\sigma_n^2 > 0$. Now, following the lead of [17], the mth CLR for problem (2.11) can be written as

$$\widehat{\Lambda}_1(\boldsymbol{Z}; m) = \log\left\{ \frac{\left[\dfrac{1}{K(N-m)} \displaystyle\sum_{i=m+1}^{N} \gamma_i\right]^{-K(N-m)}}{\displaystyle\prod_{k=1}^{m} \left(\frac{\gamma_k}{K}\right)^K \left[\frac{1}{NK}\sum_{h=1}^{N} \gamma_h\right]^{-NK}} \right\}, \tag{2.12}$$

where $\boldsymbol{Z} = [\boldsymbol{z}_1, \dots, \boldsymbol{z}_K]$ and $\gamma_1 \geq \dots \geq \gamma_N \geq 0$ are the eigenvalues of $\boldsymbol{Z}\boldsymbol{Z}^\dagger$.

2.3 Detection Architectures and CFAR Properties

All the radar applications presented in Section 2.2 have in common that the problem is to detect a deterministic signal of unknown length in complex Gaussian disturbance and that the detection problem is posed as a composite multiple hypothesis test. In this section, we define the general multiple hypothesis testing problem that includes the previous problems as special cases and provide a design solution for this kind of problem. In order to do this, the data matrix's columns are represented as statistically independent random vectors with shared distribution and parameters.

Thus, to generalize the above problems, let $f_0(\boldsymbol{Z};\boldsymbol{\theta}_{r,0},\boldsymbol{\theta}_s)$ (the subscript r refers to the unknown parameter of interest whereas s the nuisance's ones), and $f_m(\boldsymbol{Z};\boldsymbol{\theta}_{r,1}^m,\boldsymbol{\theta}_s,p_{r,m})$ be the joint PDFs of $\boldsymbol{Z}$, under H_0 and $H_{1,m}$, $m=1,\dots,M$, respectively, where:

- $M\in\mathbb{N}$ is an upper bound on the number of alternative hypotheses that is supposed known or fixed according to the operating and/or computational requirements;
- $\boldsymbol{\theta}_{r,0}\in\mathbb{R}^{p_{r,0}\times 1}$ and $\boldsymbol{\theta}_{r,1}^m\in\mathbb{R}^{p_{r,m}\times 1}$, with $p_{r,1}\leq\cdots\leq p_{r,M}$ being the parameter vectors of interest under H_0 and $H_{1,m}$, respectively;
- $\boldsymbol{\theta}_s\in\mathbb{R}^{p_s\times 1}$ represents the so-called nuisance parameters' vector, that does not enter into the decision process [19].

Thus, the general testing problem is formed by multiple, possibly nested, alternative hypotheses, i.e.,

$$\begin{cases} H_0: & \boldsymbol{Z}\sim f_0(\boldsymbol{Z};\boldsymbol{\theta}_{r,0},\boldsymbol{\theta}_s),\\ H_{1,1}: & \boldsymbol{Z}\sim f_{1,1}(\boldsymbol{Z};\boldsymbol{\theta}_{r,1}^1,\boldsymbol{\theta}_s,p_{r,1}),\\ \vdots & \vdots\\ H_{1,M}: & \boldsymbol{Z}\sim f_{1,M}(\boldsymbol{Z};\boldsymbol{\theta}_{r,1}^M,\boldsymbol{\theta}_s,p_{r,M}), \end{cases} \tag{2.13}$$

where:

$$f_0(\boldsymbol{Z};\boldsymbol{\theta}_{r,0},\boldsymbol{\theta}_s)=\prod_{k=1}^{K} g_0(\boldsymbol{z}_k;\boldsymbol{\theta}_{r,0},\boldsymbol{\theta}_s), \tag{2.14}$$

and

$$f_{1,m}(\boldsymbol{Z};\boldsymbol{\theta}_{r,1}^m,\boldsymbol{\theta}_s,p_{r,m})=\prod_{k=1}^{K} g_{1,m}(\boldsymbol{z}_k;\boldsymbol{\theta}_{r,1}^m,\boldsymbol{\theta}_s,p_{r,m}), \tag{2.15}$$

with $g_0(\boldsymbol{z}_k;\boldsymbol{\theta}_{r,0},\boldsymbol{\theta}_s)$ and $g_{1,m}(\boldsymbol{z}_k;\boldsymbol{\theta}_{r,1}^m,\boldsymbol{\theta}_s,p_{r,m})$ being the PDFs of $\boldsymbol{z}_k$ under H_0 and $H_{1,m}$, respectively.

Decision architectures for multiple alternative hypotheses have been developed in [10], wherein the adaptivity with respect to the full set of parameters is gradually obtained from the case where all parameters are known. The decision statistic results in the sum of the CLR based on available data and a penalty term that takes into account the number of unknown parameters. Such a term emerges from appropriate approximations of the KLD [20] between a candidate PDF and the true PDF (also known as the relative entropy); the KLD can be seen as the information loss when the true distribution is used to approximate the candidate one [21].

Specifically, the design framework relies on the KLD between $f_{1,m}$, the true data distribution, and $\overline{f}$, the candidate one, that can be expressed as [22]

$$D(\overline{f}\|f_{1,m}) = -h(\overline{f}) + h(\overline{f},f_{1,m}), \tag{2.16}$$

under $H_{1,m}$, and

$$D(\overline{f}\|f_0) = -h(\overline{f}) + h(\overline{f},f_0), \tag{2.17}$$

under H_0, where $h(\overline{f})$ is the differential entropy of $\overline{f}$, $h(\overline{f},f_{1,m})$ is the cross entropy between $\overline{f}$ and $f_{1,m}$, and $h(\overline{f},f_0)$ is the cross entropy between $\overline{f}$ and f_0.

2.3.1 KLIC-Based Detection Architectures

Considering the most general case where all the parameters are unknown, we summarize the design framework proposed in [10]. A possible strategy to deal with multiple alternative hypotheses consists in selecting the most likely hypothesis by replacing the unknown number of parameters with its ML estimate. In the case of nested models, this approach would fail due to an overfitting problem which arises from the fact that the log-likelihood function monotonically increases with model order [1]. In [10], it has been shown that, under some mild assumptions, we can start from the following criterion

$$D(\overline{f}\|f_0) \underset{H_0}{\overset{H_1}{\gtrless}} \min_m D(\overline{f}\|f_{1,m}), \tag{2.18}$$

and suitably approximating estimators of the cross entropy through series expansions, and it is possible to come up with the following general decision criterion [10]

$$\max_{m\in\{1,\dots,M\}} \left\{\widehat{\Lambda}_1(\boldsymbol{Z};m) - q(m)\right\} \underset{H_0}{\overset{H_{1,\hat{m}}}{\gtrless}} \eta, \tag{2.19}$$

where η is a threshold that is introduced to control the false alarm probability P_{fa},[3] $\widehat{\Lambda}_1(\boldsymbol{Z};m) = \log f_{1,m}(\boldsymbol{Z};\widehat{\boldsymbol{\theta}}^m_{r,1},\widehat{\boldsymbol{\theta}}_{s,1},p_{r,m}) - \log f_0(\boldsymbol{Z};\widehat{\boldsymbol{\theta}}_{r,0},\widehat{\boldsymbol{\theta}}_{s,0})$ represents the CLR assuming $p_{r,m}+p_s$ as the model order, with $\widehat{\boldsymbol{\theta}}^m_{r,1},\widehat{\boldsymbol{\theta}}_{s,1},\widehat{\boldsymbol{\theta}}_{r,0},\widehat{\boldsymbol{\theta}}_{s,0}$ being the ML estimates of $\boldsymbol{\theta}^m_{r,1},\boldsymbol{\theta}_{s,1},\boldsymbol{\theta}_{r,0},\boldsymbol{\theta}_{s,0}$, respectively. $\hat{m}$ is found by maximizing the left-hand side of (2.19), and $q(m)$ is a penalty term depending on

- $p_{r,m}$, the number of unknown parameters of interest under $H_{1,m}$;
- p_s, the number of nuisance parameters;
- $\mathcal{T}$, the total number of samples.

---

3 The false alarm probability is defined as the probability of rejecting H_0 when the latter is true.

Specifically, the expression of $q(m)$ is given by

$$q(m) = \begin{cases} p_m, & \text{(a)} \\ \dfrac{1+\rho}{2} p_m, \ \rho > 1, & \text{(b)} \\ \dfrac{p_m}{2} \log(\mathcal{T}), & \text{(c)} \end{cases}, \tag{2.20}$$

where $p_m = p_{r,m} + p_s$, and, (a), (b), and (c) are related to AIC, GIC, and BIC, respectively. It is important to highlight that data distribution under H_0 does not depend on m, and as a consequence, threshold setting can be accomplished as for conventional binary hypothesis tests. Finally, $\widehat{\Lambda}_1(\boldsymbol{Z}; m)$ is the CLR under the assumption that the model order is p_m. It would make sense to consider (where possible) alternative decision statistics that exhibit the same asymptotic behavior, for sufficiently high $\mathcal{T}$. For example, the decision statistics of the Rao or Wald test can be used in place of the CLR [19]. Another possible alternative to (2.19) would be to substitute $\widehat{\Lambda}_1(\boldsymbol{Z}; m)$ with the decision statistics obtained derived applying ad hoc[4] GLRT-based design procedures [8, 13, 14, 23, 24]. As a result, decision rule (2.19) represents the foundation for the design of detection architectures that address the problem of one null hypothesis against several competing hypotheses. It is demonstrated in [10] that, after considering a suitable prior to the number of parameters, (2.19) may also be regarded as the outcome of the joint application of the Bayesian and ML frameworks.

Moreover, the false alarm number limitation is a constraint intended to prevent the terrible consequences that the latter may have. Therefore, detection thresholds are specified by system engineers to ensure very low values for the P_{fa}. Since, in general, a detection system does not grant the CFAR property, it must be demonstrated before that claim on the receiver can be made.

2.3.2 Bayesian and ML Framework

Decision rule (2.19) can be viewed as the result of the joint application of the Bayesian and ML frameworks after assigning a suitable prior to the number of parameters. In the following, a conventional binary hypothesis test is considered, and the number of parameters of interest under H_1 is considered as a discrete random variable with a proper prior promoting low-dimensional models. This is done in order to mitigate the ML natural inclination of overestimating the model size. A possible prior for p_r is

$$f(p_r) = \frac{1}{\sum_{p_r \in \Omega_r} e^{-g(p_r)}} e^{-g(p_r)}, \quad p_r \in \{p_{r,1}, \dots, p_{r,M}\}, \tag{2.21}$$

4 In the sense that they break the ML principle.

where $g(\cdot)$ is a positive and increasing known function of p_r and A is a normalization constant such that

$$\sum_{p_r \in \Omega_r} e^{-g(p_r)} < +\infty. \tag{2.22}$$

In the hypothesis that p_r is a continuous random variable, the joint PDF of $\boldsymbol{Z}$ and p_r (under H_1) can be expressed as

$$f_1(\boldsymbol{Z}, p_r; \boldsymbol{\theta}_{p_r}, \boldsymbol{\theta}_{p_s}) = f_1(\boldsymbol{Z}; \boldsymbol{\theta}_{r,1}, \boldsymbol{\theta}_s | p_r) f(p_r), \tag{2.23}$$

where $f(p_r)$ denotes the PDF of p_r. The following decision rule

$$\frac{\max\limits_{p_r \in \Omega_r} \max\limits_{\boldsymbol{\theta}_{r,1}, \boldsymbol{\theta}_s} f_1(\boldsymbol{Z}; \boldsymbol{\theta}_{r,1}, \boldsymbol{\theta}_s | p_r) f(p_r)}{\max\limits_{\boldsymbol{\theta}_{r,0}, \boldsymbol{\theta}_s} f_0(\boldsymbol{Z}; \boldsymbol{\theta}_{r,0}, \boldsymbol{\theta}_s)} \underset{H_0}{\overset{H_1}{\gtrless}} \eta, \tag{2.24}$$

is considered, where the numerator's maximization can be viewed as a joint ML and maximum a posteriori estimation procedure [1, 25]. Thus,

$$\max_{p_r \in \Omega_r} \max_{\boldsymbol{\theta}_{r,1}, \boldsymbol{\theta}_s} \log f_1(\boldsymbol{Z}; \boldsymbol{\theta}_{r,1}, \boldsymbol{\theta}_s | p_r) f(p_r) - \max_{\boldsymbol{\theta}_{r,0}, \boldsymbol{\theta}_s} \log f_0(\boldsymbol{Z}; \boldsymbol{\theta}_{r,0}, \boldsymbol{\theta}_s) \underset{H_0}{\overset{H_1}{\gtrless}} \eta, \tag{2.25}$$

Maximizing with respect to $\boldsymbol{\theta}_{r,1}$, $\boldsymbol{\theta}_{r,0}$, and $\boldsymbol{\theta}_s$, given p_r, we have to solve the following optimization problem

$$\max_{p_r \in \Omega_r} \left\{ \log \frac{f_1(\boldsymbol{Z}; \widehat{\boldsymbol{\theta}}_{r,1}, \widehat{\boldsymbol{\theta}}_{s,1} | p_r)}{f_0(\boldsymbol{Z}; \widehat{\boldsymbol{\theta}}_{r,0}, \widehat{\boldsymbol{\theta}}_{s,0})} + \log f(p_r) \right\} \underset{H_0}{\overset{H_1}{\gtrless}} \eta, \tag{2.26}$$

where $\widehat{\boldsymbol{\theta}}_{s,0}$ and $\widehat{\boldsymbol{\theta}}_{s,1}$ denote the ML estimates of $\boldsymbol{\theta}_s$, under H_0 and H_1, respectively. It is important to notice that (2.26) is equivalent to (2.19) when $g(p_r) = q(p_r)$. However, several alternatives for $g(p_r)$ are possible that would produce different penalization terms.

2.3.3 CFAR Property

In radar, as well as in other application fields, the design of architectures that can ensure the CFAR property is of great importance because it allows accurate target detection even in situations where the spectral properties of the interference (or undesired components) are unknown or highly variable. Actually, limiting and maintaining low false alarms is a preventive measure meant to avert the terrible consequences that they may otherwise have. Therefore, detection thresholds are specified by system engineers to ensure very low values for P_{fa} [26–30]. Since a generic detection technique generally cannot guarantee the CFAR property, it is necessary to show that, under the null hypothesis, its decision statistic is independent of the interference parameters. Nonetheless, there are instances in which the decision statistic is functionally invariant to those transformations that modify the nuisance parameters, which do not enter the decision process, and, at the same time, preserve the hypothesis testing problem. As a consequence, under the null hypothesis, such statistics follow a distribution that is independent of the nuisance parameters ensuring the CFAR property [14, 24, 31]. The propositions presented in [10] enable us to determine when (2.19) is invariant with respect to a given group of transformations and satisfies the CFAR property.

Let us assume that each binary hypothesis testing problem

$$\begin{cases} H_0: & \boldsymbol{Z} \sim f_0(\boldsymbol{Z}; \boldsymbol{\theta}_{r,0}, \boldsymbol{\theta}_s), \\ H_{1,m}: & \boldsymbol{Z} \sim f_{1,m}(\boldsymbol{Z}; \boldsymbol{\theta}_{r,1}^m, \boldsymbol{\theta}_s, p_{r,m}), \end{cases} \tag{2.27}$$

for all $m \in \{1, \dots, M\}$, is invariant with respect to the group of transformations. Thus, the CLR for the m-th testing problem, namely $\widehat{\Lambda}_1(\boldsymbol{Z}; m)$, is also invariant to the same transformation group [32, 33], namely,

$$\widehat{\Lambda}_1(l(\boldsymbol{Z}); m) = \widehat{\Lambda}_1(\boldsymbol{Z}; m), \quad \forall m \in \{1, \dots, M\}, \tag{2.28}$$

where $l(\boldsymbol{Z})$ belongs to the transformations' group for which the hypothesis testing problem is invariant. As a consequence,

$$\max_{m\in\{1,\dots,M\}} \left\{\widehat{\Lambda}_1(l(\boldsymbol{Z}); m) - q(m)\right\} = \max_{m\in\{1,\dots,M\}} \left\{\widehat{\Lambda}_1(\boldsymbol{Z}; m) - q(m)\right\}, \tag{2.29}$$

and the left-hand side of (2.19) is invariant with respect to the group transformations. Moreover, observe that if (2.19) is invariant, then, by *Theorem 6.2.1* of [32], its decision statistic can be expressed as a function of the maximal invariant statistics. Next, the distribution of a maximal invariant statistic depends on the induced maximal invariant. Finally, the left-hand side of (2.19) is a function of the maximal invariant statistic, whose distribution does not depend on the specific value of the parameter vector under H_0. Then, (2.19) ensures the CFAR property.

2.4 Performance Analysis for Application Examples

In this section, the decision criterion (2.19) is applied to the radar applications treated in Section 2.2. (2.19) will be referred to as the multiple hypothesis penalized-likelihood ratio test (MHP-LRT) in the following application examples. Furthermore, we will refer to as MHP-LRT(x) with $x \in \{\mathrm{a}, \mathrm{b}, \mathrm{c}\}$ based on the value of $q(m)$ used in (2.20).

A natural competitor for MHP-LRT is the so-called multifamily likelihood ratio test (MFLRT) [2], expressed as

$$\max_{m\in\{1,\dots,M\}} \left\{ \left[g_m(\boldsymbol{Z}) - p_{r,m}\left(\log \frac{g_m(\boldsymbol{Z})}{p_{r,m}} + 1\right)\right] u\left(\frac{g_m(\boldsymbol{Z})}{p_{r,m}} - 1\right)\right\} \underset{H_0}{\overset{H_{1,\hat{m}}}{\gtrless}} \eta, \tag{2.30}$$

with $g_m(\boldsymbol{Z}) = 2\widehat{\Lambda}_1(\boldsymbol{Z}, m)$, $u(\cdot)$, the Heaviside step function, and $\hat{m}$, the maximizer for the left-hand side of (2.30).

It is important to note that the ML technique is used by decision rule (2.19) to obtain estimates of the unknown parameters. However, in some cases, finding such estimates requires untractable mathematics; hence, alternative heuristic solutions, such as the so-called two-step GLRT design approach [14, 24], might be pursued instead of the plain GLRT. This criterion entails estimating the unknown parameters and replacing them with the plain GLRT after assuming that a subset of them are known. For the sake of comparison, the effectiveness of (2.19) is evaluated when $\widehat{\Lambda}_1(\boldsymbol{Z}; m)$ is substituted with another *ad hoc* statistic, which is referred to as $\widehat{\Lambda}_2(\boldsymbol{Z}; m)$. We will call the final decision schemes modified-MHP-LRT (M-MHP-LRT).

Moreover, the heuristic two-stage detectors (TS-Ds), obtained by cascading a MOS-based estimation stage, is also used for performance comparison. As depicted in Figure 2.4, the TS-D architecture is realized by cascading the MOS rule, for the estimation of the unknown number of range bins contaminated by targets or NLJs, and the detection stage exploiting the estimates provided by the first stage.

While the proposed design technique also reflects a theoretically founded design criterion, it is evident that the two-stage architectures created by cascading a MOS rule and a decision scheme might comply with overcoming bad performance of GLRT in the situation of nested hypothesis only. Furthermore, it is crucial to emphasize that, in contrast to the two-stage architectures where the penalty term is simply utilized for classification, using the proposed detectors' decision statistics, the penalty term that is directly included in the decision process. Because of this, it is anticipated that the suggested detectors and their two-stage equivalents will have comparable classification performance; nonetheless, we are unable to declare an exact equivalency in terms of detection performances. Thus, it is crucial to compare the detection capabilities of the proposed detectors with those of the two-stage architectures to ascertain whether or not the diversity

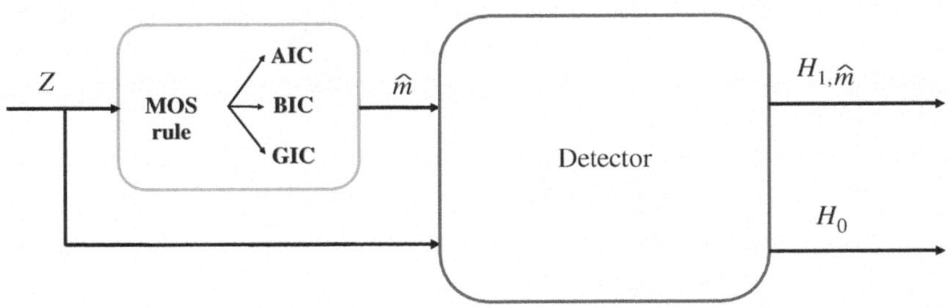

Figure 2.4 Two-stage detection architecture.

in the decision statistics yields different behaviors. Decision schemes obtained by applying the two-step GLRT design procedure differ from those based on the plain GLRT and, hence, a performance comparison becomes compelling and inevitable as corroborated by the huge amount of contributions that can be found in the related literature [14, 15, 24, 34, 35]. Notice also that in some cases, the two-step GLRT and the plain GLRT give rise to detectors that are asymptotically equivalent. With the above remarks in mind, it is important considering the two-stage architectures in terms of comparison, in addition to the MFLRT that is a natural competitor as it relies on a theoretically founded design criterion as well.

Additional competitors based upon theoretically founded design criteria can be obtained following the lead of [36]. However, this approach exploits the ML principle and a generalized Neyman–Pearson criterion [37–39] and leads to architectures whose probability of false alarm is difficult to control due to the fact that the number of thresholds required to set the probability of false alarm depends on the number of alternative hypotheses, and that the infinite number of threshold combinations returning the same value of the probability of false alarm and different detection performance. As a consequence, when the number of alternative hypotheses is high, the computational load required to set the thresholds becomes prohibitive and, in addition, a performance criterion is required to select one of the infinite threshold combinations corresponding to the preassigned probability of false alarm. Finally, this approach might provide poor performance in the case of nested hypotheses since the criterion selects the most plausible hypothesis on the basis of the difference between the decision statistic of the GLRT and the respective threshold associated to a given alternative hypothesis. For the above reasons, such a comparison is neither straightforward nor fair.

In the following, the considered performance metric is the probability of correct detection defined as $P_{d|m} = P(H_{1,m}|H_{1,m})$, $m \in \{1, \dots, M\}$. The detection threshold is estimated by means of $100/P_{fa}$ Monte Carlo trials, assuming $P_{fa} = 10^{-4}$. We adopt the classical definition for P_{fa}, i.e., the probability of rejecting H_0 when the latter is true. As stated above, the value of $\hat{m}$, which is an estimate of the model order under H_0, does not play any role in threshold setting, because, under H_0, data distribution does not depend on m and the decision statistic might return any admissible value for $\hat{m} \in \{1, \dots, M\}$. For this reason, it becomes clear that the threshold value depends on the set $\{1, \dots, M\}$ that is used in the last maximization of (2.19) to come up with $\hat{m}$.

To better clarify this point, let us assume that we want to estimate η by means of Monte Carlo trials. Thus, we follow the steps listed below:

1. for each Monte Carlo trial generate data under H_0 (observe that distribution under H_0 does not depend on m);
2. compute the left-hand side of (2.19) (observe that the selected value for m might be arbitrary but constrained to belong to $\{1, \dots, M\}$);

3. record the value of the decision statistic for each Monte Carlo trial;
4. select the empirical quantile returning the preassigned P_{fa}.

From the above list, it turns out that only the range of variation of m (namely, the set of alternative hypotheses under consideration) can affect the threshold selection.

In addition, we would like to highlight that m in the proposed framework and applications is not known, whereas the set $\{1, \dots, M\}$ is known. For instance, for the range-spread targets' case, $\Omega_{p,c}$ is defined as the set of consecutive range bins containing useful signal components whose cardinality c and position p are known under H_m with $m = (p-1)M + c$, but the actual hypothesis (i.e., the true m) is unknown. Thus, we estimate m (namely, the hypothesis) by computing

$$\widehat{\Lambda}_1(\boldsymbol{Z}; m) - q(m), \tag{2.31}$$

for each m, then we select $\hat{m}$ returning the maximum value of (2.31) and compare this value with the detection threshold.

Finally, we would like to highlight that the CFAR analysis for each detection problem is currently missing. As a matter of fact, the decision statistic of the newly proposed architectures is the maximum of correlated statistics whose distribution, in most cases, is not easily tractable from a mathematical point of view. As an alternative, we can resort to the *invariance principle* and, hence, we have to find a suitable group of transformations that includes those acting on nuisance parameters for each model order. Nevertheless, in the following, a preliminary sensitivity analysis for the problems at hand is presented.

2.4.1 Range-Spread Targets

For the case of range-spread targets, the expression of the decision statistic for the two-step GLRT design procedure, as well as for the heuristic two-stage architectures, is [15]

$$\widehat{\Lambda}_2(\boldsymbol{Z}; m) = \sum_{l \in \Omega_{p,c}} \frac{|\boldsymbol{v}^\dagger(\theta_T)\boldsymbol{S}_K^{-1}\boldsymbol{z}_l|^2}{\boldsymbol{v}^\dagger(\theta_T)\boldsymbol{S}_K^{-1}\boldsymbol{v}(\theta_T)}, \tag{2.32}$$

where $\boldsymbol{S}_K = (1/K)\boldsymbol{S}$, with $K \geq N$, is the sample covariance matrix of the secondary data. As for penalty terms, $p_m = 2c + N^2$ for the MHP-LRT, $p_m = 2c = p_{r,m}$ for the M-MHP-LRT, while $\mathcal{T} = 2(L+K)N$ for the MHP-LRT, and $\mathcal{T} = 2LN$ for the M-MHP-LRT.

Interference vectors $\boldsymbol{n}_l$, $l \in \Omega_T$, and $\boldsymbol{n}_k$, $k \in \Omega_S$, are generated as circular complex normal random vectors with zero mean and covariance matrix

$$\boldsymbol{\Sigma} = \sigma_n^2 \boldsymbol{I} + \sigma_c^2 \boldsymbol{\Sigma}_c, \tag{2.33}$$

where $\mathbf{\Sigma}_c(n,m) = \rho_c^{|n-m|}$ with $\rho_c = 0.95$, σ_c^2 is the clutter power set according to the clutter-to-noise ratio (CNR) that is equal to 20 dB, and σ_n^2 is the thermal noise power, which is set to 1 without loss of generality. Moreover, the simulation parameters are $N = 16, K = 32, L = 10, M = 4$, and the target is considered located at the range bins $\{4, 5\}$ with an SINR defined as

$$\text{SINR} = \sum_{l\in\{4,5\}} |\alpha_l|^2 \boldsymbol{v}^\dagger(0^\circ)\mathbf{\Sigma}^{-1}\boldsymbol{v}(0^\circ), \tag{2.34}$$

where $\boldsymbol{v}(0^\circ) = [1, 1, \dots, 1]^{\mathrm{T}}/\sqrt{N} \in \mathbb{C}^{N\times 1}$. Target power in different range bins is the same, namely, $|\alpha_4|^2 = |\alpha_5|^2$, whereas the phase is uniformly generated in $[0, 2\pi)$ for each α_l, $l = 4, 5$. The number of alternative hypotheses is 28 and the true hypothesis is indexed by 14.

The curves of the other decision rules saturate to values lower than 1, even though the MHP-LRT(c) and BIC-TS-D achieve $P_{d|14} = 0.98$. It turns out that the MHP-LRT(b) and GIC-TS-D with $\rho = 15$ ensure the best performance for SINR ≥ 20. This is because only their curves achieve $P_{d|14} = 1$.

In Figure 2.5, $P_{d|14}$ is plotted versus SINR. It turns out that the MHP-LRT(b) and GIC-TS-D with $\rho = 15$ ensure the best performance for SINR ≥ 20 since only their curves achieve $P_{d|14} = 1$, the MHP-LRT(c) and BIC-TS-D achieve $P_{d|14} = 0.98$, and the other decision rules saturate to values lower than 1. For SINR < 20 dB, the MHP-LRT(c) and BIC-TS-D almost share the same behavior and outperform the other architectures. Specifically, a gain of 1 dB for the MHP-LRT(c) and BIC-TS-D

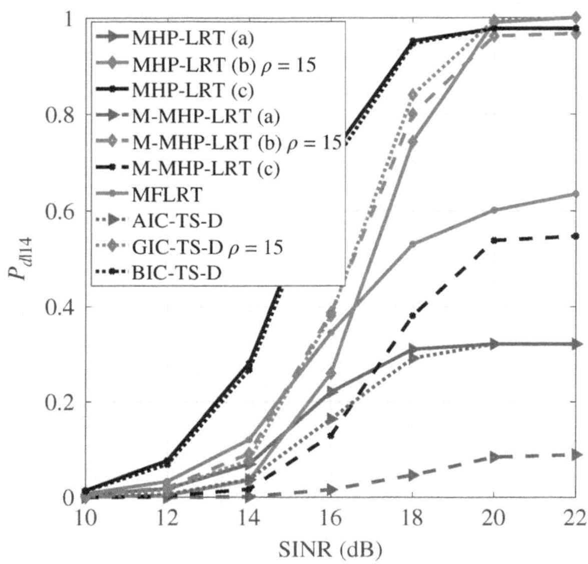

Figure 2.5 $P_{d|14}$ versus SINR for range-spread targets.

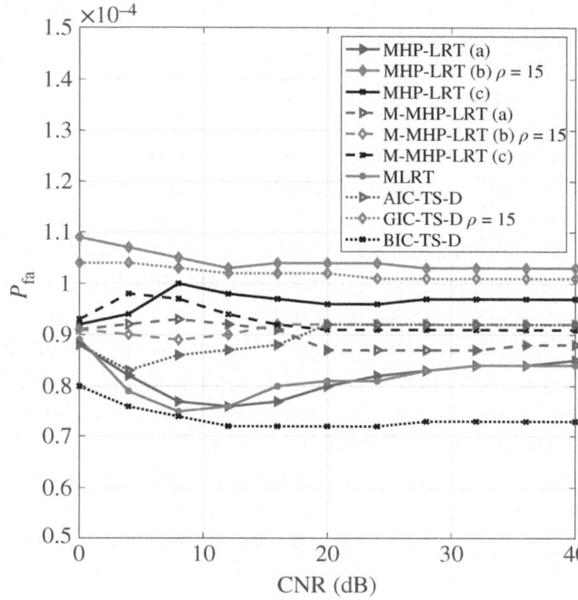

Figure 2.6 P_{fa} versus the noise power assuming the nominal $P_{\text{fa}} = 10^{-4}$ for range-spread targets.

over the MHP-LRT(b) and GIC-TS-D is found at $P_{d|14} = 0.8$. There is a notable decrease in the M-MHP-LRT(c) performance: for SINR > 20 dB, the returned $P_{d|14}$ value is only 0.55. MFLRT performs better than M-MHP-LRT(c). As expected, due to AIC's inclination to overestimate the model order, the worst performance is obtained using MHP-LRT(a), M-MHP-LRT(a) and AIC-TS-D.

Finally, the CFAR analysis, for the nominal $P_{\text{fa}} = 10^{-4}$, is shown in Figure 2.6. Results point out a certain degree of robustness for the considered range of the CNR values variation.

2.4.2 Multiple-Point Targets

The two-step GLRT statistic is given by

$$\hat{\Lambda}_2(\boldsymbol{Z}; m) = \sum_{l \in \Omega_m^e} \frac{|\boldsymbol{v}^\dagger(\theta_T)\boldsymbol{S}_K^{-1}\boldsymbol{z}_l|^2}{\boldsymbol{v}^\dagger(\theta_T)\boldsymbol{S}_K^{-1}\boldsymbol{v}(\theta_T)}. \tag{2.35}$$

As for the number of unknowns, $p_m = 3m + N^2$ and $\mathcal{T} = 2LN + 2NK$ for the MHP-LRT, whereas $p_m = 3m = p_{r,m}$ and $\mathcal{T} = 2LN$ for the M-MHP-LRT.

Circular complex normal random vectors with zero mean and covariance matrix given by (2.33) are generated for the interference vectors $\boldsymbol{n}_l, l \in \Omega_T$, and $\boldsymbol{n}_k, k \in \Omega_S$. The clutter power set according to the CNR is equal to 40 dB and σ_n^2 is set to 1. Five targets are randomly generated at each Monte Carlo trial within a window of length $L = 30$, which coincides with the upper bound on the number of targets, M. The SINR is defined as

$$\text{SINR} = P_\alpha^{\text{av}} \boldsymbol{v}^\dagger(0°)\boldsymbol{\Sigma}^{-1}\boldsymbol{v}(0°), \tag{2.36}$$

where P_α^{av} is the average power of the point-like targets, and it is set ranging from 7 to 17 dB. The modulus of the target responses is given by $|\alpha_l| = P_\alpha^{\text{av}}/5, l = 1, \dots, 5$, whereas their phases are uniformly generated in $[0, 2\pi]$. The tuning parameters are five for both GIC-TS-D, MHP-LRT(c) and 12 for M-MHP-LRT(c) (the first integer values assuring that $P_{d|5}$ reaches 1 for high SINR values). The amount of training vectors is $K = 48$, while $N = 16$.

The performance in terms of $P_{d|5}$ as functions of SINR are reported in Figure 2.7 for the MHP-LRT and M-MHP-LRT (a), (b), and (c) together with the MFLRT and TS-Ds. It is possible to notice that MHP-LRT detector with penalties (b) and (c) show good performance reaching $P_{d|5} = 1$ at near 16 dB. The GIC-TS-D is also capable to reach $P_{d|5} = 1$ at SINR = 17 dB. On the other hand, MHP-LRT(b) performs slightly better than the analogous two-stage architecture for SINR > 13 dB. As for M-MHP-LRT (b) and (c), they share almost the same performances as GIC-TS-D and BIC-TSD, respectively. The MHP-LRT (a), M-MHP-LRT (a), and MFLRT have the worst performance.

The results of the CFAR analysis for the nominal $P_{fa} = 10^{-4}$ are shown in Figure 2.8, where a good degree of robustness is appreciable for the considered CNR values.

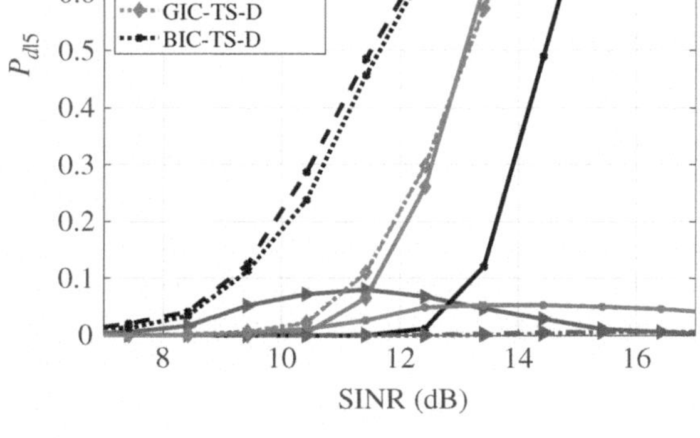

Figure 2.7 $P_{d|5}$ versus SINR for multiple-point targets.

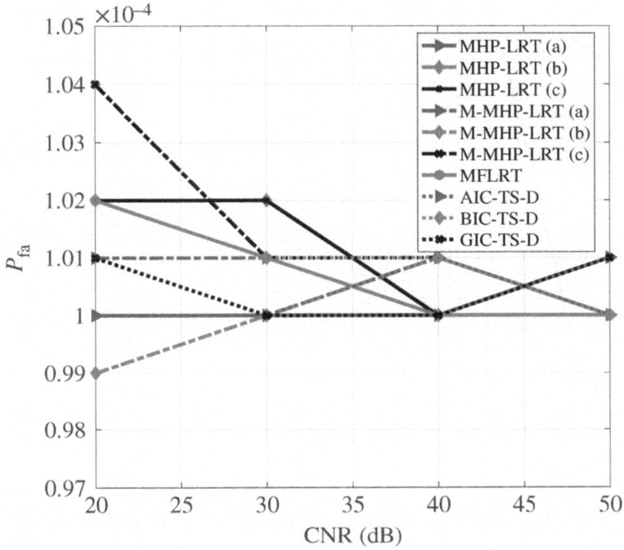

Figure 2.8 P_{fa} versus the noise power assuming the nominal $P_{fa} = 10^{-4}$ for multiple-point targets.

2.4.3 Multiple NLJs

The considered simulated scenario $N = 16, K = 32$, and $M = 6$. Data are simulated using circular complex normal random vectors with zero mean and covariance matrix as

$$\boldsymbol{\Sigma} = \sigma_n^2 \boldsymbol{I} + \text{JNR} \sum_{i=1}^{3} \boldsymbol{v}(\theta_i)\boldsymbol{v}^{\dagger}(\theta_i), \tag{2.37}$$

where $\sigma_n^2 = 1$, the jammers share the same power which is equal to the jammer-to-noise ratio (JNR), and the AOAs are $\theta_1 = 10°, \theta_2 = 20°$, and $\theta_3 = -15°$. It is supposed that $H_{1,3}$ is in force, i.e., we are in the presence of three NLJs sharing the same power and illuminating the radar from three different AOAs with values $10°, 20°$, and $-15°$. As for the computation of the penalty term we have $\mathcal{T} = 2KN$ and the number of unknowns is $p_m = m(2N - m) + 1$ ($p_{r,m} = m(2N - m)$). As for the BIC-like architectures, we consider the approximation for large K resulting to $\log(K)$ in place of $\log \mathcal{T}$.

For all the considered detection architectures, under the above assumptions, $P_{d|3}$ is plotted as a function of the JNR in Figure 2.9. The overlapping curves in the figure demonstrate how the MHP-LRTs and their two-stage counterparts share the same performance. Moreover, the MHP-LRT(a) and AIC-TS-D ensure the best

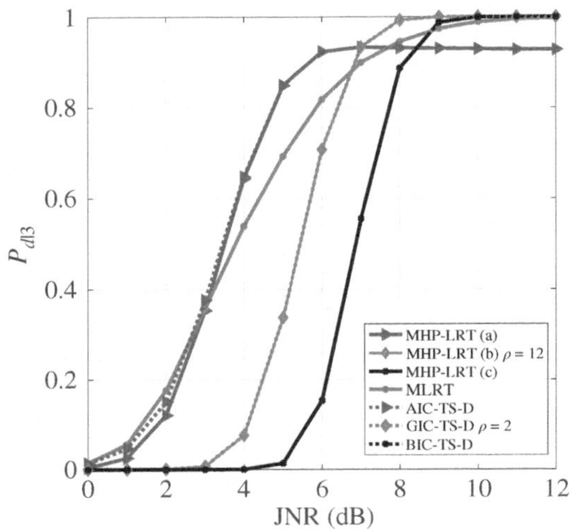

Figure 2.9 $P_{d|3}$ versus JNR.

performance for JNR ≤ 7 dB. However, they are not capable of achieving $P_{d|3} = 1$ at least for the considered parameter values. On the contrary, the MHP-LRT(b) with $\rho = 2$ and MHP-LRT(c) and their respective two-stage counterparts can reach $P_{d|3} = 1$. The MHP-LRT(b) with $\rho = 2$ outperforms MHP-LRT(c) with a gain of about 1 dB at $P_{d|3} = 0.8$. The curves of MFLRT are in between those of MHP-LRT(a) and MHP-LRT(b) with $\rho = 2$ and intersect with them.

The results for the CFAR analysis at nominal $P_{fa} = 10^{-4}$ are shown in Figure 2.10. For this case, a very good degree of robustness is observed for the considered range of noise power variation.

2.4.4 Discussion

The detection performances of the KLIC detectors are compared with those of the two-stage architectures to ascertain whether or not the diversity in the decision statistics yields different behaviors. We would like to highlight that, besides the considered competitors, a further term of comparison is the generalized Neyman–Pearson criterion [37, 38], that is applied in [39] and [36] in the context of target tracking and ECCM, respectively. However, we did not consider such an approach because of its inherent difficulty in controlling the false alarm rate (in fact, the number of thresholds depends on the number of alternative hypotheses) along with the fact that in the case of nested hypotheses it might return poor performance (in fact, it does not include any *balancing* term that accounts for the number of unknowns). The performance of the MFLRT, a natural competitor that is also based on a theoretically founded design approach, is also assessed.

The examined numerical examples have shown that the KLIC detectors are capable of outperforming the MFLRT and, in certain situations, also the heuristic

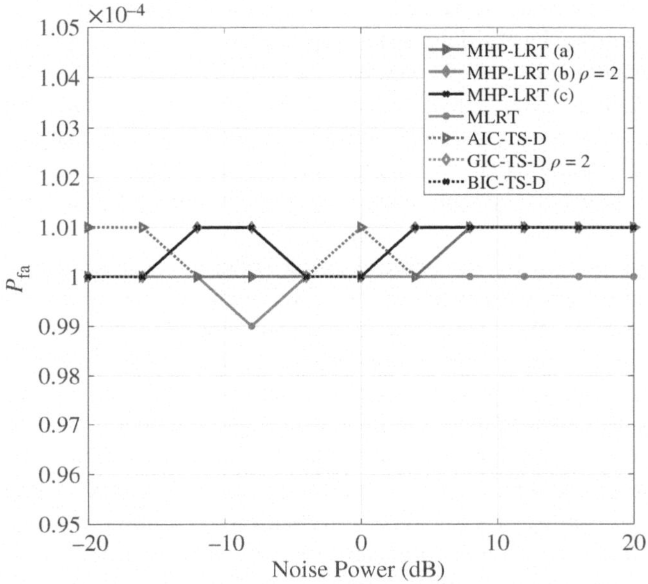

Figure 2.10 P_{fa} versus the noise power assuming the nominal $P_{fa} = 10^{-4}$.

two-stage architectures also provide better performance than the latter. The CFAR experimental analysis pointed out a certain degree of robustness for the considered range of parameter variations.

2.5 Conclusions

In this chapter, a design criterion, to solve multiple alternative hypotheses tests, based on the minimization of the KLD is presented. This criterion gives rise to decision schemes which inherit the penalty term of MOS rules.

These approaches enable the handling of multiple hypothesis tests, which frequently emerge and pose significant challenges in estimating unknown parameters. Three examples of radar applications include detecting range-spread objects with unknown size and position, identifying an unknown number of multiple point-like targets, and detecting an unknown number of NLJs.

The derived test solves this kind of problem, which contains only one null hypothesis and several alternative hypotheses. The decision rule compares the null hypothesis with a known number of alternative hypotheses. The unknown actual number of the true alternative hypothesis does not play an important role in threshold setting because, under the null hypothesis, the data distribution does not depend on it. For this reason, the decision rule solves the problem of the high computational burden raised by multiple nested hypotheses.

In all the scenarios examined, the decision strategies involve combining the CLR derived from available data with a penalty term that accounts for the number of unknown parameters. This term is determined through suitable approximations of the KLD between a candidate PDF and the true one. Thanks to the invariance principle, the proposed decision systems share the CFAR property given certain restrictions.

We applied the proposed decision schemes to three different radar detection problems also in comparison with heuristic two-stage architectures, and with a more natural competitor, the MFLRT, a theoretically founded framework. Alternative theoretically grounded competitors are found in [36]. However, the computational burden associated with threshold estimation can become exceedingly prohibitive in such cases, as these methods would necessitate a multitude of thresholds corresponding to the number of alternative hypotheses. Moreover, this approach, exploiting a generalized Neyman–Pearson criterion, would lead to architectures whose probability of false alarm is difficult to control because the number of thresholds required to set the probability of false alarm depends on the number of alternative hypotheses, as well as because of the infinite number of threshold combinations returning the same value of the probability of false alarm and different detection performance. Therefore, when the number of alternative hypotheses is high, the computational load required to set the thresholds becomes prohibitive and, in addition, a performance criterion is required to select one of the infinite threshold combinations corresponding to the preassigned probability of false alarm. Finally, this approach might provide poor performance in the case of nested hypotheses since, the criterion selects the most plausible hypothesis based on the difference between the decision statistic and the respective threshold associated with a given alternative hypothesis.

Performance analysis demonstrated that the KLIC detectors provide valuable tools for radar designers and operators to enhance radar performance in the presence of interference. These advancements contribute to the continued evolution of radar technology and its applications in various domains.

Potential future research directions could include investigating the CFAR property of the decision schemes discussed here within the framework of invariance. Additionally, it would be beneficial to evaluate the performance of these schemes by replacing the CLR with asymptotically equivalent statistics. Furthermore, exploring the derivation of alternative decision rules based on regularized likelihood ratios, which depend on various priors for the number of parameters, could also be valuable.

References

1 Stoica, P. and Selen, Y. (2004). Model-order selection: a review of information criterion rules. *IEEE Signal Processing Magazine* 21(4): 36–47.

2 Kay, S.M. (2005). The multifamily likelihood ratio test for multiple signal model detection. *IEEE Signal Processing Letters* 12(5): 369–371.

3 Bandiera, F., Orlando, D., and Ricci, G. (2006). CFAR detection of extended and multiple point-like targets without assignment of secondary data. *IEEE Signal Processing Letters* 13(4): 240–243.

4 Bandiera, F. and Ricci, G. (2006). Adaptive detection and interference rejection of multiple point-like radar targets. *IEEE Transactions on Signal Processing* 54 (12): 4510–4518.

5 Melvin, W.L. and Scheer, J.A. (2013). *Principles of Modern Radar: Advanced Techniques*. Edison, NJ: Publishing S.

6 Farina, A. (2008). ECCM Techniques. In: *Radar Handbook* (ed. M.I. Skolnik). McGraw-Hill.

7 Adamy, D. (2001). *EW101: A First Course in Electronic Warfare*. Norwood, MA: Artech House.

8 Yan, L., Addabbo, P., Hao, C. et al. (2019). New ECCM techniques against noise-like and/or coherent interferers. *IEEE Transactions on Aerospace and Electronic Systems* 56(2): 1172–1188.

9 Selén, Y. (2007). *Model Selection and Sparse Modeling*. Department of Information Technology, Uppsala University.

10 Addabbo, P., Han, S., Biondi, F. et al. (2021). Adaptive radar detection in the presence of multiple alternative hypotheses using Kullback-Leibler information criterion-Part I: detector designs. *IEEE Transactions on Signal Processing* 69: 3730–3741.

11 Addabbo, P., Han, S., Biondi, F. et al. (2021). Adaptive radar detection in the presence of multiple alternative hypotheses using Kullback-Leibler information criterion-Part II: applications. *IEEE Transactions on Signal Processing* 69: 3742–3754.

12 Richards, M.A., Scheer, J.A., and Holm, W.A. (2010). *Principles of Modern Radar: Basic Principles*. Raleigh, NC: Publishing S.

13 Abramovich, Y.I. and Johnson, B.A. (2008). GLRT-based detection-estimation for undersampled training conditions. *IEEE Transactions on Signal Processing* 56(8): 3600–3612.

14 Robey, F.C., Fuhrmann, D.R., Kelly, E.J., and Nitzberg, R. (1992). A CFAR adaptive matched filter detector. *IEEE Transactions on Aerospace and Electronic Systems* 28(1): 208–216.

15 Conte, E., De Maio, A., and Ricci, G. (2001). GLRT-based adaptive detection algorithms for range-spread targets. *IEEE Transactions on Signal Processing* 49 (7): 1336–1348.

16 Addabbo, P., Liu, J., Orlando, D., and Ricci, G. (2020). Novel parameter estimation and radar detection approaches for multiple point-like targets: designs and comparisons. *IEEE Signal Processing Letters* 27: 1789–1793.

17 Carotenuto, V., Hao, C., Orlando, D. et al. (2018). Detection of multiple noise-like jammers for radar applications. *2018 5th IEEE International Workshop on Metrology for AeroSpace (MetroAeroSpace)*, 328–333.

18 Yan, L., Hao, C., Addabbo, P. et al. (2019). An improved adaptive radar detector based on two sets of training data. *2019 IEEE Radar Conference (RadarConf)*, 1–6.

19 Kay, S.M. (1998). *Fundamentals of Statistical Signal Processing: Detection Theory*, vol. 2. Prentice-Hall.

20 Kullback, S. and Leibler, R.A. (1951). On information and sufficiency. *Annals of Mathematical Statistics* 22(1): 79–86.

21 Burnham, K.P. and Anderson, D.R. (2002). *Model Selection And Multimodel Inference, A Practical Information-Theoretic Approach*, 2e. New York: Springer.

22 Cover, T.M. and Thomas, J.A. (2012). *Elements of Information Theory*. Wiley.

23 Hao, C., Orlando, D., Foglia, G., and Giunta, G. (2016). Knowledge-based adaptive detection: joint exploitation of clutter and system symmetry properties. *IEEE Signal Processing Letters* 23(10): 1489–1493.

24 Bandiera, F., Orlando, D., and Ricci, G. (2009). *Advanced Radar Detection Schemes Under Mismatched Signal Models*, Synthesis Lectures on Signal Processing No 8. San Rafael, CA: Morgan & Claypool Publishers.

25 Yeredor, A. (2000). The joint MAP-ML criterion and its relation to ML and to extended least-squares. *IEEE Transactions on Signal Processing* 48(12): 3484–3492.

26 Farina, A. and Studer, F.A. (1986). A review of CFAR detection techniques in radar systems. *Microwave Journal* 29: 115.

27 Rohling, H. (1983). Radar CFAR thresholding in clutter and multiple target situations. *IEEE Transactions on Aerospace and Electronic Systems* 4: 608–621.

28 Barkat M, Himonas SD, Varshney PK. CFAR detection for multiple target situations. *IEE Proceedings F (Radar and Signal Processing)*, volume 136. IET; 1989. p. 193-209.

29 Roman, J.R., Rangaswamy, M., Davis, D.W. et al. (2000). Parametric adaptive matched filter for airborne radar applications. *IEEE Transactions on Aerospace and Electronic Systems* 36(2): 677–692.

30 Gini, F. and Greco, M. (2002). Covariance matrix estimation for CFAR detection in correlated heavy tailed clutter. *Signal Processing* 82(12): 1847–1859.

31 Kelly, E.J. (1986). An adaptive detection algorithm. *IEEE Transactions on Aerospace and Electronic Systems* 2: 115–127.

32 Lehmann, E.L. and Romano, J.P. (2008). *Testing Statistical Hypotheses*, Springer Texts in Statistics. New York: Springer.

33 Eaton, M.L. (1983). *Multivariate Statistics: A Vector Space Approach*, Lecture Notes-Monograph Series. Institute of Mathematical Statistics.

34 Bandiera, F., Besson, O., Orlando, D. et al. (2007). GLRT-based direction detectors in homogeneous noise and subspace interference. *IEEE Transactions on Signal Processing* 55(6): 2386–2394.

35 Abramovich, Y.I., Spencer, N.K., and Gorokhov, A.Y. (2007). Modified GLRT and AMF framework for adaptive detectors. *IEEE Transactions on Aerospace and Electronic Systems* 43(3): 1017–1051.

36 Bandiera F, Farina A, Orlando D, Ricci G. Detection algorithms to discriminate between radar targets and ECM signals. *IEEE Transactions on Signal Processing* 2010 58(12): 5984–5993.

37 Stuller, J. (1975). Generalized likelihood signal resolution. *IEEE Transactions on Information Theory* 21(3): 276–282.

38 Thomas, J. and Wolf, J. (1962). On the statistical detection problem for multiple signals. *IRE Transactions on Information Theory* 8(4): 274–280.

39 Buzzi, S., Lops, M., and Venturino, L. (2005). Track-before-detect procedures for early detection of moving target from airborne radars. *IEEE Transactions on Aerospace and Electronic Systems* 41(3): 937–954.

3

Information-Theoretic Approaches to Radar Target Enumeration

Lei Huang[1] *and Hing Cheung So*[2]

[1] *State Key Laboratory of Radio Frequency Heterogeneous Integration, Shenzhen University, Shenzhen, China*
[2] *Department of Electrical Engineering, City University of Hong Kong, Hong Kong, China*

3.1 Introduction

Array radar, such as multiple-input multiple-output (MIMO) radar and phased-array radar, offers extensive degrees of freedom in the spatial domain. This substantially augments the performance metrics of direction-of-arrival (DOA) estimation, especially in terms of identifiability, resolution, and accuracy. To address the challenge of resolving closely spaced targets situated at a considerable distance from the radar, subspace-based DOA estimation techniques become crucial, as they are capable of providing super-resolution in direction finding. This, in turn, underscores the importance of radar target enumeration (also known as source enumeration) because it serves as a prerequisite for implementing subspace-based DOA methods. However, as the number of antennas increases substantially, potentially equating to or surpassing the number of available snapshots, the covariance matrix becomes ill-conditioned. This phenomenon results in a pronounced dispersion of eigenvalues, thereby causing traditional source enumeration methods ineffective. Consequently, this impedes the correct decomposition of signal and noise subspaces, rendering subspace-based DOA estimation techniques inoperative.

In recent decades, the study of super-resolution DOA estimation has evolved considerably. Research has transitioned from traditional ℓ_2-norm (Euclidean) space approaches, such as the multiple signal classification (MUSIC) [1], estimation of signal parameters via rotational invariance techniques (ESPRIT) [2], method of direction estimation (MODE) [3, 4] and principal-eigenvector utilization for modal analysis (PUMA) [5, 6], to ℓ_p-norm (non-Euclidean) space

Information-Theoretic Radar Signal Processing, First Edition. Edited by Yujie Gu and Yimin D. Zhang.

methodologies where $p \geq 0$ but $p \neq 2$. Examples of the latter include the ℓ_1-based singular value decomposition (ℓ_1-SVD) method [7], Bayesian compressive sensing DOA (BCS-DOA) approach [8], and ℓ_p-MUSIC algorithm [9]. Nevertheless, an inherent drawback of these techniques lies in their dependence of the prior knowledge of the number of source signals, which, however, is usually unknown in advance and thus needs to be estimated. As a result, the field of array processing has identified source enumeration as a fundamental area of research [10–14]. Various algorithms for source enumeration exist, ranging from approaches based on hypothesis testing [15–17] to those guided by information-theoretic criterion (ITC) [18–21]. Fundamentally, these strategies are often designed for classical asymptotic settings, where the number of antennas remains constant while the number of snapshots increases without limit. Under such conditions, the Gaussian model proves to be an effective approximation, and ITCs having their roots in likelihood become the default methods for estimating the number of sources, especially when the number of snapshots is adequately large.

It is important to point out that Anderson's results [22] attain optimality in the maximum likelihood (ML) framework only within the classical asymptotic region, wherein the number of snapshots (n) increases toward infinity and the sensor size (m) remains unchanged. Such a restriction indicates that ITCs may not deliver optimal outcomes in the so-called general asymptotic setting [23–29], a context in which both m and n grow at a comparable pace. Under these conditions, the efficacy of ML estimators diminishes. The general asymptotic scenario is of considerable importance. For example, the tactical employment of waveform diversity in MIMO radar configurations [30–32] can lead to a virtually expanded sensor array, where sensor size may come close to, or even surpass, the number of snapshots. Additionally, Ref. [33] has suggested that the general asymptotic case offers a more accurate framework for describing practical scenarios. In order to improve the precision of ITC-based source enumeration under these parameters, the random matrix theory (RMT)-based Akaike information criterion (AIC) has been introduced in [12]. While RMT-AIC is purported to be effective for source enumeration in the general asymptotic setting, its consistency remains an open question. Consequently, there is a burgeoning interest in developing consistent methodologies for source enumeration in the general asymptotic context.

In this chapter, we first formulate the minimum description length (MDL) criterion by applying linear shrinkage (LS) techniques [23] and assuming Gaussian distribution for the observed data. This gives rise to an effective LS-MDL algorithm for source enumeration specifically tailored to the general asymptotic setting [11]. In this context, the number of sources d remains fixed, while $m, n \to \infty$ and $m/n \to c$, where c is a constant and $0 < c < \infty$. Importantly, our proposed covariance matrix estimator differs markedly from existing LS-based methods [23–26, 29]. Prior approaches are either for estimating the true covariance matrix [23–25, 29] or for predicated on *a priori* knowledge [26]. In contrast, our method is adept at leveraging consistent estimators of both the first- and second-order moments of the noise eigenvalues to calculate the shrinkage coefficient. As a result, we achieve a more reliable estimate of the covariance matrix of the noise subspace components. Nevertheless, the existing techniques can only depend on the sample covariance matrix (SCM).

Beyond the LS-MDL methodology, we also address the complex issue of source enumeration through the unique attributes of LS coefficients [17]. The efficacy of LS methods in improving the accuracy of covariance matrix estimation has already been well documented in the literature [23, 26, 28]. Utilizing the properties of the LS coefficients, and considering the asymptotic Gaussian nature of the noise shrinkage coefficients, we introduce both threshold-like testing and heuristic approaches for accurate source enumeration. The threshold-like strategy exploits the distribution of noise shrinkage coefficients to establish the decision-making boundary. Conversely, the heuristic approach is innovatively designed around a consistent relationship among noise shrinkage coefficients which are generally much greater than their signal counterparts. This leads to the introduction of shrinkage coefficient-based detectors (SCDs) [17]. Note that the threshold-like SCD, denoted as SCD_{thre}, is able to accurately identify the theoretical threshold even without the need for noise variance information. This makes its detection capabilities robust against uncertainties in noise power estimation, hence achieving constant false alarm rate (CFAR) properties. Furthermore, the iterative procedure for noise variance estimation, as outlined in [27], is obviated, which substantially reduces computational complexity. On the other hand, heuristic testing enables the effective separation of noise shrinkage coefficients from signal shrinkage coefficients. This gives rise to a novel heuristic shrinkage coefficient-based detector, denoted as SCD_{heur}.

The rest of the chapter is organized as follows. The source enumeration problem is formulated in Section 3.2. The LS-MDL approach is presented in Section 3.3. Moreover, the consistency of the LS-MDL criterion is discussed and the numerical results are provided to confirm the superiority of the LS-MDL technique. The SCD methods are addressed and the simulation results are presented in Section 3.4. Finally, the conclusion is drawn in Section 3.5.

3.2 Problem Formulation

In general, the observation space for array data can be divided into two orthogonal components, that is, the signal and noise subspaces. This separation is accomplished through the eigenvalue decomposition (EVD) of the covariance matrix. Accurate categorization of eigenvectors into these subspaces requires an exact estimate of the number of sources. Within the context of DOA estimation, the standard signal model is

$$\boldsymbol{y}_t = \boldsymbol{A}\boldsymbol{s}_t + \boldsymbol{w}_t,\ t = 1, \dots, n, \tag{3.1}$$

where

$$\begin{aligned}
\boldsymbol{y}_t &= [y_1(t), \dots, y_m(t)]^{\mathrm{T}} \in \mathbb{C}^{m\times 1},\\
\boldsymbol{A} &= [\boldsymbol{a}(\theta_1), \dots, \boldsymbol{a}(\theta_d)] \in \mathbb{C}^{m\times d},\\
\boldsymbol{s}_t &= [s_1(t), \dots, s_d(t)]^{\mathrm{T}} \in \mathbb{C}^{d\times 1},\\
\boldsymbol{w}_t &= [w_1(t), \dots, w_m(t)]^{\mathrm{T}} \in \mathbb{C}^{m\times 1}
\end{aligned}$$

are the observed sample vector, steering matrix, signal vector, and noise vector, respectively. Here, $\boldsymbol{a}(\theta_i)$ represents the steering vector where θ_i is the DOA of the ith source, $(\cdot)^{\mathrm{T}}$ denotes the transpose operator, m stands for the antenna size, n signifies the number of snapshots, and d is the *unknown* number of sources with the condition $d < \min(m, n)$. Meanwhile, the assumption here is that the incoherent signals follow a zero-mean complex Gaussian distribution, expressed as $\boldsymbol{s}_t \sim \mathcal{CN}(\boldsymbol{0}_d, \boldsymbol{\Sigma}_s)$ where $\boldsymbol{0}_d$ is the $d \times 1$ zero vector, $\boldsymbol{\Sigma}_s \triangleq \mathbb{E}[\boldsymbol{s}_t\boldsymbol{s}_t^{\mathrm{H}}] \in \mathbb{C}^{d\times d}$ is a full rank diagonal matrix with $(\cdot)^{\mathrm{H}}$ being the conjugate transpose and $\mathbb{E}[\cdot]$ denoting the mathematical expectation. Here, $\mathcal{CN}(\boldsymbol{v}, \boldsymbol{\Sigma})$ represents the complex Gaussian distribution characterized by mean $\boldsymbol{v}$ and covariance $\boldsymbol{\Sigma}$, and $\sim$ signifies "distributed as." Moreover, the noise $\boldsymbol{w}_t$ is presumed to be an independent and identically distributed (i.i.d.) complex Gaussian vector with zero mean and unknown covariance $\tau\boldsymbol{I}_m$, that is, $\boldsymbol{w}_t \sim \mathcal{CN}(\boldsymbol{0}_m, \tau\boldsymbol{I}_m)$ where $\boldsymbol{I}_m$ is the $m \times m$ identity matrix. Furthermore, the noise is considered independent of the signals. The primary objective of our study is to determine the number of sources d based on the noisy observations $\{\boldsymbol{y}_t\}_{t=1}^n$.

Based on our assumptions, the observed samples are taken as the independent Gaussian vector, that is, $\boldsymbol{y}_t \sim \mathcal{CN}(\boldsymbol{0}_m, \boldsymbol{\Sigma})$ where $\boldsymbol{\Sigma}$ represents the population covariance matrix, which is computed as

$$\boldsymbol{\Sigma} = \mathbb{E}\left[\boldsymbol{y}_t\boldsymbol{y}_t^{\mathrm{H}}\right] = \boldsymbol{A}\boldsymbol{\Sigma}_s\boldsymbol{A}^{\mathrm{H}} + \tau\boldsymbol{I}_m. \tag{3.2}$$

Recall that the signals in this context are incoherent and the condition $d < \min(m, n)$ holds, which imply that $\boldsymbol{\Sigma}_s$ is nonsingular and $\boldsymbol{A}$ has a full column rank. Let λ_i and $\boldsymbol{u}_i$, $i = 1, \dots, m$, represent the population eigenvalues and the associated population eigenvectors of $\boldsymbol{\Sigma}$, respectively. These eigenvalues λ_i are sorted in a nonincreasing manner, as follows:

$$\lambda_1 \geq \cdots \geq \lambda_d \geq \lambda_{d+1} = \cdots = \lambda_m = \tau. \tag{3.3}$$

It is intuitively simple to implement a threshold mechanism to distinguish between the d largest eigenvalues and the $(m - d)$ smallest ones, thereby determining the number of sources d. Upon achieving this, identifying the signal subspace and the noise subspace becomes relatively straightforward. The subspaces are, respectively, defined as

$$\boldsymbol{U}_{\mathcal{S}} = [\boldsymbol{u}_1, \dots, \boldsymbol{u}_d],$$
$$\boldsymbol{U}_{\mathcal{N}} = [\boldsymbol{u}_{d+1}, \dots, \boldsymbol{u}_m].$$

However, in a practical scenario, only the SCM is readily available which is computed as $\boldsymbol{S} = (1/n)\sum_{t=1}^{n} \boldsymbol{y}_t \boldsymbol{y}_t^{\mathrm{H}}$, and its EVD is

$$\boldsymbol{S} = \boldsymbol{E}\boldsymbol{L}\boldsymbol{E}^{\mathrm{H}}, \tag{3.4}$$

where $\boldsymbol{L} = \mathrm{diag}(\ell_1, \dots, \ell_m)$ with ℓ_i, $i = 1, \dots, m$, being the sample eigenvalue and $\mathrm{diag}(\boldsymbol{a})$ representing a diagonal matrix whose diagonal elements are the same as those of $\boldsymbol{a}$. $\boldsymbol{E} = [\boldsymbol{e}_1, \dots, \boldsymbol{e}_m]$ with $\boldsymbol{e}_i$, $i = 1, \dots, m$, denoting the corresponding sample eigenvector. Here, the sample eigenvalues are arranged in a nonincreasing order, expressed as

$$\ell_1 \geq \cdots \geq \ell_d \geq \ell_{d+1} \geq \cdots \geq \ell_m. \tag{3.5}$$

In the scenarios of high signal-to-noise-ratio (SNR), the signal variance substantially outweighs the noise variance, leading to $\ell_d \gg \ell_{d+1}$. Utilizing RMT [34–39], specifically, the $(d+1)$th sample eigenvalue, ℓ_{d+1}, approximately tends to the Tracy–Widom distribution [37], and a threshold can be calculated based on the limiting distribution of ℓ_{d+1}. Candidate models with sample eigenvalues falling below this threshold are subsequently eliminated, thereby yielding an estimate for the number of sources. This constitutes the foundational principle of threshold-like and RMT-based detection methods [27, 40]. However, in low SNR conditions, the variance of the signal is often masked by the noise, creating substantial barriers to precise source enumeration. In fact, these methods are restricted to use a single eigenvalue's information for detection, neglecting the potential information carried by the complete set of sample eigenvalues. Consequently, they may fail to provide a reliable estimate of the number of sources in low SNR environments. Additionally, the absence of a closed-form expression for the inverse of the Tracy–Widom distribution in turn requires a numerical evaluation procedure for threshold determination, adding computational overhead to the detection process. Lastly, these methods often rely on subjective criteria, such as the confidence level, to make decisions, which may not be straightforwardly implemented in real-world applications [10, 41, 42]. To address these problems, the LS-based MDL criterion has been developed in [11], offering an adaptive approach for detecting the number of sources in environments where parameters change dynamically.

According to the MDL principle, when confronted with a particular dataset and an assortment of probabilistic models, the optimal model to choose is the one that minimizes the description length for the dataset, leading to a metric that can be quantitatively assessed. In general, given an observation dataset $\boldsymbol{Y} = \{\boldsymbol{y}_t\}_{t=1}^{n}$ and a corresponding probabilistic model $f(\boldsymbol{Y}|\boldsymbol{\mu})$ where $\boldsymbol{\mu}$ represents the unknown parameter vector, the minimum code length needed for data encoding through this model can be expressed asymptotically as [43, 44]:

$$\mathcal{L}\{\boldsymbol{y}_t|\hat{\boldsymbol{\mu}}\} = -\log f(\boldsymbol{Y}|\hat{\boldsymbol{\mu}}) + \frac{1}{2}\nu \log n, \tag{3.6}$$

where $\hat{\boldsymbol{\mu}}$ represents the ML estimate of $\boldsymbol{\mu}$ and ν is the number of free parameters within $\hat{\boldsymbol{\mu}}$. In [45], Williams has demonstrated that the parameter vector should be constructed as $\boldsymbol{\mu} = \left[\boldsymbol{u}_1^{\mathrm{T}}, \dots, \boldsymbol{u}_d^{\mathrm{T}}\right]^{\mathrm{T}}$. This is because the first d eigenvectors and eigenvalues are just the weights for different signal components. Omitting these eigenvalues does not affect the definition of the signal subspace. Given k presumed source signals, it follows from [22] that the ML estimate for $\boldsymbol{\mu}$ is

$$\hat{\boldsymbol{\mu}}^{(k)} = \left[\boldsymbol{e}_1^{\mathrm{T}}, \dots, \boldsymbol{e}_k^{\mathrm{T}}\right]^{\mathrm{T}}. \tag{3.7}$$

Consequently, the MDL corresponding to k presumed source signals can be expressed as

$$\mathcal{L}\{\boldsymbol{y}_t|\hat{\boldsymbol{\mu}}^{(k)}\} = -\log f(\boldsymbol{Y}|\hat{\boldsymbol{\mu}}^{(k)}) + \frac{1}{2}\nu_k \log n, \tag{3.8}$$

where ν_k is the number of free parameters in $\hat{\boldsymbol{\mu}}^{(k)}$. Minimizing (3.8) with respect to k yields the estimate for the number of sources. As indicated in [42, 46], the log-likelihood term in (3.8) can be interpreted as a measure of information extracted from the observed data given the signal model. Conversely, the second term in (3.8), frequently termed the penalty function, is a function of the number of model's free parameters. When more parameters are incorporated into the model, the information gain typically decreases monotonically as the model more accurately fits the data. On the other hand, as the number of parameters increases, the error in the estimated parameters correspondingly grows, leading to an increased penalty function. Once the hypothesized number of sources aligns with the true model order, the diminishing information gain intersects with the escalating penalty function. This intersection point provides the optimal estimate for the number of sources.

3.3 LS-MDL Approach

3.3.1 Minimum Description Length

We note from (3.1) that in a noise-free environment, $\boldsymbol{y}_t$ resides in the column space of $\boldsymbol{A}$, commonly termed the signal subspace. In the presence of noise, however, $\boldsymbol{y}_t$ may diverge into its orthogonal complement, known as the noise subspace. This geometrical framework has been effectively exploited in high-resolution DOA estimation algorithms. To facilitate tractable MDL calculations within this framework, we introduce a unitary coordinate transformation matrix $\boldsymbol{U} = [\boldsymbol{u}_1, \ldots, \boldsymbol{u}_m]$, defined as $\boldsymbol{U} \triangleq [\boldsymbol{U}_k, \boldsymbol{U}_{m-k}]$. This transformation maps the first k coordinates with the signal subspace and the remaining $(m-k)$ coordinates with the noise subspace. For further details on how to construct $\boldsymbol{U}$, interested readers can refer to [18, 41, 47]. In the sequel, $\boldsymbol{U}_k \in \mathbb{C}^{m\times k}$ and $\boldsymbol{U}_{m-k} \in \mathbb{C}^{m\times(m-k)}$ represent the signal subspace and the noise subspace, respectively. With this unitary coordinate transformation, the observations $\{\boldsymbol{y}_t\}$ can be partitioned as

$$\boldsymbol{U}^{\mathrm{H}}\boldsymbol{y}_t = \begin{bmatrix} \boldsymbol{U}_k^{\mathrm{H}} \\ \boldsymbol{U}_{m-k}^{\mathrm{H}} \end{bmatrix} \boldsymbol{y}_t \triangleq \begin{bmatrix} \boldsymbol{y}_{\mathcal{S},t}^{(k)} \\ \boldsymbol{y}_{\mathcal{N},t}^{(k)} \end{bmatrix}, \tag{3.9}$$

where

$$\boldsymbol{y}_{\mathcal{S},t}^{(k)} = \boldsymbol{U}_k^{\mathrm{H}}\boldsymbol{y}_t, \tag{3.10}$$

$$\boldsymbol{y}_{\mathcal{N},t}^{(k)} = \boldsymbol{U}_{m-k}^{\mathrm{H}}\boldsymbol{y}_t \tag{3.11}$$

are the signal and noise subspace components, respectively. Since $\boldsymbol{U}$ is a unitary matrix, encoding the observation set $\{\boldsymbol{y}_t\}$ amounts to the encoding lengths for the signal subspace components $\{\boldsymbol{y}_{\mathcal{S},t}^{(k)}\}$ and the noise subspace components $\{\boldsymbol{y}_{\mathcal{N},t}^{(k)}\}$. As demonstrated in [11], the overall description length for encoding $\{\boldsymbol{y}_t\}$ can be asymptotically represented as

$$\mathcal{L}\left\{\boldsymbol{y}_{\mathcal{S},t}^{(k)}, \boldsymbol{y}_{\mathcal{N},t}^{(k)} | \boldsymbol{U}\right\} = n(m-k)\log \frac{\frac{1}{m-k}\operatorname{tr}\left[\boldsymbol{S}_{nn}^{(k)}\right]}{\left(\det\left[\boldsymbol{S}_{nn}^{(k)}\right]\right)^{\frac{1}{m-k}}} + \frac{1}{2}k(k-1)\log n, \tag{3.12}$$

where $\operatorname{tr}[\cdot]$ is the trace, $\det[\cdot]$ is the determinant, and $\boldsymbol{S}_{nn}^{(k)} = \frac{1}{n}\sum_{t=1}^{n}\boldsymbol{y}_{\mathcal{N},t}^{(k)}\left(\boldsymbol{y}_{\mathcal{N},t}^{(k)}\right)^{\mathrm{H}}$ is the SCM of the noise subspace components.

Note that minimizing the MDL as outlined in (3.12) yields an accurate estimation of the number of sources when $n \to \infty$ while m remains constant. This is due to the fact that $\boldsymbol{S}_{nn}^{(k)}$ is a consistent estimator of $\boldsymbol{\Sigma}_{nn}^{(k)} \triangleq \mathbb{E}\left[\boldsymbol{y}_{\mathcal{N},t}^{(k)}\left(\boldsymbol{y}_{\mathcal{N},t}^{(k)}\right)^{\mathrm{H}}\right]$, which is the population covariance matrix for the noise subspace components, under this classical asymptotic setting. However, in the more general asymptotic scenario where $m, n \to \infty$ and $m/n \to c$, $\boldsymbol{S}_{nn}^{(k)}$ has a notably high variance and thus becomes an inaccurate estimator for $\boldsymbol{\Sigma}_{nn}^{(k)}$. This inaccuracy compromises the effectiveness of the MDL criterion in (3.12). To ameliorate this issue, one viable approach is to recast the MDL criterion through the minimization of Kullback–Leibler (KL) divergence. However, this poses a big challenge, as it is difficult to determine the KL divergence in the situation where $m, n \to \infty$ and $m/n \to c$. Alternatively, we can enhance the estimation of $\boldsymbol{\Sigma}_{nn}^{(k)}$ in (3.12) for better performance of source enumeration. To achieve this, we utilize the LS technique in conjunction with Gaussian assumptions to derive a consistent estimator for $\boldsymbol{\Sigma}_{nn}^{(k)}$. The LS technique is advantageous as it offers a simple explicit formula for computing the covariance matrix, thereby simplifying the implementation of the proposed source enumerator. Moreover, it is important to recognize that $\boldsymbol{\Sigma}_{nn}^{(k)}$ adopts an identity structure, that is, $\tau \boldsymbol{I}_{m-k}$, for $k \geq d$, but not applicable when $k < d$. Leveraging this identity structure, we develop a precise estimator for $\boldsymbol{\Sigma}_{nn}^{(k)}$ in Section 3.3.2. Utilizing the refined estimation in the MDL calculation, ending up with an LS-MDL criterion, we are able to substantially improve the detection performance under the general asymptotic cases.

3.3.2 Estimation of Noise Subspace Covariance Matrix

Recall that the population covariance matrix of $\boldsymbol{y}_{\mathcal{N},t}^{(k)}$ is

$$\boldsymbol{\Sigma}_{nn}^{(k)} = \boldsymbol{U}_{m-k}^{\mathrm{H}} \boldsymbol{\Sigma} \boldsymbol{U}_{m-k} \triangleq \boldsymbol{Q} \boldsymbol{\Lambda}_{m-k} \boldsymbol{Q}^{\mathrm{H}}, \tag{3.13}$$

where $\boldsymbol{\Lambda}_{m-k} = \mathrm{diag}(\lambda_{k+1}, \dots, \lambda_m)$, $\boldsymbol{Q} = [\boldsymbol{q}_{k+1}, \dots, \boldsymbol{q}_m]$ with λ_i $(i = k+1, \dots, m)$ and $\boldsymbol{q}_i \in \mathbb{C}^{(m-k)\times 1}$ $(i = k+1, \dots, m)$ respectively being the eigenvalue and eigenvector of $\boldsymbol{\Sigma}_{nn}^{(k)}$. It is crucial to note that the eigenvalues of $\boldsymbol{\Sigma}_{nn}^{(k)}$ are identical when $k \geq d$ but distinct when $k < d$. Therefore, given $\boldsymbol{S}_{nn}^{(k)}$, our objective is to derive an optimal estimator for $\boldsymbol{\Sigma}_{nn}^{(k)}$ in the sense of minimum mean square error (MMSE) by means of the LS technique along with the i.i.d. Gaussian assumption of $\boldsymbol{y}_{\mathcal{N},t}^{(k)}$. To achieve this, we consider a constrained minimization problem constructed by the mean square error (MSE):

$$\begin{aligned} \min_{\alpha^{(k)}} \ & \mathbb{E}\left[\left\|\boldsymbol{R}^{(k)} - \boldsymbol{\Sigma}_{nn}^{(k)}\right\|_F^2\right] \\ \text{s.t.} \ & \boldsymbol{R}^{(k)} = \alpha^{(k)} \boldsymbol{P}^{(k)} + \left(1 - \alpha^{(k)}\right) \boldsymbol{S}_{nn}^{(k)}, \end{aligned} \tag{3.14}$$

where $\|\cdot\|_F$ is the Frobenius norm, $\alpha^{(k)} \in [0, 1]$ denotes the shrinkage coefficient [24] and $\boldsymbol{P}^{(k)} = \mathrm{tr}\left[\boldsymbol{\Sigma}_{nn}^{(k)}\right] / (m-k) \boldsymbol{I}_{m-k}$.

Notice that $\boldsymbol{S}_{nn}^{(k)}$ is an unbiased ML estimator of $\boldsymbol{\Sigma}_{nn}^{(k)}$ when m is fixed and $n \to \infty$. Nevertheless, this estimator is plagued by a large variance and is often ill-conditioned as $m \to \infty$, which has been thoroughly discussed in [23–25, 29]. In contrast, the structured estimator $\boldsymbol{P}^{(k)}$ manages to mitigate the variance at the cost of an increased bias. The estimated $\boldsymbol{R}^{(k)}$ in (3.14) is derived through shrinking $\boldsymbol{S}_{nn}^{(k)}$ toward $\boldsymbol{P}^{(k)}$, balancing the need for low bias with the demand for low variance. As a result, minimizing the MSE in (3.14) generates an accurate estimator for $\boldsymbol{\Sigma}_{nn}^{(k)}$ under the general asymptotic conditions.

Utilizing $\mathbb{E}\left[\boldsymbol{S}_{nn}^{(k)}\right] = \boldsymbol{\Sigma}_{nn}^{(k)}$, $\|\boldsymbol{A}\|_F^2 = \mathrm{tr}\left[\boldsymbol{A}\boldsymbol{A}^{\mathrm{H}}\right]$, and $\|\boldsymbol{A} + \boldsymbol{B}\|_F^2 = \|\boldsymbol{A}\|_F^2 + \|\boldsymbol{B}\|_F^2 + 2\mathrm{tr}\left[\Re\left(\boldsymbol{A}\boldsymbol{B}^{\mathrm{H}}\right)\right]$ with $\boldsymbol{A}, \boldsymbol{B} \in \mathbb{C}^{m\times m}$ and $\Re(\cdot)$ denoting the real part of complex number ·, the MSE cost function in (3.14) can be easily rewritten as

$$\begin{aligned} f\left(\alpha^{(k)}\right) &\triangleq \mathbb{E}\left[\left\|\alpha^{(k)}\boldsymbol{P}^{(k)} + \left(1-\alpha^{(k)}\right)\boldsymbol{S}_{nn}^{(k)} - \boldsymbol{\Sigma}_{nn}^{(k)}\right\|_F^2\right] \\ &= \mathbb{E}\left[\left\|\alpha^{(k)}\left(\boldsymbol{P}^{(k)} - \boldsymbol{\Sigma}_{nn}^{(k)}\right) + \left(1-\alpha^{(k)}\right)\left(\boldsymbol{S}_{nn}^{(k)} - \boldsymbol{\Sigma}_{nn}^{(k)}\right)\right\|_F^2\right] \\ &= \left(\alpha^{(k)}\right)^2\left\|\boldsymbol{P}^{(k)} - \boldsymbol{\Sigma}_{nn}^{(k)}\right\|_F^2 + \left(1-\alpha^{(k)}\right)^2\mathbb{E}\left[\left\|\boldsymbol{S}_{nn}^{(k)} - \boldsymbol{\Sigma}_{nn}^{(k)}\right\|_F^2\right]. \end{aligned} \tag{3.15}$$

By equating the derivative of $f(\alpha^{(k)})$ to zero and carrying out some algebraic operations, we arrive at the oracle estimation for $\alpha^{(k)}$:

$$\alpha_O^{(k)} = \frac{\mathbb{E}\left[\mathrm{tr}\left[\boldsymbol{S}_{nn}^{(k)}(\boldsymbol{S}_{nn}^{(k)})^{\mathrm{H}}\right]\right] - \mathrm{tr}\left[\boldsymbol{\Sigma}_{nn}^{(k)}(\boldsymbol{\Sigma}_{nn}^{(k)})^{\mathrm{H}}\right]}{\mathbb{E}\left[\mathrm{tr}\left[\boldsymbol{S}_{nn}^{(k)}(\boldsymbol{S}_{nn}^{(k)})^{\mathrm{H}}\right]\right] - \frac{\mathrm{tr}^2\left[\boldsymbol{\Sigma}_{nn}^{(k)}\right]}{m-k}}. \tag{3.16}$$

Accordingly, the oracle estimate of $\boldsymbol{\Sigma}_{nn}^{(k)}$ is

$$\boldsymbol{R}_O^{(k)} = \alpha_O^{(k)}\boldsymbol{P}^{(k)} + \left(1-\alpha_O^{(k)}\right)\boldsymbol{S}_{nn}^{(k)}. \tag{3.17}$$

The equivalence of eigenvectors between the shrunken and sample covariance matrices, as derived from (3.17), ensures the invariance of the parameter vector in MDL computations. However, it is important to note that the oracle estimate $\boldsymbol{R}_O^{(k)}$ in (3.17) is computationally infeasible. The reason is that $\alpha_O^{(k)}$ and $\boldsymbol{P}^{(k)}$ are contingent upon $\boldsymbol{\Sigma}_{nn}^{(k)}$, while $\boldsymbol{S}_{nn}^{(k)}$ is dependent on $\boldsymbol{U}_{m-k}$. Recalling that $\hat{\tau} = (1/(m-k))\sum_{i=k+1}^{m}\ell_i$ and $\boldsymbol{E}_{m-k}$ are respectively the ML estimates of τ and $\boldsymbol{U}_{m-k}$, it follows that

$$\hat{\boldsymbol{P}}^{(k)} = \hat{\tau}\boldsymbol{I}_{m-k}, \tag{3.18}$$

$$\hat{\boldsymbol{S}}_{nn}^{(k)} = \mathrm{diag}(\ell_{k+1}, \dots, \ell_m) \tag{3.19}$$

are the estimates of $\boldsymbol{P}^{(k)}$ and $\boldsymbol{S}_{nn}^{(k)}$, and from (3.16) that

$$\hat{\alpha}_O^{(k)} = \frac{\mathbb{E}\left[\sum_{i=k+1}^{m}\ell_i^2\right] - \sum_{i=k+1}^{m}\lambda_i^2}{\mathbb{E}\left[\sum_{i=k+1}^{m}\ell_i^2\right] - \frac{\left(\sum_{i=k+1}^{m}\lambda_i\right)^2}{m-k}} \tag{3.20}$$

is the estimate of $\alpha_O^{(k)}$. Nevertheless, (3.20) still involves the population eigenvalues. Indeed, for the Gaussian observations, invoking the results in [29, 48], i.e.,

$$\mathbb{E}\left[\mathrm{tr}\left[\hat{\boldsymbol{S}}_{nn}^{(k)}\right]\right] = \mathrm{tr}\left[\boldsymbol{\Sigma}_{nn}^{(k)}\right], \tag{3.21}$$

$$\mathbb{E}\left[\mathrm{tr}\left[(\hat{\boldsymbol{S}}_{nn}^{(k)})^2\right]\right] = \frac{n+1}{n}\,\mathrm{tr}\left[(\boldsymbol{\Sigma}_{nn}^{(k)})^2\right] + \frac{1}{n}\mathrm{tr}^2(\boldsymbol{\Sigma}_{nn}^{(k)}), \tag{3.22}$$

we have

$$\sum_{i=k+1}^{m}\lambda_i = \mathbb{E}\left[\sum_{i=k+1}^{m}\ell_i\right], \tag{3.23}$$

$$\sum_{i=k+1}^{m} \lambda_i^2 = \frac{1}{n+1}\left(n\mathbb{E}\left[\sum_{i=k+1}^{m} \ell_i^2\right] - \left(\mathbb{E}\left[\sum_{i=k+1}^{m} \ell_i\right]\right)^2 \right). \tag{3.24}$$

Substituting (3.23) and (3.24) into (3.20) yields

$$\hat{\alpha}_O^{(k)} = \frac{\mathbb{E}\left[\sum_{i=k+1}^{m} \ell_i^2\right] + \left(\mathbb{E}\left[\sum_{i=k+1}^{m} \ell_i\right]\right)^2}{(n+1)\left(\mathbb{E}\left[\sum_{i=k+1}^{m} \ell_i^2\right] - \frac{\left(\mathbb{E}\left[\sum_{i=k+1}^{m} \ell_i\right]\right)^2}{m-k}\right)}. \tag{3.25}$$

Nonetheless, the estimate is not immediately usable in real-world applications, as it demands knowledge of both $\mathbb{E}\left[\sum_{i=k+1}^{m} \ell_i^2\right]$ and $\mathbb{E}\left[\sum_{i=k+1}^{m} \ell_i\right]$. In fact, as demonstrated in [11], when $m, n \to \infty$ and $m/n \to c$, we have

$$\frac{1}{m-k}\sum_{i=k+1}^{m} \ell_i \xrightarrow{\text{m.s.}} \mathbb{E}\left[\frac{1}{m-k}\sum_{i=k+1}^{m} \ell_i\right], \tag{3.26a}$$

$$\frac{1}{m-k}\sum_{i=k+1}^{m} \ell_i^2 \xrightarrow{\text{m.s.}} \mathbb{E}\left[\frac{1}{m-k}\sum_{i=k+1}^{m} \ell_i^2\right], \tag{3.26b}$$

where $\xrightarrow{\text{m.s.}}$ signifies convergence in mean square. Following this, when these consistent estimates are substituted into (3.25), the consistent estimate of $\hat{\alpha}_O^{(k)}$ is obtained by

$$\hat{\alpha}^{(k)} = \frac{\sum_{i=k+1}^{m} \ell_i^2 + \left(\sum_{i=k+1}^{m} \ell_i\right)^2}{(n+1)\left(\sum_{i=k+1}^{m} \ell_i^2 - \frac{\left(\sum_{i=k+1}^{m} \ell_i\right)^2}{m-k}\right)}. \tag{3.27}$$

Since $\hat{\alpha}^{(k)}$ has the potential to exceed 1, we opt for $\beta^{(k)} = \min(\hat{\alpha}^{(k)}, 1)$ [23] instead of using $\hat{\alpha}^{(k)}$, to serve as the estimated shrinkage coefficient. Therefore, the covariance matrix is finally estimated as the following diagonal matrix:

$$\hat{\boldsymbol{R}}^{(k)} = \beta^{(k)}\hat{\boldsymbol{P}}^{(k)} + \left(1-\beta^{(k)}\right)\hat{\boldsymbol{S}}_{nn}^{(k)} \triangleq \operatorname{diag}\left(\rho_{k+1}^{(k)}, \dots, \rho_m^{(k)}\right) \tag{3.28}$$

with

$$\rho_i^{(k)} = \beta^{(k)}\hat{\tau}^{(k)} + \left(1-\beta^{(k)}\right)\ell_i,\ i = k+1, \dots, m. \tag{3.29}$$

It is implied in (3.29) that the noise eigenvalues are estimated by shrinking the noise sample eigenvalues to the estimated noise variance.

Proposition 3.1 For $k \geq d$, $\hat{\boldsymbol{R}}^{(k)}$ is the consistent estimate of $\boldsymbol{\Sigma}_{nn}^{(k)}$ as m increases to infinity at the same speed as n. That is,

$$\mathbb{E}\left(\left\|\hat{\boldsymbol{R}}^{(k)} - \boldsymbol{\Sigma}_{nn}^{(k)}\right\|_F^2\right) \to 0 \text{ as } m, n \to \infty \text{ and } m/n \to c. \tag{3.30}$$

Proof: The proof of Proposition 3.1 can be found in [11]. Recall the fact that the ML estimate, $\boldsymbol{S}_{nn}^{(k)}$, is also the consistent estimate of $\boldsymbol{\Sigma}_{nn}^{(k)}$ as $n \to \infty$ while m remains constant. Consequently, one can obtain the consistent estimator either by minimizing the MSE in (3.14) or by maximizing the likelihood function, given sufficiently large snapshots. This thereby allows us to replace $\boldsymbol{S}_{nn}^{(k)}$ with $\hat{\boldsymbol{R}}^{(k)}$ in (3.12) for source enumeration. However, under the general asymptotic conditions, $\boldsymbol{S}_{nn}^{(k)}$ becomes a poor estimator for $\boldsymbol{\Sigma}_{nn}^{(k)}$, necessitating the use of $\hat{\boldsymbol{R}}^{(k)}$ to improve the performance in source enumeration.

3.3.3 Source Enumeration via LS-MDL Criterion

For the purpose of detecting the number of sources, we substitute $\boldsymbol{S}_{nn}^{(k)}$ in (3.12) with the consistent estimator of $\boldsymbol{\Sigma}_{nn}^{(k)}$, that is, $\hat{\boldsymbol{R}}^{(k)}$. This results in a new variant of the MDL criterion, referred to as LS-MDL:

$$\begin{aligned}
\text{LS-MDL}(k) &= n\,(m-k)\log\frac{\frac{1}{m-k}\,\mathrm{tr}\left[\hat{\boldsymbol{R}}^{(k)}\right]}{\left(\det\left[\hat{\boldsymbol{R}}^{(k)}\right]\right)^{\frac{1}{m-k}}} + \frac{1}{2}k(k-1)\log n \\
&= n\,(m-k)\log\frac{\frac{1}{m-k}\sum\limits_{i=k+1}^{m}\rho_i^{(k)}}{\left(\prod\limits_{i=k+1}^{m}\rho_i^{(k)}\right)^{\frac{1}{m-k}}} + \frac{1}{2}k(k-1)\log n. \qquad (3.31)
\end{aligned}$$

By minimizing (3.31) with respect to k, an estimate for the number of sources is given by

$$\hat{d} = \arg\min_{k=0,\ldots,\bar{m}-1} \text{LS-MDL}(k), \qquad (3.32)$$

where $\bar{m}$ is the upper limit of the number of sources. Typically, $\bar{m}$ can be set as $\min(m,n)-1$, just like that in [12]. In practical applications, however, d is usually a relatively small number, while both m and n may asymptotically approach infinity. Under such circumstances, it becomes imperative to set $\bar{m}$ as a fixed value considerably smaller than $\min(m,n)$. The summary of the LS-MDL method for source enumeration is presented in Table 3.1.

Table 3.1 LS-MDL algorithm.

Step 1: Perform EVD on $\boldsymbol{S}$ and obtain the sample eigenvalues $\ell_1, \ldots, \ell_m$.
Step 2: Estimate the noise variance by $\hat{\tau}^{(k)} = (1/(m-k))\sum_{i=k+1}^{m}\ell_i$ and calculate the shrinkage coefficient as $\beta^{(k)} = \min(\hat{\alpha}^{(k)}, 1)$ with $\hat{\alpha}^{(k)}$ being given in (3.27).
Step 3: The shrunken estimates of λ_i, $i = k+1, \ldots, m$, are calculated as $\rho_i^{(k)} = \beta^{(k)}\hat{\tau}^{(k)} + \left(1-\beta^{(k)}\right)\ell_i$.
Step 4: The number of sources is estimated by minimizing the LS-MDL criterion of (3.31).

3.3.4 Consistency of LS-MDL Criterion

The consistency of the LS-MDL criterion is stated in the following proposition.

Proposition 3.2 The LS-MDL criterion given in (3.31) has strong consistency. That is, as $m, n \to \infty$ and $m/n \to c$, the LS-MDL criterion detects the number of sources with probability one.

Proof: See [11].

3.3.5 Numerical Examples

3.3.5.1 Performance of Source Enumeration

In this experiment, we evaluate the performance of the LS-MDL criterion in comparison to existing ITCs. Specifically, we consider the RMT-AIC, corrected AIC (AIC_{c_3}) [49], exponentially embedded families (EEF) [50, 51], classical MDL, and AIC [10]. For self-contained presentation, their functions are tabulated in Table 3.2.

Table 3.2 Criterion functions of RMT-AIC, AIC_{c_3}, EEF, MDL, and AIC.

Criterion	Function
RMT-AIC	$\hat{d} = \arg\min_{k\in[0,\check{m}-1]} \left\{ \frac{n^2}{2}\left[\frac{\frac{1}{m-k}\sum_{i=k+1}^{m}\ell_i^2}{\left(\frac{1}{m-k}\sum_{i=k+1}^{m}\ell_i\right)^2} - \left(1+\frac{m}{n}\right)\right]^2 + 2k \right\}$
AIC_{c_3}	$\hat{d} = \arg\min_{k\in[0,\check{m}-1]} \left\{ 2n(m-k)\log\frac{\frac{1}{m-k}\sum_{i=k+1}^{m}\ell_i}{\left(\prod_{i=k+1}^{m}\ell_i\right)^{1/(m-k)}} + \frac{2(\nu_k+1)(\check{m}+\nu_k+2)}{\check{m}-\nu_k-2} - \frac{2\nu_k}{\check{m}-\nu_k} \right\}$ $\nu_k = k(2m-k), \check{m} = 2mn$
EEF	$\hat{d} = \arg\max_{k\in[0,\check{m}-1]} \left\{ L_{G_k} - \nu_k\left[\log\left(\frac{L_{G_k}}{\nu_k}\right)+1\right]\right\}\cdot u\left(\frac{L_{G_k}}{\nu_k}-1\right)$ $\nu_k = k(2m-k)$ $u(\gamma) = 1$ for $\gamma \geq 0$ and $u(\gamma) = 0$ for $\gamma < 0$ $L_{G_k} = -2n\left[\log\left(\prod_{i=1}^{k}\ell_i\right) + (m-k)\log\left(\frac{1}{m-k}\sum_{i=k+1}^{m}\ell_i\right) - m\log\left(\frac{1}{m}\sum_{i=1}^{m}\ell_i\right)\right]$
MDL	$\hat{d} = \arg\min_{k\in[0,\check{m}-1]} \left\{ n(m-k)\log\frac{\frac{1}{m-k}\sum_{i=k+1}^{m}\ell_i}{\left(\prod_{i=k+1}^{m}\ell_i\right)^{1/(m-k)}} + \frac{1}{2}k(2m-k)\log n \right\}$
AIC	$\hat{d} = \arg\min_{k\in[0,\check{m}-1]} \left\{ 2n(m-k)\log\frac{\frac{1}{m-k}\sum_{i=k+1}^{m}\ell_i}{\left(\prod_{i=k+1}^{m}\ell_i\right)^{1/(m-k)}} + 2k(2m-k) \right\}$

We consider a uniform linear array configured with elements separated by half a wavelength, capturing three narrowband stationary Gaussian signals with the DOAs: $[2.3°, 9.8°, 17.5°]$. Given that both m and n can be large, we set $\bar{m}$ to be a fixed value of 10 for all criteria, unless otherwise specified, to ensure a fair comparison among the aforementioned schemes. Additionally, all numerical results are based on 2000 independent runs.

We consider a uniform linear array configured with elements separated by half a wavelength, capturing three narrowband stationary Gaussian signals with the DOAs: $[2.3°, 9.8°, 17.5°]$. Given that both m and n can be large, we set $\bar{m}$ to be a fixed value of 10 for all criteria, unless otherwise specified, to ensure a fair comparison among the aforementioned schemes. Additionally, all numerical results are based on 2000 independent runs.

Figure 3.1(a) depicts the empirical probabilities of error detection versus SNR for a scenario where the number of snapshots is smaller than the number of sensors,

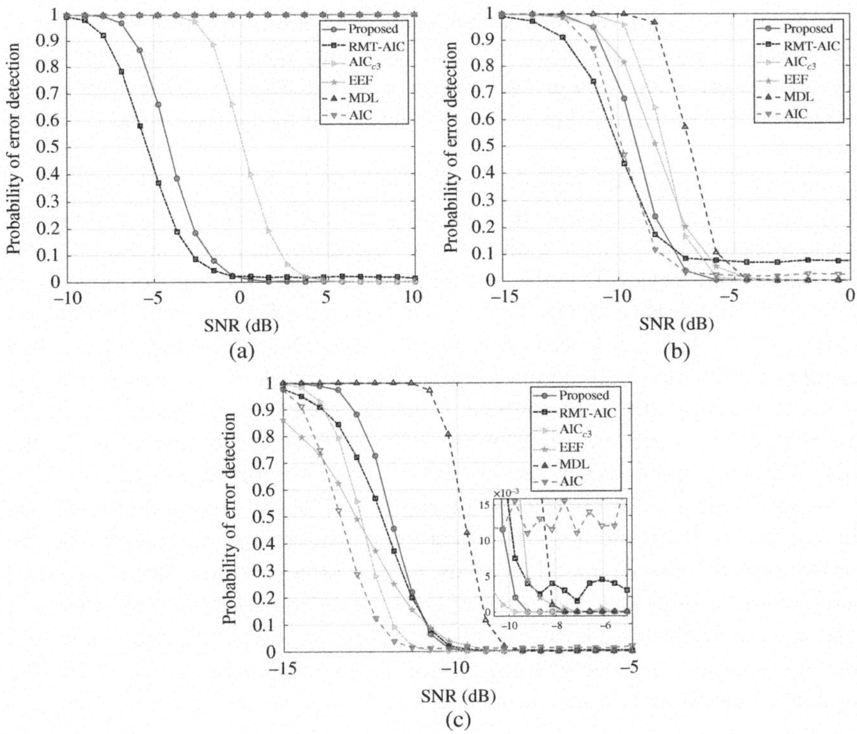

Figure 3.1 Probability of error detection versus SNR. (a) $m = 30$ and $n = 10$, (b) $m = 30$ and $n = 30$ and (c) $m = 30$ and $n = 90$.

specifically, $m/n = 3$. In this sample-starving case, only the proposed criterion, along with AIC_{c_3} and RMT-AIC methods, are capable of identifying the number of sources. The proposed method notably provides a consistent estimate of the number of sources, a feature not exhibited by the RMT-AIC approach. Furthermore, the proposed method substantially outperforms AIC_{c_3} in terms of error detection probability. When the number of snapshots expands to m, Figure 3.1(b) shows that the proposed criterion, along with the AIC_{c_3}, EEF, and MDL methods, can correctly detect the number of sources. Moreover, the AIC_{c_3}, EEF, and MDL methods require higher SNRs to reach the detection accuracy achieved by the proposed criterion. However, the RMT-AIC and AIC methods fail to provide reliable detection. They cannot offer consistent detection of the number of sources even though the SNR becomes large enough. Figure 3.1(c) depicts the scenario where n substantially exceeds m. In this case, the ML estimate becomes optimal, and the AIC_{c_3}, EEF, and MDL criteria exhibit good detection performance. Despite this, a substantial performance discrepancy still exists between our proposed criterion and MDL method. It is noteworthy that the AIC method, being an inconsistent estimator, never achieves a zero error detection probability, regardless of SNR levels. Similarly, the RMT-AIC and AIC_{c_3} approaches do not converge to zero in their error detection probabilities even at high SNRs. This behavior is substantiated by the magnified graph section presented in the right-hand corner of Figure 3.1(c).

Figure 3.2 depicts the probability of error detection versus the number of snapshots. The results confirm that our proposed method effectively enumerates the sources for $m = 3n$, $m = n$, and $m = n/3$. Meanwhile, Figure 3.2(a) indicates that while both RMT-AIC and EEF methods can detect the number of sources, their performance is suboptimal compared to the proposed scheme. Specifically, RMT-AIC fails to reach a zero error detection probability, and the EEF method requires approximately 10 additional snapshots to reach the detection accuracy of the proposed method. In contrast, other methods entirely fail to accurately enumerate the sources in this scenario. Figure 3.2(b) reveals that although the MDL and AIC_{c_3} methods can provide accurate source enumeration when $m = n$, their performance is still markedly inferior to that of the proposed criterion. In the condition where the number of snapshots exceeds the number of sensors, that is, $m = n/3$, the ML-based methods, such as MDL and EEF, become optimal and provide accurate source enumeration. However, as depicted in the magnified section in the right-hand corner of Figure 3.2(c), the RMT-AIC, AIC_{c_3}, and AIC methods still do not converge to zero error detection probability, even when the number of snapshots is sufficiently large.

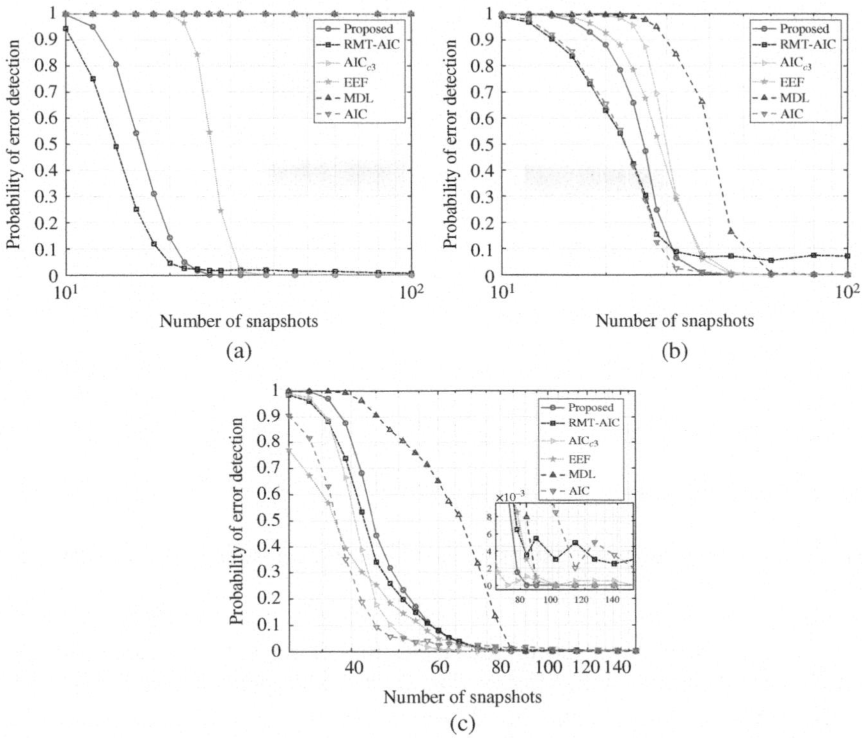

Figure 3.2 Probability of error detection versus number of snapshots. SNR = −8 dB. (a) $m = 3n$, (b) $m = n$ and (c) $m = n/3$.

3.4 SCD Approaches

Recall that only the SCM $\boldsymbol{S}$ can be accessed in practice. The noise sample eigenvalues of $\boldsymbol{S}$ satisfy $\ell_{d+1} \geq \cdots \geq \ell_m$ where the inequalities hold with probability one. When $n \to \infty$ and m remains constant, the ML estimates ℓ_i $(i = 1, \ldots, m)$ serve as efficient estimators for λ_i $(i = 1, \ldots, m)$. However, these sample eigenvalues tend to be inaccurate in a general asymptotic context. It has been established in [37] that these noise sample eigenvalues asymptotically follow the half-circle distribution or the Marčenko–Pastur distribution. This is illustrated in Figure 3.3, where $g(x)$ represents the probability density function of the random variable x.

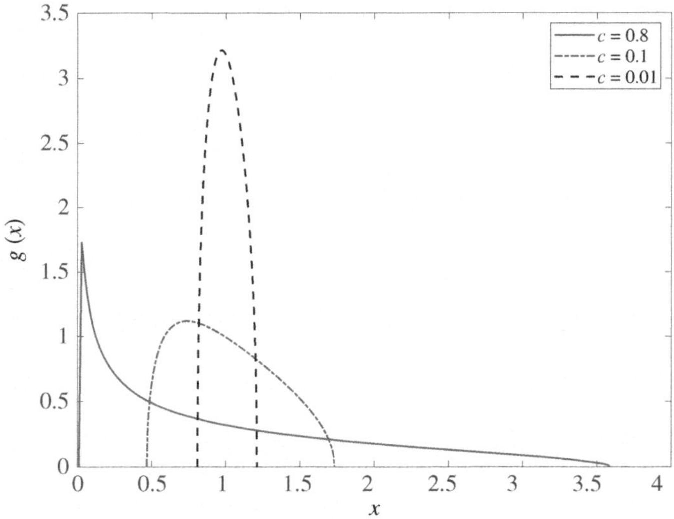

Figure 3.3 Marčenko–Pastur distribution $g(x)$ for eigenvalues x of noise-only SCM with unity noise variance.

Consequently, relying solely on the eigenvalues of the SCM for source enumeration is insufficient for achieving satisfactory detection performance in this case. One possible approach to enhance detection is to first accurately determine the distribution of the sample eigenvalues using the Marčenko–Pastur distribution. Subsequently, threshold-like testing methods such as RMT-based threshold testing (TT) [27] can be applied for source enumeration. However, this strategy suffers performance degradation owing to the inaccuracy in the noise variance estimation. Another possible approach is the use of ITC, including the AIC and MDL methods, which are good candidates for adaptive source enumeration. While these approaches are effective under the classical asymptotic conditions, they are less efficient in the general asymptotic scenario. This inefficiency arises because the probability model cannot be precisely approximated when the number of snapshots is relatively small and potentially comparable to antenna size.

To address this issue and provide reliable source number detection in general asymptotic scenarios under lower SNRs, the coefficients $\hat{\alpha}^{(k)}(k = 0, \dots, m-1)$ are leveraged for source enumeration. It should be emphasized that the behavior of the shrinkage coefficient diverges significantly when $k < d$ because the signal sample eigenvalues, that is, $\ell_k, \dots, \ell_d$, do not conform to the limiting values stipulated in (3.26), thereby providing a reliable metric for source enumeration. In contrast to the LS-MDL approach [11] which leverages LS to refine eigenvalue estimates, the methodologies formulated in this section are based on the distribution of the shrinkage coefficient and the disparity between the signal and noise shrinkage coefficients. Given that this distribution can be determined with high accuracy, and the divergence between the signal and noise shrinkage coefficients is substantial, the source enumerators proposed herein exhibit superior performance.

3.4.1 Threshold Testing for Source Enumeration

Let us define $\mathcal{H}_0$ and $\mathcal{H}_1$ to represent the hypotheses of having no more than k sources and a minimum of $k+1$ sources, respectively. Under the hypothesis $\mathcal{H}_0$, $\ell_{k+1}, \dots, \ell_m$ are the noise sample eigenvalues. In accordance with (3.27), $\hat{\alpha}^{(k)}$ $(k = d, \dots, m-1)$ is identified as the noise shrinkage coefficient, as it is wholly determined by the true noise eigenvalues. Conversely, $\hat{\alpha}^{(k)}$ $(k = 0, \dots, d-1)$ is designated as the signal shrinkage coefficient due to its predominant influence from the signal eigenvalues. The fluctuation of $\hat{\alpha}^{(k)}$ under $\mathcal{H}_0$ can be analyzed to establish a decision threshold in line with a prescribed probability of false alarm. Additionally, it is noteworthy that the value of $\hat{\alpha}^{(k)}$ under the hypothesis $\mathcal{H}_1$ is consistently smaller compared to its value under $\mathcal{H}_0$. As a result, the task of source enumeration effectively boils down to the following decision

$$\hat{\alpha}^{(k)} \underset{\mathcal{H}_1}{\overset{\mathcal{H}_0}{\gtrless}} \eta_k, \quad k = 0, \dots, m-1, \tag{3.33}$$

where η_k is the prescribed decision threshold, the determination of which is imperative and must be independent of the noise variance in order to achieve the CFAR property. Accurate threshold calculation for source enumeration necessitates a comprehensive understanding of the distribution of the shrinkage coefficients in the scenarios where $k \geq d$ which corresponds to the null hypothesis $\mathcal{H}_0$. To achieve this objective, we need the following results.

Proposition 3.3 The shrinkage coefficient $\hat{\alpha}^{(k)}$ with $k \geq d$ asymptotically obeys the Gaussian distribution as $m, n \to \infty$ and $m/n \to c$. That is,

$$\hat{\alpha}^{(k)} \xrightarrow{\mathcal{D}} \mathcal{N}(\nu_k, \sigma_k^2), \tag{3.34}$$

where $\xrightarrow{\mathcal{D}}$ denotes convergence in distribution and

$$\nu_k = \frac{c + (m - k + 1)}{c(n+1)}, \tag{3.35}$$

$$\sigma_k^2 = \frac{2(m-k+1)^2}{c^2(n+1)^2(m-k)^2}. \tag{3.36}$$

Proof: See [17].

Considering a specified level of false alarm denoted as ε, it is inferred from Proposition 3.3 that

$$\begin{aligned}\Pr\left(\mathcal{H}_1 \middle| \mathcal{H}_0\right) &= \Pr\left(\frac{\hat{\alpha}^{(k)} - \nu_k}{\sigma_k} < \tilde{\eta}_k \middle| \mathcal{H}_0\right) \\ &= 1 - Q(\tilde{\eta}_k) \triangleq \varepsilon,\end{aligned} \tag{3.37}$$

where $\tilde{\eta}_k$ serves as the threshold for the Gaussian distribution and $Q(\tilde{\eta}_k) = \int_{\tilde{\eta}_k}^{\infty} 1/\sqrt{2\pi}\,\exp(-t^2/2)dt$. Consequently, the threshold for the shrinkage coefficient is established as

$$\begin{aligned}\eta_k &= Q^{-1}(1-\varepsilon)\sigma_k + \nu_k \\ &= \frac{\sqrt{2}(m-k+1)}{c(n+1)(m-k)}Q^{-1}(1-\varepsilon) + \frac{c+(m-k+1)}{c(n+1)}.\end{aligned} \tag{3.38}$$

It has been demonstrated in [12, 39] that the limiting parameter $c = \lim_{m,n\to\infty} m/n$ can be substituted with $c_m \triangleq m/n$ for real-world applications. Consequently, one can employ the SCD_{thre} methodology to derive an estimator of the number of sources, specifically,

$$\hat{d} = \max_{k=0,\ldots,\bar{m}-1} \{i_k\}, \tag{3.39}$$

where $\bar{m} = \min(m, n)$ and

$$i_k = \begin{cases} k, & \hat{\alpha}^{(k)} < \eta_k, \\ 0, & \hat{\alpha}^{(k)} \geq \eta_k. \end{cases} \tag{3.40}$$

3.4.2 Heuristic Method for Source Enumeration

While the SCD_{thre} technique is able to accurately determine the theoretical threshold for detecting the number of sources, it does not efficiently leverage the substantial disparity between the signal and noise shrinkage coefficients. This disparity is empirically evident, as illustrated in Figure 3.4. Such intuition is corroborated by (3.14), which aligns well with the observed results in Figure 3.4. Specifically, the signal shrinkage coefficient is a diminutive value approximating zero, while the noise shrinkage coefficient is close to one. In the scenarios where $k < d$, the unknown true covariance matrix $\mathbf{\Sigma}_{nn}^{(k)}$ is diagonal but with different elements. To minimize the MSE in (3.14), a minimal weight should be assigned to the identity matrix, thereby indicating that $\alpha^{(k)}$ ought to be a small value. Conversely, when $k \geq d$, $\mathbf{\Sigma}_{nn}^{(k)}$ exhibits an identity structure, necessitating a large weight for $\mu_k \boldsymbol{I}_{m-k}$ in (3.14). As a result, $\alpha^{(k)}$ is close to one. Utilizing this substantial gap between the signal and noise shrinkage coefficients offers the potential for significant improvement in detection performance.

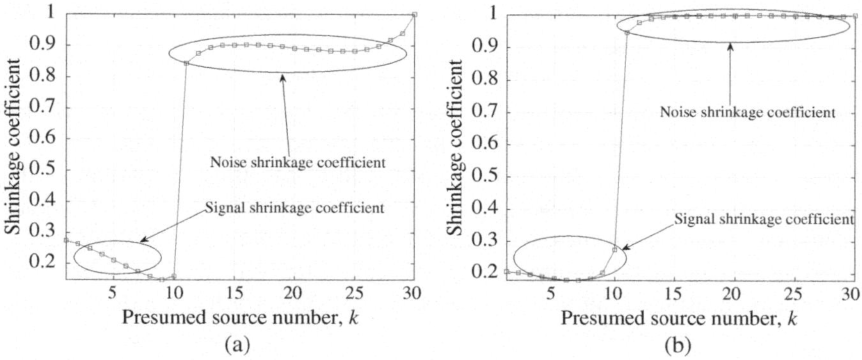

Figure 3.4 Shrinkage coefficient versus presumed source number. Ten sources with DOAs: $[0°, 6.67°, 13.33°, 20.00°, 26.67°, 33.33°, 40.00°, 46.67°, 53.33°, \text{and } 60.00°]$. (a) $m = 30, n = 50, d = 10$, and SNR = 0 dB. (b) $m = 30, n = 100, d = 10$, and SNR = −5 dB.

Proposition 3.4 As $m, n \to \infty$ and $m/n \to c$, the noise shrinkage coefficients $\hat{\alpha}^{(k)}$ $(k = d, \dots, m-1)$ tend to one almost surely (a.s.):

$$\hat{\alpha}^{(k)} = 1, \quad k = d, \dots, m-1, \tag{3.41}$$

whereas the signal shrinkage coefficients $\hat{\alpha}^{(k)}$ $(k = 0, \dots, d-1)$ are almost surely much smaller than one in large SNR region $(\ell_i/\tau \gg 1,\ i = 1, \dots, d)$ and strictly lower than one in small SNR region $(\ell_i/\tau \ll m,\ i = 1, \dots, d)$, namely,

$$\hat{\alpha}^{(k)} \begin{cases} \ll 1, & \ell_i/\tau \gg 1,\ i = 1, \dots, d, \\ < 1, & \ell_i/\tau \ll m,\ i = 1, \dots, d. \end{cases} \tag{3.42}$$

Proof: See [17].

In accordance with Proposition 3.4 that, when k varies from 0 to $m-1$, the noise shrinkage coefficients approach one asymptotically, while the signal shrinkage coefficients are asymptotically much smaller than one under large SNRs and strictly less than one under low SNRs. Consequently, paralleling the findings in [16, 52], the number of sources can be ascertained using the SCD_{heur}:

$$\hat{d} = \max_{k=0,\dots,\bar{m}-1} \{j_k\}, \tag{3.43}$$

where

$$j_k = \begin{cases} k, & \hat{\alpha}^{(k)} < \frac{D_n}{\bar{m}} \sum_{i=1}^{\bar{m}} \hat{\alpha}^{(i)}, \\ 0, & \hat{\alpha}^{(k)} \geq \frac{D_n}{\bar{m}} \sum_{i=1}^{\bar{m}} \hat{\alpha}^{(i)}. \end{cases} \tag{3.44}$$

and $D_n = \kappa / \log(n)$ with κ representing a constant value approximating $\log(n)$ which is the natural logarithm of n.

Remark 3.1 It is noteworthy that both the $\mathrm{SCD}_{\mathrm{heur}}$ and $\mathrm{SCD}_{\mathrm{thre}}$ algorithms involve the computations of the SCM and EVD, which are around $\mathcal{O}(m^2 n + m^3)$ flops. Furthermore, as delineated by (3.40) and (3.44), the formulation of the source enumerators for both $\mathrm{SCD}_{\mathrm{heur}}$ and $\mathrm{SCD}_{\mathrm{thre}}$ solely depends on the shrinkage coefficient, which is, in turn, a straightforward function of sample eigenvalues, as explicated in (3.27). Consequently, the computational overhead associated with the realization of source enumerators is negligible, confirming that the total computational complexity for either $\mathrm{SCD}_{\mathrm{heur}}$ or $\mathrm{SCD}_{\mathrm{thre}}$ remains approximately $\mathcal{O}(m^2 n + m^3)$ flops.

3.4.3 Numerical Examples

3.4.3.1 Accuracy of Theoretical Threshold for $\mathrm{SCD}_{\mathrm{thre}}$

We conduct an evaluation to assess the accuracy of the theoretical threshold for the $\mathrm{SCD}_{\mathrm{thre}}$. For comparative analysis, the empirical results for the RMT-TT [27] are also included. To establish a performance benchmark, the true threshold is numerically determined by repeatedly calculating $\hat{\alpha}^{(d)}$ under the assumption that the true number of sources d is known, which is based on a large number of independent trials. The true threshold is the one corresponding to a desired probability of false alarm: $P_{\mathrm{fa}} = \epsilon$. In this experiment, we conduct 50,000 independent runs, resulting in 50,000 $\{\hat{\alpha}^{(d)}\}$. Subsequent to organizing them in an ascending order, the $(P_{\mathrm{fa}} \cdot 50,000)$th coefficient is selected as the true threshold. The theoretical threshold for the $\mathrm{SCD}_{\mathrm{thre}}$ is obtained by employing (3.38), which is independent of noise variance τ. However, the theoretical threshold for RMT-TT [27] relies on the estimated noise variance $\hat{\tau}_{\mathrm{RMT}}$ that is determined by the following nonlinear system of equations:

$$\hat{\tau}_{\mathrm{RMT}} - \frac{1}{m-k}\left[\sum_{i=k+1}^{m} \ell_i + \sum_{i=1}^{k}(\ell_i - \hat{\rho}_i)\right] = 0, \tag{3.45}$$

$$\hat{\rho}_i^2 - \hat{\rho}_i\left(\ell_i + \hat{\tau}_{\mathrm{RMT}} - \hat{\tau}_{\mathrm{RMT}}\frac{m-k}{n}\right) + \ell_i\hat{\tau}_{\mathrm{RMT}} = 0. \tag{3.46}$$

The relative error is defined as Error $\triangleq (|\eta_{\mathrm{theo}} - \eta_{\mathrm{simu}}|)/\eta_{\mathrm{simu}} \times 100\%$ with η_{theo} and η_{simu} being the theoretical and simulated thresholds, respectively. The noise is an i.i.d. Gaussian process with zero mean and unknown variance τ.

The empirical results pertaining to threshold comparison are tabulated in Table 3.3, which specifies a configuration with 30 antennas, 40 snapshots, and the DOAs of five sources at $[1.2°, 5.8°, 11.7°, 17°, 20.3°]$. The results demonstrate

Table 3.3 Thresholds for $\mathrm{SCD}_{\mathrm{thre}}$ and RMT-TT at $m = 30$, $n = 40$, $c = 0.75$, SNR = 0 dB, and DOAs of $[1.2°, 5.8°, 11.7°, 17°, 20.3°]$.

Method	**$\mathrm{SCD}_{\mathrm{thre}}$**				**RMT-TT**			
P_{fa}	0.1	0.05	0.01	0.001	0.1	0.05	0.01	0.001
η_{theo}	0.8086	0.7912	0.7586	0.7221	3.3788	3.4283	3.5324	3.6646
η_{simu}	0.8540	0.8369	0.8066	0.7712	2.8454	2.9131	3.0500	3.2054
Error (%)	5.3102	5.4548	5.9439	6.3672	18.7452	17.6827	15.8158	14.3229

Table 3.4 Thresholds for $\mathrm{SCD}_{\mathrm{thre}}$ and RMT-TT at $m = 60$, $n = 80$, $c = 0.75$, SNR = 0 dB, and DOAs of $[1.2°, 5.8°, 11.7°, 17°, 20.3°]$.

Method	**$\mathrm{SCD}_{\mathrm{thre}}$**				**RMT-TT**			
P_{fa}	0.1	0.05	0.01	0.001	0.1	0.05	0.01	0.001
η_{theo}	0.9038	0.8952	0.8790	0.8609	3.4746	3.5067	3.5743	3.6601
η_{simu}	0.9256	0.9170	0.9007	0.8833	3.1495	3.1954	3.2898	3.3964
Error (%)	2.3557	2.3781	2.4111	2.5401	10.3207	9.7411	8.6501	7.7652

that the $\mathrm{SCD}_{\mathrm{thre}}$ algorithm outperforms the RMT-TT in terms of the accuracy of theoretical threshold determination. When the number of snapshots is increased to 80 while other parameters remain unchanged, the performance of the $\mathrm{SCD}_{\mathrm{thre}}$ algorithm substantially surpasses that of the RMT-TT, as highlighted in Table 3.4. Specifically, the $\mathrm{SCD}_{\mathrm{thre}}$ yields a relative error as low as approximately 2%, whereas the RMT-TT has an error on the order of 10% at $P_{\mathrm{fa}} = 0.1$. Furthermore, when the numbers of antennas and snapshots are both sufficiently large but grow at a uniform rate $c = 0.75$, e.g., $m = 90$ and $n = 120$, $\mathrm{SCD}_{\mathrm{thre}}$ continues to demonstrate the superior accuracy compared to the RMT-TT, as indicated in Table 3.5. Notice that the RMT-TT can approach its asymptotic limit of detection, i.e., $\tau\sqrt{n/m}$ [27], only when the noise variance τ is known *a priori*. In the scenarios where τ is unknown and must be estimated, the RMT-TT no longer reaches this asymptotic limit, thereby losing its optimality. This confirms the comparative inferiority of the RMT-TT algorithm relative to the $\mathrm{SCD}_{\mathrm{thre}}$.

3.4.3.2 Detection Performance

We now turn our attention to an evaluation of the detection performance of the SCDs. For the purpose of a comprehensive comparison, we include empirical results from state-of-the-art algorithms in source enumeration, specifically,

Table 3.5 Thresholds for SCD_{thre} and RMT-TT at $m = 90, n = 120, c = 0.75$, SNR = 0 dB and DOAs of $[1.2°, 5.8°, 11.7°, 17°, 20.3°]$.

Method	SCD_{thre}				RMT-TT			
P_{fa}	0.1	0.05	0.01	0.001	0.1	0.05	0.01	0.001
η_{theo}	0.9357	0.9300	0.9192	0.9072	3.4733	3.4979	3.5497	3.6153
η_{simu}	0.9498	0.9441	0.9336	0.9224	3.2526	3.2878	3.3588	3.4369
Error (%)	1.4777	1.4957	1.5373	1.6509	6.7865	6.3909	5.6811	5.1910

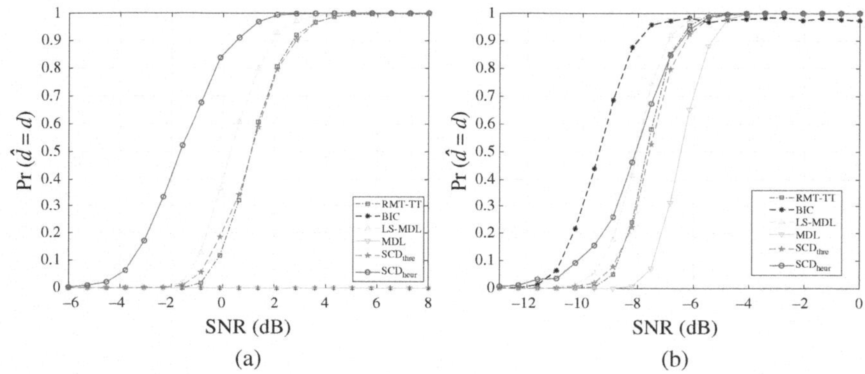

Figure 3.5 Probability of correct detection versus SNR. $\kappa = \log(n)$ and DOAs are $[1.2°, 5.8°, 11.7°, 17°, 20.3°]$ for $d = 5$. (a) $m = 30, n = 15$. (b) $m = 30, n = 40$.

the RMT-TT [27], Bayesian information criterion (BIC) [20], LS-MDL [11], and MDL [10]. Simulation results are based on 1000 independent runs. The probability of false alarm is preset at 0.001 for both the RMT-TT and SCD_{thre}, and $\kappa = \log(n)$ is assumed for the SCD_{heur}.

The empirical probabilities of correct detection versus SNR are depicted in Figure 3.5 where the number of antennas is 30 and the DOAs of five sources are $[1.2°, 5.8°, 11.7°, 17°, 20.3°]$. As illustrated in Figure 3.5(a), the classical MDL and BIC algorithms are ineffective, where the number of snapshots is 15 which is significantly fewer than the antenna size. This is because both algorithms are designed for the classical asymptotic regime where snapshots substantially outnumber antennas. Moreover, the SCD_{thre} performs similarly to the RMT-TT, but both lag behind the LS-MDL algorithm. Furthermore, the SCD_{heur}, owing to its ability to efficiently leverage the distance information between the signal and noise shrinkage coefficients, excels relative to other source enumeration methods. When the number of snapshots increases to exceed the antenna size,

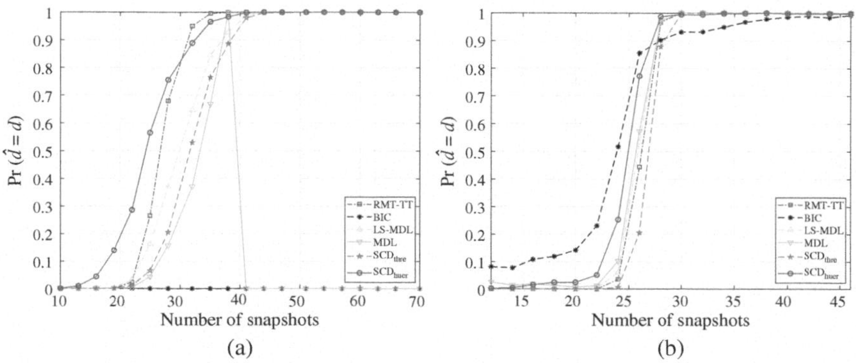

Figure 3.6 Probability of correct detection versus number of snapshots. $\kappa = \log(n)$ and DOAs are $[1.2°, 5.8°, 11.7°, 17°, 20.3°]$ for $d = 5$. (a) $m/n = 1.6$, SNR $= -5$ dB. (b) $m/n = 0.6$, SNR $= 5$ dB.

that is, $m/n = 3/4$, all tested algorithms work properly. Nevertheless, the BIC algorithm, while operates well at low SNRs, fails to achieve a probability of correct detection of one, even at high SNRs. As a threshold-like detector, the SCD_{thre} attains detection accuracy comparable to the RMT-TT. The performance of the SCD_{heur} is analogous to the LS-MDL but surpasses the threshold-like and MDL methods. The superiority of the SCD_{heur} can be rationalized particularly when the number of antennas exceeds the number of snapshots. Although LS-MDL enhances noise covariance matrix estimates via the LS technique, thereby generating an efficient MDL variant for the general asymptotic setting, it neglects the significant disparity between signal and noise shrinkage coefficients. In contrast, the SCD_{heur} algorithm effectively capitalizes on this distinction to enhance source enumeration, leading to a marked improvement in detection accuracy.

Figure 3.6 illustrates the empirical probability of correct detection versus the number of snapshots. From Figure 3.6(a), it is evident that both the MDL and BIC methods fail to correctly detect the number of sources when the number of snapshots is smaller than the antenna size. In this scenario, the RMT-TT algorithm outperforms both the SCD_{thre} and the LS-MDL methods. Nonetheless, it is not as accurate as the SCD_{heur} approach. Conversely, when $m/n = 3/5$, all the methods succeed in accurately enumerating the sources. With the exception of the BIC method, the performance across all algorithms is similar in terms of detection accuracy. Specifically, the SCD_{heur} exhibits a slight edge over other techniques, whereas the SCD_{thre} is marginally less effective. It is worth noting that the BIC algorithm converges to a probability of correct detection of one at a significantly slower rate compared to the other algorithms. Nevertheless, it maintains

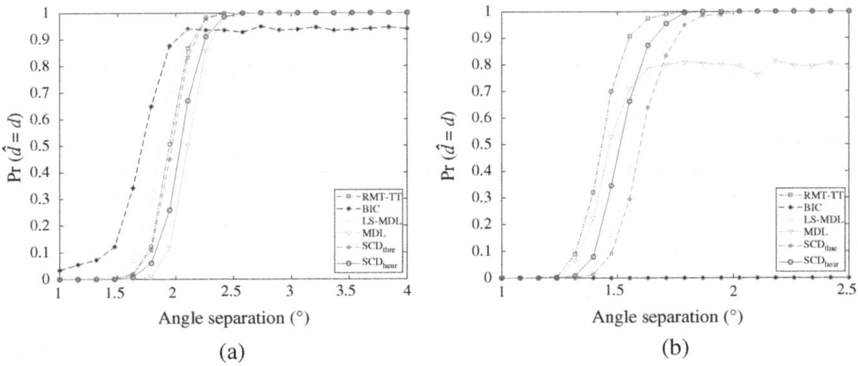

Figure 3.7 Probability of correct detection versus angle separation Δ. $[\theta_1, \theta_2, \theta_3] = [0, \Delta, 2\Delta]$ and $\kappa = \log(n)$. (a) $m = 25$, $n = 30$ and SNR $= 3$ dB. (b) $m = 50$, $n = 30$ and SNR $= -3$ dB.

relatively higher accuracy when operating with small numbers of snapshots and antennas.

Figure 3.7 presents the correct detection probabilities versus the angular separation, where the DOAs of the three sources are set as $[0, \Delta, 2\Delta]°$. As observed from Figure 3.7(a), all methods successfully identify the true number of sources when the number of snapshots exceeds that of the antennas. However, the BIC algorithm does not achieve a probability of correct detection of one, even when the angular separation is large enough. Among the methods studied, LS-MDL delivers the highest accuracy, while the MDL method performs the least. The SCD$_{\text{thre}}$ exhibits detection accuracy closely aligned with that of the RMT-TT, whereas the SCD$_{\text{heur}}$ is marginally less accurate. Conversely, in the scenarios where the number of snapshots is fewer than the antenna size, as shown in Figure 3.7(b), both the MDL and BIC algorithms cannot work properly. Under this condition, the RMT-TT method outshines all other techniques in terms of detection accuracy.

To investigate the identifiability performance of the methods under consideration, the empirical probabilities of correct detection versus the number of sources are depicted in Figure 3.8. In this context, the DOAs of $d \in [0, 15]$ sources are set as $[\theta_1, \dots, \theta_d] = \boldsymbol{\theta}(1 : d)$ with

$$\boldsymbol{\theta} = \underbrace{[0°, 3.57°, 7.14°, 10.71°, \dots, 50°]}_{15}. \tag{3.47}$$

Figure 3.8(a) reveals that both SCD$_{\text{heur}}$ and SCD$_{\text{thre}}$ exhibit superior identifiability compared to other methods when $m = 30$ and $n = 20$. In the scenario where $m = 30$ and $n = 60$, all the evaluated algorithms successfully identify the correct number of sources. Although the identifiability of SCD$_{\text{heur}}$ remains above that

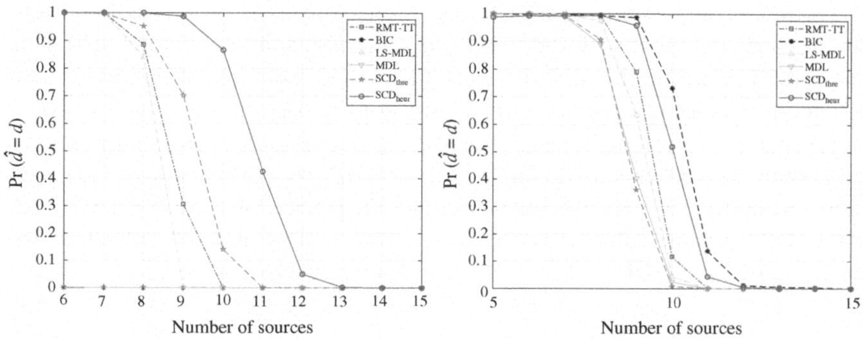

(a) (b)

Figure 3.8 Probability of correct detection versus number of sources. $\kappa = \log(n)$ and DOAs are $[\theta_1, \ldots, \theta_d] = \theta(1 : d)$. (a) $m = 30$, $n = 20$ and SNR = 3 dB. (b) $m = 30$, $n = 60$, and SNR = −5 dB.

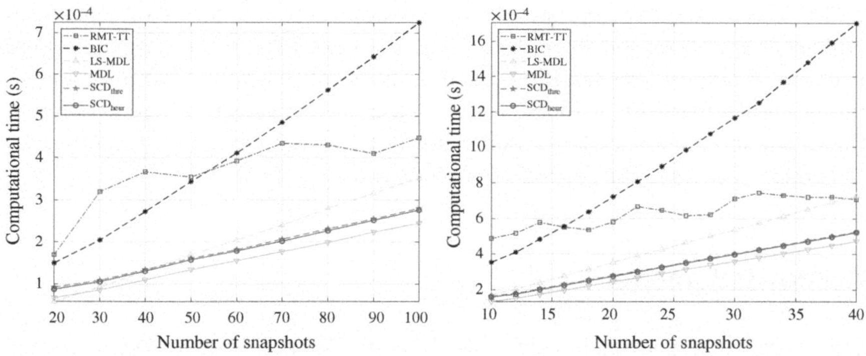

(a) (b)

Figure 3.9 Computational cost versus number of snapshots. $\kappa = \log(n)$ and DOAs are $[1.2°, 5.8°, 11.7°, 17°, 20.3°]$ for $d = 5$. (a) $m/n = 0.1$ and SNR = 0 dB. (b) $m/n = 0.5$, and SNR = 5 dB.

of most other methods, it is slightly surpassed by the BIC algorithm in this setting.

Figure 3.9 illustrates the computational costs of various algorithms versus the number of snapshots. The graph indicates that the proposed SCDs are computationally less demanding than the RMT-TT, MDL, LS-MDL, and BIC algorithms, particularly as both m and n scale up. Interestingly, the computational expenditure for the SCD methods closely mirrors that of the MDL method. This similarity can be attributed to the fact that the computational complexity for the MDL method largely stems from the calculations of the SCM and EVD, which are also the primary components in the SCD algorithms. The observed simulation results are in good alignment with the complexity analysis conducted for the SCD techniques. It is worth noting that the RMT-TT method relies on an iterative process for noise variance estimation, and the iterations persist until a predefined stopping criterion is met. Furthermore, the BIC approach incurs greater computational costs than the MDL, LS-MDL, SCD_{heur}, and SCD_{thre} algorithms due to the necessity of computing an extra likelihood term for the eigenvalues. Consequently, both the RMT-TT and BIC algorithms impose a greater computational burden than the SCD_{heur}, SCD_{thre}, MDL, and LS-MDL, as substantiated by Figure 3.9.

3.5 Conclusion

Utilizing the LS technique, we have derived three approaches to accurate source enumeration in the general asymptotic situation. It is demonstrated that the LS-MDL is superior to the state-of-the-art approaches in terms of robustness and accuracy. Moreover, since the SCD_{thre} does not rely on the estimated noise variance, it outperforms the RMT-TT in terms of robustness against noise uncertainty. Unlike the SCD_{thre}, the SCD_{heur} can use the distance information between the signal and noise shrinkage coefficients and thus is much more accurate than the other schemes in the general asymptotic case.

Acknowledgments

This work was supported in part by the National Science Fund for Distinguished Young Scholars under Grant 61925108, the Key Project of International Cooperation and Exchanges of the National Natural Science Foundation of China under Grant 62220106009, the project of Shenzhen Peacock Plan Teams under Grant KQTD202108 11090051046, and Shenzhen University 2035 Program for Excellent Research.

References

1 Schmidt, R. (1986). Multiple emitter location and signal parameter estimation. *IEEE Transactions on Antennas and Propagation* 34(3): 276–280.

2 Roy, R., Paulraj, A., and Kailath, T. (1986). ESPRIT–A subspace rotation approach to estimation of parameters of cisoids in noise. *IEEE Transactions on Acoustics, Speech, and Signal Processing* 34(5): 1340–1342.

3 Stoica, P. and Sharman, K.C. (1990). Novel eigenanalysis method for direction estimation. *IEE Proceedings F (Radar and Signal Processing)* 137: 19–26.
4 Qian, C., Huang, L., Cao, M. et al. (2017). PUMA: an improved realization of mode for DOA estimation. *IEEE Transactions on Aerospace and Electronic Systems* 53(5): 2128–2139.
5 Sun, W., So, H.C., Chan, F.K.W., and Huang, L. (2013). Tensor approach for eigenvector-based multi-dimensional harmonic retrieval. *IEEE Transactions on Signal Processing* 61(13): 3378–3388.
6 Qian, C., Huang, L., Sidiropoulos, N.D., and So, H.C. (2016). Enhanced PUMA for direction-of-arrival estimation and its performance analysis. *IEEE Transactions on Signal Processing* 64(16): 4127–4137.
7 Malioutov, D., Cetin, M., and Willsky, A.S. (2005). A sparse signal reconstruction perspective for source localization with sensor arrays. *IEEE Transactions on Signal Processing* 53(8): 3010–3022.
8 Carlin, M., Rocca, P., Oliveri, G. et al. (2013). Directions-of-arrival estimation through Bayesian compressive sensing strategies. *IEEE Transactions on Antennas and Propagation* 61(7): 3828–3838.
9 Zeng, W.-J., So, H.-C., and Huang, L. (2013). ℓ_p-music: robust direction-of-arrival estimator for impulsive noise environments. *IEEE Transactions on Signal Processing* 61(17): 4296–4308.
10 Wax, M. and Kailath, T. (1985). Detection of signals by information theoretic criteria. *IEEE Transactions on Acoustics, Speech, and Signal Processing* 33(2): 387–392.
11 Huang, L. and So, H.-C. (2013). Source enumeration via MDL criterion based on linear shrinkage estimation of noise subspace covariance matrix. *IEEE Transactions on Signal Processing* 61(19): 4806–4821.
12 Nadakuditi, R.R. and Edelman, A. (2008). Sample eigenvalue based detection of high-dimensional signals in white noise using relatively few samples. *IEEE Transactions on Signal Processing* 56(7): 2625–2638.
13 Ke, X., Zhao, Y., and Huang, L. (2021). On accurate source enumeration: a new Bayesian information criterion. *IEEE Transactions on Signal Processing* 69: 1012–1027.
14 Wax, M. and Adler, A. (2022). Detection of the number of signals in uniform arrays by invariant-signal-subspace matching. *IEEE Transactions on Signal Processing* 70: 1270–1281.
15 Williams, D.B. and Johnson, D.H. (1990). Using the sphericity test for source detection with narrow-band passive arrays. *IEEE Transactions on Acoustics, Speech, and Signal Processing* 38(11): 2008–2014.
16 Huang, L., Long, T., and Wu, S. (2008). Source enumeration for high-resolution array processing using improved Gerschgorin radii without eigendecomposition. *IEEE Transactions on Signal Processing* 56(12): 5916–5925.

17 Huang, L., Qian, C., So, H.C., and Fang, J. (2015). Source enumeration for large array using shrinkage-based detectors with small samples. *IEEE Transactions on Aerospace and Electronic Systems* 51(1): 344–357.

18 Huang, L., Wu, S., and Li, X. (2007). Reduced-rank MDL method for source enumeration in high-resolution array processing. *IEEE Transactions on Signal Processing* 55(12): 5658–5667.

19 Huang, L., Long, T., Mao, E., and So, H.-C. (2009). MMSE-based mdl method for robust estimation of number of sources without eigendecomposition. *IEEE Transactions on Signal Processing* 57(10): 4135–4142.

20 Lu, Z. and Zoubir, A.M. (2012). Generalized Bayesian information criterion for source enumeration in array processing. *IEEE Transactions on Signal Processing* 61(6): 1470–1480.

21 Liu, K., Da Costa, J.P.C.L., So, H.C. et al. (2016). Detection of number of components in CANDECOMP/PARAFAC models via minimum description length. *Digital Signal Processing* 51: 110–123.

22 Anderson, T.W. (1963). Asymptotic theory for principal component analysis. *The Annals of Mathematical Statistics* 34(1): 122–148.

23 Ledoit, O. and Wolf, M. (2004). A well-conditioned estimator for large-dimensional covariance matrices. *Journal of Multivariate Analysis* 88(2): 365–411.

24 Ledoit, O. and Wolf, M. (2003). Improved estimation of the covariance matrix of stock returns with an application to portfolio selection. *Journal of Empirical Finance* 10(5): 603–621.

25 Ledoit, O. and Wolf, M. (2004). Honey, I shrunk the sample covariance matrix. *Journal of Portfolio Management* 30(1): 110–119.

26 Stoica, P., Li, J., Zhu, X., and Guerci, J.R. (2008). On using a priori knowledge in space-time adaptive processing. *IEEE Transactions on Signal Processing* 56(6): 2598–2602.

27 Kritchman, S. and Nadler, B. (2009). Non-parametric detection of the number of signals: hypothesis testing and random matrix theory. *IEEE Transactions on Signal Processing* 57(10): 3930–3941.

28 Du, L., Li, J., and Stoica, P. (2010). Fully automatic computation of diagonal loading levels for robust adaptive beamforming. *IEEE Transactions on Aerospace and Electronic Systems* 46(1): 449–458.

29 Chen, Y., Wiesel, A., Eldar, Y.C., and Hero, A.O. (2010). Shrinkage algorithms for MMSE covariance estimation. *IEEE Transactions on Signal Processing* 58(10): 5016–5029.

30 Niu, R., Blum, R.S., Varshney, P.K., and Drozd, A.L. (2012). Target localization and tracking in noncoherent multiple-input multiple-output radar systems. *IEEE Transactions on Aerospace and Electronic Systems* 48(2): 1466–1489.

31 Xu, L., Li, J., and Stoica, P. (2008). Target detection and parameter estimation for MIMO radar systems. *IEEE Transactions on Aerospace and Electronic Systems* 44(3): 927–939.

32 Gorji, A.A., Tharmarasa, R., Blair, W.D., and Kirubarajan, T. (2012). Multiple unresolved target localization and tracking using colocated MIMO radars. *IEEE Transactions on Aerospace and Electronic Systems* 48(3): 2498–2517.

33 Mestre, X. and Lagunas, M.Á. (2008). Modified subspace algorithms for DoA estimation with large arrays. *IEEE Transactions on Signal Processing* 56(2): 598–614.

34 Jonsson, D. (1982). Some limit theorems for the eigenvalues of a sample covariance matrix. *Journal of Multivariate Analysis* 12(1): 1–38.

35 Jonsson, D. (1998). On fluctuations of random Hermitian matrices. *Duke Mathematical Journal* 91: 151–203.

36 Johnstone, I.M. (2001). On the distribution of the largest eigenvalue in principal components analysis. *The Annals of Statistics* 29(2): 295–327.

37 Tracy, C.A. and Widom, H. (1996). On orthogonal and symplectic matrix ensembles. *Communications in Mathematical Physics* 177: 727–754.

38 Bai, Z.D. (2008). Methodologies in spectral analysis of large dimensional random matrices, a review. In: *Advances in Statistics* (ed. Z. Chen, J.-T. Zhang, and F. Hu), 174–240. World Scientific.

39 Bai, Z.D. and Silverstein, J.W. (2004). CLT for linear spectral statistics of a large dimensional sample covariance matrix. *Annals of Probability* 32: 553–605.

40 Nadakuditi, R.R. and Silverstein, J.W. (2010). Fundamental limit of sample generalized eigenvalue based detection of signals in noise using relatively few signal-bearing and noise-only samples. *IEEE Journal of Selected Topics in Signal Processing* 4(3): 468–480.

41 Wax, M. and Ziskind, I. (1989). Detection of the number of coherent signals by the MDL principle. *IEEE Transactions on Acoustics, Speech, and Signal Processing* 37(8): 1190–1196.

42 Wong, K.M., Zhang, Q.-T., Reilly, J.P., and Yip, P.C. (1990). On information theoretic criteria for determining the number of signals in high resolution array processing. *IEEE Transactions on Acoustics, Speech, and Signal Processing* 38(11): 1959–1971.

43 Rissanen, J. (1978). Modeling by shortest data description. *Automatica* 14(5): 465–471.

44 Rissanen, J. (2007). *Information and Complexity in Statistical Modeling*. Springer Science & Business Media.

45 Williams, D.B. (1994). Counting the degrees of freedom when using AIC and MDL to detect signals. *IEEE Transactions on Signal Processing* 42(11): 3282–3284.

46 Akaike, H. (1974). A new look at the statistical model identification. *IEEE Transactions on Automatic Control* 19(6): 716–723.

47 Wax, M. (1991). Detection and localization of multiple sources via the stochastic signals model. *IEEE Transactions on Signal Processing* 39(11): 2450–2456.

48 Letac, G. and Massam, H. (2004). All invariant moments of the Wishart distribution. *Scandinavian Journal of Statistics* 31(2): 295–318.

49 Seghouane, A.-K. (2010). Asymptotic bootstrap corrections of AIC for linear regression models. *Signal Processing* 90(1): 217–224.

50 Kay, S. (2005). Exponentially embedded families-new approaches to model order estimation. *IEEE Transactions on Aerospace and Electronic Systems* 41(1): 333–345.

51 Xu, C. and Kay, S. (2008). Source enumeration via the EEF criterion. *IEEE Signal Processing Letters* 15: 569–572.

52 Wu, H.-T., Yang, J.-F., and Chen, F.-K. (1995). Source number estimators using transformed Gerschgorin radii. *IEEE Transactions on Signal Processing* 43(6): 1325–1333.

4

Information-Theoretic Compressive Sensing for Time Delay Estimation

Yujie Gu[1], Nathan A. Goodman[2], and Yimin D. Zhang[3]

[1]*Advanced Safety & User Experience, Aptiv, Agoura Hills, CA, USA*
[2]*Advanced Radar Research Center, School of Electrical and Computer Engineering, University of Oklahoma, Norman, OK, USA*
[3]*Department of Electrical and Computer Engineering, Temple University, Philadelphia, PA, USA*

4.1 Introduction

Time delay estimation plays a fundamental role in various defense and commercial applications, including radar and sonar ranging, source and target localization, remote sensing, and network synchronization [1–6]. Furthermore, it represents a classic nonlinear parameter estimation challenge and plays a pivotal role in comprehending the fundamentals of parameter estimation. Unlike passive radar systems, which process target reflections from noncooperative sources of illumination, active radar systems transmit modulated radio frequency (RF) signals, such as linear frequency modulated (LFM) waveforms. They then correlate the received signals reflected from targets with the known transmit waveform using a matched filter to maximize the output signal-to-noise-ratio (SNR). Subsequently, time delay estimation is achieved by identifying the peaks of the matched filter output. Given that the improvement of time delay resolution necessitates an increase in waveform bandwidth, signals with increasingly higher bandwidth are being adopted to enhance the accuracy of time delay estimation for ranging and other radar tasks.

Nevertheless, these signals require a high sampling rate for analog-to-digital converters (ADCs), posing heightened challenges in terms of hardware complexity, system cost, and power consumption. Additionally, data throughput from the ADC output to the subsequent processor and/or storage subsystem can present a bottleneck. Motivated by these facts, in this chapter, we introduce a compressive sampling scheme to offer enhanced estimation performance compared to the

Information-Theoretic Radar Signal Processing, First Edition. Edited by Yujie Gu and Yimin D. Zhang.

classical Nyquist sampling scheme operating at the same ADC sampling rate. Unlike the Nyquist scheme, where the signal bandwidth must not exceed the ADC sampling rate, compressive sampling allows for signal bandwidth to surpass the ADC sampling rate. We analyze the trade-off between resolution improvement and SNR loss in sub-Nyquist RF systems, highlighting their respective advantages and disadvantages.

Unlike the Nyquist–Shannon sampling theorem, compressive sensing (CS) theory [7] demonstrates that signals sampled at sub-Nyquist rates can be accurately reconstructed given side information on certain sparsity conditions. Such sub-Nyquist sampling essentially involves thinning out Nyquist-sampled results through projection using sensing kernels. The rationale behind this lies in signal sparsity, which effectively confines the signal to a low-dimensional manifold within a bandlimited signal space. To ensure robust signal recovery, it is crucial for the measurement kernels to exhibit incoherence with the sparse representation basis of the signal [7, 8]. Random measurement kernels, such as the Gaussian and Bernoulli kernels, are known for meeting this incoherence requirement to ensure accurate signal recovery, especially when the representation basis remains fixed [9]. However, random projections suffer an SNR loss which is approximately proportional to the compression factor [10, 11], rendering them less attractive for radar sensing applications. Hence, there is a need to devise a framework capable of leveraging available prior knowledge to alleviate such performance degradation. For instance, in target tracking, one can utilize tracking information to initialize prior distributions on target properties, such as estimated target range, and then optimize sensing kernels based on a suitable performance metric.

In this chapter, we depart from the traditional use of random sensing kernels in CS theory and applications [7–9, 12–14]. Instead, we introduce and apply an information-theoretic metric [15–18] to optimize CS kernels tailored for time delay estimation in compressive radar applications. To this end, we model the time delay as a random variable characterized by using a known probability density function (PDF). By leveraging a linear compressive measurement model, we approximate the measurement PDF as a Gaussian mixture model through discretization of the prior distribution of the time delay. The Gaussian mixture model possesses the capability to serve as a universal approximator for arbitrary densities, provided that the number of components in the mixture is sufficiently large. Given the mutual dependence between the compressive measurement and the time delay, we adopt the mutual information maximization criterion to optimize the sensing kernel. We then derive the approximate gradient of mutual information with respect to the sensing matrix, which is subsequently employed in a gradient-based search method to find the optimal sensing kernel over the Stiefel manifold. Our simulation results validate that the optimized sensing kernel suffers less SNR loss compared to random sensing kernels, resulting in significantly enhanced performance.

To evaluate the performance of time delay estimation, we derive two typical performance bounds. First, we derive the Cramér–Rao bound (CRB) for time delay estimation under compressive sampling [19, 20]. Unlike the CRB analysis in random CS [21–24], our CRB analyses distinctly illustrate a tradeoff between resolution improvement and SNR loss in compressive radar systems. Since the CRB is tight only in the asymptotic region, we further derive the globally tight Ziv–Zakai bound (ZZB) for time delay estimation from the compressive measurement [25, 26]. Different from the CRB which is based on the covariance inequality, the ZZB correlates the mean square error (MSE) in the estimation problem with the probability of error (POE) in a binary hypothesis testing problem. By incorporating both prior information and compressive measurement, the ZZB furnishes a globally tight bound on the MSE over a wide range of SNR [27–39]. Subsequently, we employ the minimum mean square error (MMSE) estimator, leveraging its closed-form expression tailored for the Gaussian mixture approximation to the prior distribution. This estimator serves as a benchmark for comparing performance across various compression strategies. The relationship between mutual information and MMSE, as elucidated in [40], validates the utility of mutual information as a useful optimization metric. With this relationship in consideration, both the Bayesian CRB (BCRB) and the ZZB are harnessed to evaluate the performance of time delay estimation.

It is worth mentioning that the information-theoretic sensing kernel optimization techniques and performance analyses discussed in this chapter are applicable to a broad spectrum of sensing and communication problems, including radar target profiling and recognition [41–43], through-the-wall radar imaging [44], direction-of-arrival (DOA) estimation [45–48], target localization [49], frequency-hopping spread spectrum (FHSS) signals detection [50], frequency-hopping pattern recognition [51], and channel estimation in massive multiple-input multiple-output (MIMO) systems [52, 53]. Therefore, the compressive measurement scheme developed in this chapter serves as a foundational framework for addressing these diverse challenges, aiding in the development and comprehension of the aforementioned results.

In this chapter, we first present the radar signal model under the framework of compressive measurement, Gaussian mixture approximation, and MMSE estimation in Section 4.2. An information-theoretic approach is introduced in Section 4.3 to optimize CS kernels for time delay estimation. We then derive the BCRB and the ZZB for time delay estimation, respectively, in Sections 4.4 and 4.5. In Section 4.6, numerical results are provided to evaluate random and optimized sensing kernels by comparing pairwise distances, output SNRs, estimation error distributions, and

Table 4.1 List of notations.

Notation	Description	Notation	Description
$a, \boldsymbol{a}, \boldsymbol{A}$	Scalar, vector, and matrix	$\|\cdot\|_2$	Euclidean norm
$\mathbb{C}$	Set of complex numbers	$\mathrm{rank}(\cdot)$	Rank of a matrix
$\Re[\cdot]$	Real part of a complex number	$\Pr(\cdot)$	Probability
$(\cdot)^*$	Complex conjugation	$\mathbb{E}\{\cdot\}$	Statistical expectation
$[\cdot]^{\mathrm{T}}$	Transpose	$\nabla_{\boldsymbol{\Phi}}\{\cdot\}$	Gradient with respect to $\boldsymbol{\Phi}$
$[\cdot]^{\mathrm{H}}$	Hermitian transpose	$\mathcal{O}(\cdot)$	Algorithm complexity
$\vert\cdot\vert$	Determinant of a matrix	$\boldsymbol{I}$	Identity matrix
$(\cdot)^{-1}$	Inverse of a matrix	$\boldsymbol{0}$	Zero vector

the root mean square error (RMSE) with BCRB and ZZB. We make our conclusions in Section 4.7. The notations used in this chapter are summarized in Table 4.1.

4.2 Compressive Measurement Model

In an active sensing system, the noisy received signal $r(t)$ reflected from a point target can be modeled using complex baseband notation as

$$r(t) = \alpha\psi(t - \tau) + n(t), \tag{4.1}$$

where $\alpha \in \mathbb{C}$ represents the signal amplitude, $\psi(t)$ denotes the radar waveform, τ is an unknown time delay, and $n(t)$ signifies the additive Gaussian noise. Given that the time delay is 1-sparse for a point target, the received signal $r(t)$ can be sparsely represented by introducing a waveform basis function parameterized by the time delay τ. According to the CS theory, this sparse signal can be reconstructed with a comparable accuracy at a sub-Nyquist sampling rate. (We argue that, under certain circumstances, such reconstruction may provide significantly *better* results compared to a uniform Nyquist sampling system operating at the same sampling rate.) In other words, we only need to capture M linear projections of the noisy received signal $r(t)$ onto a collection of measurement kernels $\{\phi_m(t), m = 1, \ldots, M\}$, where M is substantially smaller than the number of measurements sampled at the Nyquist rate. Among them, the mth compressed measurement corresponds to a projection of the noisy received signal $r(t)$ onto the mth measurement kernel $\phi_m(t)$, represented as

$$y_m = \int_0^T \phi_m(t) r(t)\, dt. \tag{4.2}$$

By partitioning the integral duration of the received signal, T, into L ($L = T/\Delta$) discrete intervals of width Δ, we can approximate y_m as

$$\begin{aligned} y_m &\approx \Delta \sum_{l=1}^{L} \phi_m[l] r[l] \\ &= \Delta \sum_{l=1}^{L} \phi_m[l] \left\{ \alpha \psi[l;\tau] + n[l] \right\}. \end{aligned} \tag{4.3}$$

When Δ is selected to be much smaller than the reciprocal of the bandwidths of both the radar waveform and the measurement kernel, the signal undergoes negligible variation within a discrete interval, ensuring a high accuracy of the approximation.

A system block diagram of the compressive radar receiver is depicted in Figure 4.1. Here, the received signal undergoes analog multiplications with the measurement kernels, followed by analog integrations to complete the linear projection between the received signal and each measurement kernel. Finally, ADCs are employed to obtain the compressive measurements. In the matrix–vector form, the compressive measurement $\boldsymbol{y} = \left[y_1, y_2, \dots, y_M\right]^{\mathrm{T}} \in \mathbb{C}^M$ can be expressed as

$$\boldsymbol{y} = \boldsymbol{\Phi} \boldsymbol{r} = \boldsymbol{\Phi} \left[\alpha \boldsymbol{\psi}(\tau) + \boldsymbol{n} \right], \tag{4.4}$$

where $\boldsymbol{\Phi} = \left[\boldsymbol{\phi}_1^{\mathrm{T}}, \boldsymbol{\phi}_2^{\mathrm{T}}, \dots, \boldsymbol{\phi}_M^{\mathrm{T}}\right]^{\mathrm{T}} \in \mathbb{C}^{M\times L}$ represents the CS matrix with the mth measurement (row) vector $\boldsymbol{\phi}_m \triangleq \left[\phi_m[1], \phi_m[2], \dots, \phi_m[L]\right] \in \mathbb{C}^{1\times L}$, $\boldsymbol{\psi}(\tau) \triangleq [\psi[1;\tau], \psi[2;\tau], \dots, \psi[L;\tau]]^{\mathrm{T}} \in \mathbb{C}^L$ represents the τ-delayed radar waveform, and $\boldsymbol{n} \triangleq [n[1], n[2], \dots, n[L]]^{\mathrm{T}} \sim \mathcal{CN}(\boldsymbol{0}, \boldsymbol{C}_{\boldsymbol{nn}})$ denotes the zero-mean complex Gaussian noise independent of the time delay τ. Since scaling the measurement kernel results in proportional scaling of both the received signal and the additive noise simultaneously without affecting the input SNR, we can assume, without loss of generality, that the CS matrix $\boldsymbol{\Phi}$ is row orthonormal. This implies that each measurement kernel is normalized, and the M measurement kernels are

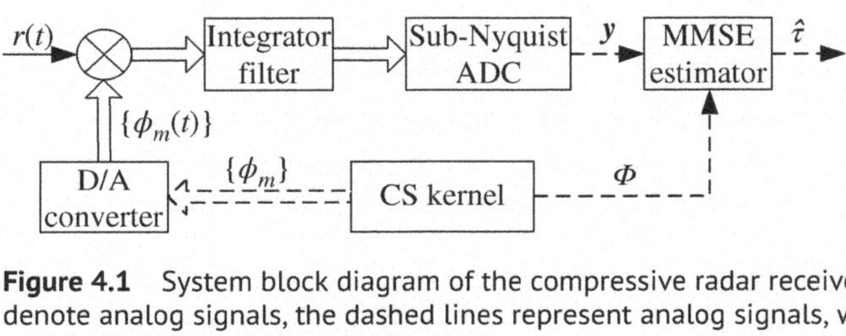

Figure 4.1 System block diagram of the compressive radar receiver. The solid lines denote analog signals, the dashed lines represent analog signals, while the hollow arrows signify parallel processing chains.

mutually orthogonal, i.e., $\forall\, m, m' \in \{1, \ldots, M\}$,

$$\int_0^T \phi_m(t)\phi_{m'}^*(t)dt = \begin{cases} 0, & m \neq m', \\ 1, & m = m'. \end{cases} \tag{4.5}$$

There are two commonly used models of the compressive measurement $\boldsymbol{y}$ depending on the signal amplitude α, namely,

- Deterministic/conditional model: α is deterministic unknown;
- Stochastic/unconditional model: α is random unknown.

In the deterministic model, the compressive measurement $\boldsymbol{y}$ follows a complex Gaussian distribution with a mean vector $\bar{\boldsymbol{y}} = \alpha\boldsymbol{\Phi}\boldsymbol{\psi}(\tau)$ and a covariance matrix $\boldsymbol{C}_{\boldsymbol{yy}} = \boldsymbol{\Phi}\boldsymbol{C}_{\boldsymbol{nn}}\boldsymbol{\Phi}^{\mathrm{H}}$. Hence, the conditional PDF of $\boldsymbol{y}$ given τ is given by

$$f(\boldsymbol{y}|\tau) = \frac{1}{\pi^M \left|\boldsymbol{C}_{\boldsymbol{yy}}\right|} e^{-(\boldsymbol{y}-\bar{\boldsymbol{y}})^{\mathrm{H}}\boldsymbol{C}_{\boldsymbol{yy}}^{-1}(\boldsymbol{y}-\bar{\boldsymbol{y}})}. \tag{4.6}$$

In the stochastic model, we assume that the signal amplitude α follows a zero-mean complex Gaussian distribution as $\alpha \sim \mathcal{CN}(0, \sigma_\alpha^2)$. Consequently, the compressive measurement $\boldsymbol{y}$ follows a zero-mean complex Gaussian distribution, i.e., $\boldsymbol{y} \sim \mathcal{CN}(\boldsymbol{0}, \boldsymbol{C}_{\boldsymbol{yy}|\tau})$, where $\boldsymbol{C}_{\boldsymbol{yy}|\tau} = \boldsymbol{\Phi}\left[\sigma_\alpha^2\boldsymbol{\psi}(\tau)\boldsymbol{\psi}^{\mathrm{H}}(\tau) + \boldsymbol{C}_{\boldsymbol{nn}}\right]\boldsymbol{\Phi}^{\mathrm{H}}$. Hence, the conditional PDF of $\boldsymbol{y}$ given τ is given by

$$f(\boldsymbol{y}|\tau) = \frac{1}{\pi^M \left|\boldsymbol{C}_{\boldsymbol{yy}|\tau}\right|} e^{-\boldsymbol{y}^{\mathrm{H}}\boldsymbol{C}_{\boldsymbol{yy}|\tau}^{-1}\boldsymbol{y}}. \tag{4.7}$$

4.2.1 Gaussian Mixture Approximation

The probability distribution of the time delay τ, denoted as $f(\tau)$, is assumed to be known *a priori* by the radar receiver. This assumption is reasonable, particularly in scenarios like radar target tracking, where radar parameter estimates are continuously updated with new measurements. According to the law of total probability, the PDF of the compressive measurement $\boldsymbol{y}$ can be written as

$$\begin{aligned} f(\boldsymbol{y}) &= \int f(\boldsymbol{y}|\tau)f(\tau)\,d\tau \\ &\approx \sum_{k\in\mathcal{K}} f(\boldsymbol{y}|\tau_k)f(\tau_k)\Delta\tau, \end{aligned} \tag{4.8}$$

where $\mathcal{K} = \{1, 2, \ldots, K\}$ denotes the set of indices for bins resulting from dividing the support of $f(\tau)$ into bins of width $\Delta\tau$. By discretizing $f(\tau)$ such that $p_k = f(\tau_k)\Delta\tau$ and $\sum_{k\in\mathcal{K}} p_k = 1$, as depicted in Figure 4.2, $f(\boldsymbol{y})$ becomes

$$f(\boldsymbol{y}) \approx \sum_{k\in\mathcal{K}} p_k f(\boldsymbol{y}|\tau_k). \tag{4.9}$$

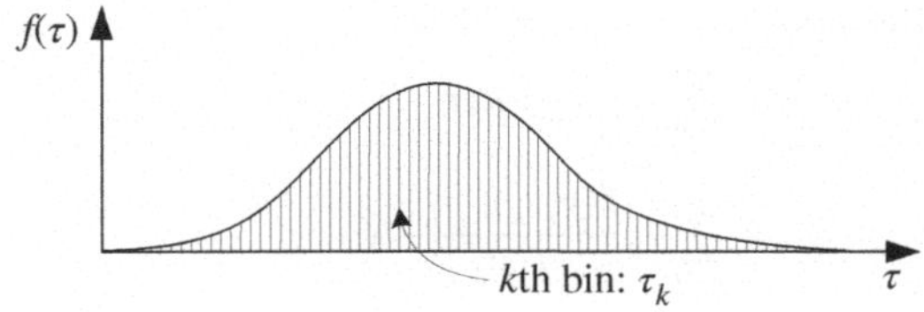

Figure 4.2 Discretization of the PDF of the time delay τ.

For time delay τ_k, i.e., the kth bin in Figure 4.2, the corresponding compressive measurement is given by

$$\begin{aligned} \boldsymbol{y}_k &= \boldsymbol{y}|_{\tau=\tau_k} \\ &= \boldsymbol{\Phi}\left[\alpha\boldsymbol{\psi}(\tau_k) + \boldsymbol{n}\right] \\ &= \alpha\boldsymbol{\Phi}\boldsymbol{\psi}(\tau_k) + \boldsymbol{\Phi}\boldsymbol{n}. \end{aligned} \tag{4.10}$$

Under either the deterministic or the stochastic model, the PDF of $\boldsymbol{y}$ is a weighted sum of K Gaussian distributions, forming a Gaussian mixture distribution. Specifically, set $\mathcal{D} = \{p_k, \bar{\boldsymbol{y}}_k, \boldsymbol{C}_{\boldsymbol{yy}}; k \in \mathcal{K}\}$ with $\bar{\boldsymbol{y}}_k = \alpha\boldsymbol{\Phi}\boldsymbol{\psi}(\tau_k)$ and $\boldsymbol{C}_{\boldsymbol{yy}} = \boldsymbol{\Phi}\boldsymbol{C}_{\boldsymbol{nn}}\boldsymbol{\Phi}^{\mathrm{H}}$ defines the Gaussian mixture parameters of $\boldsymbol{y}$ in the deterministic model, whereas set $\mathcal{S} = \{p_k, \boldsymbol{0}, \boldsymbol{C}_{\boldsymbol{yy}|k}; k \in \mathcal{K}\}$ with $\boldsymbol{C}_{\boldsymbol{yy}|k} = \boldsymbol{\Phi}\left[\sigma_\alpha^2\boldsymbol{\psi}(\tau_k)\boldsymbol{\psi}^{\mathrm{H}}(\tau_k) + \boldsymbol{C}_{\boldsymbol{nn}}\right]\boldsymbol{\Phi}^{\mathrm{H}}$ defines the Gaussian mixture parameters of $\boldsymbol{y}$ in the stochastic model. If the bin size $\Delta\tau$ is chosen to be infinitesimal, the approximate density of the compressive measurement $\boldsymbol{y}$ converges toward the true density. However, this convergence comes at the expense of an infinite number of components in the mixture distribution. Therefore, selecting an appropriate bin size $\Delta\tau$ is important for Gaussian mixture approximation. This choice ensures that the algorithm complexity remains manageable for the subsequent sensing kernel optimization and MMSE estimation.

4.2.2 MMSE Estimation

Since the MSE is a widely accepted metric for evaluating estimator performance, the MMSE estimator is a popular choice among various parameter estimators. The MMSE estimate of the time delay τ is given by

$$\hat{\tau} = \mathbb{E}_{\tau|\boldsymbol{y}}\{\tau\} = \int \tau f(\tau|\boldsymbol{y})\,d\tau. \tag{4.11}$$

While there is no closed-form expression for $\hat{\tau}$ corresponding to the full PDF in (4.8), by using Bayes' theorem, the MMSE estimate can be approximated using the same discretized PDF as

$$\begin{aligned}\hat{\tau} &\approx \sum_{k\in\mathcal{K}} \tau_k f(\tau_k|\boldsymbol{y})\Delta\tau \\ &= \sum_{k\in\mathcal{K}} \tau_k \frac{f(\boldsymbol{y}|\tau_k)f(\tau_k)}{f(\boldsymbol{y})}\Delta\tau \\ &= \frac{1}{f(\boldsymbol{y})}\sum_{k\in\mathcal{K}} p_k \tau_k f(\boldsymbol{y}|\tau_k). \end{aligned} \tag{4.12}$$

The aforementioned approximation holds true when the bin size $\Delta\tau$ is substantially smaller than the MSE of the nonapproximated MMSE estimator. Essentially, this implies that the error introduced by quantization via the approximation is negligible compared to the error arising from measurement noise. As the input SNR increases, finer bins become necessary to keep the impact of quantization error on the outcome to be negligible. Another perspective is that there must be a sufficient number of bins for the posterior PDF of τ, i.e., $f(\tau|\boldsymbol{y})$ in (4.11), to be smoothly represented. With a higher input SNR, the posterior PDF becomes sharper, thereby necessitating narrower discrete bins for ensuring smoothness.

Evidently, the MMSE estimate relies on the CS matrix $\boldsymbol{\Phi}$ through the compressive measurement $\boldsymbol{y}$. Unlike randomly generated sensing kernels commonly used in the CS literature, we optimize $\boldsymbol{\Phi}$ by leveraging *a priori* knowledge of the time delay. This optimization strategy enables to enhance the accuracy of time delay estimation from the compressive measurement.

4.3 Compressive Sensing Kernel Optimization

In this section, we adopt the mutual information maximization criterion to optimize the CS kernel for time delay estimation. Mutual information quantifies the dependence between two random variables, making it a suitable criterion for this optimization task. The optimization problem is formulated as

$$\max_{\boldsymbol{\Phi}\in \mathrm{St}(L,M)} I(\boldsymbol{y};\tau), \tag{4.13}$$

where $I(\boldsymbol{y};\tau) = h(\boldsymbol{y}) - h(\boldsymbol{y}|\tau)$ represents the Shannon mutual information between the compressive measurement $\boldsymbol{y}$ and the time delay τ. Here, $h(\boldsymbol{y}) = \mathbb{E}\{-\ln[f(\boldsymbol{y})]\}$ and $h(\boldsymbol{y}|\tau) = \mathbb{E}\{-\ln[f(\boldsymbol{y}|\tau)]\}$ denote the differential entropy of $\boldsymbol{y}$ and the conditional differential entropy of $\boldsymbol{y}$ given τ, respectively. The set

$$\mathrm{St}(L,M) = \{\boldsymbol{\Phi} \in \mathbb{C}^{M\times L} : \boldsymbol{\Phi}\boldsymbol{\Phi}^{\mathrm{H}} = \boldsymbol{I}\} \tag{4.14}$$

stands for the Stiefel manifold comprising all row-orthonormal sensing kernels defined in (4.5). Consequently, the information-theoretic sensing kernel optimization problem (4.13) aims to seek the optimal CS kernel that maximizes the mutual information between the compressive measurement and the parameter of interest (in this case, time delay) over the Stiefel manifold.

Since it is computationally prohibitive to compute the mutual information directly for optimizing the CS matrix due to its high dimensionality, we resort to a gradient ascent search approach to maximize the mutual information. For this purpose, we compute the gradient of the mutual information $I(\mathbf{y};\tau)$ with respect to the sensing matrix $\boldsymbol{\Phi}$ as

$$\nabla_{\boldsymbol{\Phi}} I(\mathbf{y};\tau) = \nabla_{\boldsymbol{\Phi}} h(\mathbf{y}) - \nabla_{\boldsymbol{\Phi}} h(\mathbf{y}|\tau). \tag{4.15}$$

However, the differential entropy lacks a closed-form expression for most PDFs of interest in practical radar applications, including the Gaussian mixture distribution approximated in Section 4.2.1. Hence, we substitute the Gaussian mixture distribution (4.9) into the definition of differential entropy and expand its logarithmic term using the Taylor series. This approximation of the differential entropy enables us to compute its gradient with respect to the CS matrix for gradient-based kernel optimization.

By employing a first-order Taylor series expansion to the logarithm of the Gaussian mixture distribution (4.9), we approximate the differential entropy of the compressive measurement as

$$\begin{aligned} h(\mathbf{y}) &\approx -\int \sum_{k'\in\mathcal{K}} p_{k'} f(\mathbf{y}|\tau_{k'}) \ln\left[\sum_{k\in\mathcal{K}} p_k f(\mathbf{y}|\tau_k)\right] d\mathbf{y} \\ &= -\sum_{k'\in\mathcal{K}} p_{k'} \int f(\mathbf{y}|\tau_{k'}) \ln\left[\sum_{k\in\mathcal{K}} p_k f(\mathbf{y}|\tau_k)\right] d\mathbf{y} \\ &\approx -\ln\left[\sum_{k\in\mathcal{K}} p_k f(\mathbf{y}_0|\tau_k)\right], \end{aligned} \tag{4.16}$$

where $\mathbf{y}_0 = \mathbb{E}\{\mathbf{y}\}$ is the mean value of $\mathbf{y}$ which serves as the point of expansion for the Taylor series. A detailed derivation is available in the Appendix of [42].

4.3.1 Mutual Information Gradient: Deterministic Model

For the deterministic model, the conditional PDF of $\mathbf{y}$ given τ_k is given by

$$f(\mathbf{y} = \mathbf{y}_0|\tau_k) = \frac{1}{\pi^M \left|\mathbf{C}_{\mathbf{yy}}\right|} e^{-|\alpha|^2[\boldsymbol{\psi}(\tau_k)-\bar{\boldsymbol{\psi}}]^{\mathrm{H}} \boldsymbol{\Phi}^{\mathrm{H}} \mathbf{C}_{\mathbf{yy}}^{-1} \boldsymbol{\Phi}[\boldsymbol{\psi}(\tau_k)-\bar{\boldsymbol{\psi}}]}, \tag{4.17}$$

where $\mathbf{y}_0 = \alpha\boldsymbol{\Phi}\bar{\boldsymbol{\psi}}$ represents the mean of $\mathbf{y}$, and $\bar{\boldsymbol{\psi}} = \int_0^T f(\tau)\boldsymbol{\psi}(\tau)\,d\tau \approx \sum_{k\in\mathcal{K}} p_k \boldsymbol{\psi}(\tau_k)$ denotes the mean of $\boldsymbol{\psi}(\tau)$ over the prior distribution of τ. Hence, the gradient of the differential entropy $h(\mathbf{y})$ with respect to the CS matrix $\boldsymbol{\Phi}$ is approximated as [20]

$$\nabla_{\boldsymbol{\Phi}} h(\boldsymbol{y}) \approx \frac{|\alpha|^2 \sum_{k\in\mathcal{K}} p_k e^{-|\alpha|^2[\boldsymbol{\psi}(\tau_k)-\bar{\boldsymbol{\psi}}]^{\mathrm{H}}\boldsymbol{\Phi}^{\mathrm{H}}\boldsymbol{C}_{yy}^{-1}\boldsymbol{\Phi}[\boldsymbol{\psi}(\tau_k)-\bar{\boldsymbol{\psi}}]} \times \boldsymbol{C}_{yy}^{-1}\boldsymbol{\Phi}\left[\boldsymbol{\psi}(\tau_k)-\bar{\boldsymbol{\psi}}\right]\left[\boldsymbol{\psi}(\tau_k)-\bar{\boldsymbol{\psi}}\right]^{\mathrm{H}}[\boldsymbol{I}-\boldsymbol{\Phi}^{\mathrm{H}}\boldsymbol{C}_{yy}^{-1}\boldsymbol{\Phi}\boldsymbol{C}_{nn}]}{\sum_{k\in\mathcal{K}} p_k e^{-|\alpha|^2[\boldsymbol{\psi}(\tau_k)-\bar{\boldsymbol{\psi}}]^{\mathrm{H}}\boldsymbol{\Phi}^{\mathrm{H}}\boldsymbol{C}_{yy}^{-1}\boldsymbol{\Phi}[\boldsymbol{\psi}(\tau_k)-\bar{\boldsymbol{\psi}}]}}. \tag{4.18}$$

It is noted that, when the time delay τ is given, the only random contribution to the compressive measurement $\boldsymbol{y}$ is due to the noise $\boldsymbol{n}$. Therefore, the conditional differential entropy $h(\boldsymbol{y}|\tau)$ simplifies to

$$h(\boldsymbol{y}|\tau) = h(\boldsymbol{\Phi n}|\tau) = h(\boldsymbol{\Phi n}) = \ln\left|\boldsymbol{\Phi}\boldsymbol{C}_{nn}\boldsymbol{\Phi}^{\mathrm{H}}\right| + M\ln(\pi e), \tag{4.19}$$

because $\boldsymbol{\Phi n} \sim \mathcal{CN}(\boldsymbol{0}, \boldsymbol{\Phi}\boldsymbol{C}_{nn}\boldsymbol{\Phi}^{\mathrm{H}})$ is independent of τ. Hence, the gradient of the conditional differential entropy $h(\boldsymbol{y}|\tau)$ with respect to the CS matrix $\boldsymbol{\Phi}$ is

$$\nabla_{\boldsymbol{\Phi}} h(\boldsymbol{y}|\tau) = \nabla_{\boldsymbol{\Phi}}\left\{\ln\left|\boldsymbol{\Phi}\boldsymbol{C}_{nn}\boldsymbol{\Phi}^{\mathrm{H}}\right|\right\} = \left(\boldsymbol{\Phi}\boldsymbol{C}_{nn}\boldsymbol{\Phi}^{\mathrm{H}}\right)^{-1}\boldsymbol{\Phi}\boldsymbol{C}_{nn}. \tag{4.20}$$

Substituting (4.18) and (4.20) into (4.15), we can obtain the approximate mutual information gradient for the deterministic model.

In the case where the additive noise $\boldsymbol{n}$ in (4.4) is Gaussian and white (i.e., $\boldsymbol{C}_{nn} = \sigma_n^2\boldsymbol{I}$), the approximate mutual information gradient is simplified as

$$\nabla_{\boldsymbol{\Phi}} I(\boldsymbol{y};\tau) \approx \frac{\frac{|\alpha|^2}{\sigma_n^2}\sum_{k\in\mathcal{K}} p_k e^{-\frac{|\alpha|^2}{\sigma_n^2}[\boldsymbol{\psi}(\tau_k)-\bar{\boldsymbol{\psi}}]^{\mathrm{H}}\boldsymbol{\Phi}^{\mathrm{H}}\boldsymbol{\Phi}[\boldsymbol{\psi}(\tau_k)-\bar{\boldsymbol{\psi}}]}\boldsymbol{\Phi}\left[\boldsymbol{\psi}(\tau_k)-\bar{\boldsymbol{\psi}}\right]\left[\boldsymbol{\psi}(\tau_k)-\bar{\boldsymbol{\psi}}\right]^{\mathrm{H}}[\boldsymbol{I}-\boldsymbol{\Phi}^{\mathrm{H}}\boldsymbol{\Phi}]}{\sum_{k\in\mathcal{K}} p_k e^{-\frac{|\alpha|^2}{\sigma_n^2}[\boldsymbol{\psi}(\tau_k)-\bar{\boldsymbol{\psi}}]^{\mathrm{H}}\boldsymbol{\Phi}^{\mathrm{H}}\boldsymbol{\Phi}[\boldsymbol{\psi}(\tau_k)-\bar{\boldsymbol{\psi}}]}} - \boldsymbol{\Phi}, \tag{4.21}$$

where $\frac{|\alpha|^2}{\sigma_n^2}$ denotes the input SNR of the target signal.

4.3.2 Mutual Information Gradient: Stochastic Model

For the stochastic model with $\alpha \sim \mathcal{CN}(0, \sigma_\alpha^2)$, the gradient of the differential entropy $h(\boldsymbol{y})$ with respect to the CS matrix $\boldsymbol{\Phi}$ is approximated as [41, 42, 45]

$$\nabla_{\boldsymbol{\Phi}} h(\boldsymbol{y}) \approx \frac{\sum_{k\in\mathcal{K}} p_k \left|\boldsymbol{C}_{yy|k}\right|^{-1}\boldsymbol{C}_{yy|k}^{-1}\boldsymbol{\Phi}\left[\sigma_\alpha^2\boldsymbol{\psi}(\tau_k)\boldsymbol{\psi}^{\mathrm{H}}(\tau_k)+\boldsymbol{C}_{nn}\right]}{\sum_{k\in\mathcal{K}} p_k\left|\boldsymbol{C}_{yy|k}\right|^{-1}}, \tag{4.22}$$

because $f(\boldsymbol{y}=\boldsymbol{y}_0=\boldsymbol{0}|\tau_k) = \frac{1}{\pi^M\left|\boldsymbol{C}_{yy|k}\right|}$.

According to (4.7), the conditional differential entropy of the compressive measurement $\boldsymbol{y}$ given the time delay τ is approximated as [20]

$$
\begin{aligned}
h(\mathbf{y}|\tau) &\approx -\iint f(\mathbf{y},\tau)\ln\left[f(\mathbf{y}_0|\tau)\right]\,d\mathbf{y}\,d\tau \\
&= -\int \ln\left[f(\mathbf{y}_0|\tau)\right] f(\tau)\,d\tau \\
&\approx \sum_{k\in\mathcal{K}} p_k \ln\left|\mathbf{C}_{\mathbf{yy}|k}\right| + M\ln\pi,
\end{aligned} \tag{4.23}
$$

where the first approximation is based on the first-order Taylor series expansion of $\ln\left[f(\mathbf{y}|\tau)\right]$ around the mean of $\mathbf{y}$, i.e., $\mathbf{y}_0 = \mathbf{0}$. Correspondingly, the approximate gradient of the conditional differential entropy $h(\mathbf{y}|\tau)$ with respect to the CS matrix $\boldsymbol{\Phi}$ is

$$
\nabla_{\boldsymbol{\Phi}} h(\mathbf{y}|\tau) \approx \sum_{k\in\mathcal{K}} p_k \mathbf{C}_{\mathbf{yy}|k}^{-1} \boldsymbol{\Phi}\left[\sigma_\alpha^2 \boldsymbol{\psi}(\tau_k)\boldsymbol{\psi}^{\mathrm{H}}(\tau_k) + \mathbf{C}_{\mathbf{nn}}\right]. \tag{4.24}
$$

Substituting (4.22) and (4.24) into (4.15), we obtain the approximate mutual information gradient for the stochastic model with $\alpha \sim \mathcal{CN}(0,\sigma_\alpha^2)$.

In the case where the additive noise $\mathbf{n}$ is Gaussian and white, the approximate mutual information gradient can be simplified as

$$
\begin{aligned}
\nabla_{\boldsymbol{\Phi}} I(\mathbf{y};\tau) \approx{}& \frac{\sum_{k\in\mathcal{K}} p_k \left|\mathbf{C}_{\mathbf{yy}|k}\right|^{-1} \left[\frac{\mathbf{C}_{\mathbf{yy}|k}}{\sigma_{\mathbf{n}}^2}\right]^{-1} \boldsymbol{\Phi}\left[\frac{\sigma_\alpha^2}{\sigma_{\mathbf{n}}^2}\boldsymbol{\psi}(\tau_k)\boldsymbol{\psi}^{\mathrm{H}}(\tau_k) + \mathbf{I}\right]}{\sum_{k\in\mathcal{K}} p_k \left|\mathbf{C}_{\mathbf{yy}|k}\right|^{-1}} \\
&- \sum_{k\in\mathcal{K}} p_k \left[\frac{\mathbf{C}_{\mathbf{yy}|k}}{\sigma_{\mathbf{n}}^2}\right]^{-1} \boldsymbol{\Phi}\left[\frac{\sigma_\alpha^2}{\sigma_{\mathbf{n}}^2}\boldsymbol{\psi}(\tau_k)\boldsymbol{\psi}^{\mathrm{H}}(\tau_k) + \mathbf{I}\right],
\end{aligned} \tag{4.25}
$$

where $\frac{\mathbf{C}_{\mathbf{yy}|k}}{\sigma_{\mathbf{n}}^2} = \boldsymbol{\Phi}\left[\frac{\sigma_\alpha^2}{\sigma_{\mathbf{n}}^2}\boldsymbol{\psi}(\tau_k)\boldsymbol{\psi}^{\mathrm{H}}(\tau_k) + \mathbf{I}\right]\boldsymbol{\Phi}^{\mathrm{H}}$, and $\frac{\sigma_\alpha^2}{\sigma_{\mathbf{n}}^2}$ represents the input SNR of the target signal.

4.3.3 Gradient Search Optimization

With the approximate mutual information gradient ((4.21) for the deterministic model or (4.25) for the stochastic model), the CS matrix is updated in a mutual information gradient ascent direction according to

$$
\tilde{\boldsymbol{\Phi}} = \boldsymbol{\Phi} + \rho\nabla_{\boldsymbol{\Phi}} I(\mathbf{y};\tau), \tag{4.26}
$$

where $\rho > 0$ is a small step size which can be either fixed or adaptive. This rule updates the CS matrix in a direction that increases the mutual information between the compressed measurements $\mathbf{y}$ and the parameter of interest τ. The updated CS matrix $\tilde{\boldsymbol{\Phi}}$ (with row orthonormalization) is used instead of $\boldsymbol{\Phi}$ in the subsequent iteration to compute the approximate mutual information gradient and is then updated using (4.26) to achieve an iterative search procedure for the

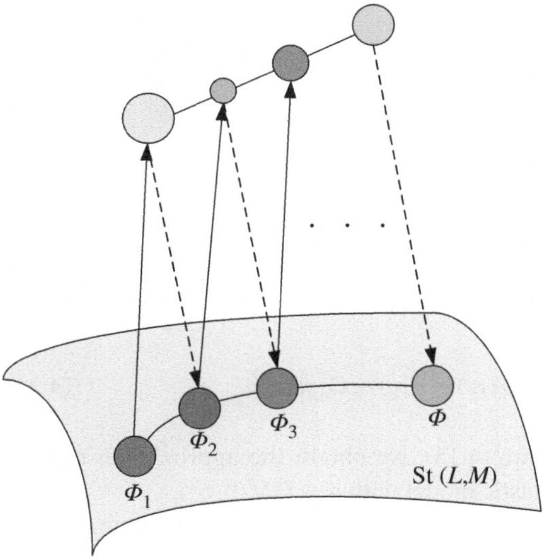

Figure 4.3 An illustration depicting the iterative process of the CS kernel optimization. The solid arrows lines denote the updating of the sensing matrix, while the dashed arrow lines denote the row orthonormalization.

optimal CS kernel that maximizes the mutual information depicted in (4.13). The iterative process of the CS kernel optimization is illustrated in Figure 4.3.

In summary, the information-theoretic CS matrix optimization performs the following iterative process:

Step 1: Randomly initialize a row-orthonormal CS matrix $\boldsymbol{\Phi}$;
Step 2: Compute the approximate mutual information gradient with respect to the CS matrix, $\nabla_{\boldsymbol{\Phi}} I(\boldsymbol{y}; \tau)$, using either (4.21) or (4.25), evaluated at the current iteration of the sensing matrix;
Step 3: Update the sensing matrix as $\tilde{\boldsymbol{\Phi}}$ via (4.26);
Step 4: Orthonormalize the rows of $\tilde{\boldsymbol{\Phi}}$, set $\boldsymbol{\Phi}$ to the result, and return to *Step 2*.

The iterative process continues until a stopping criterion is met, such as reaching a maximum number of iterations or when a convergence criterion is satisfied (e.g., the norm difference of the mutual information gradient is less than a predefined threshold [42]).

The row orthonormalization operation in *Step 4* projects the updated CS matrix $\tilde{\boldsymbol{\Phi}}$ back to the Stiefel manifold $\mathrm{St}(L, M)$ defined in (4.14). This operation, known as the orthogonal Procrustes problem [54], can be efficiently solved using singular value decomposition (SVD). Denoting the SVD of $\tilde{\boldsymbol{\Phi}}$ as $\tilde{\boldsymbol{\Phi}} = \boldsymbol{U}\boldsymbol{\Lambda}\boldsymbol{V}^{\mathrm{H}}$, where $\boldsymbol{U}$ and $\boldsymbol{V}$ are unitary matrices and $\boldsymbol{\Lambda}$ is a diagonal matrix containing the singular values, the row-orthonormal matrix closest to $\tilde{\boldsymbol{\Phi}}$ can be obtained as $\boldsymbol{U}\boldsymbol{I}_{M\times L}\boldsymbol{V}^{\mathrm{H}}$, where $\boldsymbol{I}_{M\times L}$ is an $M \times L$ rectangular matrix with all main diagonal elements being 1 and all other elements being 0. This matrix effectively preserves the orthonormality of the rows of $\boldsymbol{U}$ and $\boldsymbol{V}^{\mathrm{H}}$, ensuring that the resulting matrix remains on the Stiefel manifold after projection.

The computational complexity of computing the mutual information gradient, as observed from (4.21) and (4.25), is $\mathcal{O}(KML^2)$ for both deterministic and stochastic models where $M \ll L$. Therefore, the overall computational complexity of the proposed sensing kernel optimization is $\mathcal{O}(GKML^2)$, where G represents the number of iterations.

It is worth noting that the CS matrix optimization proposed in this chapter is not an adaptive process and can be performed offline. However, it can be easily extended to an adaptive manner by updating the PDF of the parameter of interest iteratively, as described in references [47, 48]. This extension allows for the optimization process to be carried out when the prior distribution about τ is unavailable or inaccurate, and to adapt to changes in the underlying data distribution, thereby broadening the applicability and improving performance in dynamic environments.

4.4 Bayesian Cramér–Rao Bound

In this section, we derive the BCRB for time delay estimation within the CS framework. According to [55], the Bayesian version of the Fisher information, denoted as J_B, can be decomposed into two components:

$$J_B = J_D + J_P, \tag{4.27}$$

where

$$J_D = \mathbb{E}_{\boldsymbol{y},\tau}\left\{-\frac{d^2 \ln f(\boldsymbol{y}|\tau)}{d\tau^2}\right\} \tag{4.28}$$

describes the information derived from the compressive measurement $\boldsymbol{y}$, whereas

$$J_P = \mathbb{E}_{\tau}\left\{-\frac{d^2 \ln f(\tau)}{d\tau^2}\right\} \tag{4.29}$$

represents the information contributed solely by the prior knowledge (i.e., the prior PDF $f(\tau)$ described in Section 4.2.1).

4.4.1 BCRB: Deterministic Model

From (4.6), the log-likelihood function conditioned on the time delay τ is expressed as

$$\ln f(\boldsymbol{y}|\tau) = -[\boldsymbol{y} - \alpha\boldsymbol{\Phi}\boldsymbol{\psi}(\tau)]^{\mathrm{H}}\boldsymbol{C}_{\boldsymbol{yy}}^{-1}[\boldsymbol{y} - \alpha\boldsymbol{\Phi}\boldsymbol{\psi}(\tau)] - \ln\left|\boldsymbol{C}_{\boldsymbol{yy}}\right| - M\ln\pi, \tag{4.30}$$

where the last two terms are independent of τ because $\boldsymbol{C}_{yy} = \boldsymbol{\Phi}\boldsymbol{C}_{nn}\boldsymbol{\Phi}^{\mathrm{H}}$. Note that, because $\mathbb{E}_{y|\tau}\{\boldsymbol{y}\} = \alpha\boldsymbol{\Phi}\boldsymbol{\psi}(\tau)$, the regularity condition

$$\mathbb{E}_{y|\tau}\left\{\frac{d\ln f(\boldsymbol{y}|\tau)}{d\tau}\right\} = \mathbb{E}_{y|\tau}\left\{2\Re\left[\alpha[\boldsymbol{y} - \alpha\boldsymbol{\Phi}\boldsymbol{\psi}(\tau)]^{\mathrm{H}}\boldsymbol{C}_{yy}^{-1}\boldsymbol{\Phi}\frac{d\boldsymbol{\psi}(\tau)}{d\tau}\right]\right\} = 0 \tag{4.31}$$

is satisfied. Hence, for a specific time delay τ, the standard Fisher information is given by [20]

$$J_F(\tau) = \mathbb{E}_{y|\tau}\left\{-\frac{d^2\ln f(\boldsymbol{y}|\tau)}{d\tau^2}\right\} = 2|\alpha|^2\frac{d\boldsymbol{\psi}^{\mathrm{H}}(\tau)}{d\tau}\boldsymbol{\Phi}^{\mathrm{H}}\boldsymbol{C}_{yy}^{-1}\boldsymbol{\Phi}\frac{d\boldsymbol{\psi}(\tau)}{d\tau}. \tag{4.32}$$

By taking its expected value, we obtain the information gained from the compressive measurement as

$$J_D = \mathbb{E}_\tau\left\{J_F(\tau)\right\} = 2|\alpha|^2\mathbb{E}_\tau\left\{\frac{d\boldsymbol{\psi}^{\mathrm{H}}(\tau)}{d\tau}\boldsymbol{\Phi}^{\mathrm{H}}\boldsymbol{C}_{yy}^{-1}\boldsymbol{\Phi}\frac{d\boldsymbol{\psi}(\tau)}{d\tau}\right\}. \tag{4.33}$$

With the prior information J_P (4.29), the deterministic BCRB is expressed as [20]

$$\mathrm{BCRB} = J_B^{-1} = \frac{1}{2|\alpha|^2\mathbb{E}_\tau\left\{\frac{d\boldsymbol{\psi}^{\mathrm{H}}(\tau)}{d\tau}\boldsymbol{\Phi}^{\mathrm{H}}\boldsymbol{C}_{yy}^{-1}\boldsymbol{\Phi}\frac{d\boldsymbol{\psi}(\tau)}{d\tau}\right\} + J_P}. \tag{4.34}$$

In the case where the additive noise $\boldsymbol{n}$ is Gaussian and white, the deterministic BCRB is simplified as

$$\mathrm{BCRB} = \frac{1}{2\frac{|\alpha|^2}{\sigma_n^2}\mathbb{E}_\tau\left\{\left\|\frac{d[\boldsymbol{\Phi}\boldsymbol{\psi}(\tau)]}{d\tau}\right\|_2^2\right\} + J_P}. \tag{4.35}$$

A close look at the expression of (4.35) would provide us insights into the BCRB. First, the BCRB depends on the input SNR, $\frac{|\alpha|^2}{\sigma_n^2}$. That is, at a high input SNR, the BCRB experiences a linear decrease with the input SNR due to the predominance of the information gained from the measurement. This behavior is typical in nonlinear parameter estimation problems operating in the asymptotic region. Second, the term $\mathbb{E}_\tau\left\{\left\|\frac{d[\boldsymbol{\Phi}\boldsymbol{\psi}(\tau)]}{d\tau}\right\|_2^2\right\}$ in the BCRB encompasses the expectation of the norm square of the derivative of the compressed signal concerning the time delay, integrated over the prior distribution of the time delay. Given that this derivative characterizes the rate of change and escalates linearly with increased waveform bandwidth, the BCRB typically exhibits a quadratic decrease with the signal bandwidth. The specific value of the derivative term depends on the interplay between the measurement kernel and the transmit waveform, as well as on the prior distribution of τ. However, the overall trend remains consistent with the traditional BCRB for time delay estimation such that, in the asymptotic region, the estimation performance experiences a linear enhancement with the input SNR and a quadratic improvement with the signal bandwidth. Unlike random projections, an optimized sensing kernel possesses the capability to encode τ-dependent fluctuations in the measurements, which evolve more rapidly with τ than what would be observable due to the waveform alone. Naturally, these fluctuations must be distinct across all conceivable values of τ to ensure the unique recovery of τ. While random projections tend to establish a straightforward one-to-one mapping from τ to the measurement, an information-theoretic sensing kernel has the potential to introduce distinct encodings, which are enhanced for the most probable values of τ based on the prior distribution.

Moreover, the norm square of these derivatives is also contingent upon the dimension of $\boldsymbol{\Phi}$, i.e., the number of rows in $\boldsymbol{\Phi}$. A larger number of measurements proportionally amplifies the norm square of these derivatives. Conversely, reducing the number of measurements leads to a decrease in energy collection, resulting in an SNR loss [11, 56, 57]. Generally, there is an equilibrium between resolution enhancement and SNR loss. For a constant sampling rate, increasing the waveform bandwidth enhances the BCRB by the *square* of the bandwidth increment, while simultaneously inducing a *linear* decline in the BCRB due to SNR loss. It is noteworthy that SNR loss depreciates roughly linearly with the compression ratio. These insights are pertinent when operating in the asymptotic region with a sufficiently high input SNR, where *resolution* predominantly limits estimation precision rather than *SNR*. In this region, the importance of increasing the native resolution outweighs the SNR loss. Therefore, it is often beneficial to tolerate a certain degree of SNR loss from compression to gain enhanced native resolution in measurements.

4.4.2 BCRB: Stochastic Model

From (4.7), the log-likelihood function conditioned on the time delay τ is expressed as

$$\ln f(\boldsymbol{y}|\tau) = -\boldsymbol{y}^{\mathrm{H}}\boldsymbol{C}_{\boldsymbol{yy}|\tau}^{-1}\boldsymbol{y} - \ln\left|\boldsymbol{C}_{\boldsymbol{yy}|\tau}\right| - M\ln\pi, \tag{4.36}$$

where the third term remains constant regardless of τ. Note that, because

$$\mathbb{E}_{\boldsymbol{y}|\tau}\left\{\boldsymbol{y}^{\mathrm{H}}\boldsymbol{C}_{\boldsymbol{yy}|\tau}^{-1}\frac{d\boldsymbol{C}_{\boldsymbol{yy}|\tau}}{d\tau}\boldsymbol{C}_{\boldsymbol{yy}|\tau}^{-1}\boldsymbol{y}\right\} = 2\sigma_\alpha^2\Re\left[\boldsymbol{\psi}^{\mathrm{H}}(\tau)\boldsymbol{\Phi}^{\mathrm{H}}\boldsymbol{C}_{\boldsymbol{yy}|\tau}^{-1}\boldsymbol{\Phi}\frac{d\boldsymbol{\psi}(\tau)}{d\tau}\right],$$

the required regularity condition

$$\begin{aligned}&\mathbb{E}_{\boldsymbol{y}|\tau}\left\{\frac{d\ln f(\boldsymbol{y}|\tau)}{d\tau}\right\}\\&= \mathbb{E}_{\boldsymbol{y}|\tau}\left\{\boldsymbol{y}^{\mathrm{H}}\boldsymbol{C}_{\boldsymbol{yy}|\tau}^{-1}\frac{d\boldsymbol{C}_{\boldsymbol{yy}|\tau}}{d\tau}\boldsymbol{C}_{\boldsymbol{yy}|\tau}^{-1}\boldsymbol{y} - 2\sigma_\alpha^2\Re\left[\boldsymbol{\psi}^{\mathrm{H}}(\tau)\boldsymbol{\Phi}^{\mathrm{H}}\boldsymbol{C}_{\boldsymbol{yy}|\tau}^{-1}\boldsymbol{\Phi}\frac{d\boldsymbol{\psi}(\tau)}{d\tau}\right]\right\} = 0\end{aligned} \tag{4.37}$$

is satisfied. Hence, for a specific time delay τ, the standard Fisher information is given by [20]

$$J_F(\tau) = \mathbb{E}_{y|\tau}\left\{-\frac{d^2 \ln f(y|\tau)}{d\tau^2}\right\} = 2\sigma_\alpha^2 \Re\left[\boldsymbol{\psi}^{\mathrm{H}}(\tau)\boldsymbol{\Phi}^{\mathrm{H}}\boldsymbol{C}_{yy|\tau}^{-1}\frac{d\boldsymbol{C}_{yy|\tau}}{d\tau}\boldsymbol{C}_{yy|\tau}^{-1}\boldsymbol{\Phi}\frac{d\boldsymbol{\psi}(\tau)}{d\tau}\right]. \tag{4.38}$$

By taking its expected value, we obtain the information gained from the compressive measurement as

$$J_D = \mathbb{E}_\tau\left\{J_F(\tau)\right\} = -2\sigma_\alpha^2 \mathbb{E}_\tau\left\{\Re\left[\boldsymbol{\psi}^{\mathrm{H}}(\tau)\boldsymbol{\Phi}^{\mathrm{H}}\frac{d\boldsymbol{C}_{yy|\tau}^{-1}}{d\tau}\boldsymbol{\Phi}\frac{d\boldsymbol{\psi}(\tau)}{d\tau}\right]\right\}. \tag{4.39}$$

With the prior information J_P (4.29), the stochastic BCRB is expressed as [20]

$$\mathrm{BCRB} = J_B^{-1} = \frac{1}{-2\sigma_\alpha^2 \mathbb{E}_\tau\left\{\Re\left[[\boldsymbol{\Phi}\boldsymbol{\psi}(\tau)]^{\mathrm{H}}\frac{d\boldsymbol{C}_{yy|\tau}^{-1}}{d\tau}\frac{d[\boldsymbol{\Phi}\boldsymbol{\psi}(\tau)]}{d\tau}\right]\right\} + J_P}. \tag{4.40}$$

In the stochastic model, the Fisher information is described in terms of the covariance matrix rather than its mean counterpart. As a result, the stochastic BCRB presented in (4.40) differs from the deterministic BCRB depicted in (4.34).

On the other hand, when the input SNR is very low, both the deterministic and stochastic BCRBs tend toward the reciprocal of the prior information J_P. This convergence occurs because the first term in the denominator of either (4.34) or (4.40) approaches zero. There are two typical distributions to characterize the time delay τ. When τ follows a uniform distribution, J_P equals zero, rendering the BCRB infinite in the *a priori* performance region. Consequently, it becomes unsuitable for evaluating the MSE of the estimator. When τ follows a normal distribution with variance σ_τ^2, $J_P = \frac{1}{\sigma_\tau^2}$ such that $\mathrm{BCRB} = \sigma_\tau^2$ which is tight in the *a priori* performance region. It is noted that the BCRB does not exhibit the same degree of tightness within the threshold region, situated between the *a priori* performance region and the asymptotic region.

4.5 Ziv–Zakai Bound

As analyzed above, the BCRB achieves tightness generally only in the asymptotic region. To overcome this limitation, in this section, we derive the ZZB to provide a globally tight bound for evaluating time delay estimation from the compressive measurement.

Unlike the BCRB that relates the lower bound of the variance of an estimator to the Fisher information, the ZZB establishes a connection between the MSE in the estimation problem and the POE in a binary hypothesis testing problem [27].

Mathematically, the ZZB for time delay estimation with $\tau \in [0, T]$ is given as [29]

$$\mathrm{ZZB} = \frac{1}{2}\int_0^T \xi \int_0^{T-\xi} \left[f_\tau(\varphi) + f_\tau(\varphi+\xi)\right] \mathrm{P}_{\min}(\varphi, \varphi+\xi) d\varphi \, d\xi, \tag{4.41}$$

where $f_\tau(\varphi) \triangleq f(\tau)|_{\tau=\varphi}$ represents the prior probability of the time delay τ when $\tau = \varphi$, and

$$\begin{aligned}\mathrm{P}_{\min}(\varphi, \varphi+\xi) &= \Pr(\mathcal{H}_0)\Pr\left(\mathcal{L}_\tau \le \gamma|\tau=\varphi\right) + \Pr(\mathcal{H}_1)\Pr\left(\mathcal{L}_\tau > \gamma|\tau=\varphi+\xi\right)\\ &\triangleq \Pr(\mathcal{H}_0)\Pr\left(\mathcal{L}_\varphi \le \gamma\right) + \Pr(\mathcal{H}_1)\Pr\left(\mathcal{L}_{\varphi+\xi} > \gamma\right)\end{aligned} \tag{4.42}$$

is the minimum POE (MPOE) of the following binary hypothesis testing problem

$$\begin{aligned}\mathcal{H}_0 &: \tau = \varphi, & \Pr(\mathcal{H}_0) &= \frac{f_\tau(\varphi)}{f_\tau(\varphi)+f_\tau(\varphi+\xi)},\\ \mathcal{H}_1 &: \tau = \varphi+\xi, & \Pr(\mathcal{H}_1) &= \frac{f_\tau(\varphi+\xi)}{f_\tau(\varphi)+f_\tau(\varphi+\xi)}\end{aligned} \tag{4.43}$$

via the following maximum a posteriori probability (MAP) criterion [58]

$$\begin{aligned}&\text{Decide } \mathcal{H}_0 : \tau = \varphi \quad &\text{if } \Pr(\mathcal{H}_0)f(\boldsymbol{y}|\varphi) > \Pr(\mathcal{H}_1)f(\boldsymbol{y}|\varphi+\xi),\\ &\text{Decide } \mathcal{H}_1 : \tau = \varphi+\xi \quad &\text{if } \Pr(\mathcal{H}_0)f(\boldsymbol{y}|\varphi) \le \Pr(\mathcal{H}_1)f(\boldsymbol{y}|\varphi+\xi).\end{aligned} \tag{4.44}$$

Here, $\Pr(\mathcal{H}_0) + \Pr(\mathcal{H}_1) = 1$,

$$\mathcal{L}_\tau = \ln \frac{f(\boldsymbol{y}|\varphi)}{f(\boldsymbol{y}|\varphi+\xi)} \tag{4.45}$$

is the log-likelihood ratio test (LRT) with $f(\boldsymbol{y}|\varphi) \triangleq f(\boldsymbol{y}|\tau)|_{\tau=\varphi}$, and

$$\gamma = \ln \frac{\Pr(\mathcal{H}_1)}{\Pr(\mathcal{H}_0)} = \ln \frac{f_\tau(\varphi+\xi)}{f_\tau(\varphi)} \tag{4.46}$$

is the log-probability ratio.

Given that $f(\tau)$ is the *a priori* known, as described in Section 4.2.1, we can evaluate the ZZB through deriving the MPOE. To calculate $\Pr\left(\mathcal{L}_\varphi \le \gamma\right)$ and $\Pr\left(\mathcal{L}_{\varphi+\xi} > \gamma\right)$, which are needed for deriving $\mathrm{P}_{\min}(\varphi, \varphi+\xi)$ in (4.42), we first find the distributions of $\mathcal{L}_\varphi$ and $\mathcal{L}_{\varphi+\xi}$ in the following two subsections.

4.5.1 ZZB: Deterministic Model

From (4.6), the LRT is given as

$$\mathcal{L}_\tau = \left(\boldsymbol{y} - \bar{\boldsymbol{y}}_{\varphi+\xi}\right)^{\mathrm{H}} \boldsymbol{C}_{\boldsymbol{yy}}^{-1} \left(\boldsymbol{y} - \bar{\boldsymbol{y}}_{\varphi+\xi}\right) - \left(\boldsymbol{y} - \bar{\boldsymbol{y}}_\varphi\right)^{\mathrm{H}} \boldsymbol{C}_{\boldsymbol{yy}}^{-1} \left(\boldsymbol{y} - \bar{\boldsymbol{y}}_\varphi\right), \tag{4.47}$$

where $\bar{\boldsymbol{y}}_\varphi \triangleq \bar{\boldsymbol{y}}|_{\tau=\varphi} = \alpha\boldsymbol{\Phi}\boldsymbol{\psi}(\varphi)$. Hence, the LRT $\mathcal{L}_\tau$ is a function of the time delay τ via the compressive measurement $\boldsymbol{y}$. Given $\tau = \varphi$, the LRT is written as

$$\begin{aligned}\mathcal{L}_\varphi &= \left(\boldsymbol{y}_\varphi - \bar{\boldsymbol{y}}_{\varphi+\xi}\right)^{\mathrm{H}} \boldsymbol{C}_{\boldsymbol{yy}}^{-1} \left(\boldsymbol{y}_\varphi - \bar{\boldsymbol{y}}_{\varphi+\xi}\right) - \left(\boldsymbol{y}_\varphi - \bar{\boldsymbol{y}}_\varphi\right)^{\mathrm{H}} \boldsymbol{C}_{\boldsymbol{yy}}^{-1} \left(\boldsymbol{y}_\varphi - \bar{\boldsymbol{y}}_\varphi\right)\\ &= \left(\bar{\boldsymbol{y}}_\varphi - \bar{\boldsymbol{y}}_{\varphi+\xi}\right)^{\mathrm{H}} \boldsymbol{C}_{\boldsymbol{yy}}^{-1} \left(\bar{\boldsymbol{y}}_\varphi - \bar{\boldsymbol{y}}_{\varphi+\xi}\right) + 2\Re\left[\left(\bar{\boldsymbol{y}}_\varphi - \bar{\boldsymbol{y}}_{\varphi+\xi}\right)^{\mathrm{H}} \boldsymbol{C}_{\boldsymbol{yy}}^{-1}\boldsymbol{\Phi}\boldsymbol{n}\right],\end{aligned} \tag{4.48}$$

where $\boldsymbol{y}_\varphi \triangleq \boldsymbol{y}|_{\tau=\varphi} = \boldsymbol{\Phi}\left[\alpha\boldsymbol{\psi}(\varphi) + \boldsymbol{n}\right]$. Since the only random variable in $\mathcal{L}_\varphi$ is the Gaussian noise $\boldsymbol{n} \sim \mathcal{CN}(\boldsymbol{0}, \boldsymbol{C}_{nn})$, it is straightforward to prove that $\mathcal{L}_\varphi$ follows a Gaussian distribution as $\mathcal{L}_\varphi \sim \mathcal{N}(\bar{\mathcal{L}}_\varphi, 2\bar{\mathcal{L}}_\varphi)$, where

$$\begin{aligned}\bar{\mathcal{L}}_\varphi &= \left(\bar{\boldsymbol{y}}_\varphi - \bar{\boldsymbol{y}}_{\varphi+\xi}\right)^{\mathrm{H}} \boldsymbol{C}_{yy}^{-1} \left(\bar{\boldsymbol{y}}_\varphi - \bar{\boldsymbol{y}}_{\varphi+\xi}\right) \\ &= |\alpha|^2[\boldsymbol{\Phi}\boldsymbol{\psi}(\varphi) - \boldsymbol{\Phi}\boldsymbol{\psi}(\varphi+\xi)]^{\mathrm{H}} \boldsymbol{C}_{yy}^{-1} [\boldsymbol{\Phi}\boldsymbol{\psi}(\varphi) - \boldsymbol{\Phi}\boldsymbol{\psi}(\varphi+\xi)]\,. \end{aligned} \tag{4.49}$$

Similarly, given $\tau = \varphi + \xi$, it is observed that

$$\mathcal{L}_{\varphi+\xi} = -\left(\bar{\boldsymbol{y}}_\varphi - \bar{\boldsymbol{y}}_{\varphi+\xi}\right)^{\mathrm{H}} \boldsymbol{C}_{yy}^{-1} \left(\bar{\boldsymbol{y}}_\varphi - \bar{\boldsymbol{y}}_{\varphi+\xi}\right) + 2\Re\left[\left(\bar{\boldsymbol{y}}_\varphi - \bar{\boldsymbol{y}}_{\varphi+\xi}\right)^{\mathrm{H}} \boldsymbol{C}_{yy}^{-1}\boldsymbol{\Phi}\boldsymbol{n}\right] \tag{4.50}$$

follows a Gaussian distribution as $\mathcal{L}_{\varphi+\xi} \sim \mathcal{N}(-\bar{\mathcal{L}}_\varphi, 2\bar{\mathcal{L}}_\varphi)$. Hence, the MPOE can be written as

$$\mathrm{P}_{\min}(\varphi, \varphi+\xi) = \Pr(\mathcal{H}_0) Q\left(\frac{\bar{\mathcal{L}}_\varphi - \gamma}{\sqrt{2\bar{\mathcal{L}}_\varphi}}\right) + \Pr(\mathcal{H}_1) Q\left(\frac{\bar{\mathcal{L}}_\varphi + \gamma}{\sqrt{2\bar{\mathcal{L}}_\varphi}}\right), \tag{4.51}$$

where $Q(x) = \frac{1}{\sqrt{2\pi}} \int_x^\infty e^{-\frac{v^2}{2}} dv$ is the tail distribution function of the standard normal distribution. Correspondingly, the deterministic ZZB is given as [25]

$$\begin{aligned}\mathrm{ZZB} = \frac{1}{2} \int_0^T \xi \int_0^{T-\xi} & [f_\tau(\varphi) + f_\tau(\varphi+\xi)] \\ & \left[\Pr(\mathcal{H}_0) Q\left(\frac{\bar{\mathcal{L}}_\varphi - \gamma}{\sqrt{2\bar{\mathcal{L}}_\varphi}}\right) + \Pr(\mathcal{H}_1) Q\left(\frac{\bar{\mathcal{L}}_\varphi + \gamma}{\sqrt{2\bar{\mathcal{L}}_\varphi}}\right)\right] d\varphi\, d\xi, \end{aligned} \tag{4.52}$$

which can be formulated as an explicit function of the CS kernel $\boldsymbol{\Phi}$ via $\bar{\mathcal{L}}_\varphi$ in (4.49).

Specifically, when τ follows a uniform distribution, i.e., $f_\tau(\varphi) = f_\tau(\varphi + \xi)$, the MPOE in (4.51) is simplified as

$$\mathrm{P}_{\min}(\varphi, \varphi+\xi) = Q\left(\sqrt{\frac{\bar{\mathcal{L}}_\varphi}{2}}\right), \tag{4.53}$$

because $\Pr(\mathcal{H}_0) = \Pr(\mathcal{H}_1) = \frac{1}{2}$ and $\gamma = 0$. Consequently, the deterministic ZZB degrades to

$$\mathrm{ZZB} = \frac{1}{T} \int_0^T \xi \int_0^{T-\xi} Q\left(\sqrt{\frac{\bar{\mathcal{L}}_\varphi}{2}}\right) d\varphi\, d\xi. \tag{4.54}$$

In the case where the additive noise $\boldsymbol{n}$ is Gaussian and white, the deterministic ZZB is explicitly expressed as

$$\mathrm{ZZB} = \frac{1}{T}\int_0^T \xi \int_0^{T-\xi} Q\left(\sqrt{\frac{1}{2}\frac{|\alpha|^2}{\sigma_{\boldsymbol{n}}^2}\|\boldsymbol{\Phi}\boldsymbol{\psi}(\varphi) - \boldsymbol{\Phi}\boldsymbol{\psi}(\varphi+\xi)\|_2^2}\right) d\varphi d\xi, \quad (4.55)$$

where $\|\boldsymbol{\Phi}\boldsymbol{\psi}(\varphi) - \boldsymbol{\Phi}\boldsymbol{\psi}(\varphi+\xi)\|_2$ denotes the pairwise distance between compressed waveforms delayed by φ and $\varphi+\xi$. Interestingly, the operation $\boldsymbol{\Phi}\boldsymbol{\psi}(\varphi) - \boldsymbol{\Phi}\boldsymbol{\psi}(\varphi+\xi)$ represents the difference form of the differential operation $\frac{d[\boldsymbol{\Phi}\boldsymbol{\psi}(\tau)]}{d\tau}$ in (4.35).

The ZZB expression given in (4.55) entails some nuanced interpretations compared to the straightforward depiction of the BCRB outlined in (4.35). First, the dependence of the ZZB on the input SNR is captured through the term $\frac{|\alpha|^2}{\sigma_{\boldsymbol{n}}^2}$. Because the Q-function is monotonically decreasing, a higher input SNR corresponds to a lower ZZB, indicating higher estimation accuracy. Conversely, when the input SNR is very low, the ZZB converges to $\frac{1}{12}T^2$, representing the variance of the uniform distribution $\tau \sim [0, T]$. Thus, the ZZB provides a tight bound in the *a priori* performance region, even when τ is uniformly distributed. Second, the dependence of the ZZB on the pairwise distances between compressed waveforms with different time delays is encapsulated in the term $\|\boldsymbol{\Phi}\boldsymbol{\psi}(\varphi) - \boldsymbol{\Phi}\boldsymbol{\psi}(\varphi+\xi)\|_2, \forall\varphi \in [0, T-\xi], \xi \in [0, T]$. Because of the monotonicity of the Q-function, a larger pairwise distance results in a lower ZZB, indicating more accurate estimation. The information-theoretic sensing kernel, in contrast to random measurement kernels, enhances these pairwise distances by leveraging the prior distribution of τ. Moreover, the tradeoff between resolution enhancement due to waveform bandwidth increment and SNR loss due to compressive measurement mirrors what we observed in Section 4.4.1. Specifically, for a constant sampling rate, the ZZB decreases quadratically with the bandwidth increment while increasing linearly with the SNR loss. This underscores how the resolution benefit from increasing waveform bandwidth outweighs the SNR loss from compressed measurement.

4.5.2 ZZB: Stochastic Model

From (4.7), the LRT is

$$\mathcal{L}_\tau = c + \boldsymbol{y}^{\mathrm{H}}\left(\boldsymbol{C}_{\boldsymbol{yy}|\varphi+\xi}^{-1} - \boldsymbol{C}_{\boldsymbol{yy}|\varphi}^{-1}\right)\boldsymbol{y}, \quad (4.56)$$

where $c = \ln\frac{|\boldsymbol{C}_{\boldsymbol{yy}|\varphi+\xi}|}{|\boldsymbol{C}_{\boldsymbol{yy}|\varphi}|}$ is a constant given φ and ξ. Because $\boldsymbol{y} \sim \mathcal{CN}(\boldsymbol{0}, \boldsymbol{C}_{\boldsymbol{yy}|\tau})$, we apply a whitening transform $\boldsymbol{z} = \boldsymbol{C}_{\boldsymbol{yy}|\tau}^{-\frac{1}{2}}\boldsymbol{y}$ to yield a whitened random vector $\boldsymbol{z} \sim \mathcal{CN}(\boldsymbol{0}, \boldsymbol{I})$. In this case, $\mathcal{L}_\tau$ can be represented as a quadratic function of $\boldsymbol{z}$ as

$$\mathcal{L}_\tau = c + \boldsymbol{z}^{\mathrm{H}}\boldsymbol{\Gamma}_\tau\boldsymbol{z}, \quad (4.57)$$

where

$$\boldsymbol{\Gamma}_\tau = \boldsymbol{C}_{yy|\tau}^{\frac{1}{2}} \left(\boldsymbol{C}_{yy|\varphi+\xi}^{-1} - \boldsymbol{C}_{yy|\varphi}^{-1} \right) \boldsymbol{C}_{yy|\tau}^{\frac{1}{2}} \tag{4.58}$$

is a Hermitian matrix. By performing eigendecomposition of $\boldsymbol{\Gamma}_\tau$ as $\boldsymbol{\Gamma}_\tau = \boldsymbol{U}_\tau \boldsymbol{\Lambda}_\tau \boldsymbol{U}_\tau^{\mathrm{H}}$ with $\boldsymbol{U}_\tau$ denoting a unitary matrix containing the eigenvectors of $\boldsymbol{\Gamma}_\tau$ and $\boldsymbol{\Lambda}_\tau$ being a diagonal matrix containing the eigenvalues of $\boldsymbol{\Gamma}_\tau$, $\mathcal{L}_\tau$ can be further written as

$$\mathcal{L}_\tau = c + \boldsymbol{w}^{\mathrm{H}} \boldsymbol{\Lambda}_\tau \boldsymbol{w}, \tag{4.59}$$

where $\boldsymbol{w} = \boldsymbol{U}_\tau^{\mathrm{H}} \boldsymbol{z} \sim \mathcal{CN}(\boldsymbol{0}, \boldsymbol{I})$. Hence, the LRT $\mathcal{L}_\tau$ can be expressed as a linear combination of chi-square random variables as

$$\mathcal{L}_\tau = c + \sum_{i=1}^{\mathrm{rank}(\boldsymbol{\Gamma}_\tau)} \frac{\lambda_{\tau,i}}{2} \chi_2^2 \triangleq c + X_\tau, \tag{4.60}$$

where $\lambda_{\tau,i}$ denotes the ith nonzero eigenvalue of $\boldsymbol{\Lambda}_\tau$ and χ_2^2 is a chi-square distributed random variable with 2 degrees of freedom (DOFs).

According to the Sherman–Morrison formula, we have

$$\boldsymbol{C}_{yy|\varphi+\xi}^{-1} - \boldsymbol{C}_{yy|\varphi}^{-1} = \boldsymbol{C}_{yy}^{-1} \boldsymbol{A} \boldsymbol{D} \boldsymbol{A}^{\mathrm{H}} \boldsymbol{C}_{yy}^{-1}, \tag{4.61}$$

where $\boldsymbol{D} = \mathrm{diag}\left(d_1, d_2\right)$ is a diagonal matrix with $d_1 = \frac{\sigma_\alpha^2}{1+\sigma_\alpha^2 \boldsymbol{\psi}^{\mathrm{H}}(\varphi)\boldsymbol{\Phi}^{\mathrm{H}} \boldsymbol{C}_{yy}^{-1} \boldsymbol{\Phi}\boldsymbol{\psi}(\varphi)} > 0$ and $d_2 = -\frac{\sigma_\alpha^2}{1+\sigma_\alpha^2 \boldsymbol{\psi}^{\mathrm{H}}(\varphi+\xi)\boldsymbol{\Phi}^{\mathrm{H}} \boldsymbol{C}_{yy}^{-1} \boldsymbol{\Phi}\boldsymbol{\psi}(\varphi+\xi)} < 0$, and $\boldsymbol{A} = [\boldsymbol{\Phi}\boldsymbol{\psi}(\varphi), \boldsymbol{\Phi}\boldsymbol{\psi}(\varphi+\xi)] \in \mathbb{C}^{M\times 2}$. Because both $\boldsymbol{C}_{yy|\tau}$ and $\boldsymbol{C}_{yy}$ are full rank, we have

$$\mathrm{rank}(\boldsymbol{\Gamma}_\tau) = \mathrm{rank}\left(\boldsymbol{C}_{yy|\varphi+\xi}^{-1} - \boldsymbol{C}_{yy|\varphi}^{-1} \right) = \mathrm{rank}\left(\boldsymbol{A}\boldsymbol{D}\boldsymbol{A}^{\mathrm{H}}\right). \tag{4.62}$$

According to Sylvester's rank inequality, we have $\mathrm{rank}(\boldsymbol{\Gamma}_\tau) = 2$ by the observation that both $\mathrm{rank}(\boldsymbol{A}) = 2$ and $\mathrm{rank}(\boldsymbol{D}) = 2$. Furthermore, by substituting (4.61) into (4.58), we have

$$\begin{aligned}\boldsymbol{\Gamma}_\tau = {} & d_1 \boldsymbol{C}_{yy|\tau}^{\frac{1}{2}} \boldsymbol{C}_{yy}^{-1} \boldsymbol{\Phi}\boldsymbol{\psi}(\varphi) \left[\boldsymbol{C}_{yy|\tau}^{\frac{1}{2}} \boldsymbol{C}_{yy}^{-1} \boldsymbol{\Phi}\boldsymbol{\psi}(\varphi) \right]^{\mathrm{H}} \\ & + d_2 \boldsymbol{C}_{yy|\tau}^{\frac{1}{2}} \boldsymbol{C}_{yy}^{-1} \boldsymbol{\Phi}\boldsymbol{\psi}(\varphi+\xi) \left[\boldsymbol{C}_{yy|\tau}^{\frac{1}{2}} \boldsymbol{C}_{yy}^{-1} \boldsymbol{\Phi}\boldsymbol{\psi}(\varphi+\xi) \right]^{\mathrm{H}},\end{aligned} \tag{4.63}$$

which is the difference between two rank-one positive semidefinite Hermitian matrices.

As mentioned above, the two nonzero eigenvalues of $\boldsymbol{\Gamma}_\tau$ contain one positive value and one negative value. Correspondingly, X_τ results in the difference of two chi-square distributions with the following cumulative distribution function (CDF) [59]:

$$F_{X_\tau}(x) = \begin{cases} 1 - \dfrac{\lambda^+}{\lambda^+ + \lambda^-} e^{-\frac{1}{\lambda^+}x}, & x \geq 0, \\ \dfrac{\lambda^-}{\lambda^+ + \lambda^-} e^{\frac{1}{\lambda^-}x}, & x < 0, \end{cases} \tag{4.64}$$

where λ^+ and λ^-, respectively, represent the modulus of the positive eigenvalue and the negative eigenvalue of $\mathbf{\Lambda}_\tau$.

By, respectively, substituting $\tau = \varphi$ and $\tau = \varphi + \xi$ into (4.58) and calculating (4.59) to (4.64), the MPOE is expressed as

$$\mathrm{P_{min}}(\varphi, \varphi + \xi) = \Pr(\mathcal{H}_0)F_{X_\varphi}(\gamma - c) + \Pr(\mathcal{H}_1)\left[1 - F_{X_{\varphi+\xi}}(\gamma - c)\right]. \tag{4.65}$$

Correspondingly, the stochastic ZZB is given as [26]

$$\begin{aligned}\mathrm{ZZB} = \frac{1}{2}\int_0^T \xi \int_0^{T-\xi} & [f_\tau(\varphi) + f_\tau(\varphi + \xi)] \\ & \left\{\Pr(\mathcal{H}_0)F_{X_\varphi}(\gamma - c) + \Pr(\mathcal{H}_1)\left[1 - F_{X_{\varphi+\xi}}(\gamma - c)\right]\right\} d\varphi\, d\xi. \end{aligned}\tag{4.66}$$

Specifically, when τ follows a uniform distribution, the stochastic ZZB (4.66) degrades to

$$\mathrm{ZZB} = \frac{1}{2T}\int_0^T \xi \int_0^{T-\xi} \left[F_{X_\varphi}(-c) + 1 - F_{X_{\varphi+\xi}}(-c)\right] d\varphi\, d\xi. \tag{4.67}$$

In summary, both the deterministic ZZB and the stochastic ZZB depend on CS kernels $\boldsymbol{\Phi}$ through the individual MPOE $\mathrm{P_{min}}(\varphi, \varphi + \xi)$.

4.6 Simulation Results

In this section, we conduct comparative analyses of time delay estimation performance between the information-theoretic CS kernels proposed in Section 4.3 and the Gaussian random CS kernels. We also evaluate the BCRB derived in Section 4.4 and the ZZB derived in Section 4.5 with respect to the input SNR and the compression ratio to demonstrate the benefit offered by the proposed compressive measurement scheme.

Without loss of generality, we assume that the ADC operates at a fixed sampling rate of $B = 1$ MHz.[1] The radar transmitter transmits an LFM waveform with a bandwidth of $CR \times B$ MHz, where $CR = \frac{L}{M} > 1$ represents the compression ratio at the compressive radar receiver. The compression ratio is a pivotal factor when either the transmitter signal bandwidth or the receiver sampling rate remains unchanged. In our simulations, we keep the sampling rate fixed at the compressive radar receiver, allowing us to assess the performance gain achieved by generating transmit waveforms with higher bandwidths. For example, when the compression ratio is $CR = 5$, the radar waveform bandwidth amounts to $5 \times 1 = 5$ MHz. Concurrently, waveforms with a 1-MHz bandwidth serve as the

1 There exists a scaling factor impacting the estimation performance by adjusting the waveform bandwidth. This observation was validated in Figure 5 of [25], where we compared the estimation performance and bounds using waveforms with bandwidths of 10 MHz and 1 GHz.

baseline for comparison, wherein the receiver sampling rate aligns with the transmit waveform bandwidth, reflecting a Nyquist sampling scenario.

In addition to the compression ratio, the input SNR of the target signal is another factor that plays an important role in both the sensing kernel optimization and the time delay estimation. We establish the relationship between the signal amplitude and the input SNR as $\alpha = \sqrt{\text{SNR}} \times \sigma_n$ in the deterministic model and $\alpha \sim \mathcal{CN}(0, \text{SNR} \times \sigma_n^2)$ in the stochastic model. In the simulations, we assume the time delay τ to fall within the range of 0–8 μs, corresponding to a target distance spanning from 0 to 1200 m. We uniformly discretize the PDF of time delay into $K = 1000$ bins for information-theoretic sensing kernel optimization, irrespective of the input SNR. Consequently, this results in 1000 Gaussian mixture components, as depicted in (4.9).

It is worth noting that, although the information-theoretic sensing kernel necessitates separate optimization for each value of the input SNR, we may perform optimization at selected input SNR values to reduce the overall complexity. In this case, we prefer to optimize the sensing kernels at a higher input SNR rather than adopting a lower SNR. Our studies indicate that, in the asymptotic region, the performance of parameter estimation exhibits relatively low sensitivity to the mismatch in the input SNR in the optimization of the sensing kernel.

4.6.1 Sensing Kernels

In Figure 4.4, we compare the information-theoretic sensing kernels with the initial Gaussian random sensing kernels. Here, the compression ratio is set to $CR = 5$, and the information-theoretic sensing kernels are optimized at an input SNR of 30 dB, with the prior distribution $\tau \sim \mathcal{N}\left(\frac{4}{B}, \left(\frac{1}{B}\right)^2\right)$. Comparing with the random sensing kernels, which exhibit an approximately uniform distribution in both the time domain and the frequency domain, the optimized sensing kernels demonstrate a higher concentration in the time domain. Moreover, they feature dominant components in their frequency spectra ranging from −2.5 to 2.5 MHz (i.e., the bandwidth of $CR \times B = 5$ MHz), thereby substantially enhancing the resolution performance.

4.6.2 Pairwise Distances

In Figure 4.5, we present the pairwise distance maps for different sensing kernels together with the corresponding slices along their main diagonals, where the parameters remain the same as those used in Section 4.6.1. The pairwise distance between compressed waveforms delayed by τ_1 and τ_2 is defined as

$$d(\tau_1, \tau_2) \triangleq \|\boldsymbol{\Phi}\boldsymbol{\psi}(\tau_1) - \boldsymbol{\Phi}\boldsymbol{\psi}(\tau_2)\|_2, \ \forall \tau_1, \tau_2, \tag{4.68}$$

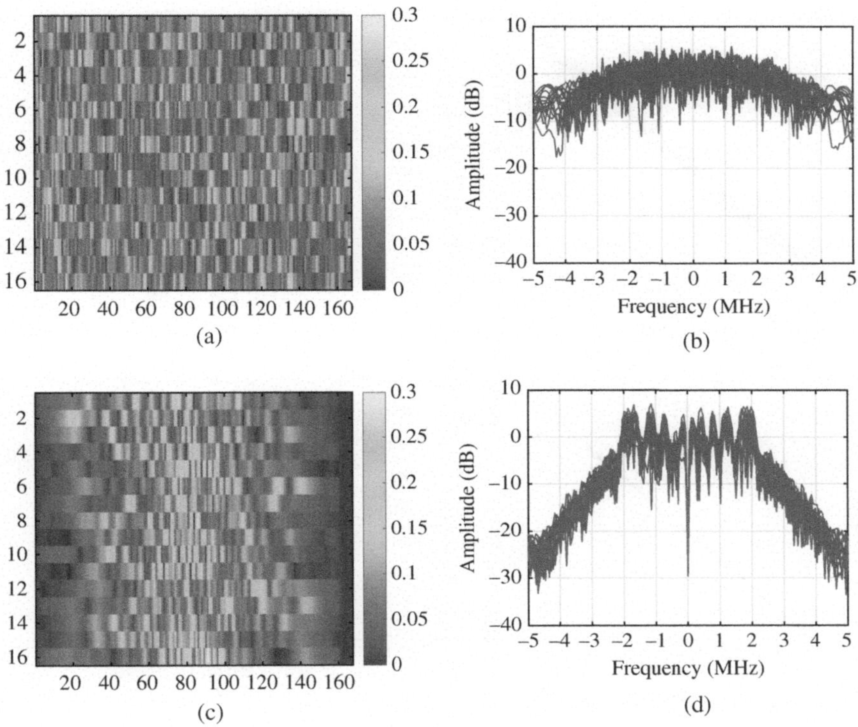

Figure 4.4 Comparison of CS kernels ($CR = 5$). (a) Magnitudes, (b) frequency spectra of the Gaussian random sensing kernels; (c) magnitudes and (d) frequency spectra of the information-theoretic sensing kernels.

where the sensing kernel $\boldsymbol{\Phi}$ becomes the sole determinant when specific time delays τ_1 and τ_2 are assigned for a given transmit waveform $\boldsymbol{\psi}(t)$. To mitigate the effects of randomness, the pairwise distance map for random sensing kernels is obtained by averaging the pairwise distances across numerous realizations of the random sensing kernel. Consequently, the pairwise distance map illustrated in Figure 4.5(c) approximates the expected distance $\mathbb{E}\{d(\tau_1, \tau_2)\}$. Ideally, the pairwise distances should be maximized because distinct time delays will map to substantially different compressed waveforms, thereby facilitating superior resolution and estimation accuracy. Notably, the anti-diagonal elements of the pairwise distance maps consistently register as zeros, regardless of the employed sensing kernel, as they represent the distances between the compressed waveforms with identical time delays. As the time delay separation increases, the pairwise distance between the corresponding compressed waveforms generally increases as well, thereby enhancing the resolution and the estimation accuracy.

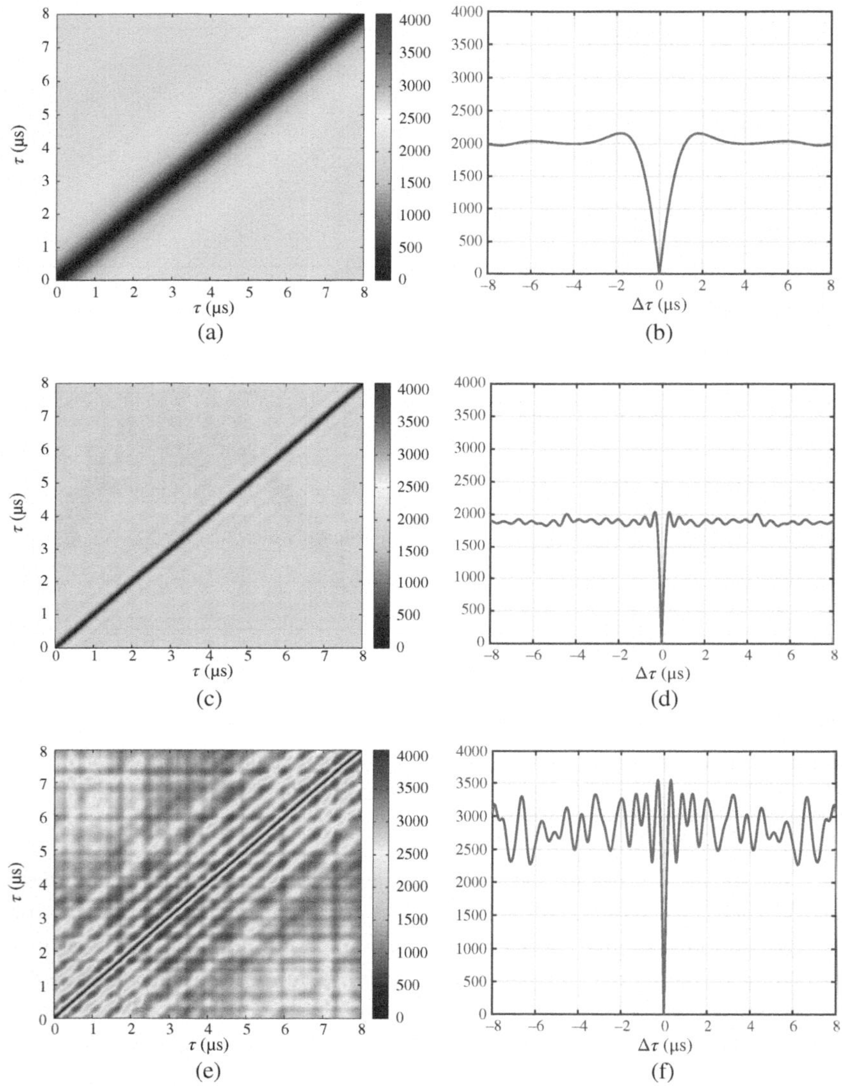

Figure 4.5 Comparison of pairwise distances. (a) and (b) Nyquist sensing kernel; (c) and (d) Gaussian random sensing kernels; (e) and (f) information-theoretic sensing kernels.

Clearly, when considering the same sampling rate, the Nyquist sampling scheme exhibits a poor resolution, as is evident from the broad null along the main diagonal of the pairwise distance map, as depicted in Figure 4.5(a) and (b). In contrast, due to the broader transmit waveform bandwidth, compressive implementations demonstrate superior native resolution, as shown in Figure 4.5(c) and (e), regardless of the utilized sensing kernels. This enhanced resolution is further underscored by the narrower nulls along the main diagonals, as observed in Figure 4.5(d) and (f). It is appealing to achieve such fundamental resolution enhancement through compressive implementations without increasing the sampling rate of the ADCs. Furthermore, upon comparing Figure 4.5(c) with Figure 4.5(e), it becomes apparent that, while both the information-theoretic sensing kernel and random sensing kernels offer the same resolution (i.e., identical null width), the optimized sensing kernel yields larger distances (and consequently better distinguishability), particularly near the center of the map, where the prior (normal) PDF of the time delay τ denotes more probable values. Additionally, at the corners of the map, the optimized sensing kernel facilitates distance reduction since these time delays are less probable to occur at the tails of the parameter's prior distribution. Overall, such pairwise distance behavior indicates improved estimation accuracy. Although random sensing kernels exhibit comparable resolution, the reduced pairwise distance for any random sensing kernel realization impairs its average performance. Finally, it is noteworthy that the pairwise distance map for the optimized sensing kernel depends on the input SNR and the prior PDF, which play critical roles in information-theoretic sensing kernel optimization.

4.6.3 Output SNRs and Estimation Error Distributions

In Figure 4.6(a), we compare the output SNRs of different sensing kernels to evaluate the SNR loss attributed to compressive sampling. The output SNR is defined as

$$\text{Output SNR} \triangleq \mathbb{E}\left\{ \frac{\|\alpha \boldsymbol{\Phi} \boldsymbol{\psi}(\tau)\|_2^2}{\|\boldsymbol{\Phi} \boldsymbol{n}\|_2^2} \right\}, \tag{4.69}$$

where the signal amplitude α is assumed deterministic, and the time delay follows a normal distribution $\tau \sim \mathcal{N}\left(\frac{4}{B}, \left(\frac{1}{B}\right)^2\right)$. Nyquist sampling exhibits no SNR loss, as its output SNR remains identical to the input SNR across all tested values. Conversely, for random CS kernels, a fixed SNR loss equivalent to the compression ratio (expressed as $10\log_{10} CR$ in dB) is evident. In contrast, the behavior differs for information-theoretic CS kernels. Specifically, at low input SNR levels (e.g., below 0 dB), the optimized sensing kernel aims to preserve SNR, i.e., minimize

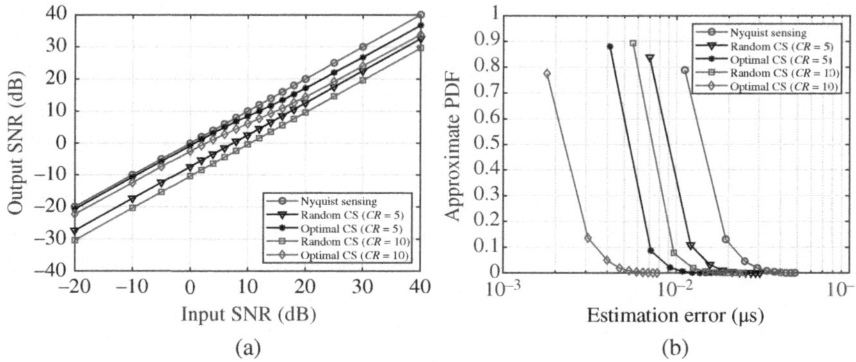

Figure 4.6 Comparison of (a) output SNRs and (b) estimation error distributions.

SNR loss. Conversely, at high input SNR levels (e.g., above 20 dB), the optimized sensing kernel compromises SNR loss to enhance resolution and discrimination across various τ values. This transition in behavior occurs within the input SNR range between 0 and 20 dB, which is often referred to as the *threshold region* in non-linear parameter estimation problems. When the input SNR is below the threshold region, it is called as the *a priori* performance region where the estimation problem is primarily noise-limited, and performance depends on the prior variance. Conversely, when the input SNR is above the threshold region, we refer it as the asymptotic region in which the system becomes resolution-limited, thus benefiting from high-resolution compressive implementations to enhance the estimation performance.

In Figure 4.6(b), we compare the approximate distributions of the estimation errors across different sensing kernels at an input SNR of 30 dB. Each result is obtained by averaging over 1000 Monte Carlo trials. It is evident that the information-theoretic CS kernels offer smaller estimation errors, despite that their average output SNR levels are lower than that of the Nyquist sensing case, as depicted in Figure 4.6(a). This fact suggests that the compressive system outperforms in the asymptotic region. These distributions corroborate that, in the asymptotic region, the compressive system yields significantly lower average estimation errors, with rare occurrences of worst-case errors. However, it is important to note that these conclusions hold true only in the asymptotic region, not in the *a priori* performance region (not depicted). In the noise-limited *a priori* performance region, estimation performance solely depends on the prior variance. The RMSE comparisons in Sections 4.6.4 and 4.6.5 will further demonstrate this claim.

4.6.4 RMSE for the Deterministic Model

In Figure 4.7, we compare the RMSE performance with respect to different sensing kernels, with the signal amplitude α being considered as a deterministic unknown. The RMSE is defined as

$$\mathrm{RMSE} \triangleq \sqrt{\frac{1}{Q}\sum_{q=1}^{Q}\left(\hat{\tau}_q - \tau_q\right)^2}, \tag{4.70}$$

where $\hat{\tau}_q$ denotes the estimate of the true time delay τ_q based on (4.12) in the qth Monte Carlo trial. For each data point (SNR), the RMSE is computed from $Q = 1000$ Monte Carlo trials.

First, it is evident that CS kernels require a higher input SNR to reach the desired asymptotic region compared to the Nyquist kernel, owing to the SNR loss they incur. This means that compensating for the SNR loss from compressive measurement necessitates additional input SNR, which increases as a higher compression ratio is adopted. However, once the input SNR surpasses a certain threshold (e.g., above 20 dB), compressively sampled high-bandwidth waveforms generally outperform Nyquist sampled low-bandwidth ones. This is exemplified by CS outperforming Nyquist sensing for the same input SNR, provided that the input SNR is approximately 20–25 dB or higher. As anticipated, information-theoretic sensing kernels notably outperform random sensing kernels in both the threshold and asymptotic regions, achieving the desired asymptotic region at a lower input SNR (around 7 dB or less). This underscores the efficacy of the information-theoretic sensing kernel optimization method delineated in this chapter.

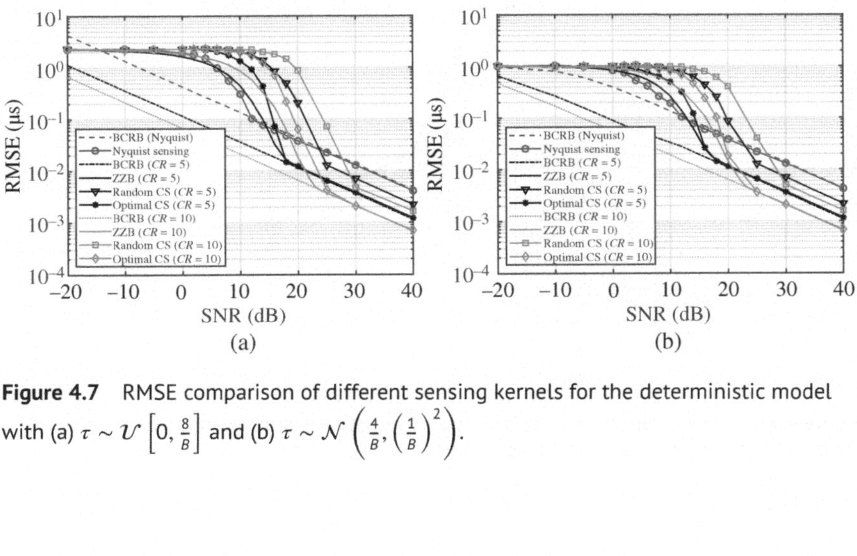

Figure 4.7 RMSE comparison of different sensing kernels for the deterministic model with (a) $\tau \sim \mathcal{U}\left[0, \frac{8}{B}\right]$ and (b) $\tau \sim \mathcal{N}\left(\frac{4}{B}, \left(\frac{1}{B}\right)^2\right)$.

Second, the BCRB consistently offers a tight lower bound on the MSE in the asymptotic region, irrespective of the compression ratio and the prior distribution. However, as the input SNR decreases, the BCRB loses validity in the *a priori* performance region ($\lim_{\text{SNR}\to 0} J_D = 0$). In this case, the parameter follows a uniform distribution which is uninformative, as shown in Figure 4.7(a). Conversely, in the *a priori* performance region where the parameter adheres to a normal distribution, the BCRB converges to its prior variance, as depicted in Figure 4.7(b). It is noteworthy that the BCRB is a loose bound in the threshold region, regardless of the prior distribution, and thus fails to accurately predict the estimation performance. In contrast, the ZZB consistently offers tight lower bounds on the MSE across both the *a priori* performance region and the asymptotic region, irrespective of the compression ratio and the prior distribution. Consequently, the ZZB serves as a global lower bound for the prediction of the estimation performance. Furthermore, unlike the BCRB, the ZZB possesses the ability to predict the threshold SNR level for the bound to transit to the asymptotic region.

In Figure 4.8, we investigate the impact of different prior distributions on the time delay estimation performance when using information-theoretic sensing kernels. Here, we keep the same actual distribution, $\tau \sim \mathcal{N}\left(\frac{4}{B}, \left(\frac{1}{B}\right)^2\right)$, for MMSE estimation, while varying the prior distribution used for sensing kernel optimization (as indicated in the legend). The results reveal that variations in the time delay estimation performance among information-theoretic sensing kernels, optimized with different prior distributions, are primarily noticeable in the threshold region.

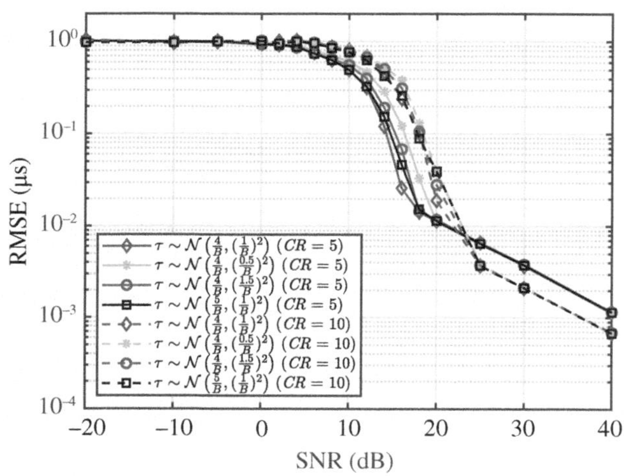

Figure 4.8 RMSE comparison of information-theoretic sensing kernels optimized with different compression ratios and prior distributions.

This discrepancy vanishes in the *a priori* performance region and the asymptotic region. Specifically, when the prior distribution fails to encompass the actual distribution, such as $\tau \sim \mathcal{N}\left(\frac{4}{B}, \left(\frac{0.5}{B}\right)^2\right)$, the information-theoretic sensing kernel yields higher estimation errors. This outcome stems from overlooking some probable values during the prior distribution-based sensing kernel optimization. Overall, in terms of parameter estimation performance, the information-theoretic sensing kernel demonstrates robustness to mismatches between the prior distribution and the actual distribution. However, in practical applications, the prior distribution should strive to cover as many scenarios as possible while maintaining sufficient tightness to yield substantial benefits.

4.6.5 RMSE for the Stochastic Model

In Figure 4.9, we analyze the RMSE performance across different sensing kernels, considering a stochastic signal amplitude α and a normal distribution for the time delay $\tau \sim \mathcal{N}\left(\frac{4}{B}, \left(\frac{1}{B}\right)^2\right)$. To capture potential catastrophic errors, we conduct a large number of $Q = 100{,}000$ Monte Carlo trials when the input SNR is 40 dB and higher. It is clear that the information-theoretic sensing kernels outperform random sensing kernels, which is consistent with the findings from Figure 4.7. However, it is noteworthy that the performance of information-theoretic sensing kernels does not surpass that of the Nyquist kernel under high input SNR conditions.

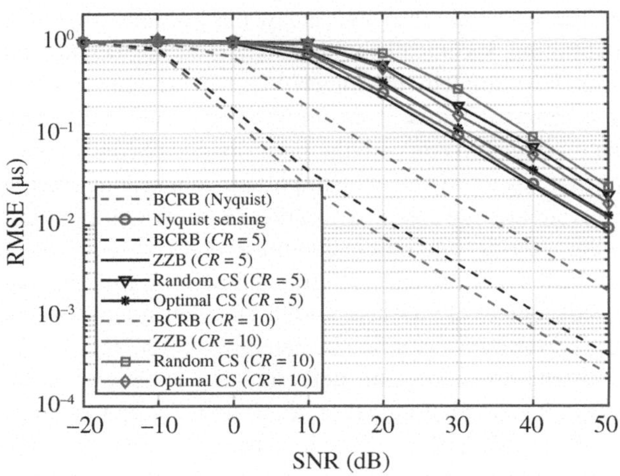

Figure 4.9 RMSE comparison of different sensing kernels for the stochastic model.

In contrast to the deterministic model, we did not observe a clear threshold point within the evaluated input SNR range for this zero-mean stochastic compressive measurement. In the high SNR region, a substantial gap between the ZZB and the BCRB is noticeable, primarily because the ZZB can encompass both small and catastrophic errors. In contrast, the BCRB, acting as a local bound, only captures local small errors and becomes loose in the presence of catastrophic errors. Consequently, the ZZB can predict the estimation performance globally regardless of the compression ratio. Through performance analysis, we find that the influence of a *zero mean* in stochastic signals can be mitigated either by reducing the stochasticity of the signals (i.e., increasing the mean-variance ratio) [25] or by increasing the number of sensing kernels [26], both of which indicate the threshold point for entering the asymptotic region.

4.7 Conclusions

In this chapter, we introduced an information-theoretic approach to optimize CS kernels for time delay estimation in a compressive radar receiver, accompanied by an analysis of fundamental behaviors through derived performance bounds. We developed an effective CS kernel optimization scheme using mutual information gradient ascent search. This approach leverages quantization of the prior distribution of the time delay parameter, resulting in a Gaussian mixture distribution of the compressive measurement. Analytical and numerical results clearly demonstrate the resolution enhancement achievable through compressive sampling, and the tradeoff between SNR loss and resolution improvement. This tradeoff tends to be disadvantageous when the expected input SNR is low but becomes attractive when the input SNR is not limiting. Specifically, at high SNR levels, the heightened native resolution of a compressive implementation enables high-resolution radar imaging, outweighing the inherent SNR loss. More importantly, we illustrated how *a priori* knowledge can be leveraged to enhance the performance of compressive systems while minimizing SNR loss. The following important conclusions emerge from our analysis: (i) the threshold input SNR of a nonlinear parameter estimation problem increases due to the SNR loss inherent in compressive RF systems; (ii) in the asymptotic region, the benefits of higher resolution afforded by compressive sampling outweigh the disadvantages of SNR loss; and (iii) the detrimental effects of SNR loss on estimation performance can be alleviated by optimizing the CS kernel based on the prior distribution of the time delay and, to a lesser extent, the expected SNR of the scenario. It is worth noting that in certain radar applications, the sparsity assumption made in this study may not hold true due to the presence of clutter. In such cases, clutter removal preprocessing is necessary before applying the proposed method.

Acknowledgments

The work of N. A. Goodman was supported in part by the Defense Advanced Research Projects Agency under Grant #N66001-10-1-4079. The work of Y. D. Zhang was supported in part by the National Science Foundation (NSF) under Grant No. ECCS-2236023 and in part by the Air Force Office of Scientific Research (AFOSR) under Award No. FA9550-23-1-0255. The authors would like to thank Dr. Zongyu Zhang for his contribution on the ZZB simulation.

References

1 Wax, M. and Leshem, A. (1997). Joint estimation of time delays and directions of arrival of multiple reflections of a known signal. *IEEE Transactions on Signal Processing* 45(10): 2477–2484.

2 Riba, J., Sala, J., and Vázquez, G. (2001). Conditional maximum likelihood timing recovery: estimators and bounds. *IEEE Transactions on Signal Processing* 49(4): 835–850.

3 Dogandzic, A. and Nehorai, A. (2001). Cramér-Rao bounds for estimating range, velocity, and direction with an active array. *IEEE Transactions on Signal Processing* 49(6): 1122–1137.

4 Sadler, B.M., Liu, N., and Xu, Z. (2010). Ziv–Zakai bounds on time delay estimation in unknown convolutive random channels. *IEEE Transactions on Signal Processing* 58(5): 2729–2745.

5 Gedalyahu, K. and Eldar, Y.C. (2010). Time-delay estimation from low-rate samples: a union of subspaces approach. *IEEE Transactions on Signal Processing* 58(6): 3017–3031.

6 Mir, H.S. and Wong, K.T. (2015). Low-rate sampling technique for range-windowed radar/sonar using nonlinear frequency modulation. *IEEE Transactions on Aerospace and Electronic Systems* 51(3): 1972–1979.

7 Donoho, D.L. (2006). Compressed sensing. *IEEE Transactions on Information Theory* 52(4): 1289–1306.

8 Candès, E.J. and Tao, T. (2006). Near-optimal signal recovery from random projections: universal encoding strategies? *IEEE Transactions on Information Theory* 52(12): 5406–5425.

9 Candès, E.J. and Wakin, M.B. (2008). An introduction to compressive sampling. *IEEE Signal Processing Magazine* 25(2): 21–30.

10 Arias-Castro, E. and Eldar, Y.C. (2011). Noise folding in compressed sensing. *IEEE Signal Processing Letters* 18(8): 478–481.

11 Davenport, M.A., Laska, J.N., Treichler, J.R., and Baraniuk, R.G. (2012). The pros and cons of compressive sensing for wideband signal acquisition: noise folding versus dynamic range. *IEEE Transactions on Signal Processing* 60(9): 4628–4642.

12 Baraniuk, R. and Steeghs, P. (2007). Compressive radar imaging. *Proceedings of IEEE Radar Conference*, 128–133. Waltham, MA.

13 Duarte, M.F., Davenport, M.A., Takhar, D. et al. (2008). Single-pixel imaging via compressive sampling. *IEEE Signal Processing Magazine* 25(2): 83–91.

14 Li, S., Zhao, G., Sun, H., and Amin, M. (2018). Compressive sensing imaging of 3-D object by a holographic algorithm. *IEEE Transactions on Antennas and Propagation* 66(12): 7295–7304.

15 Cover, T.M. and Thomas, J.A. (2006). *Elements of Information Theory*, 2e. Hoboken, NJ: Wiley-Interscience.

16 Neifeld, M.A., Ashok, A., and Baheti, P.K. (2007). Task-specific information for imaging system analysis. *Journal of the Optical Society of America. A* 24(12): B25–B41.

17 Carson, W.R., Chen, M., Rodrigues, M.R.D. et al. (2012). Communications-inspired projection design with application to compressive sensing. *SIAM Journal on Imaging Sciences* 5(4): 1185–1212.

18 Duarte-Carvajalino, J., Yu, G., Carin, L., and Sapiro, G. (2013). Task-driven adaptive statistical compressive sensing of Gaussian mixture models. *IEEE Transactions on Signal Processing* 61(3): 585–600.

19 Gu, Y. and Goodman, N.A. (2014). Compressive sensing kernel optimization for time delay estimation. *Proceedings of IEEE Radar Conference*, 1209–1213. Cincinnati, OH.

20 Gu, Y. and Goodman, N.A. (2017). Information-theoretic compressive sensing kernel optimization and Bayesian Cramér–Rao bound for time delay estimation. *IEEE Transactions on Signal Processing* 65(17): 4525–4537.

21 Babadi, B., Kalouptsidis, N., and Tarokh, V. (2009). Asymptotic Achievability of the Cramér–Rao bound for noisy compressive sampling. *IEEE Transactions on Signal Processing* 57(3): 1233–1236.

22 Niazadeh, R., Babaie-Zadeh, M., and Jutten, C. (2012). On the achievability of Cramér–Rao bound in noisy compressed sensing. *IEEE Transactions on Signal Processing* 60(1): 518–526.

23 Pakrooh, P., Scharf, L.L., Pezeshki, A., and Chi, Y. (2013). Analysis of Fisher information and the Cramer–Rao bound for nonlinear parameter estimation after compressed sensing. *Proceedings of IEEE International Conference on Acoustics, Speech, and Signal Processing (ICASSP)*, 1–5. Vancouver, Canada.

24 Ramasamy, D., Venkateswaran, S., and Madhow, U. (2014). Compressive parameter estimation in AWGN. *IEEE Transactions on Signal Processing* 62(8): 2012–2027.

25 Zhang, Z., Shi, Z., Zhou, C. et al. (2022). Ziv–Zakai bound for compressive time delay estimation. *IEEE Transactions on Signal Processing* 70: 4006–4019.

26 Zhang, Z., Shi, Z., Gu, Y. et al. (2023). Ziv–Zakai bound for compressive time delay estimation from zero-mean Gaussian signal. *IEEE Signal Processing Letters* 94: 105–113.

27 Ziv, J. and Zakai, M. (1969). Some lower bounds on signal parameter estimation. *IEEE Transactions on Information Theory* 15(3): 386–391.

28 Bellini, S. and Tartara, G. (1974). Bounds on error in signal parameter estimation. *IEEE Transactions on Communications* 22(3): 340–342.

29 Chazan, D., Zakai, M., and Ziv, J. (1975). Improved lower bounds on signal parameter estimation. *IEEE Transactions on Information Theory* 21(1): 90–93.

30 Van Trees, H.L. and Bell, K.L. (2007). *Bayesian Bounds for Parameter Estimation and Nonlinear Filtering/Tracking*. NY: Wiley.

31 Routtenberg, T. and Tabrikian, J. (2012). A general class of outage error probability lower bounds in Bayesian parameter estimation. *IEEE Transactions on Signal Processing* 60(5): 2152–2166.

32 Weiss, A. and Weinstein, E. (1983). Fundamental limitations in passive time delay estimation–Part I: narrow-band systems. *IEEE Transactions on Acoustics, Speech, and Signal Processing* 31(2): 472–486.

33 Bell, K.L., Ephraim, Y., and Van Trees, H.L. (1996). Explicit Ziv–Zakai lower bound for bearing estimation. *IEEE Transactions on Signal Processing* 44(11): 2810–2824.

34 Sadler, B.M., Huang, L., and Xu, Z. (2007). Ziv–Zakai time delay estimation bound for ultra-wideband signals. *Proceedings of IEEE International Conference on Acoustics, Speech, and Signal Processing (ICASSP)*, 549–552. Honolulu, HI.

35 Liu, N., Xu, Z., and Sadler, B.M. (2010). Ziv–Zakai time-delay estimation bounds for frequency-hopping waveforms under frequency-selective fading. *IEEE Transactions on Signal Processing* 58(12): 6400–6406.

36 Decarli, N. and Dardari, D. (2014). Ziv–Zakai bound for time delay estimation of unknown deterministic signals. *Proceedings of IEEE International Conference on Acoustics, Speech, and Signal Processing (ICASSP)*, 4673–4677. Florence, Italy.

37 Chiriac, V.M., He, Q., Haimovich, A.M., and Blum, R.S. (2015). Ziv–Zakai bound for joint parameter estimation in MIMO radar systems. *IEEE Transactions on Signal Processing* 63(18): 4956–4968.

38 Mishra, K.V. and Eldar, Y.C. (2017). Performance of time delay estimation in a cognitive radar. *Proceedings of IEEE International Conference on Acoustics, Speech, and Signal Processing (ICASSP)*, 3141–3145. New Orleans, LA.

39 Bell, K.L., Steinberg, Y., Ephraim, Y., and Van Trees, H.L. (1997). Extended Ziv–Zakai lower bound for vector parameter estimation. *IEEE Transactions on Information Theory* 43(2): 624–637.

40 Guo, D., Shamai, (S.)S., and Verdú, S. (2005). Mutual information and minimum mean-square error in Gaussian channels. *IEEE Transactions on Information Theory* 51(4): 1261–1282.

41 Gu, Y. and Goodman, N.A. (2013). Compressed sensing kernel design for radar range profiling. *Proceedings of IEEE Radar Conference*, 1–5. Ottawa, Canada.

42 Gu, Y., Goodman, N.A., and Ashok, A. (2014). Radar target profiling and recognition based on TSI-optimized compressive sensing kernel. *IEEE Transactions on Signal Processing* 62(12): 3194–3207.

43 Gu, Y. and Goodman, N.A. (2018). Impacts of hardware nonlinearities on compressive sensing performance. *Proceedings of IEEE Radar Conference*, 1579–1583. Oklahoma City, OK.

44 Gu, Y. and Goodman, N.A. (2014). Time domain CS kernel design for mitigation of wall reflections in urban radar. *Proceedings of IEEE SAM Workshop. A Coruña*, 1–4. Spain.

45 Gu, Y., Zhang, Y.D., and Goodman, N.A. (2017). Optimized compressive sensing-based direction-of-arrival estimation in massive MIMO. *Proceedings of IEEE International Conference on Acoustics, Speech, and Signal Processing (ICASSP)*, 1–5. New Orleans, LA.

46 Gu, Y. and Zhang, Y.D. (2019). Compressive sampling optimization for user signal parameter estimation in massive MIMO systems. *Digital Signal Processing* 94: 105–113.

47 Zhang, Y.D. (2022). Iterative learning for optimized compressive measurements in massive MIMO systems. *Proceedings of IEEE Radar Conference*, 1–5. New York.

48 Pavel, S.R., Zhang, Y.D., Greco, M.S., and Gini, F. (2023). Deep learning-based compressive sampling optimization in massive MIMO systems. *Proceedings of IEEE International Conference on Acoustics, Speech, and Signal Processing (ICASSP)*, 1–5. Rhodes Island, Greece.

49 Ni, T., Liu, S., Mao, Z., and Huang, Y. (2022). Information-theoretic target localization with compressed measurement using FDA radar. *Proceedings of IEEE Radar Conference*, 1–5. New York.

50 Liu, F., Marcellin, M.W., Goodman, N.A., and Bilgin, A. (2016). Compressive sampling for detection of frequency-hopping spread spectrum signals. *IEEE Transactions on Signal Processing* 64(21): 5513–5524.

51 Gu, Y. and Goodman, N.A. (2018). Information-theoretic compressive measurement for frequency hopping pattern recognition. *Proceedings of IEEE Radar Conference*, 1445–1449. Oklahoma City, OK.

52 Gu, Y. and Zhang, Y.D. (2018). Pilot design for Gaussian mixture channel estimation in massive MIMO. *Proceedings of IEEE International Conference on Acoustics, Speech, and Signal Processing (ICASSP)*, 3266–3270. Calgary, Canada.

53 Gu, Y. and Zhang, Y.D. (2019). Information-theoretic pilot design for downlink channel estimation in FDD massive MIMO systems. *IEEE Transactions on Signal Processing* 67(9): 2334–2346.

54 Schönemann, P.H. (1966). A generalized solution of the orthogonal Procrustes problem. *Psychometrika* 31(1): 1–10.

55 Van Trees, H.L. (2002). *Detection, Estimation, and Modulation Theory, Part IV: Optimum Array Processing*. New York: Wiley.

56 Pollock, B. and Goodman, N.A. (2012). An examination of the effects of sub-Nyquist sampling on SNR. *Proceedings of SPIE*, 1–4. Baltimore, MD.

57 Goodman, N.A. and Potter, L.C. (2015). Pitfalls and possibilities of radar compressive sensing. *Applied Optics* 54(8): C1–C13.

58 Kay, S.M. (1993). *Fundamentals of Statistical Signal Processing, Volume II: Detection Theory*. Upper Saddle River, NJ: Prentice-Hall PTR.

59 Bekker, A., Chen, D.G., and Ferreira, J.T. (2020). *Computational and Methodological Statistics and Biostatistics: Contemporary Essays in Advancement*. Springer.

5

Entropy-Enhanced One-Bit Compressive Sensing for DOA Estimation

Bin Liao[1], Qianhui You[1], and Peng Xiao[2]

[1] *Guangdong Key Laboratory of Intelligent Information Processing, College of Electronics and Information Engineering, Shenzhen University, Shenzhen, China*
[2] *School of Marine Engineering and Technology, Sun Yat-sen University, Zhuhai, China*

5.1 Introduction

Quantization is the process of representing continuous-valued signals with a limited set of discrete values. It is an essential concept in digital signal processing, and the accuracy of the quantization process plays a crucial role in the quality of digital signals. For signal recovery, it is desired to make the quantization error sufficiently small by adopting high-resolution analog-to-digital converters (ADCs). However, in certain applications, such as emerging large-scale antenna array systems [1], the usage of high-resolution ADCs could be impractical due to the high hardware cost and circuit power consumption [2]. Another example is that, in an impulse ultra-wideband (UWB) radar system with over 10 GHz bandwidth, ADCs with a high sampling rate of over 20 GHz are needed at its receiver. Such kinds of ADCs, especially with high-resolution quantization, may be too expensive to be commercially viable [3]. Therefore, system design with low-resolution ADCs and corresponding signal processing techniques have received increasing attention in the last few years [4–7]. In particular, extremely low-resolution (i.e., one-bit) ADCs composed of simple and cost-effective comparators, which consume negligible circuit power, have been widely studied for radar [8–11] as well as communication systems [12–14], in which direction-of-arrival (DOA) estimation is a fundamental problem. In this context, this chapter focuses on the problem of DOA estimation with one-bit array measurements.

One of the earliest works dedicated to DOA estimation with one-bit measurements was proposed by Bar-Shalom and Weiss in 2002 [15]. Using the covariance matrix estimates obtained directly from the array data, DOA estimators based on the conventional methods, such as minimum variance distortionless

Information-Theoretic Radar Signal Processing, First Edition. Edited by Yujie Gu and Yimin D. Zhang.

response (MVDR) beamformer, were studied. It was shown in [16] that if the signal-to-noise-ratio (SNR) is mildly low, the signal and noise subspaces can be straightforwardly determined through the eigendecomposition of the one-bit covariance matrix. Thus, subspace-based DOA estimation methods such as multiple signal classification (MUSIC) can be directly applied. In [17], the arcsine law was used to estimate the DOAs of strictly noncircular signals and obtain the quantization relationship between approximated covariance matrix and one-bit covariance. More recently, to apply MUSIC to one-bit DOA estimation, the work [18] presented an enhanced estimate of the normalized covariance matrix from one-bit sparse linear array measurements by exploiting the structure of the normalized covariance matrix. Similar to the infinite-resolution cases, these DOA estimators require a high number of snapshots to ensure the accuracy of the covariance matrix estimates and the number of sources must be known for subspace-based algorithms. Therefore, it is of great importance to develop techniques which require only few one-bit snapshots with unknown number of sources. In this context, one-bit compressive sensing (CS) theory has been exploited for DOA estimation in [19–21].

Unlike the ordinary CS assuming infinite-resolution quantization and using few samples to recover the signal that has a sparsity structure, one-bit CS aims to perform signal recovery by reducing the data bits per sample to the extreme case of keeping measurements' signs only, instead of reducing the number of samples. So far, a wide range of research on one-bit CS has been conducted in the literature. Based on the fixed-point continuation (FPC) algorithm, the problem of one-bit signal recovery was studied in [22], where the one-sided ℓ_2-norm was used to enforce consistency (i.e., the signs of the unquantized measurement and quantized counterpart are the same). In [23], it is shown that the one-sided ℓ_1-norm can achieve improved performance over the one-sided ℓ_2-norm. Moreover, the robustness of the algorithm against noise can be enhanced by incorporating the adaptive outlier pursuit (AOP) algorithm [24] into the FPC-ℓ_1 approach. In order to improve the performance of sparse signal recovery in noisy environments, various robust one-bit CS algorithms with weighted ℓ_1-norm minimization were presented in [25]. Besides, there have been many other one-bit CS methods developed in the past decade, such as the matching sign pursuit (MSP) [26], ℓ_1-norm minimization-based linear programming [27], binary iterative hard thresholding (BIHT) [28], Hamming support detection and coefficients recovery (HISTORY) algorithm [29], PASSIVE algorithm [30], and restricted-step shrinkage (RSS) [31]. Among them, the sparsity level should be known to implement the MSP, BIHT, and HISTORY algorithms. In contrast, this knowledge is unnecessary in the FPC, PASSIVE, and RSS algorithms. Moreover, the MSP, BIHT, FPC, and RSS algorithms were shown to be sensitive to the one-bit measurement sign flips caused by noise, whereas the PASSIVE and AOP algorithms are able to perform well in noisy environments. It is worth mentioning that deep unfolding methodologies have also been applied for one-bit CS in various works [32–34]. In general, these algorithms are applicable to DOA estimation with one-bit array measurements.

In order to determine the DOAs from one-bit array measurements using sparse representation under the CS framework with improved performance over existing approaches, this chapter introduces Schur-concave functions [35] that have the ability of inducing a vector to be sparse and incorporates this kind of function into one-bit CS as well as DOA estimation. More specifically, Schur-concave functions are the class of functions consistent with the so-called Lorentz ordering, which is a partial ordering on vectors also known as majorization. Particularly, Schur-concave functions attain their minima at sparse solutions to the best basis selection problem [36]. This suggests the suitability of Schur-concave function minimization for one-bit CS. The normalized ℓ_1-norm Shannon entropy function (ℓ_1-SEF), which is a representative of the family of Schur-concave functions, is exploited for sparse signal recovery and applied to DOA estimation with one or multiple one-bit measurement vectors. This is achieved by taking advantage of the concentration property, i.e., minimizing the ℓ_1-SEF with respect to a vector will enforce the entries of the vector to be concentrated on only few entries with relatively large magnitudes [37]. In other words, the vector tends to be sparse. It is shown that the Shannon entropy-enhanced algorithm does not require the sparsity level of the signal as an input. This implies that the knowledge of the number of sources for DOA estimation is unnecessary. More importantly, it exhibits better performance than existing algorithms that do not rely on the prior information of sparsity level. Simulation results demonstrate the outstanding performance of the ℓ_1-SEF-based one-bit CS algorithm for general sparse signal recovery, and for DOA estimation with sparse representation as well.

The remainder of this chapter is organized as follows. Some preliminaries including the signal model and the problem formation are first given in Section 5.2. In Sections 5.3 and 5.4, the one-bit CS algorithms and the entropy-enhanced one-bit CS are introduced, respectively. In Section 5.5, the representative ℓ_1-SEF-based one-bit CS as well as the case of multiple measurement vectors (MMVs) in DOA estimation are presented. Simulation results are given in Section 5.6 to assess the performance of the entropy-enhanced one-bit CS algorithm, and conclusions are drawn in Section 5.7.

5.2 Signal Model and Problem Formulation

5.2.1 Signal Model

As shown in Figure 5.1, consider K narrowband far-field signals impinging on an array with M' elements from distinct directions $\{\theta_1, \ldots, \theta_K\}$. Each radio frequency

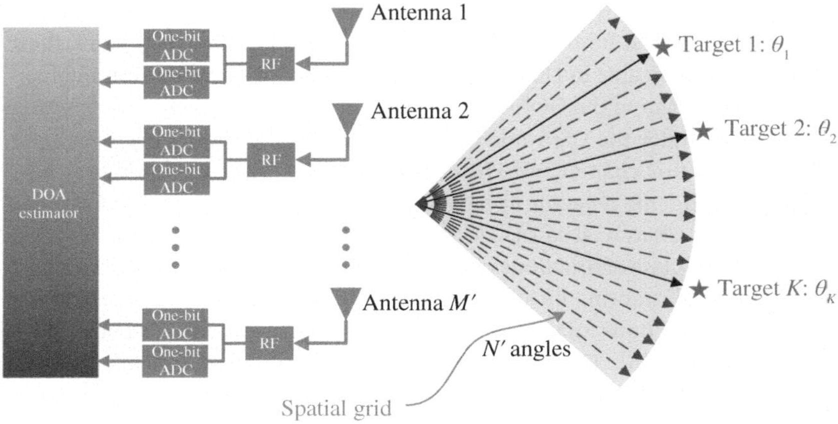

Figure 5.1 Diagram of DOA estimation with one-bit array measurements.

(RF) chain is equipped with a pair of one-bit ADCs. At the tth snapshot, the array output can be expressed as

$$\boldsymbol{y}_{1\text{-bit}}(t) = \mathcal{Q}(\boldsymbol{A}\boldsymbol{s}(t) + \boldsymbol{e}(t)), \tag{5.1}$$

where $\mathcal{Q}(\cdot)$ represents the element-wise one-bit quantizer that only keeps the sign information of the real and imaginary parts (denoted by $\Re\{\cdot\}$ and $\Im\{\cdot\}$, respectively) of a complex number α as

$$\mathcal{Q}(\alpha) = \text{sign}(\Re\{\alpha\}) + j \cdot \text{sign}(\Im\{\alpha\}) \tag{5.2}$$

with sign$(\cdot)$ being an element-wise sign function of a real-valued number. In (5.1), $\boldsymbol{s}(t)$ and $\boldsymbol{e}(t)$ denote the signal waveform and additive white Gaussian noise, respectively, and $\boldsymbol{A}$ is the steering matrix corresponding to the DOAs. Specifically, they can be expressed as

$$\boldsymbol{s}(t) = \left[s_1(t), \ldots, s_K(t)\right]^{\mathrm{T}} \in \mathbb{C}^K, \tag{5.3a}$$

$$\boldsymbol{e}(t) = \left[e_1(t), \ldots, e_{M'}(t)\right]^{\mathrm{T}} \in \mathbb{C}^{M'}, \tag{5.3b}$$

$$\boldsymbol{A} = \left[\boldsymbol{a}\left(\theta_1\right), \ldots, \boldsymbol{a}\left(\theta_K\right)\right] \in \mathbb{C}^{M' \times K}, \tag{5.3c}$$

where $\boldsymbol{a}(\theta_k) \in \mathbb{C}^{M'}$ is the steering vector corresponding to θ_k, and $(\cdot)^{\mathrm{T}}$ is the transpose operator. Without loss of generality and for simplicity, we consider a uniform linear array (ULA) with half-wavelength inter-element spacing, so that

$$\boldsymbol{a}(\theta_k) = \left[1, e^{-j\pi \sin(\theta_k)}, \ldots, e^{-j\pi(M'-1)\sin(\theta_k)}\right]^{\mathrm{T}}. \tag{5.4}$$

5.2.2 Sparse Representation

The one-bit array measurement vector can be expressed in a sparse representation model. First, a grid is constructed by discretizing the spatial range $[-\pi/2, \pi/2)$ into N' points $\{\vartheta_1, \dots, \vartheta_{N'}\}$ with $N' \gg K$, as illustrated in Figure 5.1. In this chapter, a uniform discretization is considered, i.e.,

$$\vartheta_n = (n-1)\Delta, \quad n = 1, 2, \dots, N', \tag{5.5}$$

where $\Delta = \pi/(N'-1)$ is the grid stepsize. For simplicity, the grid is assumed to be sufficiently dense, such that the true DOAs exactly lie on the sampling grids, i.e., $\{\theta_k\}_{k=1}^{K} \subset \{\vartheta_n\}_{n=1}^{N'}$. Then, one can construct a sparse vector $\tilde{\boldsymbol{s}}(t) = [\tilde{s}_1(t), \dots, \tilde{s}_{N'}(t)]^{\mathrm{T}} \in \mathbb{C}^{N'}$, which is related to $\boldsymbol{s}(t)$ as

$$\tilde{s}_n(t) = \begin{cases} s_k(t), & \text{if } \vartheta_n \in \{\theta_1, \dots, \theta_K\}, \\ 0, & \text{otherwise.} \end{cases} \tag{5.6}$$

This indicates that the signal waveform vector $\tilde{\boldsymbol{s}}(t)$ at each time instant has a K-sparse structure, i.e., the sparsity level of $\tilde{\boldsymbol{s}}(t)$ is K. As a result, the one-bit array measurement vector can be rewritten as

$$\boldsymbol{y}_{\text{1-bit}}(t) = \mathcal{Q}(\tilde{\boldsymbol{A}}\tilde{\boldsymbol{s}}(t) + \boldsymbol{e}(t)), \tag{5.7}$$

where $\tilde{\boldsymbol{A}} = [\boldsymbol{a}(\vartheta_1), \dots, \boldsymbol{a}(\vartheta_{N'})] \in \mathbb{C}^{M' \times N'}$ is an overcomplete basis in the sense that $\{\boldsymbol{a}(\theta_k)\}_{k=1}^{K} \subset \{\boldsymbol{a}(\vartheta_n)\}_{n=1}^{N'}$.

In order to convert the complex-valued measurement vector (5.7) to the real-valued domain, we introduce the following notations:

$$\boldsymbol{y}(t) = [\Re\{\boldsymbol{y}_{\text{1-bit}}^{\mathrm{T}}(t)\}, \Im\{\boldsymbol{y}_{\text{1-bit}}^{\mathrm{T}}(t)\}]^{\mathrm{T}} \in \{-1, +1\}^{2M'}, \tag{5.8a}$$

$$\boldsymbol{x}(t) = [\Re\{\tilde{\boldsymbol{s}}^{\mathrm{T}}(t)\}, \Im\{\tilde{\boldsymbol{s}}^{\mathrm{T}}(t)\}]^{\mathrm{T}} \in \mathbb{R}^{2N'}, \tag{5.8b}$$

$$\boldsymbol{\Phi} = \begin{bmatrix} \Re\{\tilde{\boldsymbol{A}}\} & -\Im\{\tilde{\boldsymbol{A}}\} \\ \Im\{\tilde{\boldsymbol{A}}\} & \Re\{\tilde{\boldsymbol{A}}\} \end{bmatrix} \in \mathbb{R}^{2M' \times 2N'}, \tag{5.8c}$$

$$\boldsymbol{v}(t) = [\Re\{\boldsymbol{e}^{\mathrm{T}}(t)\}, \Im\{\boldsymbol{e}^{\mathrm{T}}(t)\}]^{\mathrm{T}} \in \mathbb{R}^{2M'}. \tag{5.8d}$$

Then, the real-valued one-bit array measurement vector can be expressed as $\boldsymbol{y}(t) = \operatorname{sign}(\boldsymbol{\Phi}\boldsymbol{x}(t) + \boldsymbol{v}(t))$. For simplicity, we define $M \triangleq 2M'$ and $N \triangleq 2N'$, and in the remainder of this chapter, we drop the time index t for the case of single measurement vector, i.e.,

$$\boldsymbol{y} = \operatorname{sign}(\boldsymbol{\Phi}\boldsymbol{x} + \boldsymbol{v}). \tag{5.9}$$

The task is to recover the sparse vector $\boldsymbol{x} \in \mathbb{R}^N$ from the measurement vector $\boldsymbol{y} \in \mathbb{R}^M$, such that the DOAs can be determined according to the indexes of the nonzero entries in $\boldsymbol{x}$. Therefore, in what follows, the problem of sparse signal recovery is first discussed, and then the recovery algorithm is applied to estimate the DOAs.

5.2.3 Problem Formulation

According to the theory of ordinary CS for the infinite-resolution case, the fundamental optimization problem can be regarded as selecting a basis vector from the feasible set by minimizing a sparsity measure. As a matter of fact, the one-bit CS problem can be formulated in a similar manner. More concretely, in the noiseless case, the basic one-bit CS problem that recovers a sparse vector $\boldsymbol{x}$ from the one-bit measurement vector $\boldsymbol{y}$ is given by

$$\begin{aligned} \underset{\boldsymbol{x}}{\text{minimize}} \quad & \|\boldsymbol{x}\|_0 \\ \text{subject to} \quad & \boldsymbol{y} \odot (\boldsymbol{\Phi x}) \succeq \boldsymbol{0} \\ & \|\boldsymbol{x}\|_2 = 1, \end{aligned} \tag{5.10}$$

where $\odot$ represents the Hadamard product, $\succeq$ denotes element-wise inequality, and $\|\cdot\|_0$ and $\|\cdot\|_2$ are the ℓ_0-norm and ℓ_2-norm, respectively. For the sparsity measure, in this problem, the ℓ_0-norm (which equals the number of nonzero entries in a vector) is adopted as the objective function. In (5.10), the constraint $\boldsymbol{y} \odot (\boldsymbol{\Phi x}) \succeq \boldsymbol{0}$ is imposed by exploiting the fact that the product of each one-bit quantized measurement (i.e., the sign of the measurement) and its true value, namely $y_m \times [\boldsymbol{\Phi x}]_m$, must be non-negative. This is the so-called consistency constraint, which stems from the one-bit quantization process in the noiseless case:

$$\boldsymbol{y} = \text{sign}(\boldsymbol{\Phi x}). \tag{5.11}$$

It should be noted that scaling the sparse signal vector as $a\boldsymbol{x}$ with $a > 0$ does not change the signs of the measurements in $\boldsymbol{y}$. In other words, if $\boldsymbol{x}$ is consistent with the one-bit measurement vector $\boldsymbol{y}$ and minimizes the ℓ_0-norm, then so is $a\boldsymbol{x}$. This will result in amplitude ambiguity and drive the solution to zero. Therefore, an equality constraint $\|\boldsymbol{x}\|_2 = 1$ is imposed in (5.10) to avoid the trivial solution $\boldsymbol{x} = \boldsymbol{0}$ by restricting the solution to a unit sphere. By doing so, the optimization search space can be reduced, and meanwhile, it plays an important role in improving the reconstruction performance [22]. Alternatively, one may use other constraints for preventing the problem from returning to a trivial solution. For example, the following ℓ_1-norm constraint:

$$\|\boldsymbol{\Phi x}\|_1 = \nu \tag{5.12}$$

with $\nu > 0$ is adopted in [27] and [38]. Here, $\|\cdot\|_1$ denotes the ℓ_1-norm. This constraint can be used to construct a convex problem for one-bit CS, when the objective function and consistency constraint are both convex.

5.3 One-Bit CS Algorithms

The one-bit CS problem (5.10) is NP-hard due to the nonconvexity of the ℓ_0-norm objective function. Moreover, the consistency constraint does not tolerate sign flips caused by the existence of noise in the model (5.9). To tackle these issues, appropriate relaxations of the objective and consistency constraint can be applied to simplify the problem (5.10). By relaxing the ℓ_0-norm objective to a more tractable function and employing a loss function to enforce the sign consistency between the one-bit measurement and unquantized counterpart, the problem (5.10) can be recast as

$$\begin{aligned} \underset{\boldsymbol{x}}{\text{minimize}} \quad & f(\boldsymbol{x}) + \lambda \varphi(\boldsymbol{x}) \\ \text{subject to} \quad & \|\boldsymbol{x}\|_2 = 1, \end{aligned} \tag{5.13}$$

where $f(\boldsymbol{x})$ is a relaxed function of $\|\boldsymbol{x}\|_0$. The consistency constraint in (5.10) has been absorbed into the objective function of the problem (5.13) as

$$\varphi(\boldsymbol{x}) = \sum_{m=1}^{M} \mathcal{L}\left([\boldsymbol{y} \odot (\boldsymbol{\Phi x})]_m\right), \tag{5.14}$$

where $\mathcal{L}(\cdot)$ is a loss function used to ensure the data fidelity and tolerate the sign flips caused by noise as well. The regularization parameter $\lambda > 0$ balances the tradeoff between sparse promotion and data fidelity.

5.3.1 Consistency Constraint Relaxation

There exist various loss functions $\mathcal{L}(\cdot)$ with different performances and applicabilities. A simple choice is known as the linear loss function [30, 39]

$$\mathcal{L}(z) = -z, \tag{5.15}$$

where z is a real-valued number. In fact, the linear loss can be deemed as a measure for the correlation between $\boldsymbol{y}$ and $\boldsymbol{\Phi x}$. Since this function is simple and shows good performance in noisy environments, it has attracted much attention in the literature. Moreover, the linear loss can prevent the problem from returning a trivial solution $\boldsymbol{x} = \boldsymbol{0}$, such that the normalization constraint can be relaxed as $\|\boldsymbol{x}\|_2 \leq 1$. As a result, a convex problem known as PASSIVE problem can be formulated and solved analytically [30].

As an extended version of the linear loss, the pinball loss function [40] can be written as

$$\mathcal{L}(z) = \max\{\rho z, -z\}, \tag{5.16}$$

where $\rho \in [-1, 0]$. If $\rho = 0$, the pinball loss function becomes the one-sided ℓ_1-norm function [23, 28]:

$$\mathcal{L}(z) = \begin{cases} -z, & \text{if } z < 0, \\ 0, & \text{if } z \geq 0, \end{cases} \tag{5.17}$$

which is related to the hinge loss function in machine learning and has been studied in [41] and [42]. If $\rho = -1$, the pinball loss function degrades into the linear loss function. It is seen that the pinball loss provides a bridge between the hinge loss and the linear loss [40]. With a proper selection of ρ, one can expect a suitable tradeoff between the hinge loss (or one-sided ℓ_1-norm function) and the linear loss and achieve a good performance in one-bit CS.

Analogous to the one-sided ℓ_1-norm function, the one-sided ℓ_2-norm function [22, 28] is defined as

$$\mathcal{L}(z) = \begin{cases} z^2, & \text{if } z < 0, \\ 0, & \text{if } z \geq 0. \end{cases} \tag{5.18}$$

It is shown in [28] that the one-sided ℓ_1-norm function is capable of achieving a superior performance on sparse signal recovery compared to the one-sided ℓ_2-norm function.

5.3.2 Objective Function Relaxation

The ℓ_0-norm is a traditional sparsity measure in many mathematical settings. However, in practical scenarios, this measure is undesirable due to the facts that (i) its derivative does not contain information and (ii) it cannot perform well in noisy environments [43]. A simple convex relaxation of the ℓ_0-norm is known to be the ℓ_1-norm:

$$f(\boldsymbol{x}) = \|\boldsymbol{x}\|_1. \tag{5.19}$$

Unlike the ℓ_0-norm, under this sparsity measure, large entries in the vector are supposed to be more crucial than small entries. In certain cases of ordinary CS, the ℓ_1-norm can be used to find the support of the ℓ_0-norm solution, i.e., minimizing the ℓ_1-norm could be equivalent to minimizing the ℓ_0-norm [44]. Hence, the ℓ_1-norm measure has been employed for one-bit CS in various studies [22, 23, 31].

Besides the ℓ_1-norm, there exist other sparsity-promoting penalty functions that can also be applied to conduct one-bit CS. For instance, the decomposable nonconvex penalty function, known as minimax concave penalty (MCP), can be written as [45]

$$f(\boldsymbol{x}) = \sum_{n=1}^{N} g_{\mu,b}(x_n), \tag{5.20}$$

where

$$g_{\mu,b}(x_n) = \begin{cases} \mu|x_n| - \dfrac{x_n^2}{2b}, & \text{if } |x_n| \leq b\mu \\ \dfrac{b\mu^2}{2}, & \text{if } |x_n| > b\mu \end{cases} \tag{5.21}$$

with b and μ being positive regularization parameters. Hereafter, $|\cdot|$ denotes the element-wise absolute operator.

The sorted ℓ_1 penalty [46] is given by

$$f(\mathbf{x}) = \sum_{n=1}^{N} w_n x'_n, \tag{5.22}$$

where x'_n denotes the nth largest entry in $\{|x_1|, |x_2|, \dots, |x_N|\}$ and $x'_1 \geq x'_2 \geq \dots \geq x'_N$ are obtained by sorting $|x_1|, |x_2|, \dots, |x_N|$ in a descending order. The weights are typically set as

$$w_n = \begin{cases} 1, & \text{if } n > \tilde{n}, \\ e^{-\xi(\tilde{n}-n)/\tilde{n}}, & \text{if } n \leq \tilde{n}, \end{cases} \tag{5.23}$$

where $\tilde{n}$ is related to the sparsity level, and ξ controls the rate of increasing the weight w_n from 0 to 1.

The Mangasarian function and the log-det function, which belong to the family of separable concave functions, can also be used to approximate the ℓ_0-norm [38]. Specifically, the Mangasarian function is defined as

$$f(\mathbf{x}) = \sum_{n=1}^{N} \left(1 - e^{-\frac{|x_n|}{\epsilon}}\right), \tag{5.24}$$

where ϵ is a fixed number within $(0, 1)$. The log-det function is given by

$$f(\mathbf{x}) = \sum_{n=1}^{N} \frac{\log(|x_n|/\varepsilon + 1)}{\log(1/\varepsilon)}, \tag{5.25}$$

where ε is a small regularization constant. Note that the log-det function can be replaced by $f(\mathbf{x}) = \sum_{n=1}^{N} \log(|x_n| + \varepsilon)$, since the constant terms do not affect the solution of the optimization problem.

5.4 Entropy-Enhanced One-Bit CS

Unlike the aforementioned sparsity-promoting penalties, in this section, the normalized ℓ_1-SEF is investigated. The entropy is a suitable measure of diversity for sparse basis selection, and it has long been used by economists and ecologists as a measure of concentration and diversity. In information theory, entropy quantifies the extent of the data spread out over its possible values. Specifically, high entropy indicates that the data are spread out as much as possible, while low entropy means that the data are nearly all concentrated on one value. For a vector, the entropy quantifies the expected value of the information contained in it, and a low entropy implies that only a few elements are significant in the vector. As a matter of fact, the ℓ_1-SEF belongs to the class of Schur-concave functions which attain their minima at sparse solutions to the best basis selection problem [36]. In this section, one-bit CS with the family of Schur-concave functions and the representative ℓ_1-SEF will be introduced.

5.4.1 Schur-Concave Function

To begin with, we introduce the following useful definition that is related to Schur-concave function.

Definition 5.1 ***(Majorization)*** Given two vectors $\boldsymbol{x} \in \mathbb{R}^N$ and $\tilde{\boldsymbol{x}} \in \mathbb{R}^N$, let $x_{\lfloor 1 \rfloor} \geq \cdots \geq x_{\lfloor N \rfloor}$ be the decreasing rearrangement of the entries of $\boldsymbol{x}$, e.g., $x_{\lfloor 1 \rfloor} = \max\{x_1, \ldots, x_N\}$ and $x_{\lfloor N \rfloor} = \min\{x_1, \ldots, x_N\}$, and so is for $\tilde{\boldsymbol{x}}$, if

$$\begin{cases} \sum_{n=1}^{\kappa} x_{\lfloor n \rfloor} \leq \sum_{n=1}^{\kappa} \tilde{x}_{\lfloor n \rfloor}, & \forall \kappa \in \{1, \ldots, N-1\} \\ \sum_{n=1}^{N} x_{\lfloor n \rfloor} = \sum_{n=1}^{N} \tilde{x}_{\lfloor n \rfloor}, & \end{cases} \tag{5.26}$$

then we say that $\tilde{\boldsymbol{x}}$ majorizes $\boldsymbol{x}$, or $\boldsymbol{x}$ is majorized by $\tilde{\boldsymbol{x}}$, and it is denoted as $\boldsymbol{x} \prec \tilde{\boldsymbol{x}}$.

From the definition, it is known that majorization means that the sum of the κ largest entries of $\boldsymbol{x}$ is not larger than the sum of the κ largest entries of $\tilde{\boldsymbol{x}}$, unless $\kappa = N$. For example, if $\boldsymbol{x} = [0.8, 0.3, 0.2]^{\mathrm{T}}$ and $\tilde{\boldsymbol{x}} = [1, 0.3, 0]^{\mathrm{T}}$, then $\boldsymbol{x} \prec \tilde{\boldsymbol{x}}$. Therefore, majorization defines a partial ordering between two vectors $\boldsymbol{x}$ and $\tilde{\boldsymbol{x}}$ and precisely describes the concept that the components of $\boldsymbol{x}$ are "less spread out" or "more nearly equal" than the components of $\tilde{\boldsymbol{x}}$ [47]. In order to quantify majorization, Schur systematically studied the order-preserving functions, which are monotonic with respect to the ordering of majorization, also known as Schur-concave functions.

Definition 5.2 ***(Schur-concave function)*** A real-valued function ψ defined on a set $\Omega \subseteq \mathbb{R}^N$ is said to be Schur-concave, if

$$\boldsymbol{x} \prec \tilde{\boldsymbol{x}} \text{ on } \Omega \qquad \Rightarrow \qquad \psi(\boldsymbol{x}) \geq \psi(\tilde{\boldsymbol{x}}). \tag{5.27}$$

If the strict inequality $\psi(\boldsymbol{x}) > \psi(\tilde{\boldsymbol{x}})$ holds whenever $\boldsymbol{x} \prec \tilde{\boldsymbol{x}}$, but $\boldsymbol{x}$ is not a permutation of $\tilde{\boldsymbol{x}}$, then ψ is strictly Schur-concave on Ω. Moreover, if ψ is Schur-concave, then $-\psi$ is Schur-convex, and vice versa.

In the above definition, permutation means that the entries of $\tilde{\boldsymbol{x}}$ and $\boldsymbol{x}$ are the same, but their orders of the entries are different. In other words, we have $\tilde{\boldsymbol{x}} = \boldsymbol{P}\boldsymbol{x}$, where $\boldsymbol{P} \in \mathbb{R}^{N\times N}$ is a permutation matrix. The definition implies that, if $\tilde{\boldsymbol{x}}$ majorizes $\boldsymbol{x}$, the value of the Schur-concave function should be smaller when the variable is $\tilde{\boldsymbol{x}}$. Note that Schur characterized the functions that preserve the ordering $\prec$ on the non-negative orthant, and later Ostrowski observed that the restriction to this orthant is unnecessary [35].

Because Schur-concave functions preserve the ordering $\prec$, they are suitable for measuring the degree of concentration. If ψ is Schur-concave and $\psi(\tilde{\boldsymbol{x}}) \le \psi(\boldsymbol{x})$, then $\tilde{\boldsymbol{x}}$ is more concentrated than $\boldsymbol{x}$. This means that one can obtain a concentrated or sparse vector by minimizing a Schur-concave function. As a result, Schur-concave functions can be utilized as sparsity-promoting penalties. According to (5.13), replacing $f(\boldsymbol{x})$ by a certain Schur-concave function $\psi(\boldsymbol{x})$, the framework of one-bit CS with Schur-concave function minimization can be written as

$$\begin{aligned} \underset{\boldsymbol{x}}{\text{minimize}} \quad & \psi(\boldsymbol{x}) + \lambda\varphi(\boldsymbol{x}) \\ \text{subject to} \quad & \|\boldsymbol{x}\|_2 = 1. \end{aligned} \tag{5.28}$$

5.4.2 ℓ_1-Norm Shannon Entropy Function

Under the above framework, a specific Schur-concave function, i.e., the normalized ℓ_1-SEF is exploited for one-bit CS. In information theory, Shannon entropy quantifies the amount of information in a variable. Let X be a discrete random variable on a finite set $\mathcal{X} = \{x_1, \ldots, x_N\}$ and probability mass function $p(x), x \in \mathcal{X}$, the entropy of X is

$$H(X) = \sum_{x\in\mathcal{X}} p(x) \log \frac{1}{p(x)} = -\sum_{x\in\mathcal{X}} p(x) \log p(x). \tag{5.29}$$

Analogously, the ℓ_1-SEF with respect to a vector $\boldsymbol{x}$ is defined as

$$H_{\ell_1\text{-SEF}}(\boldsymbol{x}) = -\sum_{n=1}^{N} \frac{|x_n|}{\|\boldsymbol{x}\|_1} \log \frac{|x_n|}{\|\boldsymbol{x}\|_1}, \tag{5.30}$$

where $\frac{|x_n|}{\|\boldsymbol{x}\|_1}$ stands for the weight of each component in the vector, and it behaves like the probability of the occurrence of x_n. It can be readily found that

$$H_{\ell_1\text{-SEF}}(|\boldsymbol{x}|) = H_{\ell_1\text{-SEF}}(\boldsymbol{x}). \tag{5.31}$$

Notice that, because $\log 0$ does not exist, if any entries of $\boldsymbol{x}$ are equal to zero, then $H_{\ell_1\text{-SEF}}(\boldsymbol{x})$ will be inexistent. A simple but effective strategy to tackle this issue is adding a sufficiently small number to those zero entries.

In what follows, several properties of the ℓ_1-SEF function are presented. These properties play an important role in recovering sparse signal based on the ℓ_1-SEF function minimization [37, 48].

- $H_{\ell_1\text{-SEF}}(\boldsymbol{x})$ is scale-invariant, Schur-concave, as well as concave with respect to $|\boldsymbol{x}|$. This property means that the ℓ_1-SEF function belongs to the class of Schur-concave functions and can be utilized as a sparsity-promoting function.
- Defining the gradient of ℓ_1-SEF with respect to $|\boldsymbol{x}|$ as

$$\nabla H_{\ell_1\text{-SEF}}(|\boldsymbol{x}|) \triangleq \frac{\partial H_{\ell_1\text{-SEF}}(\boldsymbol{x})}{\partial |\boldsymbol{x}|}, \tag{5.32}$$

then the nth entry of $\nabla H_{\ell_1\text{-SEF}}(|\boldsymbol{x}|)$ is given by

$$\nabla H_{\ell_1\text{-SEF}}(|x_n|) = -\frac{1}{\|\boldsymbol{x}\|_1} \log |x_n| + \frac{1}{\|\boldsymbol{x}\|_1^2} \sum_{n=1}^{N} |x_n| \log |x_n|, \tag{5.33}$$

and we always have

$$\left\langle \nabla H_{\ell_1\text{-SEF}}(|\boldsymbol{x}|), |\boldsymbol{x}| \right\rangle = 0, \tag{5.34}$$

where $\langle \cdot, \cdot \rangle$ denotes the inner product operator. This property implies that the gradient of ℓ_1-SEF is orthogonal to the vector $|\boldsymbol{x}|$.
- The nth entry of $\nabla H_{\ell_1\text{-SEF}}(|\boldsymbol{x}|)$ satisfies

$$\nabla H_{\ell_1\text{-SEF}}(|x_n|) \begin{cases} < 0, & \text{if } |x_n| > \gamma, \\ = 0, & \text{if } |x_n| = \gamma, \\ > 0, & \text{if } |x_n| < \gamma, \end{cases} \tag{5.35}$$

where

$$\gamma = \exp\left(\frac{\sum_{n=1}^{N} |x_n| \log |x_n|}{\|\boldsymbol{x}\|_1} \right). \tag{5.36}$$

This property indicates that if $|x_n| > \gamma$, then minimizing the ℓ_1-SEF with respect to $\boldsymbol{x}$ will make $|x_n|$ even larger. On the other hand, if $|x_n| < \gamma$, then minimizing the ℓ_1-SEF will make $|x_n|$ even smaller. In other words, minimizing $H_{\ell_1\text{-SEF}}(\boldsymbol{x})$ promotes large-magnitude entries while suppressing small-magnitude entries of $\boldsymbol{x}$.

5.5 ℓ_1-SEF-Based One-Bit CS

5.5.1 Case of Single Measurement Vector

To begin with, we consider the case of single measurement vector. According to the properties of the ℓ_1-SEF, one can make use of this function to prompt sparsity in one-bit CS. More concretely, the problem of one-bit CS via ℓ_1-SEF minimization can be written as

$$\begin{aligned} \underset{x}{\text{minimize}} \quad & H_{\ell_1\text{-SEF}}(|\boldsymbol{x}|) + \lambda\varphi(\boldsymbol{x}) \\ \text{subject to} \quad & \|\boldsymbol{x}\|_2 = 1. \end{aligned} \tag{5.37}$$

The objective function $H_{\ell_1\text{-SEF}}(|\boldsymbol{x}|)$ will be used instead of $H_{\ell_1\text{-SEF}}(\boldsymbol{x})$ by taking advantage of the equivalence in (5.31). This allows us to explore its concavity, which facilitates solving the problem. In this section, an iterative algorithm based on FPC will be presented to tackle the problem (5.37).

Owing to the fact that the objective function is nonconvex, popular convex optimization methods are not directly applicable. In this chapter, the non-convex problem is tackled by an iterative algorithm, in which a surrogate function is constructed to convert the problem into a simpler one in each iteration. Since the ℓ_1-SEF function is concave, the tangent at any point is above the function. Thus, given the solution $|\boldsymbol{x}_{(i)}|$ in the ith iteration, we have

$$\begin{aligned} H_{\ell_1\text{-SEF}}(|\boldsymbol{x}|) &\leq H_{\ell_1\text{-SEF}}(|\boldsymbol{x}_{(i)}|) + \langle \nabla H_{\ell_1\text{-SEF}}(|\boldsymbol{x}_{(i)}|), |\boldsymbol{x}| - |\boldsymbol{x}_{(i)}|\rangle \\ &= H_{\ell_1\text{-SEF}}(|\boldsymbol{x}_{(i)}|) - \langle \nabla H_{\ell_1\text{-SEF}}(|\boldsymbol{x}_{(i)}|), |\boldsymbol{x}_{(i)}|\rangle \\ &\quad + \langle \nabla H_{\ell_1\text{-SEF}}(|\boldsymbol{x}_{(i)}|), |\boldsymbol{x}|\rangle. \end{aligned} \tag{5.38}$$

Based on this observation, by omitting the constant terms, the following surrogate function:

$$\langle \nabla H_{\ell_1\text{-SEF}}(|\boldsymbol{x}_{(i)}|), |\boldsymbol{x}|\rangle \tag{5.39}$$

is utilized to attain a solution by applying the majorization–minimization scheme [49]. More specifically, in the $(i+1)$th iteration, we replace the objective function $H_{\ell_1\text{-SEF}}(|\boldsymbol{x}|)$ in (5.37) by the surrogate function (5.39) to formulate the following optimization problem:

$$\begin{aligned} \underset{x}{\text{minimize}} \quad & \langle \nabla H_{\ell_1\text{-SEF}}(|\boldsymbol{x}_{(i)}|), |\boldsymbol{x}|\rangle + \lambda\varphi(\boldsymbol{x}) \\ \text{subject to} \quad & \|\boldsymbol{x}\|_2 = 1, \end{aligned} \tag{5.40}$$

whose solution, denoted by $\boldsymbol{x}_{(i+1)}$, will be utilized to construct the surrogate function in the next iteration. By iteratively minimizing the surrogate function, the objective ℓ_1-SEF penalty decreases and ultimately results in a sparse solution.

It should be emphasized that the FPC algorithm [50] and its variants such as iterative shrinkage-thresholding algorithms (ISTA) [51], forward–backward splitting (FBS) [52], and fast ISTA (FISTA) [53] are not straightforwardly applicable to solve the problem (5.40). This is mainly because the entries of the derivative $\nabla H_{\ell_1\text{-SEF}}(|\boldsymbol{x}_{(i)}|)$ could be either positive or negative, such that the weighted ℓ_1-norm of $\boldsymbol{x}$, i.e., $\langle \nabla H_{\ell_1\text{-SEF}}(|\boldsymbol{x}_{(i)}|), |\boldsymbol{x}|\rangle$, cannot meet the requirement of positive weights for FPC. In this case, the objective function is nonconvex, and the FPC algorithm is not directly applicable. Toward this end, the generalized FPC (GFPC) algorithm [54] can be applied. The GFPC algorithm was developed based on the following theorem.

Theorem 5.1 Consider the problem

$$\underset{\boldsymbol{x}}{\text{minimize}} \quad \langle \boldsymbol{w}, |\boldsymbol{x}| \rangle + \lambda\varphi(\boldsymbol{x}), \tag{5.41}$$

where $\varphi(\boldsymbol{x})$ is convex and differentiable with $\boldsymbol{w} \in \mathbb{R}_*^N$, $\mathbb{R}_* = (-\infty, 0) \cup (0, +\infty)$ and $\lambda > 0$. For any $\beta > 0$, $\boldsymbol{x}^\star = [x_1^\star, x_2^\star, \dots, x_N^\star]^\text{T}$ satisfies the following fixed-point equation:

$$\begin{aligned}\boldsymbol{x}^\star &= \boldsymbol{\tau}_2 \odot \text{sign}\left(\boldsymbol{x}^\star - \beta \mathbf{g}(\boldsymbol{x}^\star)\right) \\ &\quad \odot \max\left\{ \min\left(\boldsymbol{\tau}_0, \boldsymbol{\tau}_2\right) \odot |\boldsymbol{x}^\star - \beta \mathbf{g}(\boldsymbol{x}^\star)| - \frac{\beta}{\lambda}\boldsymbol{w}, \mathbf{0} \right\},\end{aligned} \tag{5.42}$$

if and only if $x_n^\star$, $\forall n \in \{1, 2, \dots, N\}$, is the optimal solution in $\mathbb{R}$ when $w_n > 0$, or the optimal solution in $(-\infty, 0]$ or $[0, +\infty)$ when $w_n < 0$, to the problem (5.41) with respect to x_n. Here, $\mathbf{g}(\boldsymbol{x})$ is the gradient function of $\varphi(\boldsymbol{x})$, and

$$\boldsymbol{\tau}_0 = \max\left(\text{sign}(\boldsymbol{w}), \mathcal{M}(\boldsymbol{x}^\star)\right) \tag{5.43a}$$

$$\boldsymbol{\tau}_1 = \text{sign}(\boldsymbol{x}^\star) \odot \text{sign}(\boldsymbol{x}^\star - \beta \mathbf{g}(\boldsymbol{x}^\star)) \tag{5.43b}$$

$$\boldsymbol{\tau}_2 = \max\left(\boldsymbol{\tau}_1, \text{sign}(\boldsymbol{w})\right) \tag{5.43c}$$

with

$$\mathcal{M}(x_n^\star) = \begin{cases} -1, & \text{if } x_n^\star = 0, \\ 1, & \text{otherwise.} \end{cases} \tag{5.44}$$

Theorem 5.1 provides a simple way to attain the solution to problem (5.41), when the weights in $\boldsymbol{w}$ are not always positive. This leads to the so-called GFPC [54]. More specifically, the solution can be achieved in an iterative manner as

$$\begin{aligned}\boldsymbol{x}^{(r+1)} &= \boldsymbol{\tau}_2^{(r)} \odot \text{sign}\left(\boldsymbol{x}^{(r)} - \beta \mathbf{g}(\boldsymbol{x}^{(r)})\right) \\ &\quad \odot \max\left\{ \min\left(\boldsymbol{\tau}_0^{(r)}, \boldsymbol{\tau}_2^{(r)}\right) \odot |\boldsymbol{x}^{(r)} - \beta \mathbf{g}(\boldsymbol{x}^{(r)})| - \frac{\beta}{\lambda}\boldsymbol{w}, \mathbf{0} \right\},\end{aligned} \tag{5.45}$$

where $\boldsymbol{\tau}_0^{(r)}$, $\boldsymbol{\tau}_1^{(r)}$, and $\boldsymbol{\tau}_2^{(r)}$ are similarly defined as (5.43) by replacing $\boldsymbol{x}^\star$ with $\boldsymbol{x}^{(r)}$, and r is the index of the iteration. In particular, for the special case of a positive weight vector $w_n > 0$, $n = 1, \dots, N$, we have $\boldsymbol{\tau}_0 = \boldsymbol{\tau}_2 = \mathbf{1}$, and the GFPC algorithm degrades to the conventional FPC method. The main steps of the GFPC algorithm for one-bit CS are summarized in Algorithm 5.1.

The iterative process (5.45) for the optimization problem (5.41) can converge to a fixed point [54]. Although only a local optimal solution can be guaranteed, the global solution is still achievable provided that the initial guess is properly

---

Algorithm 5.1: GFPC algorithm for one-bit CS.

1: **Input** Measurement vector: $\boldsymbol{y} \in \{\pm 1\}^M$, Measuring matrix: $\boldsymbol{\Phi}$.
2: **Initialize** Seed: $\boldsymbol{x}_{(0)}$ with $\|\boldsymbol{x}_{(0)}\|_2 = 1$, Descent step size: β, Relaxation parameter: λ, Outer iteration index: $i \leftarrow 0$.
3: **repeat**
4: Outer iteration index update: $i \leftarrow i + 1$.
5: Inner iteration index initialization: $r \leftarrow 0$.
6: Gradient descent: $\boldsymbol{h}^{(0)} \leftarrow \boldsymbol{x}_{(i-1)} - \beta \mathbf{g}(\boldsymbol{x}_{(i-1)})$.
7: New problem establish:
$\boldsymbol{x}_{(i)}^{(0)} \leftarrow \boldsymbol{h}^{(0)}$.
$\boldsymbol{w}_{(i)} \leftarrow \nabla H_{\ell_1\text{-SEF}}(\boldsymbol{x}_{(i)}^{(0)})$.
8: **repeat**
9: Inner iteration index update: $r \leftarrow r + 1$.
10: Compute:
$\boldsymbol{\tau}_0{}^{(r-1)} \leftarrow \max(\text{sign}(\boldsymbol{w}_{(i)}), \mathcal{M}(\boldsymbol{x}_{(i)}^{(r-1)}))$.
$\boldsymbol{\tau}_1{}^{(r-1)} \leftarrow \text{sign}(\boldsymbol{x}_{(i)}^{(r-1)}) \odot \text{sign}(\boldsymbol{h}^{(r-1)})$.
$\boldsymbol{\tau}_2{}^{(r-1)} \leftarrow \max(\boldsymbol{\tau}_1{}^{(r-1)}, \text{sign}(\boldsymbol{w}_{(i)}))$.
11: Shrinkage:
$\boldsymbol{u}_{(i)}^{(r)} \leftarrow \boldsymbol{\tau}_2{}^{(r-1)} \odot \text{sign}\left(\boldsymbol{h}^{(r-1)}\right) \odot \max\{\min(\boldsymbol{\tau}_0{}^{(r-1)}, \boldsymbol{\tau}_2{}^{(r-1)}) \odot |\boldsymbol{h}^{(r-1)}| - \frac{\beta}{\lambda}\boldsymbol{w}_{(i)}, 0\}$.
12: Normalization: $\boldsymbol{x}_{(i)}^{(r)} \leftarrow \boldsymbol{u}_{(i)}^{(r)} / \|\boldsymbol{u}_{(i)}^{(r)}\|_2$.
13: Gradient descent: $\boldsymbol{h}^{(r)} \leftarrow \boldsymbol{x}_{(i)}^{(r)} - \beta \mathbf{g}(\boldsymbol{x}_{(i)}^{(r)})$.
14: **until** Termination conditions are satisfied (Denote by $r_{\max}$ the total number of inner iterations)
15: Update $\boldsymbol{x}_{(i)}$: $\boldsymbol{x}_{(i)} \leftarrow \boldsymbol{x}_{(i)}^{(r_{\max})}$.
16: **until** Termination conditions are satisfied (Denote by $i_{\max}$ the total number of outer iterations)
17: **Output** Recovered signal: $\boldsymbol{x}^{\star} \leftarrow \boldsymbol{x}_{(i_{\max})}$.

---

selected. Since the one-bit CS problem with normalization constraint $\|\boldsymbol{x}\|_2 = 1$ is nonconvex, the recovery algorithms cannot be guaranteed to converge to the global minimum. Hence, the initialization should be carefully selected. Analogous to the conventional FPC algorithm, a good heuristic initialization for the GFPC algorithm can be chosen as $\boldsymbol{x}_{(0)} = \boldsymbol{\Phi}^{\mathrm{T}}\boldsymbol{y}$. It is experimentally found that such an initialization can always lead to excellent performance.

5.5.2 Case of Multiple Measurement Vectors

In this section, the above one-bit CS problem is extended to the case of MMVs. Assuming that L snapshots are collected, then the one-bit array measurements in the real-valued domain can be expressed as

$$\boldsymbol{Y} = Q(\boldsymbol{\Phi}\boldsymbol{X} + \boldsymbol{V}), \tag{5.46}$$

where $\boldsymbol{Y} = [\boldsymbol{y}(1), \dots, \boldsymbol{y}(L)] \in \{-1, +1\}^{M\times L}$, $\boldsymbol{X} = [\boldsymbol{x}(1), \dots, \boldsymbol{x}(L)] \in \mathbb{R}^{N\times L}$, and $\boldsymbol{V} = [\boldsymbol{v}(1), \dots, \boldsymbol{v}(L)] \in \mathbb{R}^{M\times L}$. Note that $\boldsymbol{X}$ is a sparse matrix, whose columns share the same support basis. The problem in the case of MMVs can thus be formulated as

$$\begin{aligned} &\underset{\boldsymbol{X}}{\text{minimize}} \quad H_{\ell_1\text{-SEF}}(\boldsymbol{X}) + \lambda \sum_{m=1}^{M}\sum_{l=1}^{L} \mathcal{L}\left([\boldsymbol{Y} \odot (\boldsymbol{\Phi}\boldsymbol{X})]_{m,l}\right) \\ &\text{subject to} \quad \|\boldsymbol{X}\|_F = 1, \end{aligned} \tag{5.47}$$

where $\|\cdot\|_F$ denotes the Frobenius norm, and

$$H_{\ell_1\text{-SEF}}(\boldsymbol{X}) = -\sum_{n=1}^{N}\sum_{l=1}^{L} \frac{|x_{n,l}|}{\|\boldsymbol{X}\|_{1,1}} \log \frac{|x_{n,l}|}{\|\boldsymbol{X}\|_{1,1}} \tag{5.48}$$

stands for the ℓ_1-SEF with respect to $\boldsymbol{X}$. Here, $x_{n,l}$ is the (n, l)th entry of $\boldsymbol{X}$, $\frac{|x_{n,l}|}{\|\boldsymbol{X}\|_{1,1}}$ corresponds to the weight of each element in $\boldsymbol{X}$ and behaves like the probability of the occurrence of $x_{n,l}$. $\|\cdot\|_{1,1}$ denotes the mixed $\ell_{1,1}$-norm of a matrix and is defined as

$$\|\boldsymbol{X}\|_{1,1} = \sum_{n=1}^{N}\sum_{l=1}^{L} |x_{n,l}|. \tag{5.49}$$

Similar to the case of single measurement vector, we have $H_{\ell_1\text{-SEF}}(\boldsymbol{X}) = H_{\ell_1\text{-SEF}}(|\boldsymbol{X}|)$. The derivative with respect to its (n, l)th entry of $|\boldsymbol{X}|$ can be written as

$$\nabla H_{\ell_1\text{-SEF}}(|x_{n,l}|) = -\frac{\log |x_{n,l}|}{\|\boldsymbol{X}\|_{1,1}} + \frac{1}{\|\boldsymbol{X}\|_{1,1}^2} \sum_{n=1}^{N}\sum_{l=1}^{L} |x_{n,l}| \log |x_{n,l}|. \tag{5.50}$$

Given the sparse matrix $\boldsymbol{X}_{(i)}$ obtained at the ith iteration, we can update $\boldsymbol{X}$ by solving the following problem:

$$\begin{aligned} &\underset{\boldsymbol{X}}{\text{minimize}} \quad \|\nabla H_{\ell_1\text{-SEF}}(|\boldsymbol{X}_{(i)}|) \odot |\boldsymbol{X}|\|_{1,1} + \lambda \sum_{m=1}^{M}\sum_{l=1}^{L} \mathcal{L}\left([\boldsymbol{Y} \odot (\boldsymbol{\Phi}\boldsymbol{X})]_{m,l}\right) \\ &\text{subject to} \quad \|\boldsymbol{X}\|_F = 1, \end{aligned} \tag{5.51}$$

which can be solved by the GFPC algorithm.

5.6 Simulation Results

In this section, the sparse signal recovery performance of the ℓ_1-SEF-based one-bit CS method is first investigated for the case of a single measurement vector.

Then, the performance of this approach for DOA estimation is examined for the cases of single and MMVs.

5.6.1 Performance of Sparse Signal Recovery

To investigate the performance of the ℓ_1-SEF-based one-bit CS approach, the data fidelity term is considered to be one-side ℓ_1-norm loss (5.17), i.e., $\varphi(\boldsymbol{x}) = \sum_{m=1}^{M} \max\{0, -[\boldsymbol{y} \odot \boldsymbol{\Phi x}]_m\}$. It can be derived that the derivative is given by $\mathbf{g}(\boldsymbol{x}) = \boldsymbol{\Phi}^{\mathrm{T}}\left(\mathrm{sign}(\boldsymbol{\Phi x}) - \boldsymbol{y}\right)$. The ℓ_1-SEF-based algorithm will be compared to various approaches based on the conventional FPC-ℓ_1 [23], FPC-ℓ_2 [22], and BIHT algorithms [28]. In addition, the sorted ℓ_1 penalty [46], MCP penalty [45], log-det function, and Mangasarian function penalty [38] are also examined for comparison. All the problems based on these penalties are solved in an iterative manner, and in each iteration the weighted ℓ_1-norm minimization problems are solved under the framework of GFPC.

To generate the data for simulations, in each experimental trial, the entries of measuring matrix $\boldsymbol{\Phi} \in \mathbb{R}^{M\times N}$ are randomly generated by a zero-mean Gaussian process with variance $1/M$. We set $N = 1000$; a uniform distribution is used to choose the support set, and the values of the support set are produced by an independent and identically distributed (i.i.d.) zero-mean Gaussian process with unit variance. The signal is normalized such that $\|\boldsymbol{x}\|_2 = 1$. We set $\xi = 5$ for the sorted ℓ_1 algorithm and empirically choose the parameters for log-det, Mangasarian, and MCP penalties, such that they can provide satisfactory performance. More specifically, the weights corresponding to the log-det penalty are set as $w_n = \frac{\varrho}{|x_n|+\varepsilon}$ with $\varrho = 0.1$ and $\varepsilon = 0.05$; the weights for the Mangasarian function penalty are set as $w_n = \frac{\varsigma}{\epsilon}e^{-|x_n|/\epsilon}$ with $\varsigma = 100/N$ and $\epsilon = \log(N)/M$; and the weights for the MCP penalty are set as $w_n = \mu - |x_n|/b$ if $|x_n| \leq b\mu$ and $w_n = 0$ if $|x_n| > b\mu$, with $\mu = 1$ and b being selected to ensure that $b\mu$ is equal to the Kth largest entry. Additionally, the maximum number of inner and outer iterations are set to be 20 and 100, respectively. Note that, to balance the convergence speed and recovery precision, λ is chosen as $\lambda_{(r)} = c^r$ with $c = 1.1$ and $r = 1, 2, \ldots, 20$ in the rth inner iteration. To demonstrate the recovery performance of different algorithms, the following mean square error (MSE):

$$\mathrm{MSE} = \frac{1}{Q}\sum_{q=1}^{Q} \|\boldsymbol{x}_q - \boldsymbol{x}_q^{\star}\|_2^2 \tag{5.52}$$

will be adopted as the performance metric. Here, $\boldsymbol{x}_q$ and $\boldsymbol{x}_q^{\star}$ denote the true signal vector and recovered signal vector, respectively, in the qth Monte Carlo trial, and $Q = 100$ is the total number of Monte Carlo trials.

In the first example, the sparsity level is assumed to be 8, and the MSEs for different measurement numbers are examined. In order to match the sparsity level for

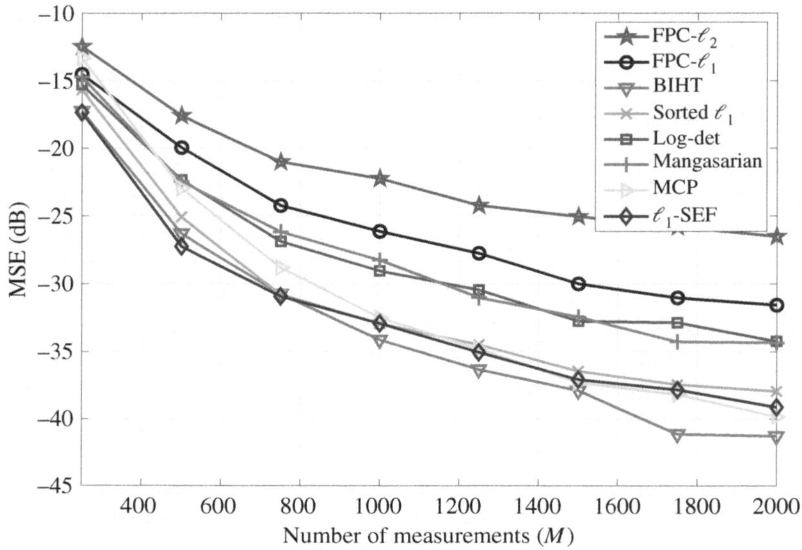

Figure 5.2 MSE versus the number of measurements M, when $N = 1000$, $K = 8$, and with no sparsity mismatch for BIHT, sorted ℓ_1, and MCP penalty.

conventional penalties, we set $\tilde{n} = 8$ for the sorted ℓ_1 penalty, and let $b\mu$ be the 8th largest entry for the MCP penalty. Figure 5.2 depicts the curves of MSE versus the number of measurements for different approaches. As expected, with the increase of the measurement number, the signal becomes sparser, and the reconstruction performance can be improved. Moreover, it can be found that the ℓ_1-SEF-based method performs similarly to the BIHT algorithm and better than other methods when the number of measurements is less than 800. The log-det and Mangasarian penalties are about 5 dB worse than the BIHT algorithm, the ℓ_1-SEF, and the sorted ℓ_1 algorithm. When the number of measurements is sufficiently large, e.g., larger than 800, the BIHT algorithm performs best. However, the knowledge of the sparsity level must be provided to achieve the outstanding performance for the BIHT algorithm, whereas the ℓ_1-SEF-based method is free of this knowledge. Further, it is seen that the sorted ℓ_1 algorithm and the ℓ_1-SEF algorithm achieve comparable performances regardless of the measurement number. On the other hand, the MCP penalty can only offer similar performance to the ℓ_1-SEF penalty when the number of measurements is relatively large, e.g., larger than 1000, and it is outperformed by the ℓ_1-SEF when the number of measurements is small.

In the second example, we set $M = 1000$ and change the sparsity level from 2 to 20. Meanwhile, we accordingly change $\tilde{n}$ from 2 and 20 for the implementation of the sorted ℓ_1 penalty-based method, and $b\mu$ is chosen as the Kth largest entry

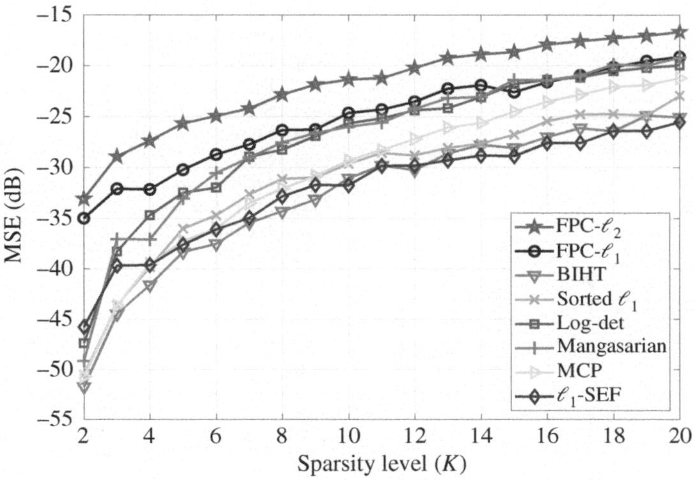

Figure 5.3 MSE versus sparsity level K, when $N = 1000$, $M = 1000$, and without sparsity mismatch for BIHT, sorted ℓ_1, and MCP penalty.

value for the MCP penalty. Figure 5.3 shows the resulting recovery performance in terms of the MSE versus sparsity level. It is observed that the performance of the ℓ_1-SEF-based approach is comparable to that of the BIHT algorithm even if the sparsity level is not provided. In particular, it performs slightly better than the BIHT algorithm in the case of $K \geq 10$. Moreover, this algorithm offers superior performance to the other sparsity-level-free algorithms. It is worth mentioning that when $K = 2$ or 3, the MCP penalty and the sorted ℓ_1 penalty give better performance than ℓ_1-SEF. However, the performance gain relies on the knowledge of the sparsity level.

To demonstrate the superiority of the ℓ_1-SEF-based approach in terms of its independence on the knowledge of sparsity level, in this example it is assumed that the sparsity level is not precisely known. More exactly, instead of the true sparsity level $K = 8$, three nominal sparsity levels, namely $K = 6$, 10, and 12 are considered for the BIHT and the sorted ℓ_1 algorithms, while other simulation settings are kept the same as the first example. The parameter $b\mu$ for the MCP penalty is set according to the nominal sparsity level. Figure 5.4 compares the MSE of the ℓ_1-SEF-based approach with unknown sparsity level to those of BIHT, sorted ℓ_1, and MCP penalties with nominal sparsity levels. The results indicate that in the presence of sparsity level mismatch, the ℓ_1-SEF-based approach offers outstanding performance because it is not affected by such mismatch. For other algorithms, the BIHT algorithm is relatively more sensitive to the precision of the sparsity level. Particularly, if the nominal level is smaller than the true value, the performance could be considerably degraded.

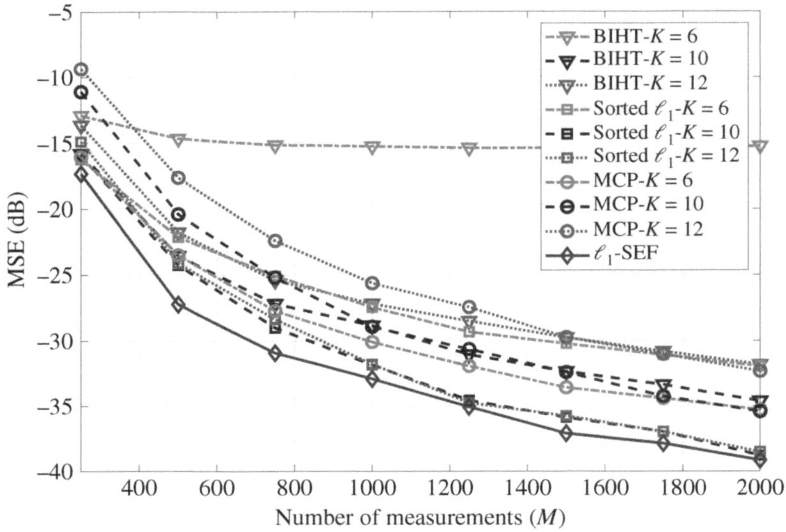

Figure 5.4 MSE versus the number of measurements M, when $N = 1000$, $K = 8$, and with sparsity mismatch for BIHT, sorted ℓ_1, and MCP penalty.

In the last example of this subsection, we further examine the influence of the sparsity level as well as the knowledge of the sparsity level on the reconstruction performance. For this purpose, we change the sparsity level K from 2 to 20, and it is assumed to be unknown for all algorithms. Since the BIHT, sorted ℓ_1, and MCP algorithms need the knowledge of sparsity level as inputs for implementation, we simply assume the sparsity level to be 8. This implies that there exist sparsity level mismatches when the true sparsity level K is not 8. The results of MSE versus sparsity level are demonstrated in Figure 5.5. It is obvious that the BIHT, sorted ℓ_1, and MCP algorithms are sensitive to the sparsity-level input, especially the BIHT algorithm. More specifically, if the presumed sparsity level (i.e., $K = 8$) is larger than the actual level, the BIHT and MCP methods offer similar performance that is worse than the sorted ℓ_1 and ℓ_1-SEF approaches. If the presumed sparsity level is smaller than the actual one, then the ℓ_1-SEF and BIHT offer the best performance and worst performance, respectively. However, when the knowledge of sparsity level is accurately available, all the methods can achieve similar performance. Overall, the ℓ_1-SEF-based approach is free of the sparsity level as an input parameter, and it is more attractive in practical applications, such as DOA estimation with unknown number of sources.

5.6.2 Performance of DOA Estimation

In this subsection, we evaluate the performance of the ℓ_1-SEF-based one-bit DOA estimation. An M'-element ULA with half-wavelength inter-element spacing is

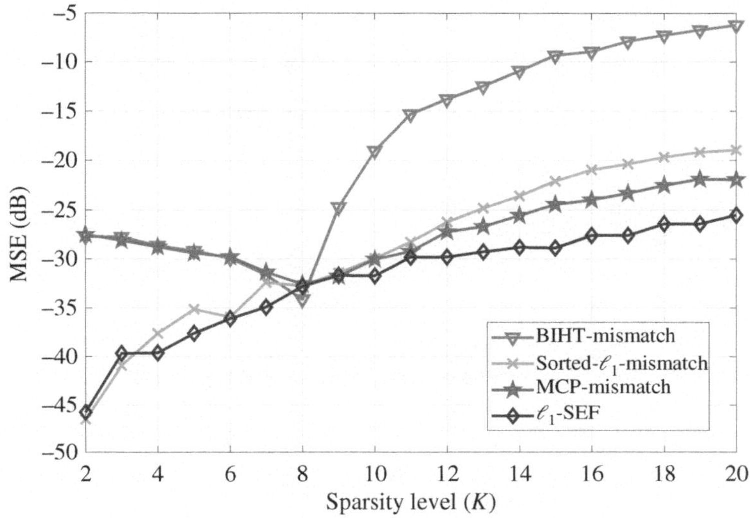

Figure 5.5 MSEs of BIHT, sorted ℓ_1, and MCP penalty versus sparsity level input. The true sparsity level is $K = 8$.

considered. The number of incoming signals is $K = 2$, and DOAs are assumed to be $\theta_1 = 1°$ and $\theta_2 = 10.4°$, unless otherwise specified. Moreover, the signals are assumed to be quadrature phase-shift keying (QPSK) modulated with zero mean and variance $\sigma_s^2 = 1$, and the noise is drawn from an i.i.d. complex Gaussian random process with zero mean and variance σ_e^2. The SNR is defined as

$$\text{SNR} = 10 \log_{10} \frac{\sigma_s^2}{\sigma_e^2}, \tag{5.53}$$

and the interval of the spatial grid is set to 0.5°. The initializations for algorithms based on ℓ_1-SEF and FPC-ℓ_1 are set as $\boldsymbol{X}_{(0)} = \boldsymbol{\Phi}^{\mathrm{T}}\boldsymbol{Y}$. Besides, the maximum number of outer and inner iterations are 60 and 500, respectively, and λ is set as $\lambda_{(r)} = c^r$ with $c = 1.25$. For comparison, the conventional subspace-based approaches, including the one-bit MUSIC [16], ESPRIT, classical CS-based method (MMV-OMP) [55], and the recent FPC-ℓ_1 [21], will also be examined.

In this example, we assume $M' = 40$, $N' = 360$, and $L = 4$, and the performance of the ℓ_1-SEF-based one-bit DOA estimator is evaluated by the mean absolute error (MAE) defined as

$$\text{MAE} = \frac{1}{QK} \sum_{q=1}^{Q} \sum_{k=1}^{K} |\hat{\theta}_{q,k} - \theta_k|, \tag{5.54}$$

where $Q = 500$ is the number of Monte Carlo trials, $\theta_{q,k}$ denotes the kth DOA estimate in the qth Monte Carlo trial, and θ_k represents the real DOA of the kth signal. Figure 5.6 compares the MAEs versus the SNR for different methods.

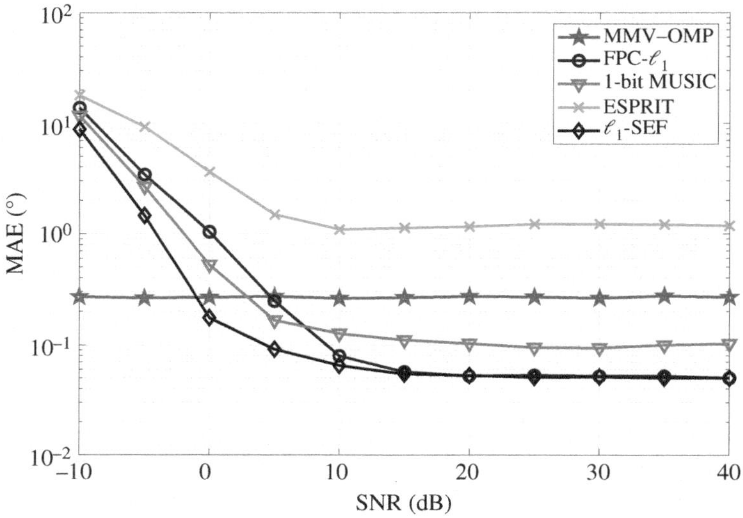

Figure 5.6 MAE versus SNR, when $M' = 40$ and $L = 4$.

It is observed from Figure 5.6 that the MAE of the ESPRIT algorithm is larger than 1° even at high SNR levels. Compared to the MMV–OMP algorithm, the performance of ℓ_1-SEF is better when SNR ≥ -1 dB. It is also noticed that the performance of the MMV–OMP algorithm does not improve with the increase of SNR although it does perform robustly at low SNR regime. In contrast, the ℓ_1-SEF algorithm outperforms all the conventional subspace-based algorithms, as well as the FPC-ℓ_1 algorithm at low SNR levels. This is mainly because ℓ_1-SEF takes advantage of its sparsity enhancement, while the FPC-ℓ_1 algorithm only focuses on the improvement of the fidelity term.

Next, we fix the SNR to be 10 dB and change the number of sensors from 30 to 50; the MAE versus the number of sensors is demonstrated in Figure 5.7. With the increase in the number of sensors, the ESPRIT algorithm still does not work properly. Moreover, it is found that the FPC-ℓ_1 outperforms the MUSIC when M' is larger than 38 and the ℓ_1-SEF achieves superior performance than all the other algorithms. A similar result is illustrated in Figure 5.8, where the MAE versus the number of snapshots is evaluated, by setting the SNR = 10 dB and varying the number of snapshots L from 1 (i.e., a single snapshot) to 8. It can be seen that the performance of all algorithms improves with the increase of snapshots, and the ℓ_1-SEF-based algorithm is superior to all the other algorithms.

In the last experiment, the resolution of ℓ_1-SEF-based algorithm is explored. We change the interval of the spatial grid to 0.1° and fix the number of snapshots and SNR to $L = 6$ and 15 dB, respectively. The DOAs of the sources are assumed

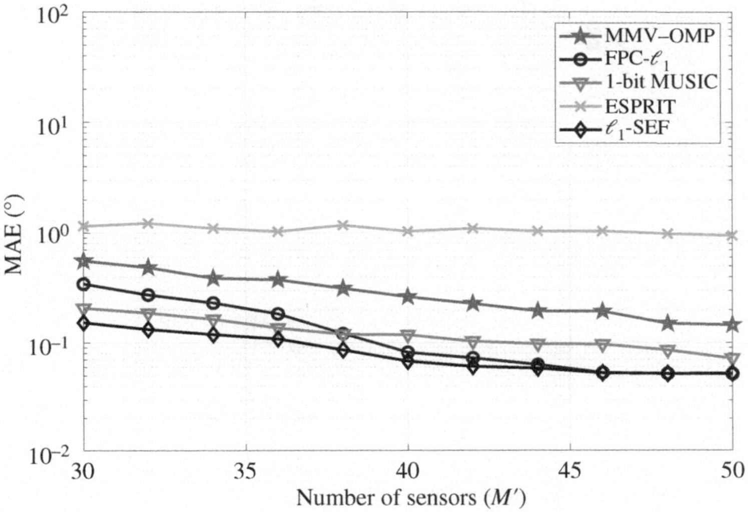

Figure 5.7 MAE versus the number of sensors M', when $L = 4$ and SNR = 10 dB.

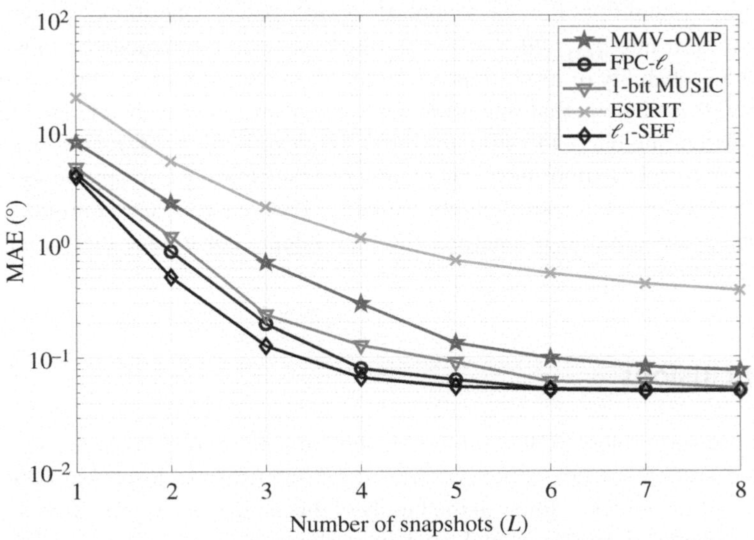

Figure 5.8 MAE versus the number of snapshots L, when $M' = 40$ and SNR = 10 dB.

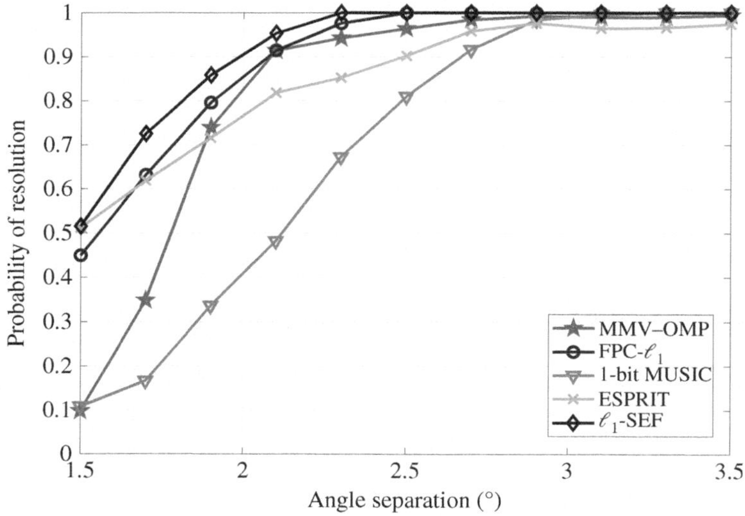

Figure 5.9 Probability of resolution versus source angle separation in degree with $M' = 40$, $L = 6$, and SNR = 15 dB.

to be $\theta_1 = 1° - \frac{\Delta\theta}{2}$ and $\theta_2 = 1° + \frac{\Delta\theta}{2}$, where $\Delta\theta$ is the angular separation between the two sources. If $\max_{k\in\{1,2\}} |\hat{\theta}_k - \theta_k| < \frac{\Delta\theta}{2}$, then the sources are deemed to be resolved. The probability of resolution versus the source separation is illustrated in Figure 5.9. It is obvious that when the angular separation is relatively small, the ℓ_1-SEF-based one-bit DOA estimator offers better resolution. Moreover, it is shown that CS-based methods have advantages over subspace-based approaches in resolution when only few snapshots are available, since relatively larger number of snapshots are required to reconstruct the high-fidelity covariance matrix for subspace extraction.

5.7 Conclusions

In this chapter, the problem of DOA estimation with one-bit array measurement(s) is considered and an entropy-enhanced one-bit CS algorithm is developed. The normalized ℓ_1-SEF, which belongs to the family of the Schur-concave function, is exploited to enhance the sparse signal recovery performance and applied to DOA estimation. Compared to the conventional approaches, the entropy-enhanced algorithm does not need the knowledge of sparsity level. In other words, the knowledge of the number of sources for DOA estimation is unnecessary. More importantly, it exhibits better performance than existing sparsity-level-free algorithms. Numerical results have demonstrated the outstanding performance of the ℓ_1-SEF-based one-bit CS algorithm for general sparse signal recovery as well as for DOA estimation with sparse representation.

Acknowledgment

This work was supported by the National Natural Science Foundation of China under Grant 62171292.

References

1 Sohrabi, F. (2018). *Hybrid Beamforming and One-Bit Precoding for Large-Scale Antenna Arrays*. University of Toronto (Canada).

2 Walden, R.H. (1999). Analog-to-digital converter survey and analysis. *IEEE Journal on Selected Areas in Communications* 17(4): 539–550.

3 Zhang, T., Ren, J., Li, J. et al. (2022). RFI mitigation for one-bit UWB radar systems. *IEEE Transactions on Aerospace and Electronic Systems* 58(2): 879–889.

4 Wang, S., Li, Y., and Wang, J. (2014). Multiuser detection in massive spatial modulation MIMO with low-resolution ADCs. *IEEE Transactions on Wireless Communications* 14(4): 2156–2168.

5 Mo, J., Alkhateeb, A., Abu-Surra, S., and Heath, R.W. (2017). Hybrid architectures with few-bit ADC receivers: achievable rates and energy-rate tradeoffs. *IEEE Transactions on Wireless Communications* 16(4): 2274–2287.

6 Liu, T., Tong, J., Guo, Q. et al. (2019). Energy efficiency of massive MIMO systems with low-resolution ADCs and successive interference cancellation. *IEEE Transactions on Wireless Communications* 18(8): 3987–4002.

7 Anokye, P., Ahiadormey, R.K., and Lee, K.-J. (2021). Full-duplex cell-free massive MIMO with low-resolution ADCs. *IEEE Transactions on Vehicular Technology* 70(11): 12179–12184.

8 Cheng, Z., Liao, B., He, Z., and Li, J. (2019). Transmit signal design for large-scale MIMO system with 1-bit DACs. *IEEE Transactions on Wireless Communications* 18(9): 4466–4478.

9 Xi, F., Xiang, Y., Chen, S., and Nehorai, A. (2020). Gridless parameter estimation for one-bit MIMO radar with time-varying thresholds. *IEEE Transactions on Signal Processing* 68: 1048–1063.

10 Deng, M., Cheng, Z., Wu, L. et al. (2022). One-bit ADCs/DACs based MIMO radar: performance analysis and joint design. *IEEE Transactions on Signal Processing* 70: 2609–2624.

11 Liu, C.-L. and Vaidyanathan, P.P. (2017). One-bit sparse array DOA estimation. *2017 IEEE International Conference on Acoustics, Speech and Signal Processing (ICASSP)*, 3126–3130. IEEE.

12 Liang, N. and Zhang, W. (2016). Mixed-ADC massive MIMO. *IEEE Journal on Selected Areas in Communications* 34(4): 983–997.

13 Li, Y., Tao, C., Seco-Granados, G. et al. (2017). Channel estimation and performance analysis of one-bit massive MIMO systems. *IEEE Transactions on Signal Processing* 65(15): 4075–4089.

14 Khalili, A., Shirani, F., Erkip, E., and Eldar, Y.C. (2021). MIMO networks with one-bit ADCs: receiver design and communication strategies. *IEEE Transactions on Communications* 70(3): 1580–1594.

15 Bar-Shalom, O. and Weiss, A.J. (2002). DOA estimation using one-bit quantized measurements. *IEEE Transactions on Aerospace and Electronic Systems* 38(3): 868–884.

16 Huang, X. and Liao, B. (2019). One-bit MUSIC. *IEEE Signal Processing Letters* 26(7): 961–965.

17 Teng, L., Wang, Q., and Chen, H. (2021). 1-bit DOA estimation algorithm for strictly non-circular sources. *IEEE Communications Letters* 25(7): 2216–2220.

18 Sedighi, S., Shankar, M.R.B., Soltanalian, M., and Ottersten, B. (2021). On the performance of one-bit DoA estimation via sparse linear arrays. *IEEE Transactions on Signal Processing* 69: 6165–6182.

19 Stöckle, C., Munir, J., Mezghani, A., and Nossek, J.A. (2015). 1-bit direction of arrival estimation based on compressed sensing. *2015 IEEE 16th International Workshop on Signal Processing Advances in Wireless Communications (SPAWC)*, 246–250. IEEE.

20 Yu, K., Zhang, Y.D., Bao, M. et al. (2016). DOA estimation from one-bit compressed array data via joint sparse representation. *IEEE Signal Processing Letters* 23(9): 1279–1283.

21 Huang, X., Xiao, P., and Liao, B. (2018). One-bit direction of arrival estimation with an improved fixed-point continuation algorithm. *2018 10th International Conference on Wireless Communications and Signal Processing (WCSP)*, 1–4. IEEE.

22 Boufounos, P.T. and Baraniuk, R.G. (2008). 1-bit compressive sensing. *2008 42nd Annual Conference on Information Sciences and Systems*, 16–21. IEEE.

23 Xiao, P., Liao, B., Huang, X., and Quan, Z. (2019). 1-bit compressive sensing with an improved algorithm based on fixed-point continuation. *Signal Processing* 154: 168–173.

24 Yan, M., Yang, Y., and Osher, S. (2012). Robust 1-bit compressive sensing using adaptive outlier pursuit. *IEEE Transactions on Signal Processing* 60(7): 3868–3875.

25 Xiao, P. and Liao, B. (2019). Robust one-bit compressive sensing with weighted ℓ_1-norm minimization. *Signal Processing* 164: 380–385.

26 Boufounos, P.T. (2009). Greedy sparse signal reconstruction from sign measurements. *2009 Conference Record of the 43rd Asilomar Conference on Signals, Systems and Computers*, 1305–1309. IEEE.

27 Plan, Y. and Vershynin, R. (2013). One-bit compressed sensing by linear programming. *Communications on Pure and Applied Mathematics* 66(8): 1275–1297.

28 Jacques, L., Laska, J.N., Boufounos, P.T., and Baraniuk, R.G. (2013). Robust 1-bit compressive sensing via binary stable embeddings of sparse vectors. *IEEE Transactions on Information Theory* 59(4): 2082–2102.

29 Sun, B., Feng, H., and Xu, X. (2016). History: an efficient and robust algorithm for noisy 1-bit compressed sensing. *IEICE Transactions on Information and Systems* 99(10): 2566–2573.

30 Zhang, L., Yi, J., and Jin, R. (2014). Efficient algorithms for robust one-bit compressive sensing. *International Conference on Machine Learning*, 820–828. PMLR.

31 Laska, J.N., Wen, Z., Yin, W., and Baraniuk, R.G. (2011). Trust, but verify: fast and accurate signal recovery from 1-bit compressive measurements. *IEEE Transactions on Signal Processing* 59(11): 5289–5301.

32 Xiao, P., Liao, B., and Deligiannis, N. (2020). DeepFPC: a deep unfolded network for sparse signal recovery from 1-bit measurements with application to DOA estimation. *Signal Processing* 176: 107699.

33 Zeng, Y., Khobahi, S., and Soltanalian, M. (2022). One-bit compressive sensing: can we go deep and blind? *IEEE Signal Processing Letters* 29: 1629–1633.

34 Khobahi, S. and Soltanalian, M. (2020). Model-based deep learning for one-bit compressive sensing. *IEEE Transactions on Signal Processing* 68: 5292–5307.

35 Marshall, A.W., Olkin, I., and Arnold, B.C. (1979). *Inequalities: Theory of Majorization and Its Applications*. Springer.

36 Kreutz-Delgado, K. and Rao, B.D. (1997). A General Approach to Sparse Basis Selection: Majorization, Concavity, and Affine Scaling. Tech. Rep. UCSD-CIE-97-7-1. San Diego, CA: University of California.

37 Huang, S. and Tran, T.D. (2018). Sparse signal recovery via generalized entropy functions minimization. *IEEE Transactions on Signal Processing* 67(5): 1322–1337.

38 Shen, L. and Suter, B.W. (2016, 2016). One-bit compressive sampling via $\ell_0$minimization. *EURASIP Journal on Advances in Signal Processing* (1): 1–16.

39 Plan, Y. and Vershynin, R. (2012). Robust 1-bit compressed sensing and sparse logistic regression: a convex programming approach. *IEEE Transactions on Information Theory* 59(1): 482–494.

40 Huang, X., Shi, L., Yan, M., and Suykens, J.A.K. (2018). Pinball loss minimization for one-bit compressive sensing: convex models and algorithms. *Neurocomputing* 314: 275–283.

41 Cucker, F. and Zhou, D.X. (2007). *Learning Theory: An Approximation Theory Viewpoint*, vol. 24. Cambridge University Press.

42 Zhang, T. (2004). Statistical analysis of some multi-category large margin classification methods. *Journal of Machine Learning Research* 5: 1225–1251.

43 Hurley, N. and Rickard, S. (2009). Comparing measures of sparsity. *IEEE Transactions on Information Theory* 55(10): 4723–4741.

44 Donoho, D.L. (2006). For most large underdetermined systems of linear equations the minimal ℓ_1-norm solution is also the sparsest solution. *Communications on Pure and Applied Mathematics: A Journal Issued by the Courant Institute of Mathematical Sciences* 59(6): 797–829.

45 Zhu, R. and Gu, Q. (2015). Towards a lower sample complexity for robust one-bit compressed sensing. *Proceedings of the 32nd International Conference on International Conference on Machine Learning*, ICML'15, volume 37, 739–747. JMLR.org.

46 Huang, X. and Yan, M. (2018). Nonconvex penalties with analytical solutions for one-bit compressive sensing. *Signal Processing* 144: 341–351.

47 Wang, J. and Palomar, D. (2017). Majorization theory and applications. *Mathematical Foundations for Signal Processing, Communications, and Networking*, 561–598. CRC Press.

48 Xiao, P., Liao, B., and Li, J. (2019). One-bit compressive sensing via Schur-concave function minimization. *IEEE Transactions on Signal Processing* 67(16): 4139–4151.

49 Lange, K., Hunter, D.R., and Yang, I. (2000). Optimization transfer using surrogate objective functions. *Journal of Computational and Graphical Statistics* 9(1): 1–20.

50 Hale, E.T., Yin, W., and Zhang, Y. (2007). A fixed-point continuation method for l1-regularized minimization with applications to compressed sensing. CAAM TR07-07. Rice University, 1–45.

51 Daubechies, I., Defrise, M., and De Mol, C. (2004). An iterative thresholding algorithm for linear inverse problems with a sparsity constraint. *Communications on Pure and Applied Mathematics: A Journal Issued by the Courant Institute of Mathematical Sciences* 57(11): 1413–1457.

52 Combettes, P.L. and Wajs, V.R. (2005). Signal recovery by proximal forward-backward splitting. *Multiscale Modeling and Simulation* 4(4): 1168–1200.

53 Beck, A. and Teboulle, M. (2009). A fast iterative shrinkage-thresholding algorithm for linear inverse problems. *SIAM Journal on Imaging Sciences* 2(1): 183–202.

54 Xiao, P., Liao, B., Tao, R., and Li, J. (2020). Generalized fixed-point continuation method: convergence and application. *IEEE Transactions on Signal Processing* 68: 5746–5758.

55 Chen, J. and Huo, X. (2006). Theoretical results on sparse representations of multiple-measurement vectors. *IEEE Transactions on Signal Processing* 54(12): 4634–4643.

6

Information-Theoretic Methods for Waveform Design in Multistatic Radar Imaging

Zacharie Idriss[1], Raghu G. Raj[1], and Ram M. Narayanan[2]

[1] *Radar Division, U.S. Naval Research Laboratory, Washington, DC, USA*
[2] *School of Electrical and Computer Engineering, The Pennsylvania State University, University Park, PA, USA*

6.1 Introduction

This chapter addresses the problem of radar waveform design for imaging targets using a multistatic radar. We explore waveform design techniques exploiting the mutual information (MI) maximization criterion based on inherent scene statistics. The radar channel can be modeled as a linear time-invariant (LTI) system as the effects of scene on the waveform are modeled as linear [1]. In this chapter, we show that by viewing the surrounding environment as an LTI system, the waveform design problem can be viewed as solving for the upper bound of the channel capacity, and the change in information between the input and output is measured using information-theoretic metrics. Therefore, extracting scene information is posed as maximizing the MI between the quantity of interest and the received waveform conditioned upon a transmit signal.

To calculate MI-related quantities described above, we incorporate a compound Gaussian (CG) prior to modeling the scene response for waveform optimization. In particular, we model the sparse wavelet coefficients as a CG distributed vector, i.e., the product of two independent random variables, one being a positive random scalar variable and the other being a multivariate vector following the zero-mean Gaussian distribution. The CG distribution provides a rich model for the overall measured scene reflectivity [2–6]. We first present an analytical characterization of the MI between the image vector described by the CG distribution and the received waveform. We then optimize the analytically derived MI to form an optimal transmit waveform. In [6, 7], it was empirically shown that with a sparse wavelet representation of natural scenes, a CG distribution is a good fit to model

Information-Theoretic Radar Signal Processing, First Edition. Edited by Yujie Gu and Yimin D. Zhang.

the image vector. This is seen by viewing the scene reflectivity function as the product of a two-dimensional (2D) wavelet basis and the corresponding wavelet coefficients, where the distribution of the underlying wavelet coefficient is well modeled using a CG distribution. Moreover, the developed tools are systematically applied to waveform optimization in a multistatic radar imaging scenario. We present mathematical bounds on the MI for the problem of multistatic waveform optimization where the CG statistics of the scene is exploited.

MI for radar waveform design and imaging has received considerable attention [2, 8–14]. The authors of [15] use MI for multiple-input multiple-output (MIMO) radar waveform design. They show that maximizing the MI at the receiver for a random target impulse response (TIR) is equivalent to minimizing the minimum mean square error (MMSE) of the estimated TIR. In [16], MI is applied to waveform optimization where the target ensemble is modeled as a Gaussian mixture model (GMM). The concept of dual-MI is used in [17] to maximize the TIR and minimize the clutter impulse response (CIR) returns respectively. The optimization of MI has also seen use in the coexistence of radar and communication systems [18]. More recently, the MI is derived for a target ensemble modeled as a Gaussian mixture in the context of matched illumination [16]. However, it is still unclear whether maximizing MI at the receiver is an optimal solution to waveform design when compared to other metrics. The authors of [19] aim to answer this question by providing a relationship between MI, signal-to-noise-ratio (SNR), and the Kullback–Leibler divergence. In this chapter, we use MI to optimize a waveform and improve image reconstruction within a multistatic radar configuration.

The remainder of this chapter is organized as follows. The system setup being considered is described in Section 6.2. The statistics of image scenes is considered in Section 6.3. A brief introduction to MI and conditional MI is presented in Section 6.4. In Section 6.5, we describe the proposed waveform design method using MI. The application of bounds is presented in Section 6.6. Section 6.7 concludes this chapter.

6.1.1 Notation

Throughout this chapter, we use boldface italic notation, such as $\boldsymbol{x}$, to denote random vectors and ordinary vectors depending on the context in which they are used. We use non-boldface italic notation, such as x, to denote random variables and scalars depending on the context. We make these distinctions clear within the context in which they are used. The range of a random vector is denoted with a boldface non-italic notation such as $\mathbf{x}$. The range of a random variable is denoted with a non-boldface non-italic notation such as x. We use capital, boldface letters, e.g., $\mathbf{X}$, to denote matrices. In our analysis all our quantities are real.

6.2 System Setup

We first present a general radar system modeled from a tomographic perspective for the case of a single monostatic radar. We then extend this to the multistatic case of M sensors for radar imaging. The multistatic radar will assume that each of the M sensors transmits sequentially. This forms $M - 1$ bistatic links and a single monostatic link for each transmitter in the sequence.

6.2.1 Radar Model

In our model, we first depict a single monostatic radar that emits a signal $x(t) : \mathbb{R} \rightarrow \mathbb{R}$. We place an energy constraint $E = T\rho$, on the transmit waveform where ρ represents the average power per sample and T corresponds to the signal's duration. As such, for a signal vector $\boldsymbol{x} = [x(1), \dots, x(T)]^{\mathrm{T}}$, its l_2-norm satisfies the following condition:

$$\|\boldsymbol{x}\|_2^2 \leq T\rho, \tag{6.1}$$

where $\|\cdot\|_2$ denotes the l_2-norm of a vector.

The radar at a location m transmits a signal onto a scene, which is defined by the scene reflectivity density matrix, $\mathbf{G} \in \mathbb{R}^{U\times V}$. The scene reflectivity density from a ground patch of a differential area is given by the (u, v)th element of $\mathbf{G}$ denoted as $g(u, v)$, for $1 \leq u \leq U$ and $1 \leq v \leq V$. U and V are referred to as the dimensions of the scene reflectivity function. The direction along the line of sight to the radar is called the "downrange." Assigning the radar position as the origin of the coordinate system, a "range bin" is defined as the reflectivity density of a ground patch centered at $r_0 = \sqrt{u_0^2 + v_0^2}$. Figure 6.1 illustrates the downrange as a dashed line intersecting the scene at the angle θ_m.

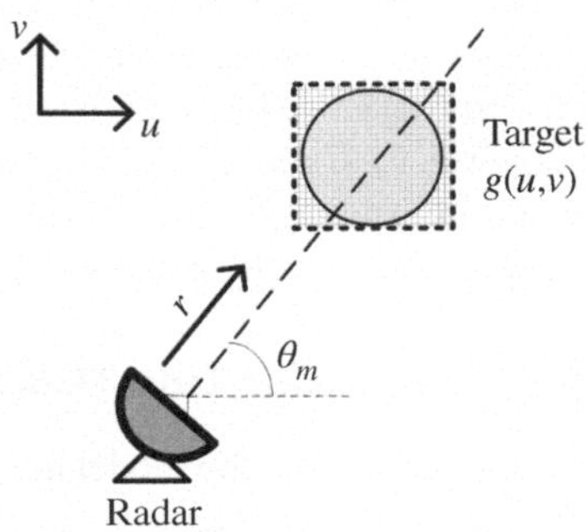

Figure 6.1 A single monostatic radar emitting a signal that intersects the target at angle θ_m. The downrange is denoted by r. We denote the horizontal and vertical axes of the target with u and v, respectively.

Therefore, the return signal from the scene reflectivity point spread function $g(u_0, v_0)$, centered at r_0 from the radar, is defined as

$$y_{r_0}(t) = g(u_0, v_0)x\left(t - \frac{2r_0}{v_p}\right) + n(t) \in \mathbb{R}, \tag{6.2}$$

where $2r_0/v_p$ denotes the two-way time delay, v_p represents the speed of wave propagation, and $n(t) \sim \mathcal{N}(0, \sigma_n^2)$ is the measurement noise.

The range bin associated with downrange r_0 is defined as the integral across radar returns from all points along a line perpendicular to the line of sight direction at r_0. This corresponds to the integration over all entries of $g(u, v)$ situated along a line defined as $r_\perp$, commonly referred to as an "iso-range contour." This can be expressed using the Radon transform of the scene evaluated at the angle θ_m and the downrange location r_0 [[20], Chapter 2] as $g(u_0, v_0) = R_{\theta_m, r_0}\{g(u, v)\}$. The operator $R_{\theta_m, r_0}\{g(u, v)\} : \mathbb{R}^N \to \mathbb{R}$ represents the Radon transform of $g(u, v)$ evaluated at θ_m and along the line perpendicular to the downrange line of sight direction at location r_0. Therefore, the model in (6.2) is rewritten as

$$y_{r_0}(t) = R_{\theta_m, r_0}\{g(u, v)\}\, x\left(t - \frac{2r_0}{v_p}\right) + n(t). \tag{6.3}$$

The complete response at time t from all L range bins, $r_1, \dots, r_L$, along the line formed by intersecting the scene at θ_m, is written as

$$y(t) = \int_r R_{\theta_m, r}\{g(u, v)\}\, x\left(t - \frac{2r}{v_p}\right) dr + n(t). \tag{6.4}$$

This can be reformulated in convolution form as:

$$y(t) = \hat{g}_{\theta_m}(t) * x(t) + n(t), \tag{6.5}$$

where $\hat{g}_{\theta_m}(t) = R_{\theta_m, r}\{g(u, v)\}$ is used for brevity, and $*$ denotes the convolution operation. This system equation can be interpreted as the response to an LTI system with an input signal $x(t)$ and an impulse response $\hat{g}_{\theta_m}(t)$. The received signal $y(t)$ is the output of this LTI system and is subsequently sampled at the receiver.

The impulse response $\hat{g}_{\theta_m}(t)$ characterizes the interaction between the transmit waveform and the scene with respect to sensing angle θ_m. Given the received signal $y(t)$, our objective is to enhance the estimation performance of the scene's impulse response, $\hat{g}_{\theta_m}(t)$, by optimizing the sensing waveform $x(t)$. To quantify how much scene information is contained within the received signal, we measure the MI between the impulse response, $\hat{g}_{\theta_m}(t)$, and the received signal, $y(t)$.

6.2.2 Multistatic Radar Model

In this subsection, the model in Section 6.2.1 is expanded to two separated radars, denoted as Radar m and Radar n, as depicted in Figure 6.2.

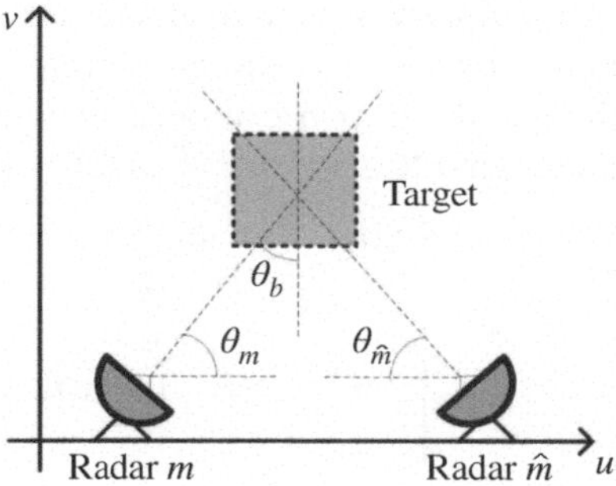

Figure 6.2 Bistatic system in which two radars are used. It is assumed that one transmits and both receive. This can be extended to a multistatic radar system comprising M radars.

Given that Radar m transmits a signal vector $\boldsymbol{x} = \left[x(t_1), \ldots, x(t_N)\right]^{\mathrm{T}} \in \mathbb{R}^N$, the received signal at Radar m, $\boldsymbol{y}_m = [y(t_1), \ldots, y(t_{N+L-1})]^{\mathrm{T}} \in \mathbb{R}^{N+L-1}$, is written using (6.5) as

$$\boldsymbol{y}_m = A_{m,m}\hat{\mathbf{g}}_{\theta_m} * \boldsymbol{x} + \boldsymbol{n}_m, \tag{6.6}$$

where the term $A_{m,m}$ is a two-way path loss coefficient, $\hat{\mathbf{g}}_{\theta_m} = [\hat{g}_{\theta_m}(t_1), \ldots, \hat{g}_{\theta_m}(t_L)]^{\mathrm{T}} \in \mathbb{R}^L$, and $\boldsymbol{n}_m = [n(t_1), \ldots, n(t_{N+L-1})]^{\mathrm{T}} \sim \mathcal{N}(\mathbf{0}, \sigma_n^2 \mathbf{I}_{N+L-1})$ is additive white Gaussian noise with $\mathbf{I}_{N+L-1}$ denoting the identity matrix of size $N + L - 1$. The impulse response $\hat{\mathbf{g}}_{\theta_m}$ is built from taking the Radon transform of the scene at L range bins located at r from the radar.

To characterize the return signal received at Radar $\hat{m}$, the bistatic signal can be approximated as a virtual monostatic situated at the bistatic bisector of the two radars [21–23]. This bisector corresponds to an angle that divides the bistatic angle in half, denoted as $\theta_b = \frac{1}{2}\left(\pi - (\theta_m + \theta_{\hat{m}})\right)$. Using θ_b, the return at Radar $\hat{m}$ can be expressed as follows:

$$\boldsymbol{y}_{\hat{m}} = A_{m,\hat{m}}\hat{\mathbf{g}}_{\theta_b} * \boldsymbol{x} + \boldsymbol{n}_{\hat{m}}. \tag{6.7}$$

To exploit the underlying CG distribution of the scene, the formulation of the system model will now incorporate a wavelet-based image model wherein we vectorize the scene reflectivity function $\mathbf{G}$ as

$$\mathrm{vec}(\mathbf{G}) = \boldsymbol{\Phi}\mathbf{c} \in \mathbb{R}^{UV}, \tag{6.8}$$

where $\boldsymbol{\Phi} \in \mathbb{R}^{UV \times D}$ is a dictionary constructed of D vectorized wavelet "atoms" used as a basis, and $\mathbf{c} \in \mathbb{R}^D$ represents the corresponding wavelet coefficients. Using the scene model presented in (6.8), the system model of (6.7) is rewritten as

$$\boldsymbol{y}_{\hat{m}} = \left(A_{m,\hat{m}}\mathbf{R}_{\theta_b}\boldsymbol{\Phi}\mathbf{c}\right) * \boldsymbol{x} + \boldsymbol{n}_{\hat{m}}, \tag{6.9}$$

where due to its linearity, the Radon transform of the scene can be implemented as a matrix operation denoted as $\mathbf{R}_{\theta_b}$. Therefore, the quantity $\mathbf{R}_{\theta_b}\boldsymbol{\Phi}\mathbf{c}$ now represents the impulse response of the scene at angle θ_b.

To write the system model for the multistatic radar system, we first denote the Radon transform of the sparse wavelet dictionary as $\boldsymbol{\Psi}_{\hat{m}} = A_{m,\hat{m}}\mathbf{R}_{\theta_b}\boldsymbol{\Phi}$. Therefore, the impulse response at radar $\hat{m}$ is given as $\boldsymbol{\Psi}_{\hat{m}}\boldsymbol{c}$. We then construct the convolution matrix $\mathbf{X} \in \mathbb{R}^{(N+L-1)\times L}$ as follows:

$$\mathbf{X} = \begin{bmatrix} x(t_1) & 0 & & \cdots & 0 \\ x(t_2) & x(t_1) & 0 & \cdots & 0 \\ \vdots & & \ddots & & \vdots \\ 0 & \cdots & 0 & x(t_N) & x(t_{N-1}) \\ 0 & \cdots & & 0 & x(t_N) \end{bmatrix}. \tag{6.10}$$

The returns at Radars m and $\hat{m}$ can be expressed as follows:

$$\boldsymbol{y}_m = \mathbf{X}\boldsymbol{\Psi}_m\boldsymbol{c} + \boldsymbol{n}_m, \tag{6.11}$$

$$\boldsymbol{y}_{\hat{m}} = \mathbf{X}\boldsymbol{\Psi}_{\hat{m}}\boldsymbol{c} + \boldsymbol{n}_{\hat{m}}. \tag{6.12}$$

It is important to note that the convolution matrix in (6.10) exhibits sparsity, with nonzero entries confined to the uppermost and lowermost diagonals. This property is harnessed through sparse coding, which reduces computational complexity and enables us to perform compressive imaging.

The model described in (6.12) can be extended to other radar systems within the multistatic array. For a total of M radar receivers with only radar i transmitting, the overall responses are represented as a system of equations as

$$\boldsymbol{y} = \overline{\mathbf{X}}\boldsymbol{\Psi}\boldsymbol{c} + \boldsymbol{n}, \tag{6.13}$$

where $\boldsymbol{y} = [\boldsymbol{y}_1^{\mathrm{T}}, \ldots, \boldsymbol{y}_M^{\mathrm{T}}]^{\mathrm{T}}$, $\overline{\mathbf{X}}$ is a block diagonal matrix with M copies of $\mathbf{X}$ along the diagonal,

$$\boldsymbol{\Psi} = [\boldsymbol{\Psi}_1^{\mathrm{T}}, \boldsymbol{\Psi}_2^{\mathrm{T}}, \ldots, \boldsymbol{\Psi}_M^{\mathrm{T}}]^{\mathrm{T}},$$

$$\boldsymbol{n} = [\boldsymbol{n}_1^{\mathrm{T}}, \boldsymbol{n}_2^{\mathrm{T}}, \ldots, \boldsymbol{n}_M^{\mathrm{T}}]^{\mathrm{T}},$$

$\boldsymbol{n}$ follows a Gaussian distribution with mean $\mathbf{0}$ and covariance matrix $\sigma_n^2\mathbf{I}$, and the multistatic TIR is given as $\boldsymbol{\Psi}\boldsymbol{c}$. Equation (6.13) forms the foundation of our multistatic radar model.

Upon reception, the estimation of the image vector $\boldsymbol{c}$ from $\boldsymbol{y}$ is well-posed by minimizing a regularized least squares cost function defined as:

$$J(\boldsymbol{c}) = \|\boldsymbol{y} - \overline{\mathbf{X}}\boldsymbol{\Psi}\boldsymbol{c}\|_2^2 + \alpha\|\boldsymbol{c}\|_p^p, \tag{6.14}$$

where $\alpha > 0$. We choose $p = 1$ to achieve sparsity of $\boldsymbol{c}$ while keeping the optimization convex. The estimated image vector $\hat{\boldsymbol{c}}$ that minimizes (6.14) is projected onto the sparse wavelet basis to form the estimated scene reflectivity function.

6.3 Statistics of Scenes

In this section, we present the CG nature of the corresponding sparse wavelet coefficients for radar scenes. It is pointed out in [1] that, for an extended target whose TIR is not known *a priori*, it makes sense to model the TIR as a finite-energy random process because the distribution of scattering centers for an extended target are well modeled using a Gaussian random process. However, it can be shown that for natural scenes, the statistics (when represented on a wavelet basis) do not behave in a Gaussian manner, but rather are characterized closely as a CG [4, 6, 7]. This fact is exploited to characterize the scene reflectivity function when deriving the MI.

We place a CG prior on the sparse wavelet coefficients $\boldsymbol{c}$ in the overall multistatic model (6.13) as follows:

$$\boldsymbol{c} = z\boldsymbol{w}, \tag{6.15}$$

where $\boldsymbol{w} \sim \mathcal{N}(\mathbf{0}, \boldsymbol{\Sigma}_w)$ is a Gaussian vector, and the multiplier z is a non-negative scalar which is represented as a discrete random variable with a probability density function (PDF) denoted as $p_z(\text{z})$. As such, the wavelet coefficient vector $\boldsymbol{c}$ is modeled as a CG vector, that is, the product of two independent random quantities, namely the random variable z and the random vector $\boldsymbol{w}$. It can be shown that the CG model subsumes many well-known distributions such as the generalized Gaussian (which includes the Laplacian distribution that is widely used in sparsity-based signal processing and compressed sensing) [34]. In particular, the CG model accommodates the sparse structure in natural images by a suitable choice of $p_z(\text{z})$ [24].

The entire set of wavelet coefficients is divided into subsets commonly referred to as neighborhoods, of size $\hat{N}$. In image processing, it is common to assume that each neighborhood shares the same underlying structure. We follow this convention and, as such, these neighborhoods are identically distributed. Consequently, the overall covariance matrix is constructed as a block diagonal matrix, where each block contains copies of a neighborhood covariance matrix denoted as $\boldsymbol{\Sigma}_{\hat{N}}$, and this construction method is in line with established practices in literature [25–27].

Within a neighborhood, the underlying multivariate zero-mean Gaussian vector becomes an $\hat{N}$-dimensional zero-mean random vector, $\boldsymbol{w}_{\hat{N}} = [w_1, \dots, w_{\hat{N}}]^{\text{T}}$, and its distribution takes the form

$$p_{\boldsymbol{w}_{\hat{N}}}(\mathbf{w}) = \frac{1}{\sqrt{(2\pi)^{\hat{N}} \det(\boldsymbol{\Sigma}_{\hat{N}})}} \exp\left\{-\frac{1}{2}\mathbf{w}^{\text{T}}\boldsymbol{\Sigma}_{\hat{N}}^{-1}\mathbf{w}\right\}, \tag{6.16}$$

where $p_{\boldsymbol{w}_{\hat{N}}}(\mathbf{w})$ denotes the joint PDF of the random vector $\boldsymbol{w}_{\hat{N}}$, and $\det(\cdot)$ returns the determinant of a matrix. The vector $\mathbf{w} = [\text{w}_1, \dots, \text{w}_{\hat{N}}]^{\text{T}}$ represents the range of the $\hat{N}$-dimensional random vector $\boldsymbol{w}_{\hat{N}}$ in the neighborhood.

Correspondingly, within a neighborhood of $\hat{N}$ wavelet coefficients, the CG distribution of $\mathbf{c}_{\hat{N}} = [c_1, \dots, c_{\hat{N}}]^{\mathrm{T}}$ is expressed as [25]

$$p_{\mathbf{c}_{\hat{N}}}(\mathbf{c}) = \int_{-\infty}^{\infty} \frac{1}{\sqrt{(2\pi)^{\hat{N}} \det(\mathrm{z}^2 \boldsymbol{\Sigma}_{\hat{N}})}} \exp\left\{ -\frac{1}{2} \left(\frac{\mathbf{c}^{\mathrm{T}} \boldsymbol{\Sigma}_{\hat{N}}^{-1} \mathbf{c}}{\mathrm{z}^2} \right) \right\} p_z(\mathrm{z})\, dz, \quad (6.17)$$

where $p_z(\mathrm{z})$ is the PDF of z with range z. The PDF depicted in (6.17) can be seen as a mixture of Gaussian distributions due to the integration and scaling of the random variable z.

It is shown that the statistical structure of a scene can be well characterized by the CG distribution [6, 25, 26]. This is exploited by projecting a natural scene onto a wavelet basis, from which the wavelet coefficients can be described as CG distributed. Because images such as natural scenes exhibit non-Gaussian behavior, models such as the power spectrum are inadequate [25]. Wavelet coefficients are highly non-Gaussian such that the marginals of the wavelet coefficients have high kurtosis (kurtosis is a measure of the tailedness of the distribution and equals three for a standard Gaussian distribution), and they show a statistical dependency in both the marginal and the joint distributions [28], although the dependent coefficients can be approximately decorrelated. Therefore, the projection of an image onto a wavelet basis accounts well for the nonlinearity in images.

In order to accurately estimate the distribution for a particular scene, the parameters of the CG distribution are estimated from the wavelet coefficients of the scene.

6.3.1 Parameter Estimation

The neighborhood covariance matrix $\boldsymbol{\Sigma}_{\hat{N}}$ can be explicitly computed from training data or is known *a priori*. Additionally, the PDF of the scalar random variable z is computed explicitly from training data [7]. Given that the wavelet coefficients of natural scenes are assumed to be well characterized by the CG distribution, it can be shown that the multiplying scalar random variable z can be estimated by maximum likelihood techniques [7]. Thus, upon the estimation of the multiplying scalar random variable, the underlying Gaussian random variable can be found. To show this, we consider a synthetic aperture radar (SAR) scene shown in Figure 6.3 with corresponding wavelet decomposition [29] shown in Figure 6.4.

For each wavelet coefficient c_0 in a subband shown in Figure 6.4, a small neighborhood of $\hat{N}$ wavelet coefficients is taken to form the CG vector $\mathbf{c}_{\hat{N}}$. From [7], the

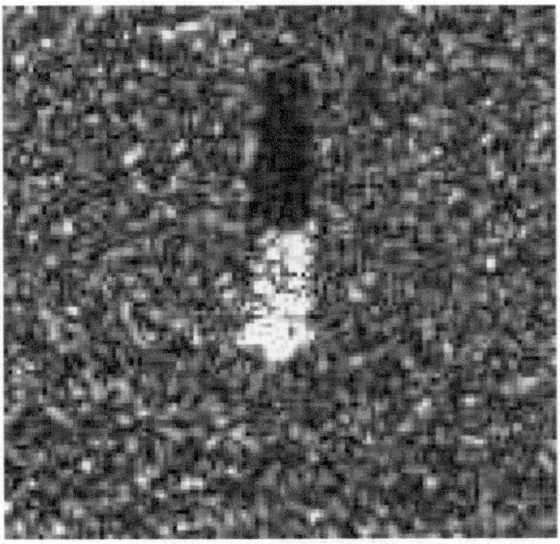

Figure 6.3 An SAR image of a tank from the MSTAR dataset.

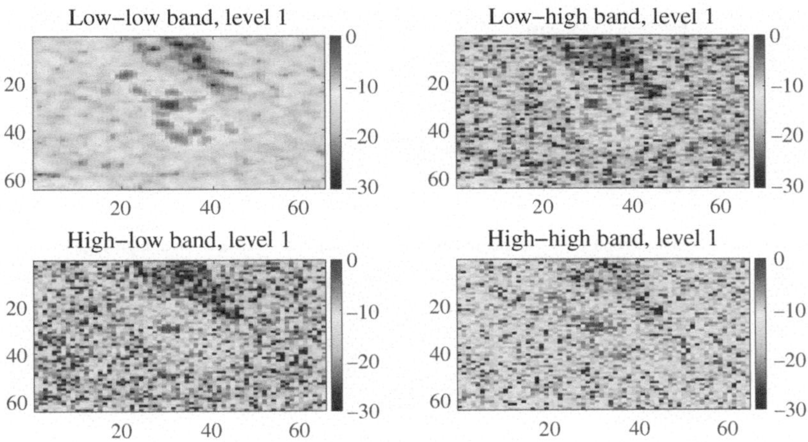

Figure 6.4 All four bands of the wavelet coefficients of Figure 6.3 at the first level/scale.

maximum likelihood estimate (MLE) for z is derived as

$$\begin{aligned}\hat{z} &= \arg\max_{z} \{\ln p(\mathbf{c}_{\hat{N}}|z)\} \\ &= \arg\max_{z} \left\{ \hat{N}\ln(z) + \frac{\mathbf{c}_{\hat{N}}^{\mathrm{T}}\boldsymbol{\Sigma}_{\hat{N}}^{-1}\mathbf{c}_{\hat{N}}}{2z^2} \right\} \\ &= \sqrt{\frac{\mathbf{c}_{\hat{N}}^{\mathrm{T}}\boldsymbol{\Sigma}_{\hat{N}}^{-1}\mathbf{c}_{\hat{N}}}{\hat{N}}}, \end{aligned} \tag{6.18}$$

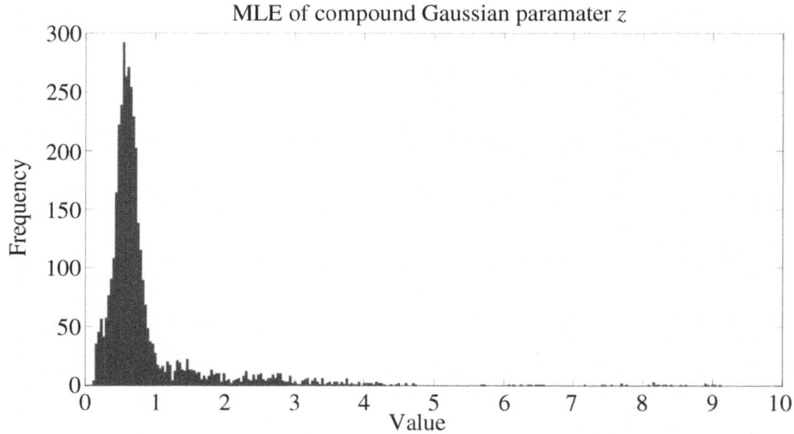

Figure 6.5 Estimated distribution of the multiplying scalar random variable z.

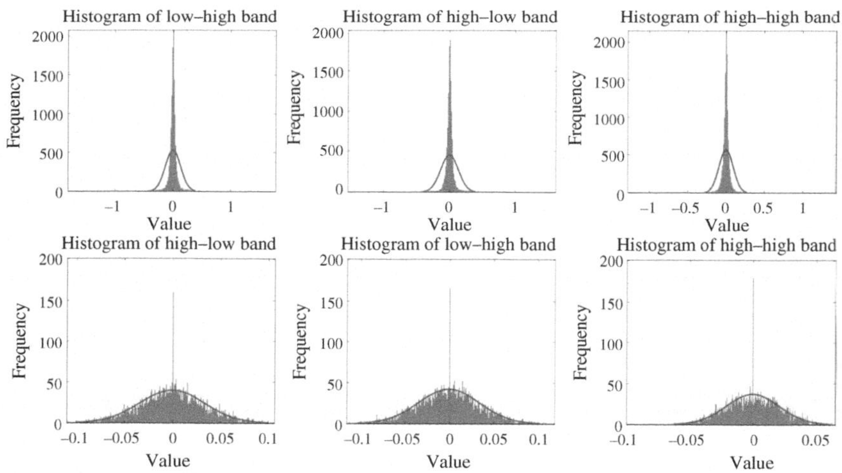

Figure 6.6 *Top*: The wavelet coefficients are shown in the histogram, and the underlying Gaussian is overlayed using a line. *Bottom*: The wavelet coefficients after dividing out the scalar random variable of the CG random variable and the underlying Gaussian.

where $p(\cdot|\cdot)$ denotes a conditional PDF, and $\ln(\cdot)$ represents the natural logarithm. For implementation, it is assumed that $\mathbf{\Sigma}_{\hat{N}}$ can be estimated directly from the data. Figure 6.5 shows the estimated distribution of the multiplier, $p_z(z)$. Figure 6.6 shows the CG nature of the wavelet coefficients before and after dividing out the multiplier z. As expected, after dividing out the multiplier z, the underlying distribution is Gaussian which validates the assumption that the underlying scene statistics of the wavelet coefficients in an SAR image are CG.

6.4 Mutual Information

6.4.1 Mutual Information for Noisy Signals

We first present a brief introduction to MI in general through a noisy signal example to motivate its application to radar waveform design. We consider a Gaussian vector characterized as

$$\boldsymbol{\eta} = \boldsymbol{\epsilon} + \boldsymbol{n}, \tag{6.19}$$

in which a zero-mean Gaussian vector $\boldsymbol{\epsilon} \in \mathbb{R}^k$ with distribution $\boldsymbol{\epsilon} \sim \mathcal{N}(\mathbf{0}, \boldsymbol{\Sigma}_\epsilon)$ is perturbed by independent white Gaussian noise $\boldsymbol{n} \sim \mathcal{N}(\mathbf{0}, \boldsymbol{\Sigma}_{\boldsymbol{n}})$ with covariance $\boldsymbol{\Sigma}_n = \sigma_n^2 \mathbf{I}_k$. The term σ_n^2 denotes the variance of each event in the random vector $\boldsymbol{n}$. Since the vector $\boldsymbol{\eta} \in \mathbb{R}^k$ is the summation of both the Gaussian source and Gaussian noise, it follows a Gaussian distribution, i.e., $\boldsymbol{\eta} \sim \mathcal{N}(\mathbf{0}, \boldsymbol{\Sigma}_\eta)$, with covariance $\boldsymbol{\Sigma}_\eta = \boldsymbol{\Sigma}_\epsilon + \boldsymbol{\Sigma}_n$. In order to reconstruct the Gaussian source, the MI can be used as a metric to extract the information of $\boldsymbol{\epsilon}$ that is contained in $\boldsymbol{\eta}$. The MI between $\boldsymbol{\epsilon}$ and $\boldsymbol{\eta}$ is described as

$$I(\boldsymbol{\epsilon}; \boldsymbol{\eta}) = h(\boldsymbol{\eta}) - h(\boldsymbol{\eta}|\boldsymbol{\epsilon}), \tag{6.20}$$

where $h(\cdot)$ is the differential entropy and $h(\cdot|\cdot)$ is the conditional differential entropy. The differential entropy of a random vector $\boldsymbol{\epsilon}$ is defined as [30]

$$h(\boldsymbol{\epsilon}) = -\int_{-\infty}^{\infty} p_\epsilon(\boldsymbol{\mathrm{\epsilon}}) \ln p_\epsilon(\boldsymbol{\mathrm{\epsilon}})\, d\boldsymbol{\mathrm{\epsilon}}, \tag{6.21}$$

where $p_\epsilon(\boldsymbol{\mathrm{\epsilon}})$ is the joint PDF of the random vector $\boldsymbol{\epsilon}$, and $\boldsymbol{\mathrm{\epsilon}}$ represents the range of $\boldsymbol{\epsilon}$. Note that the differential entropy considers continuous variables as opposed to the entropy, which is its discrete counterpart. The differential entropy of a Gaussian vector is defined as [30]

$$h(\boldsymbol{\epsilon}) = \frac{1}{2} \ln \left\{ (2\pi e)^k \det \left(\boldsymbol{\Sigma}_\epsilon \right) \right\}. \tag{6.22}$$

Because $\boldsymbol{\eta}$ has a covariance matrix $\boldsymbol{\Sigma}_\eta = \boldsymbol{\Sigma}_\epsilon + \boldsymbol{\Sigma}_n$, the differential entropy $h(\boldsymbol{\eta})$ in (6.20) is obtained as

$$h(\boldsymbol{\eta}) = \frac{1}{2} \ln \left\{ (2\pi e)^k \det (\boldsymbol{\Sigma}_\epsilon + \boldsymbol{\Sigma}_{\boldsymbol{n}}) \right\}. \tag{6.23}$$

The second term on the right-hand side of (6.20), $h(\boldsymbol{\eta}|\boldsymbol{\epsilon})$, can be obtained as

$$h(\boldsymbol{\eta}|\boldsymbol{\epsilon}) = h(\boldsymbol{\epsilon} + \boldsymbol{n}|\boldsymbol{\epsilon}) = h(\boldsymbol{n}|\boldsymbol{\epsilon}) = h(\boldsymbol{n}) = \frac{1}{2} \ln \left\{ (2\pi e)^k \det (\boldsymbol{\Sigma}_{\boldsymbol{n}}) \right\}, \tag{6.24}$$

where $h(\boldsymbol{\epsilon}|\boldsymbol{\epsilon}) = 0$, and due to the statistical independence of $\boldsymbol{\epsilon}$ and $\boldsymbol{n}$, $h(\boldsymbol{n}|\boldsymbol{\epsilon}) = h(\boldsymbol{n})$. Therefore, using (6.23) and (6.24), the MI (6.20) can be written as

$$I(\boldsymbol{\epsilon};\boldsymbol{\eta}) = \frac{1}{2}\ln\left\{(2\pi e)^k \det\left(\boldsymbol{\Sigma}_{\epsilon} + \boldsymbol{\Sigma}_{\boldsymbol{n}}\right)\right\} - \frac{1}{2}\ln\left\{(2\pi e)^k \det\left(\boldsymbol{\Sigma}_{\boldsymbol{n}}\right)\right\}. \tag{6.25}$$

Since a linear combination of time shifted and attenuated received signals can be considered as the output of an LTI system, maximizing the received information is equivalent to maximizing the channel capacity. The maximization of the MI (6.20) yields the channel capacity, i.e.,

$$C = \max_{p_{\epsilon}(\boldsymbol{\epsilon})} I(\boldsymbol{\epsilon};\boldsymbol{\eta}). \tag{6.26}$$

This can be formally defined as finding the joint PDF of the waveform that best denoises the signal at the receiver.

In the underlying radar imaging problem, we exploit MI maximization for optimized transmit waveform design, and the problem becomes one of finding the waveform that best extracts certain channel characteristics of interest.

6.4.2 Conditional MI

Now we consider a convolutional channel and examine the MI between the received signal and a transmitted waveform. In Section 6.4.1, we introduced the MI for a channel that consists of only additive Gaussian noise. In this chapter, we consider the case in which a transmit waveform probes a scene as presented in Section 6.2. In order to consider a channel in which a transmitted waveform can be both scaled and attenuated by a scene, the conditional MI is introduced. We define these channel effects on the waveform using the multistatic TIR, $\tilde{\mathbf{g}} = \boldsymbol{\Psi}\mathbf{c}$, defined in (6.13). After transmission, the waveform is scattered from the scene, and the received signal $\boldsymbol{y}$ is expressed by the convolutional channel in (6.6).

Our objective is to maximize the TIR information contained in the received signal $\boldsymbol{y}$ by optimizing the transmit waveform. Therefore, the channel capacity is described as the maximum conditional MI between the TIR $\tilde{\mathbf{g}}$, presented in Section 6.2, and the received signal $\boldsymbol{y}$, conditioned upon the transmit waveform $\boldsymbol{x}$ given as

$$C = \max_{p_{\boldsymbol{x}}(\mathbf{x})} I(\boldsymbol{y};\tilde{\mathbf{g}}|\boldsymbol{x}), \tag{6.27}$$

where the conditional MI $I(\boldsymbol{y};\tilde{\mathbf{g}}|\boldsymbol{x})$ is expressed as

$$I(\boldsymbol{y};\tilde{\mathbf{g}}|\boldsymbol{x}) = h(\boldsymbol{y}|\boldsymbol{x}) - h(\boldsymbol{y}|\tilde{\mathbf{g}},\boldsymbol{x}). \tag{6.28}$$

Informally, we optimize the conditional MI with the goal of maximizing the scene information of interest in the presence of noise for a particular transmit waveform. In order to find the waveform that best achieves this goal, we exploit the statistics of the scene by placing a prior upon the distribution of $\tilde{\mathbf{g}}$. This provides a statistical model for the TIR for the maximization of the MI.

6.5 Waveform Design Using MI

In this section, we provide the objective functions for the optimization of the upper and lower bounds of the MI defined in Section 6.4. We show that the objective functions are convex with respect to the transmit waveform. This is proven by decomposing the cost function as a convex combination of elementary convex functions where we then exploit the positive definiteness of the relevant Hessian waveform matrix to prove the convexity.

6.5.1 Upper and Lower Bounds of Differential Entropy

In Section 6.2, we presented a multistatic radar model where the scene is expressed as the product of a sparsifying basis set $\boldsymbol{\Psi}$ and a set of corresponding sparse coefficients, $\mathbf{c}$, which denotes the scene information encoded within the sparse representation. Although the multistatic radar model holds in general for any basis set representation, we exploit the fact that, when the scene is projected onto a wavelet basis, a CG prior can be placed upon the corresponding wavelet coefficients. Therefore, the CG distribution is used as a prior to model the sparse scene reflectivity wavelet coefficients. We derive the MI between the set of signals received within in a multistatic radar vector $\boldsymbol{y} = [\boldsymbol{y}_1^{\mathrm{T}}, \boldsymbol{y}_2^{\mathrm{T}}, \dots, \boldsymbol{y}_M^{\mathrm{T}}]^{\mathrm{T}}$ and the sparse scene coefficients $\mathbf{c}$, conditioned upon a transmit waveform, $\boldsymbol{x}$.

To derive the waveform that optimally estimates the sparse scene coefficients $\mathbf{c}$ we define the conditional MI as

$$I(\boldsymbol{y}; \mathbf{c}|\boldsymbol{x}) = h(\boldsymbol{y}|\boldsymbol{x}) - h(\boldsymbol{y}|\mathbf{c}, \boldsymbol{x}). \tag{6.29}$$

The second term on the right-hand side of (6.29) can be reduced to the differential entropy of the Gaussian noise term $\boldsymbol{n}$ as shown in Section 6.4.1, expressed as

$$h(\boldsymbol{y}|\mathbf{c}, \boldsymbol{x}) = h(\boldsymbol{n}) = \frac{1}{2} K \ln\left(2\pi e \sigma_n^2\right), \tag{6.30}$$

where $K = M(N + L - 1)$. Because the term in (6.30) is constant and not waveform-dependent, this term is ignored when solving for the optimal waveform $\boldsymbol{x}$. As such, we focus on defining the waveform-dependent term $h(\boldsymbol{y}|\boldsymbol{x})$ on the right-hand side of (6.29).

To incorporate the CG prior to the scene shown in (6.15) within the received signal model $\boldsymbol{y}$, we rewrite (6.13) as

$$\boldsymbol{y} = z\overline{\mathbf{X}}\boldsymbol{\Psi}\boldsymbol{w} + \boldsymbol{n}. \tag{6.31}$$

Since the matrix $\overline{\mathbf{X}}\boldsymbol{\Psi}$ is deterministic, $\boldsymbol{q} = \overline{\mathbf{X}}\boldsymbol{\Psi}\boldsymbol{w} \sim \mathcal{N}\left(\mathbf{0}, \boldsymbol{\Sigma}_q\right)$ remains a Gaussian vector with a covariance matrix

$$\boldsymbol{\Sigma}_q = \mathbb{E}\left[\boldsymbol{q}\boldsymbol{q}^{\mathrm{T}}\right] = \overline{\mathbf{X}}\boldsymbol{\Psi}\boldsymbol{\Sigma}_w\boldsymbol{\Psi}^{\mathrm{T}}\overline{\mathbf{X}}^{\mathrm{T}}, \tag{6.32}$$

where $\mathbb{E}[\cdot]$ denotes the expectation. The overall system model is rewritten as

$$\boldsymbol{y} = z\boldsymbol{q} + \boldsymbol{n}. \tag{6.33}$$

Due to the presence of the random scalar variable z in (6.33), the term $z\boldsymbol{q}$ is a CG random variable.

The CG multistatic radar model derived in (6.33) is used to rewrite the MI in (6.29) as

$$I(\boldsymbol{y}; \mathbf{c}|\boldsymbol{x}) = h(z\boldsymbol{q} + \boldsymbol{n}|\boldsymbol{x}) - \frac{1}{2}K \ln\left(2\pi e\sigma_n^2\right), \tag{6.34}$$

where again we note that the constant term $\frac{1}{2}K \ln\left(2\pi e\sigma_n^2\right)$ can be dropped when optimizing for the waveform $\boldsymbol{x}$. Consequently, maximizing the MI of the multistatic radar model $\boldsymbol{y}$ becomes a maximization of the term $h(z\boldsymbol{q} + \boldsymbol{n}|\boldsymbol{x})$.

Although Gaussian functions have a closed-form differential entropy as shown in (6.30), there is no closed-form solution for the differential entropy of the sum of a Gaussian and a CG random vector [31]. However, there exist upper and lower bounds on the entropy of Gaussian mixtures as shown in [31]. Therefore, the upper and lower bounds on the remaining waveform-dependent term in (6.34) are respectively denoted as $h_u(\boldsymbol{x})$ and $h_l(\boldsymbol{x})$, i.e.,

$$h_l(\boldsymbol{x}) \leq h(z\boldsymbol{q} + \boldsymbol{n}|\boldsymbol{x}) \leq h_u(\boldsymbol{x}). \tag{6.35}$$

We express the *upper* bound $h_u(\boldsymbol{x})$ as [31]

$$h_u(\boldsymbol{x}) \triangleq \sum_i w_i \left(-\ln w_i + \frac{1}{2}\ln\left((2\pi e)^D \det\left((\boldsymbol{\Sigma}_y)_i\right)\right)\right), \tag{6.36}$$

where $w_i = p_z(z_i)$ denotes the probability mass function (PMF) of the scalar random variable z, and the index i denotes that the sum is over the range of the random variable, z. $(\boldsymbol{\Sigma}_y)_i = z_i^2\boldsymbol{\Sigma}_q + \sigma_n^2\mathbf{I}$ represents the covariance matrix of the received signal vector of the multistatic radar model, $\boldsymbol{y} = z\boldsymbol{q} + \boldsymbol{n}$.

Similarly, we express the *lower* bound $h_l(\boldsymbol{x})$ as [31]

$$h_l(\boldsymbol{x}) = -\sum_i w_i \ln\left(\sum_j w_j \left[(2\pi)^{D/2} \det\left((\boldsymbol{\Sigma}_y)_i + (\boldsymbol{\Sigma}_y)_j\right)\right]^{-1/2}\right), \tag{6.37}$$

where the covariance matrices $(\boldsymbol{\Sigma}_y)_i$ and $(\boldsymbol{\Sigma}_y)_j$ are defined in (6.36) but summed over the separate indices i and j where the indices denote an instances with the range of the PMFs w_i and w_j, respectively, for the random variable, z.

Having defined the upper and lower bounds on the differential entropy shown in (6.36) and (6.37), respectively, we are now able to maximize the MI (6.29) between the received signal vector and the image vector. In Sections 6.5.2 and 6.5.3, we present the objective functions to maximize the upper and lower bounds, respectively.

6.5.2 Optimization of Upper Bound

An objective function is defined in order to find the waveform that optimizes the upper bound on the MI. We denote $\boldsymbol{x}_u$ to be the transmit waveform that maximizes the upper bound on the MI depicted in (6.36), i.e.,

$$\boldsymbol{x}_u = \underset{\boldsymbol{x}:\|\boldsymbol{x}\|_2^2 \leq T\rho}{\arg\max}\, h_u(\boldsymbol{x}). \tag{6.38}$$

Using the expression for $h_u(\boldsymbol{x})$ shown in (6.36) and ignoring any waveform-independent constant terms, we can rewrite (6.38) as

$$\boldsymbol{x}_u = \underset{\boldsymbol{x}:\|\boldsymbol{x}\|_2^2 \leq T\rho}{\arg\max} \sum_i w_i \ln\ \det\left((\boldsymbol{\Sigma}_y)_i\right). \tag{6.39}$$

The cost function presented in (6.39) is for a single waveform vector as by definition the MI is conditioned on a single vector representation of the waveform. However, there is a one-to-one correspondence between the waveform, $\boldsymbol{x}$, and the multistatic waveform convolution matrix, $\overline{\mathbf{X}}$. Therefore, we can expand (6.39) for a multistatic system as

$$\overline{\mathbf{X}}_{up} = \underset{\overline{\mathbf{X}}:\|\overline{\mathbf{X}}\|_F^2 \leq LMN\rho}{\arg\max} \sum_i w_i \ln\det\left(z_i^2 \overline{\mathbf{X}}\boldsymbol{\Psi}\boldsymbol{\Sigma}_w\boldsymbol{\Psi}^{\mathrm{T}}\overline{\mathbf{X}}^{\mathrm{T}} + \sigma_n^2\mathbf{I}\right), \tag{6.40}$$

where $\|\cdot\|_F$ represents the Frobenius norm. We first apply the Cholesky decomposition to the covariance matrix $\boldsymbol{\Sigma}_w = \mathbf{D}_w\mathbf{D}_w^{\mathrm{T}}$ and rewrite the determinant within (6.40), scaled by σ_n^2, as

$$\det\left(\frac{z_i^2}{\sigma_n^2}\overline{\mathbf{X}}\boldsymbol{\Psi}\mathbf{D}_w\mathbf{D}_w^{\mathrm{T}}\boldsymbol{\Psi}^{\mathrm{T}}\overline{\mathbf{X}}^{\mathrm{T}} + \mathbf{I}\right). \tag{6.41}$$

Then, using Sylvester's determinant identity, (6.40) can be rewritten as

$$\overline{\mathbf{X}}_{up} = \underset{\overline{\mathbf{X}}:\|\overline{\mathbf{X}}\|_F^2 \leq LMN\rho}{\arg\max} \sum_i w_i \ln\ \det\left(\frac{z_i^2}{\sigma_n^2}\mathbf{D}_w^{\mathrm{T}}\boldsymbol{\Psi}^{\mathrm{T}}\overline{\mathbf{X}}^{\mathrm{T}}\overline{\mathbf{X}}\boldsymbol{\Psi}\mathbf{D}_w + \mathbf{I}\right), \tag{6.42}$$

which provides an objective function that is dependent on a positive quantity, $\overline{\mathbf{X}}^{\mathrm{T}}\overline{\mathbf{X}}$, and further shows that the objective function (6.40) is concave in $\overline{\mathbf{X}}$. We then define the matrix $\mathbf{R}_{\overline{\mathbf{X}}} = \overline{\mathbf{X}}^{\mathrm{T}}\overline{\mathbf{X}}$ as an autocorrelation matrix which becomes the main variable for optimization. Note that $\mathbf{R}_{\overline{\mathbf{X}}}$ is a block diagonal matrix where each of the M identical blocks $\mathbf{R}_{\mathbf{X}} = \mathbf{X}^{\mathrm{T}}\mathbf{X}$ along the main diagonal define a sensor within the multistatic radar model. The cost function (6.42) is thus rewritten as

$$\underset{\mathbf{R}_{\overline{\mathbf{X}}}:\mathrm{tr}(\mathbf{R}_{\overline{\mathbf{X}}}) \leq LMN\rho}{\arg\max} \sum_i w_i \ln\det\left(\frac{z_i^2}{\sigma_n^2}\mathbf{D}_w^{\mathrm{T}}\boldsymbol{\Psi}^{\mathrm{T}}\mathbf{R}_{\overline{\mathbf{X}}}\boldsymbol{\Psi}\mathbf{D}_w + \mathbf{I}\right), \tag{6.43}$$

where tr(·) denotes the trace operator. Since we have recast the optimization problem to be over the autocorrelation matrix, we show in Observation 6.1 that we can extract the optimum real-valued waveform $\boldsymbol{x}$ from the matrix $\overline{\mathbf{R}}_{\mathbf{X}}$.

Observation 6.1 The transmit convolution matrix $\mathbf{X}$ is defined in (6.10) as a function of a vector $\boldsymbol{x}$ of length N and an impulse response of length L. Then, the Toeplitz autocorrelation matrix $\mathbf{R}_{\mathbf{X}} = \mathbf{X}^{\mathrm{T}}\mathbf{X} \in \mathbb{R}^{L\times L}$ can be written as

$$\mathbf{R}_{\mathbf{X}} = \begin{bmatrix} \tilde{r}_0 & \tilde{r}_1 & \cdots & \tilde{r}_{L-1} \\ \tilde{r}_1 & \tilde{r}_0 & \cdots & \tilde{r}_{L-2} \\ \vdots & & \ddots & \vdots \\ \tilde{r}_{L-1} & \tilde{r}_{L-2} & \cdots & \tilde{r}_0 \end{bmatrix}, \tag{6.44}$$

where $\tilde{r}_\tau$, $0 \le \tau \le L-1$ denotes an entry in the matrix $\mathbf{R}_{\mathbf{X}}$ and is defined as

$$\tilde{r}_\tau = \sum_{n=1}^{N} x(t_n)x(t_n - \tau)dt. \tag{6.45}$$

Note that $x(t)$ is nonzero only on the interval $0 \le n < N$.

Building on Observation 6.1, extraction of $\boldsymbol{x}$ from $\mathbf{R}_{\mathbf{X}}$ is given in Corollary 6.1. It is noted that to emphasize the dependence of $\mathbf{X}$ on $\boldsymbol{x}$ and L, we explicitly express $\mathbf{X}$ as $\mathbf{X}(\boldsymbol{x}, L)$.

Corollary 6.1 Given a Toeplitz and positive semi-definite matrix $\mathbf{R}_{\mathbf{X}} \in \mathbb{R}^{L\times L}$, a convolution matrix $\mathbf{X}$ that satisfies $\mathbf{R}_{\mathbf{X}} = \mathbf{X}^{\mathrm{T}}(\boldsymbol{x}, L)\mathbf{X}(\boldsymbol{x}, L)$ is given by $\mathbf{X}(\boldsymbol{x}_0, L)$ where $\boldsymbol{x}_0 \in \mathbb{R}^N$ is defined as

$$\boldsymbol{x}_0 = \mathcal{F}^{-1}\left\{ \sqrt{\mathcal{F}\left\{\boldsymbol{r}_{\mathbf{X}}\right\}} \right\}, \tag{6.46}$$

$\tilde{\boldsymbol{r}} \in \mathbb{R}^N$ is an autocorrelation sequence built from $\mathbf{R}_{\mathbf{X}}$, $\mathcal{F}\{\cdot\}$ is the discrete Fourier transform, and $\mathcal{F}^{-1}\{\cdot\}$ is the discrete inverse Fourier transform. Note that it is assumed that $N > L$, and $\tilde{\boldsymbol{r}}$ is zero-padded to be of length $2(N-1)+1$.

Proof: Since $\mathbf{R}_{\mathbf{X}}$ and L are given, this problem becomes equivalent to finding $\boldsymbol{x}$ to satisfy the equality $\mathbf{R}_{\mathbf{X}} = \mathbf{X}^{\mathrm{T}}(\boldsymbol{x}, L)\mathbf{X}(\boldsymbol{x}, L)$. As was described in Observation 6.1, the entries of $\mathbf{R}_{\mathbf{X}}$ correspond to a length $2(L-1)+1$ autocorrelation function $\tilde{\boldsymbol{r}}$. Since

$$\mathcal{F}\{\tilde{\boldsymbol{r}}\} = |X_0|^2, \tag{6.47}$$

where X_0 is the discrete Fourier transform for $\boldsymbol{x}_0$, and $|\cdot|$ denotes the magnitude. Therefore, by taking the square root and the inverse discrete Fourier transform of (6.47), the magnitude of $\boldsymbol{x}_0$ is obtained. For a practical implementation, a phase retrieval procedure using alternating projections can be used to find the phase information of $\boldsymbol{x}_0$ [2].

The overall optimization problem is written as in Lemma 6.1.

Lemma 6.1 The block diagonal autocorrelation matrix of the transmit waveform, $\mathbf{R}_{\overline{\mathbf{X}}} = \overline{\mathbf{X}}^{\mathrm{T}}\overline{\mathbf{X}}$, that maximizes the upper bound $h_{\mathrm{up}}(\boldsymbol{x})$ on the MI is found from the following convex optimization problem,

$$\begin{aligned} \boldsymbol{x}_0 = \arg\max_{\boldsymbol{x}} \ & \sum_i p_z(z_i) \ln \det\left(\frac{z_i^2}{\sigma_n^2}\mathbf{D}_w^{\mathrm{T}}\boldsymbol{\Psi}^{\mathrm{T}}\overline{\mathbf{R}}_{\overline{\mathbf{X}}}\boldsymbol{\Psi}\mathbf{D}_w + \mathbf{I}\right) \\ \text{subject to} \quad & \mathrm{tr}(\overline{\mathbf{R}}_{\overline{\mathbf{X}}}) \leq LMN\rho, \\ & \mathbf{R}_{\mathbf{X}}^{\mathrm{T}} = \mathbf{R}_{\mathbf{X}}, \\ & \mathbf{R}_{\mathbf{X}} \text{ is Toeplitz}, \\ & \mathbf{R}_{\mathbf{X}} \succeq 0, \end{aligned} \tag{6.48}$$

where $\mathbf{R}_{\mathbf{X}} \succeq 0$ denotes that $\mathbf{R}_{\mathbf{X}}$ is positive semidefinite.

Proof: Follows from Observation 6.1 and Corollary 6.1.

In this subsection, we have defined an objective function for the optimization of the upper bound on the MI. In the following subsection, we present an objective function for the optimization of the lower bound on the MI.

6.5.3 Optimization of Lower Bound

The lower bound on the MI (6.37) while ignoring waveform-independent constants is rewritten as

$$h_l(\boldsymbol{x}) = -\sum_i w_i \ln\left(\sum_j w_j \left[\det\left((\boldsymbol{\Sigma}_y)_i + (\boldsymbol{\Sigma}_y)_j\right)\right]^{-1/2}\right). \tag{6.49}$$

To show that this lower bound is also concave in the waveform, we first expand the covariance matrices $\boldsymbol{\Sigma}_y$ within the logarithm to yield

$$h_l(\boldsymbol{x}) = -\sum_i w_i \ln\left(\sum_j w_j \left[\det\left(\left(z_i^2 + z_j^2\right)\overline{\mathbf{X}}\boldsymbol{\Psi}\boldsymbol{\Sigma}_w\boldsymbol{\Psi}^{\mathrm{T}}\overline{\mathbf{X}}^{\mathrm{T}} + 2\sigma_n^2\mathbf{I}\right)\right]^{-1/2}\right). \tag{6.50}$$

Using the Cholesky decomposition and Sylvester's determinant theorem as in (6.40), we can rewrite (6.50) as

$$h_l(\boldsymbol{x}) = \sum_i w_i \ln\left(\sum_j w_j \left[\det\left(\frac{z_i^2 + z_j^2}{2\sigma_n^2}\mathbf{D}_w^{\mathrm{T}}\boldsymbol{\Psi}^{\mathrm{T}}\overline{\mathbf{R}}_{\overline{\mathbf{X}}}\boldsymbol{\Psi}\mathbf{D}_w + \mathbf{I}\right)\right]^{-1/2}\right). \tag{6.51}$$

To prove the convexity of the lower bound (6.51), the following Lemmas are presented.

Lemma 6.2 Let $f(t) \geq 0$ be a twice-differentiable, real, and non-negative function whose derivatives satisfy the following inequality,

$$(1 + \lambda t + f(t))f''(t) - f'(t) - \lambda \leq 0, \tag{6.52}$$

where $\lambda > 0$. Then, the following function,

$$a(t) = \ln(1 + \lambda t + f(t)), \tag{6.53}$$

is concave in $t \geq 0$.

Lemma 6.3 Recall that z is a non-negative random variable, with associated PMF $p_z(z)$. Given the constants $d_{\ell j} \geq 0$, $\lambda_\ell \geq 0$ for $\ell = 1, 2, \ldots, B$, the following function, $b(t)$, is convex in $t \geq 0$,

$$b(t) = \ln\left(\sum_j p_z(z_j)\left[z_j^{2B}\prod_{\ell=1}^{B}\left(1 + d_{\ell j} + \lambda_\ell t\right)\right]^{-1}\right). \tag{6.54}$$

Proof: To begin the proof, a simpler version of the equation is studied. Defining $\tilde{b}(t)$ as $b(t)$ with $d_{\ell j}$ changed to be d_ℓ, we have

$$\tilde{b}(t) = \ln\left(\sum_j p_z(z_j)z_j^{-2B}\right) - \sum_{\ell=1}^{B}\ln\left(1 + d_\ell + \lambda_\ell t\right). \tag{6.55}$$

It can be shown that (6.55) is convex in t [32]. Thus, by mapping (6.54) to (6.55) the convexity follows. This is shown by using the following definition of d_ℓ that depends on t

$$d_\ell(t) = \frac{1}{\sum_j \beta_j [1 + d_{\ell j} + \lambda_\ell t]^{-1}} - (1 + \lambda_\ell t), \tag{6.56}$$

where $\beta_j = \frac{p_z(z_j)}{z_j^{2D}}\left[\sum_j \frac{p_z(z_j)}{z_j^{2D}}\right]^{-1}$. It can be shown that substituting $d_\ell(t)$ for d_ℓ in (6.55) gives $h(t) = \tilde{h}(t)$; thus, the two expressions are equivalent. Therefore, $\tilde{h}(t)$ can be rewritten as

$$\tilde{b}(t) = \ln\left(\sum_j p_z(z_j)z_j^{-2D}\right) - \sum_{\ell=1}^{B}\ln\left(1 + d_\ell(t) + \lambda_\ell t\right). \tag{6.57}$$

Since it can be verified that $d_\ell(t) \geq 0$, $d'_\ell(t) \geq 0$ and $d''_\ell(t) \leq 0$, by examining the second term of (6.57) and Lemma 6.2 the full expression is convex.

Lemma 6.4 The function, $f(\mathbf{B}) : \mathbb{R}^{B\times B} \to \mathbb{R}$ given as

$$f(\mathbf{B}) = -\sum_i p_z(z_i) \ln\left(\sum_j p_z(z_j)\left[\det\left(\mathbf{I} + (z_i^2 + z_j^2)\mathbf{B}\right)\right]^{-1}\right), \tag{6.58}$$

is concave for a positive-semidefinite **B**.

Proof: To prove concavity, the convexity of $-f(\mathbf{B})$ is proven. Since a non-negative weighted sum of convex functions is convex, it remains to show the convexity of

$$\eta(\mathbf{B}) = \ln\left(\sum_j p_z(z_j)\left[\det\left(\mathbf{I} + (z_i^2 + z_j^2)\mathbf{B}\right)\right]^{-1}\right). \tag{6.59}$$

For simplicity, we let $\tilde{z}_j^2 = z_i^2 + z_j^2$ as it is now assumed that the index i is fixed. To prove the convexity of this expression, a commonly used method of checking the convexity using one variable is employed. This is done by seeing if $\eta(\mathbf{X} + t\mathbf{V})$ is convex in t, for any positive semidefinite matrices **X** and **V**. Expanding out $\eta(\mathbf{X} + t\mathbf{V})$ results in

$$\begin{aligned}
\eta(\mathbf{B} + t\mathbf{V}) &= \ln\left(\sum_j p_z(z_j)\left[\det\left(\mathbf{I} + \tilde{z}_j^2(\mathbf{B} + t\mathbf{V})\right)\right]^{-1}\right) \\
&= -\ln\det(\mathbf{B}) + \ln\left(\sum_j p_z(z_j)\left[\det\left(\mathbf{B} + \tilde{z}_j^2(\mathbf{I} + t\mathbf{B}^{-1/2}\mathbf{V}\mathbf{B}^{-1/2})\right)\right]^{-1}\right) \\
&= -\ln\det(\mathbf{B})\ln\left(\sum_j p_z(z_j)\left[\prod_{\ell=1}^{D}\left(\tilde{z}_j^2(1 + \lambda_l t) + \lambda_{x,\ell}\right)\right]^{-1}\right),
\end{aligned} \tag{6.60}$$

where $\lambda_l \geq 0$ and $\lambda_{x,\ell} \geq 0$ are l-th the eigenvalues of $\mathbf{B}^{-1/2}\mathbf{V}\mathbf{B}^{-1/2}$ and **B**, respectively. Pulling out the $\tilde{z}_j^2$ term from the product results in

$$\eta(\mathbf{B} + t\mathbf{V}) = -\ln\det(\mathbf{B}) + \ln\left(\sum_j p_z(z_j)\tilde{z}_j^{-2D}\left[\prod_{\ell=1}^{D}\left(1 + \lambda_l t + \frac{\lambda_{x,\ell}}{\tilde{z}_j^2}\right)\right]^{-1}\right), \tag{6.61}$$

where $d_{\ell,j} = \frac{\lambda_{x,\ell}}{\tilde{z}_j^2}$. Therefore, the second term is of the form as shown in Lemma 6.3 and is convex in t.

Therefore, as per Lemma 6.4 it can be seen that the lower bound is convex.

In Section 6.6, we will present the application of the bounds to the SAR image shown in Figure 6.3. Specifically, we will present performance evaluations on image reconstruction using the optimized waveforms within a multistatic radar image architecture and compare the results with a linear frequency modulated (LFM) waveform and a simple rectangular pulse.

6.6 Application of Bounds

The upper bound is optimized using convex techniques. As the cost function (6.40) is convex, the solution is guaranteed to reach a global optimum using convex optimization techniques. We will show both the result of waveform optimization using the lower and upper bounds and present the numerical results of applying the waveforms on a test scene.

We consider a sensor geometry in which the possible sensor locations lie on a semicircle of radius $R = 2$ around the scene shown in Figure 6.3. Therefore, we generate M locations on the semicircle to form the multistatic aperture, where M is instantiated in Section 6.6.2 and 6.6.3. We compare our waveforms to both an LFM waveform with an equal bandwidth, and a rectangular pulse. We measure the degree of improvement over other waveforms using the mean square error (MSE) between the ground truth scene and the reconstructed scene corresponding to each transmit waveform.

6.6.1 Scene Reconstruction

As is shown in (6.8), the vectorized scene reflectively function can be decomposed into the product of a sparse basis set and a sparse vector of wavelet coefficients modeled as a CG vector. To test the optimized waveforms, the signals are transmitted to the scene shown in Figure 6.3. We use a sparsity-driven regularization radar imaging technique to estimate the sparse vector of wavelet coefficients $\hat{\mathbf{c}}$ that make up the scene reflectivity function [33]. Upon the estimation of $\hat{\mathbf{c}}$, we reconstruct the final scene by projecting $\hat{\mathbf{c}}$ onto the wavelet basis $\boldsymbol{\Phi}$, as shown in (6.8), to provide an estimate of the scene $\tilde{\mathbf{g}}_{\text{est}}$.

To perform scene reconstruction, we first define the transmit waveform convolution matrix $\mathbf{X}$, from either the optimized waveform or any other desired waveform under test. For the signal model shown in (6.13), the sparse image vector $\mathbf{c}$ is the only unknown and is estimated using the least squares cost function shown in (6.14) with $p = 1$, given as

$$\hat{\mathbf{c}} = \arg\min_{c \in \mathbb{R}} \left\| \mathbf{y} - \overline{\mathbf{X}}\boldsymbol{\Psi}\mathbf{c} \right\|_2^2 + \alpha \left| \mathbf{c} \right| . \tag{6.62}$$

The truncated Newton interior-point method is used to solve (6.62) [34]. Once the coefficients $\hat{\mathbf{c}}$ are estimated, the received scene reflectivity function is $\tilde{\mathbf{g}}_{\text{est}} = \boldsymbol{\Phi}\hat{\mathbf{c}}$. $\alpha = 1/2$ is used in the simulations. Note that the regularization term, $\alpha \left| \mathbf{c} \right|$, forces the scene reconstruction to be sparse.

6.6.2 Scene Reconstruction Performance

The upper bound on the MI shown in (6.40) is optimized to produce a time-domain waveform that maximizes the MI between the scene and the received signal at the radar. The optimized waveform is shown in Figure 6.7.

The LFM waveform and the rectangular pulse used for comparison, as respectively shown in Figures 6.8 and 6.9, are generated using the same waveform length and bandwidth as the optimized waveform. In all our simulations, we keep the parameters of the waveforms consistent. Specifically, we consider each waveform to be of length $N = 31$ with average power $\rho = 62$, over a pulse duration $T = 1 \times 10^{-6}$ seconds, and a bandwidth of 400 MHz.

We consider two multistatic radar imaging scenarios. In the first case, the multistatic radar has sensors placed randomly around an area of interest. We considered $M = 15$ randomly placed sensors at a radius r from the center of the scene. In the

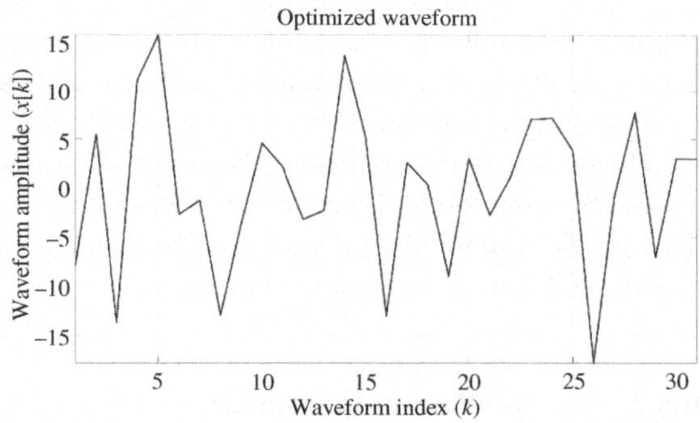

Figure 6.7 The optimized waveform formed by finding the optimal solution to (6.40).

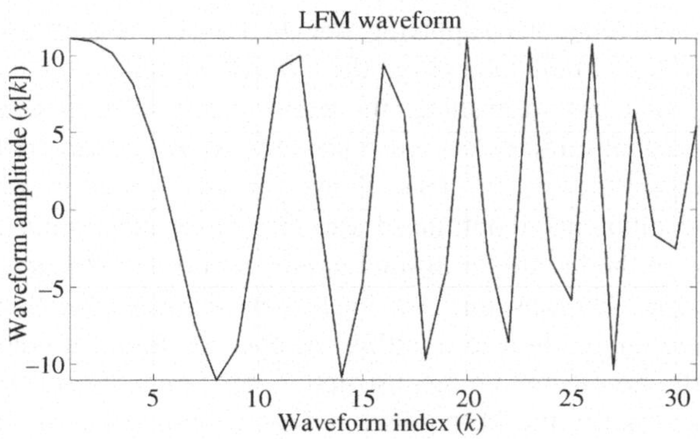

Figure 6.8 LFM waveform used for comparison.

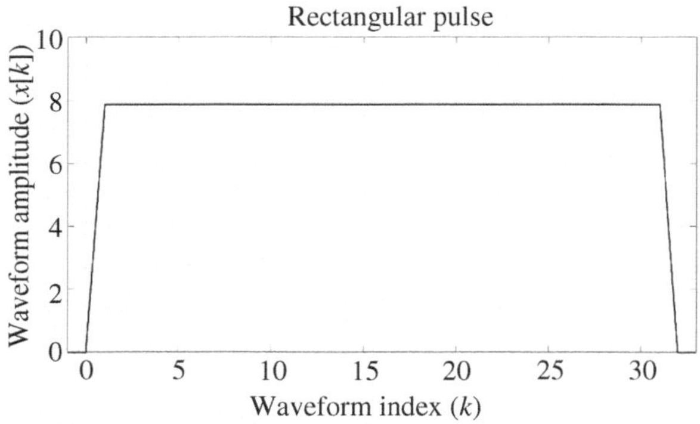

Figure 6.9 Rectangular pulse used for comparison.

second case, $M = 5$ sensors are considered. In each multistatic case, each sensors transmits sequentially, while all M sensors receive.

In order to evaluate the performance of each waveform, we run Monte Carlo trials for randomly extracted patches in the scene. Each extracted patch is of size 8×8 to estimate the CG parameters as shown in Section 6.3.1. With the estimated CG distribution, the waveform is optimized and transmitted onto each patch with an added noise with variance σ_n^2. The reconstruction performance of that patch is evaluated using the MSE between the ground truth and the reconstructed scene. This provides bounds on the average performance of each waveform when it is transmitted from a multistatic radar system.

6.6.3 Random Sensor Placement

We compare the optimized waveform with the LFM waveform and a rectangular pulse function by applying them to the scene in Figure 6.3 which is a sample image of a T-72 tank taken from the MSTAR dataset.

In order to evaluate the performance of each waveform as the noise level gradually increases, we repeat the Monte Carlo trials for each waveform for $\sigma_n^2 = [0.1, 1, 5, 10]$. This allows for an evaluation of the performance of the traditional and optimized waveforms over various operating conditions.

In the Monte Carlo simulations, we keep the parameters of each waveform the same. The waveform parameters are mentioned in Section 6.6.2. $M = 15$ sensors are randomly placed within the aperture for each noise level. Figure 6.10 shows the average scene reconstruction performance after transmitting the optimized waveform, the LFM pulse, and the rectangular pulse for each noise level. Thousand Monte Carlo trails over random patches provide average metric values for each waveform. As can be seen in Figure 6.10, the optimized waveform outperforms both the LFM and the rectangular pulse waveform at each noise level. The

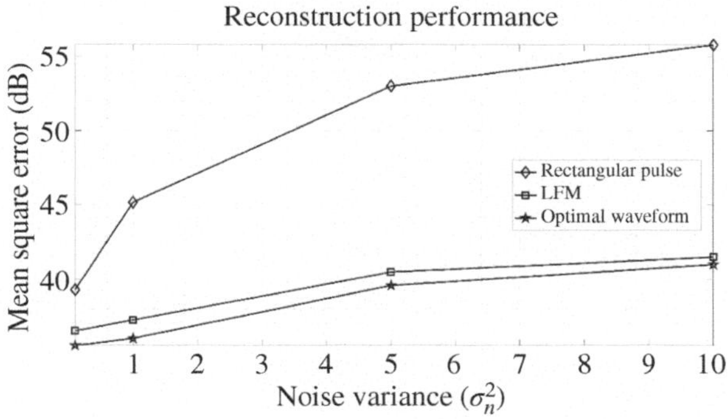

Figure 6.10 Image reconstruction performance with $M = 15$ randomly placed sensors.

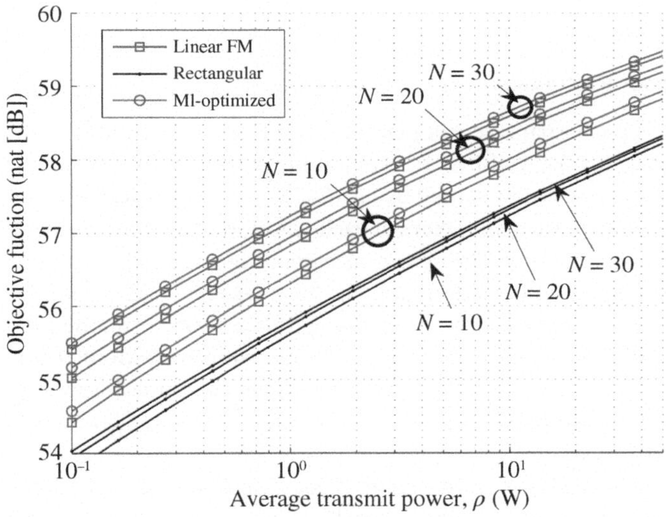

Figure 6.11 The objective function (6.40) evaluated for an optimal waveform and compared with an LFM waveform and rectangular pulse waveform. N denotes the waveform length and ρ the average power.

performance of the LFM waveform is close to the optimized waveform with the rectangular pulse having the poorest reconstruction error.

Figure 6.11 shows the performance of the objective function (6.40) for varying power levels ρ and waveform lengths, N, where $M = 15$. We compare the objective function evaluated for the optimized waveform, the LFM waveform, and the rectangular pulse. As can be seen in Figure 6.11, the rectangular pulse

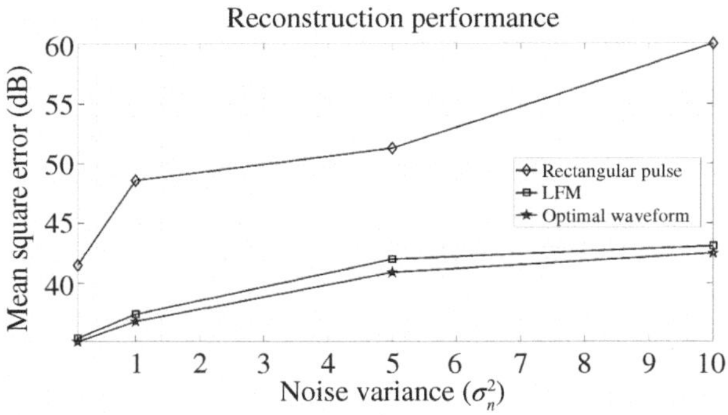

Figure 6.12 Image reconstruction performance with $M = 5$ sensors.

performance is poorly compared to the LFM, and the optimized waveform performs the best.

6.6.4 Sparse Sensor Placement

We now consider the case of sparse sensor placement. Compared to the multistatic radar considered in Section 6.6.3 with $M = 15$ randomly placed sensors within the aperture, the area of interest is now sensed using $M = 5$ uniformly placed sensors. As in Section 6.6.3, we keep the parameters of each waveform the same throughout the Monte Carlo simulations. The sparse placement of the smaller number of sensors limits the angular information available for image reconstruction. The average MSE of the reconstructed scene using the optimized waveform, the LFM pulse, and a rectangular pulse is shown in Figure 6.12, using 1000 Monte Carlo trials. As can be seen in Figure 6.12, the optimized waveform outperforms both the LFM and the rectangular pulse by returning lower reconstruction errors over varying degrees of noise variance.

The optimized waveform shown in this chapter does not discriminate between undesirable scene effects such as "clutter," and the target of interest. In our work [2], we present a waveform that incorporates knowledge of clutter and target characteristics for waveform optimization.

6.7 Conclusion

In this chapter, we presented a formulation for optimizing the waveform for a multistatic radar that maximizes the MI between the received signal and the scene of interest in the presence of noise. We place a prior distribution over the scene as a means of exploiting the underlying structure of natural scenes. The CG distribution is shown to model image statistics of natural scenes well when projected onto a wavelet basis. Thus, a sparse prior can be placed over the wavelet coefficients of natural scenes to optimize a waveform.

In order to optimize the waveform, the MI criterion is used. We model the effects of the scene as an LTI system, and the core challenge becomes determining the upper limit of the channel's capacity. We assess the change in information between the input and output using information-theoretic metrics. Therefore, to extract scene information we maximize the MI between the quantity of interest and the received waveform, given a transmit waveform.

We derived the upper and lower bounds on the MI that can be optimized to find an optimal transmit signal. We observe that by optimizing the bounds on the MI the average image reconstruction performance of the optimized waveform has a lower MSE than traditional waveforms such as the LFM waveform and the rectangular pulse. Monte Carlo results show that the optimized waveforms perform better over varying noise levels for both random and uniform sensor placement scenarios.

References

1 Gini, F., De Maio, A., and Patton, L. (ed.) (2012). *Waveform Design and Diversity for Advanced Radar Systems*. Stevenage: The Institution of Engineering and Technology.

2 Idriss, Z., Raj, R.G., and Narayanan, R.M. (2021). Waveform optimization for multistatic radar imaging using mutual information. *IEEE Transactions on Aerospace and Electronic Systems* 57(4): 2410–2425.

3 Sangston, K.J. and Farina, A. (2016). Coherent radar detection in compound-Gaussian clutter: clairvoyant detectors. *IEEE Aerospace and Electronic Systems Magazine* 31(11): 42–63.

4 Idriss, Z., Raj, R.G., and Narayanan, R.M. (2019). A compound Gaussian-based waveform design approach for enhanced target detection in multistatic radar imaging. *Proceedings of SPIE Conference on Radar Sensor Technology XXIII*, volume 11003. Baltimore, MD.

5 Sangston, K.J., Gini, F., Greco, M.V., and Farina, A. (1999). Structures for radar detection in compound Gaussian clutter. *IEEE Transactions on Aerospace and Electronic Systems* 35(2): 445–458.

6 Chance, Z., Raj, R.G., and Love, D.J. (2011). Information-theoretic structure of multistatic radar imaging. *Proceedings of 2011 IEEE Radar Conference (RADAR)*, 853–858. Kansas City, MO.

7 Wainwright, M.J. and Simoncelli, E. (1999). Scale mixtures of Gaussians and the statistics of natural images. *Advances in Neural Information Processing Systems* 12: 855–861.

8 Bell, M.R. (1993). Information theory and radar waveform design. *IEEE Transactions on Information Theory* 39(5): 1578–1597.

9 Guo, D., Shamai, S., and Verdú, S. (2005). Mutual information and minimum mean-square error in Gaussian channels. *IEEE Transactions on Information Theory* 51(4): 1261–1282.

10 Romero, R.A., Bae, J., and Goodman, N.A. (2011). Theory and application of SNR and mutual information matched illumination waveforms. *IEEE Transactions on Aerospace and Electronic Systems* 47(2): 912–927.

11 Leshem, A., Naparstek, O., and Nehorai, A. (2007). Information theoretic adaptive radar waveform design for multiple extended targets. *IEEE Journal of Selected Topics in Signal Processing* 1(1): 42–55.

12 Tang, B. and Li, J. (2019). Spectrally constrained MIMO radar waveform design based on mutual information. *IEEE Transactions on Signal Processing* 67(3): 821–834.

13 Wang, L., Wang, L., Zeng, Y., and Wang, M. (2017). Jamming power allocation strategy for MIMO radar based on MMSE and mutual information. *IET Radar, Sonar and Navigation* 11(7): 1081–1089.

14 Farina, A. and Studer, F.A. (1991). Detection with high resolution radar - great promise, big challenge. *Microwave Journal* 34: 263–273.

15 Yang, Y. and Blum, R. (2007). MIMO radar waveform design based on mutual information and minimum mean-square error estimation. *IEEE Transactions on Aerospace and Electronic Systems* 43(1): 330–343.

16 Gu, Y. and Goodman, N.A. (2019). Information-theoretic waveform design for Gaussian mixture radar target profiling. *IEEE Transactions on Aerospace and Electronic Systems* 55(3): 1528–1536.

17 Xin, F., Wang, B., Xu, Z., and Chen, X. (2019). A novel algorithm for optimal waveform design based on dual mutual information for radar systems. *Journal of Physics: Conference Series* 1324: 012022.

18 Bica, M., Huang, K., Koivunen, V., and Mitra, U. (2016). Mutual information based radar waveform design for joint radar and cellular communication systems. *2016 IEEE International Conference on Acoustics, Speech and Signal Processing (ICASSP)*, 3671–3675. Shanghai, China.

19 Zhu, Z., Kay, S., and Raghavan, R.S. (2017). Information-theoretic optimal radar waveform design. *IEEE Signal Processing Letters* 24(3): 274–278.

20 Jakowatz, C.V. Jr., Wahl, D.E., Eichel, P.H. et al. (1996). *Spotlight-Mode Synthetic Aperture Radar: A Signal Processing Approach*. New York: Kluwer Academic Publishers.

21 Crispin, J.W., Goodrich, R.F., and Siegel, K.M. (1959). A Theoretical Method for the Calculation of Radar Cross Section of Aircraft and Missiles, Rep. 2591-1-H. Ann Arbor, MI: University of Michigan Radiation Laboratory.

22 Willis, N.J. (2005). *Bistatic Radar*. Raleigh, NC: SciTech Publishing.

23 Kell, R.E. (1965). On the derivation of bistatic RCS from monostatic measurements. *Proceedings of the IEEE* 53(8): 983–988.

24 Lyons, C., Raj, R.G., and Cheney, M. (2023). A compound Gaussian least squares algorithm and unrolled network for linear inverse problems. *IEEE Transactions on Signal Processing* 71: 4303–4316.

25 Portilla, J., Strela, V., Wainwright, M., and Simoncelli, E. (2003). Image denoising using scale mixtures of Gaussians in the wavelet domain. *IEEE Transactions on Image Processing* 12(11): 1338–1351.

26 Wainwright, M., Simoncelli, E., and Willsky, A. (2001). Random cascades on wavelet trees and their use in analyzing and modeling natural images. *Applied and Computational Harmonic Analysis* 11(1): 89–123.

27 Raj, R.G. (2016). A hierarchical Bayesian-map approach to inverse problems in imaging. *Inverse Problems* 32(7): 075003.

28 Zetzsche, C., Barth, E., and Wegmann, B. (1993). Nonlinear aspects of primary vision: entropy reduction beyond decorrelation. *Proceedings of SID*, volume XXIV (ed. J. Morreale), 933–936. Society for Information Display.

29 Mallat, S. (2008). *A Wavelet Tour of Signal Processing: The Sparse Way*. Elsevier Science.

30 Cover, T.M. and Thomas, J.A. (2006). *Elements of Information Theory, 2e*. Hoboken, NJ: Wiley-Interscience.

31 Huber, M.F., Bailey, T., Durrant-Whyte, H., and Hanebeck, U.D. (2008). On entropy approximation for Gaussian mixture random vectors. *Proceedings of IEEE International Conference on Multisensor Fusion and Integration for Intelligent Systems*, 181–188. Seoul, South Korea.

32 Boyd, S. and Vandenberghe, L. (2004). *Convex Optimization*. Cambridge: Cambridge University Press.

33 Fang, Y., Cheney, M., and Roecker, S. (2010). Multistatic radar imaging from sparse measurements. *Proceedings of 2010 IEEE Radar Conference*, 485–490. Washington, DC.

34 Kim, S., Koh, K., Lustig, M. et al. (2007). An interior-point method for large-scale ℓ_1-regularized least squares. *IEEE Journal of Selected Topics in Signal Processing* 1(4): 606–617.

7

Statistical Information Theory in SAR and PolSAR Image Analysis

Alejandro C. Frery[1] *and Abraão D. C. Nascimento*[2]

[1] *School of Mathematics and Statistics, Victoria University of Wellington, Wellington, New Zealand*
[2] *Departamento de Estatística, Universidade Federal de Pernambuco, Recife, Brazil*

7.1 Introduction

The goal of this chapter is to propose a way to unify two areas, namely (i) the processing and analysis of synthetic aperture radar (SAR) and polarimetric SAR (PolSAR) images and (ii) information-theoretic tools. We do this by presenting seemingly unrelated problems and showing that they can be successfully addressed within the same framework: Tools from the application of information theory, complemented by statistics.

Society relies on the information provided by remote sensing. Solutions to critical problems are related to remotely sensed data, e.g., oil spill detection [1] and forest fire detection [2]. Microwave remote sensing is one of the most important sources of information. Active microwave sensors have as main advantages the following characteristics: they operate independent of the time of day or weather conditions, they are capable of providing high spatial image resolution, and they provide a variety of information about the scene.

Such sensors send electromagnetic pulses with different polarizations to a typically distant target; then, the returning echoes are detected and registered. This process is characterized by the use of coherent illumination, which adds a grainy noise to the images called "speckle." This noise is multiplicative in nature and its intensity does not follow the Gaussian law. Therefore, special modeling is required for the processing and analysis of such images.

There is extensive literature on techniques for SAR image processing and analysis that use a variety of approaches to solve relevant problems. Among others, advances in modeling with statistical information theory have led to several interesting solutions to the problems of edge detection, noise reduction,

Information-Theoretic Radar Signal Processing, First Edition. Edited by Yujie Gu and Yimin D. Zhang.

classification, parameter estimation, change detection, and image enhancement procedures. In addition, these approaches provide unique insights into the properties of the data and the information they convey.

The basic idea of information theory was developed by Hartley [3], who introduced a logarithmic measure of information in communication. The terms "information" and "entropy" were formally defined mathematically by Shannon [4] in the context of data communication. Since then, the formulation and application of information and entropy measures have become an active area of research in various fields. Let us mention a few that are important in the context of this chapter. Nadarajah and Zografos [5, 6] have derived closed formulas for Shannon and Rényi entropies for various univariate, bivariate, and multivariate distributions. Among these information-theoretic tools, Shannon entropy was applied to PolSAR images. In the context of PolSAR images, Morio et al. [7] analyzed such an entropy to characterize polarimetric targets under the complex, circular, and multidimensional Gaussian distribution. This measure was compared with the von Neumann entropy for classification purposes [8, 9].

After its beginnings, the entropy concept evolved into the concept of divergence developed by Kullback–Leibler [see Ref. [10], for details]. Recently, Basseville [11] presented an important and complete survey of divergence measures.

An image can be understood as a set of regions. The pixel values in each region are observations of random variables that follow a certain distribution. Therefore, stochastic dissimilarity measures can be used as image processing tools, as they are able to evaluate the difference between distributions describing different image regions [12]. For a systematic and comprehensive treatment of dissimilarity measures, see Ali and Silvey [13] and Csiszár [14], and as a result, the class of h-ϕ-divergences was proposed by Salicrú et al. [15]. We will use the notation "υ–ϕ" in this text.

In this chapter we try to achieve the goal of linking information theory with the processing of SAR and PolSAR data. The remainder of this chapter is organized as follows. Section 7.2, where we build *ab initio* the statistical properties of SAR and PolSAR images, is grounded in physical aspects of the interaction between matter and electromagnetic waves. In Section 7.3, we apply elements of information theory to problems encountered in processing and understanding SAR and PolSAR images. In Section 7.4, we illustrate the use of information-theoretic-based tools for contrast analysis, edge detection, noise reduction, and classification.

7.2 Statistical Models for SAR and PolSAR Imagery

Active microwave sensors emit signals in the direction of the target and capture the returning information. The spatial resolution of such microwave imaging devices depends on the antenna aperture, and to obtain satisfactory results, they often simulate an artificially large aperture, hence the term "synthetic." The reader is referred to the books by Ulaby and Long [16] and Frery et al. [[17], Chapter 1] for more details.

Suppose the radar beam illuminates an area with N point-like scatterers that return part of the incident energy. The returned complex-valued field from this area is

$$S = \sum_{n=1}^{N} S_n = \sum_{n=1}^{N} A_n \exp\{i\phi_n\} = \underbrace{\sum_{n=1}^{N} A_n \cos\phi_n}_{\Re(S)} + \mathrm{i}\underbrace{\sum_{n=1}^{N} A_n \sin\phi_n}_{\Im(S)}, \tag{7.1}$$

where S_n is the complex envelope representing the echo from the nth scatterer, which has an amplitude A_n and phase ϕ_n, and $\mathrm{i} = \sqrt{-1}$ is the imaginary unit, and $\Re(\cdot)$ and $\Im(\cdot)$ denote the real and imaginary parts, respectively. We need to be aware of each pair (A_n, ϕ_n) for every pixel to know S. This is not feasible except under laboratory conditions. In practice, we must rely on physically based statistical assumptions that lead to expressive models.

According to Yue et al. [18], $\Re(S)$ and $\Im(S)$ are independent and identically distributed $\mathcal{N}(0, \sigma^2/2)$ random variables. An immediate consequence is that the amplitude, i.e., the norm $\|S\| = \sqrt{SS^*}$, where $(\cdot)^*$ denotes the conjugate operator, and the intensity $Z' = \|S\|^2$ of this signal are Rayleigh and exponentially distributed, respectively. The average (or mean) intensity over L looks from a target is calculated by

$$Z = \frac{1}{L}\sum_{\ell=1}^{L} Z'_\ell, \tag{7.2}$$

where Z'_ℓ denotes the intensity of the ℓth look.

If the number of elementary backscatterers N among the resolution cells changes, it must be described as a random variable. Thus, the return is no longer Gaussian. Delignon and Pieczynski [19] discussed the single-look case. In the following, we discuss this derivation for the multilook case. Specifically, we assume that:

(i) The random number of scatterers N in each elementary cell follows a Poisson distribution.
(ii) Its expected value $\mathbb{E}(N)$ is a continuous random variable Λ with density f.
(iii) The conditional model $Z \mid \Lambda = \lambda_0$ (for Λ as a random variable for the Poisson parameter) follows for λ_0, which is large enough, a gamma distribution with shape L and scale $(\lambda_0\sigma^2)^{-1}L$, so that σ^2 is the common variance of the amplitude of the individual scatterers.

(iv) The density of the unconditional intensity Z is

$$\begin{aligned} g(x) &= \int_0^\infty f_{Z|\Lambda=\lambda_0}(x) f(\lambda_0) d\lambda_0 \\ &= \int_0^\infty \frac{\left[L/(\lambda_0\sigma^2)\right]^L x^{L-1}}{\Gamma(L)} \exp\left\{-\frac{Lx}{\lambda_0\sigma^2}\right\} f(\lambda_0) d\lambda_0 \\ &= \int_0^\infty \frac{1}{\lambda_0} \underbrace{f_{\Gamma(L,L/\sigma^2)}(x/\lambda_0)}_{\text{unidimensional speckle}} \underbrace{f_X(\lambda_0)}_{\text{backscatter}} d\lambda_0. \end{aligned} \tag{7.3}$$

The last equation represents the density of a random variable defined by the products of two other independent random variables. Physically, one follows the gamma distribution denoted as $\Gamma(L, L/\sigma^2)$, and the other is used to account for the speckle variability created by the target. This approach is called "the multiplicative model," to which we will return later to extend the discussion to its polarimetric counterpart. This law is also known as compound Gaussian model.

The most commonly used distribution for the backscatter X is the reciprocal gamma law, $X \sim \Gamma^{-1}(-\alpha, (-\alpha-1)\mu)$, whose probability density function is

$$\begin{aligned} f_X(x; \mu, \alpha) = \frac{1}{\Gamma(-\alpha)} \frac{1}{(-\alpha-1)\mu} \left[\frac{(-\alpha-1)\mu}{x}\right]^{\alpha+1} \\ \exp\left[-\frac{(-\alpha-1)\mu}{x}\right] \mathbb{1}_{(0,\infty)}(x), \end{aligned} \tag{7.4}$$

where $\mathbb{1}_C(x)$ is the indicator function in the set C, $\Gamma(x) = \int_0^\infty x^{a-1} e^{-x} dx$ and $\mu > 0$ and $\alpha < 0$ represent scale and shape parameters, respectively.

PolSAR systems register amplitude and phase of the backscattered signals from linearly received and transmit combinations, resulting in four polarization channels: S_{HH}, S_{HV}, S_{VH}, and S_{VV} (H for horizontal polarization and V for vertical polarization). If the conditions of the reciprocity theorem [[16], Chapter 3] are satisfied, then $S_{HV} = S_{VH}$. Thus, the *single-look PolSAR* data is given by:

$$\begin{bmatrix} S_{HH} & S_{HV} \\ S_{HV} & S_{VV} \end{bmatrix}. \tag{7.5}$$

Data of the form (7.2) has low signal-to-noise-ratio. The *multilook PolSAR* processing reduces the noise by computing the following average:

$$\boldsymbol{S} = \sum_{n=1}^{N} \frac{1}{L} \sum_{\ell=1}^{L} \begin{bmatrix} S_{HH,n}^{(\ell)} \\ S_{HV,n}^{(\ell)} \\ S_{VV,n}^{(\ell)} \end{bmatrix} \left[\left(S_{HH,n}^{(\ell)}\right)^* \ \left(S_{HV,n}^{(\ell)}\right)^* \ \left(S_{VV,n}^{(\ell)}\right)^* \right], \tag{7.6}$$

where $S_{B,n}^{(\ell)} \in \mathbb{C}$ is the scattering in channel $B \in \{HH, HV, VV\}$ at the nth individual scatterer and ℓth look, and $\left(S_{B,n}^{(\ell)}\right)^*$ is its complex conjugate. Based on the results of Goodman [20] and Hagedorn et al. [21], we have evidence that the scaled complex Wishart distribution can adequately represent the PolSAR return in homogeneous scenarios.

Taking the multiplicative model as the generator mechanism for PolSAR distributions, each returned matrix associated with an image entry is the product of two independent random variables describing the terrain and speckle influences. For example, consider the m-dimensional scaled complex Wishart model (say $\boldsymbol{Y} \sim \mathcal{CW}_m$) discussed by Nascimento et al. [22] to describe the multilook multidimensional speckle noise for m polarization channels with density $f_{\boldsymbol{Y}}$ and $X \in \mathbb{R}^+$ as a random variable for the terrain with density f_X. Then, the return model $\boldsymbol{Z} = \boldsymbol{Y}X$ holds from the general expression [23]

$$f_{\boldsymbol{Z}}(\boldsymbol{Z}) = \int_0^\infty x^{-m^2} \underbrace{f_{\boldsymbol{Y}}(\boldsymbol{Z}/x)}_{\text{Multilook speckle}} \underbrace{f_X(x)}_{\text{Backscatter}} \, dx, \tag{7.7}$$

where $\boldsymbol{Z} \in \boldsymbol{\Omega}_+ := \{\boldsymbol{Z} \in \mathbb{C}^{m\times m} : \boldsymbol{Z} = \boldsymbol{Z}^{\mathrm{H}}\}$ and $(\cdot)^{\mathrm{H}}$ is the conjugate transpose or Hermitian operator. Equation (7.7) represents a general density of a distribution for multilook PolSAR data. From the work of Khan and Guida [24], we can see that in the single-look complex case, this expression becomes

$$f_{\boldsymbol{z}}(\boldsymbol{z}) = \int_0^\infty x^{-m} \underbrace{f_{\boldsymbol{y}}(\boldsymbol{z}/\sqrt{x})}_{\text{Single-look speckle}} f_X(x) dx, \tag{7.8}$$

where $\boldsymbol{y}$ is a random vector that follows the complex normal law and $\boldsymbol{z} \in \mathbb{C}$.

7.3 SIT: Statistical Information Theory

In Sections 7.3.1 and 7.3.2, we introduce distance and entropy measures embedded in their statistical frameworks that can be used to solve problems related to PolSAR modeled images. We will use the $\mathcal{G}_I^0$ and the scaled complex Wishart laws as basic distributions to illustrate their use in SAR and PolSAR data, respectively. The idea is to first consider the derived υ–ϕ-statistics as preprocessing tools for the analysis of PolSAR images and then apply them in methods for boundary detectors.

7.3.1 Entropy Measures

Let $f_{\boldsymbol{Z}}(\boldsymbol{Z}; \boldsymbol{\theta})$ be a probability density function with parameter vector $\boldsymbol{\theta}$ characterizing the distribution of the random matrix $\boldsymbol{Z} \in \boldsymbol{\Omega}_+$. The (υ, ϕ)-entropy relative to $\boldsymbol{Z}$ is defined by

$$h_\phi^\upsilon(\boldsymbol{\theta}) = \upsilon\left(\int_A \phi\left(f_{\boldsymbol{Z}}(\boldsymbol{Z}; \boldsymbol{\theta})\right) d\boldsymbol{Z}\right), \tag{7.9}$$

Table 7.1 (v, ϕ)-distances and their functions.

(v, ϕ)-distance	$v(y)$	$\phi(x)$
Shannon [25]	y	$-x \log x$
Restricted Tsallis (order $\beta \in \mathbb{R}_+ : \beta \neq 1$) [26]	y	$\frac{x^\beta - x}{1-\beta}$
Rényi (order $\beta \in \mathbb{R}_+ : \beta \neq 1$) [27]	$\frac{\log y}{1-\beta}$	x^β

where $\mathcal{A} \subseteq \mathbf{\Omega}_+$ and either $\phi : [0, \infty) \to \mathbb{R}$ concave and $v : \mathbb{R} \to \mathbb{R}$ is increasing, or ϕ is convex and v is decreasing. The differential element $d\boldsymbol{Z}$ is given by

$$d\boldsymbol{Z} = \prod_{i=1}^{m} dZ_{ii} \underbrace{\prod_{i,j=1}^{m}}_{i<j} d\Re\{Z_{ij}\} d\Im\{Z_{ij}\}, \tag{7.10}$$

where Z_{ij} is the (i,j)th entry of the matrix $\boldsymbol{Z}$ [20]. Table 7.1 shows the specification of the three entropies: Shannon, Rényi, and restricted Tsallis.

The following result derived by Pardo et al. [28] paves the way for the proposal of asymptotic statistical inference methods based on entropy.

Lemma 7.1 ***Convergence in distribution*** Let $\widehat{\boldsymbol{\theta}} = [\widehat{\theta_1}, \widehat{\theta_2}, \dots, \widehat{\theta_p}]^{\mathrm{T}}$ be the maximum likelihood (ML) estimator of the parameter vector $\boldsymbol{\theta} = [\theta_1, \theta_2, \dots, \theta_p]^{\mathrm{T}}$ based on a random sample of size N under the model $f(\boldsymbol{Z}; \boldsymbol{\theta})$, where $(\cdot)^{\mathrm{T}}$ is the transpose operator. Then

$$\sqrt{N}\left[h_v^\phi(\widehat{\boldsymbol{\theta}}) - h_v^\phi(\boldsymbol{\theta})\right] \xrightarrow[N\to\infty]{\mathcal{D}} \mathcal{N}(0, \sigma_h^2(\boldsymbol{\theta})), \tag{7.11}$$

where $\mathcal{N}(\mu, \sigma^2)$ is the Gaussian distribution with mean μ and variance σ^2, "$\xrightarrow{\mathcal{D}}$" denotes convergence in distribution,

$$\sigma_h^2(\boldsymbol{\theta}) = \boldsymbol{\delta}^{\mathrm{T}}\left[\boldsymbol{J}(\boldsymbol{\theta})\right]^{-1}\boldsymbol{\delta}, \tag{7.12}$$

where $\boldsymbol{J}(\boldsymbol{\theta}) = \mathbb{E}\left[-\partial^2 \log f_{\boldsymbol{Z}}(\boldsymbol{Z}; \boldsymbol{\theta})/\partial\boldsymbol{\theta}^2\right]$ is the Fisher information matrix, and $\boldsymbol{\delta} = \left[\delta_1\ \delta_2\ \cdots\ \delta_p\right]^{\mathrm{T}}$ such that $\delta_i = \partial h_v^\phi(\boldsymbol{\theta})/\partial\theta_k$ for $k = 1, 2, \dots, p$.

In the following, we present a methodology for hypothesis testing and confidence intervals based on entropy.

Let $\boldsymbol{Z}_i$ be a positive-definite random matrix with probability density function defined over $\mathcal{A}$ with parameter vector $\boldsymbol{\theta}_k$ for $k = 1, 2, \dots, r$, where r is the number of populations to be studied. We are interested in testing the following hypotheses: For o being a prespecified entropy,

$$\begin{cases} \mathcal{H}_0 : h_v^{\phi}(\boldsymbol{\theta}_1) = h_v^{\phi}(\boldsymbol{\theta}_2) = \cdots = h_v^{\phi}(\boldsymbol{\theta}_r) = o, \\ \mathcal{H}_1 : h_v^{\phi}(\boldsymbol{\theta}_k) \neq h_v^{\phi}(\boldsymbol{\theta}_\ell) \text{ for some } k \text{ and } \ell. \end{cases} \tag{7.13}$$

In other words, statistical evidence is sought to assess whether at least one of the r regions of a PolSAR image has a different entropy than the other regions.

Salicrú et al. [25] have demonstrated an asymptotic distribution under the null hypothesis for statistics:

$$S_{\phi}^{v}(\widehat{\boldsymbol{\theta}}_1, \widehat{\boldsymbol{\theta}}_2, \dots, \widehat{\boldsymbol{\theta}}_r) = \sum_{k=1}^{r} \frac{N_k \left(h_v^{\phi}(\widehat{\boldsymbol{\theta}}_k) - \overline{o} \right)^2}{\sigma_h^2(\widehat{\boldsymbol{\theta}}_k)}, \tag{7.14}$$

where

$$\overline{o} = \left[\sum_{k=1}^{r} \frac{N_k}{\sigma_h^2(\widehat{\boldsymbol{\theta}}_k)} \right]^{-1} \sum_{k=1}^{r} \frac{N_k h_v^{\phi}(\widehat{\boldsymbol{\theta}}_k)}{\sigma_h^2(\widehat{\boldsymbol{\theta}}_k)}. \tag{7.15}$$

We are now in a position to state the following result:

- Let N_k, $k = 1, 2, \dots, r$, be sufficiently large. If $S_{\phi}^{v}(\widehat{\boldsymbol{\theta}}_1, \widehat{\boldsymbol{\theta}}_2, \dots, \widehat{\boldsymbol{\theta}}_r) = s$, then the null hypothesis $\mathcal{H}_0$ can be rejected at a level α if $\Pr\left(\chi_{r-1}^2 > s\right) \leq \alpha$.

7.3.2 Stochastic Distances

Suppose that $\boldsymbol{X}$ and $\boldsymbol{Y}$ are random matrices with densities $f_{\boldsymbol{X}}(\boldsymbol{Z}; \boldsymbol{\theta}_1)$ and $f_{\boldsymbol{Y}}(\boldsymbol{Z}; \boldsymbol{\theta}_2)$, where $\boldsymbol{\theta}_1$ and $\boldsymbol{\theta}_2$ are parameter vectors. The densities are assumed to have a common support given by the cone of Hermitian positive definite matrices $\mathcal{A}$. The (v, ϕ)-divergence between $f_{\boldsymbol{X}}$ and $f_{\boldsymbol{Y}}$ is defined by

$$D_{\phi}^{v}(\boldsymbol{X}, \boldsymbol{Y}) = v\left(\int_{\mathcal{A}} \phi\left(\frac{f_{\boldsymbol{X}}(\boldsymbol{Z}; \boldsymbol{\theta}_1)}{f_{\boldsymbol{Y}}(\boldsymbol{Z}; \boldsymbol{\theta}_2)} \right) f_{\boldsymbol{Y}}(\boldsymbol{Z}; \boldsymbol{\theta}_2) d\boldsymbol{Z} \right), \tag{7.16}$$

where $v : (0, \infty) \to [0, \infty)$ is a strictly increasing function with $v(0) = 0$ and $\phi : (0, \infty) \to [0, \infty)$ is a convex function such that $0\,\phi(0/0) = 0$ and $0\,\phi(x/0) = \lim_{x\to\infty} \phi(x)/x$.

Appropriate choices of v and ϕ yield known divergences. Among them, the following have been studied: (i) the Kullback–Leibler (KL) divergence [29], (ii) the Rényi divergence, (iii) the Bhattacharyya divergence [30], and (iv) the Hellinger divergence [12]. Since the triangular inequality is not necessarily satisfied, not every divergence measure is a metric [31]. Moreover, the symmetry property is not satisfied for some of these divergence measures. Nevertheless, such instruments are mathematically suitable for comparing the distribution of

Table 7.2 (υ, ϕ)-distances and their functions.

(υ, ϕ)-distance	$\upsilon(y)$	$\phi(x)$
Kullback–Leibler	$y/2$	$(x-1)\log x$
Rényi (order β)	$\frac{1}{\beta-1}\log((\beta-1)y+1)$, $0 \le y < \dfrac{1}{1-\beta}$	$\dfrac{x^{1-\beta}+x^{\beta}-\beta(x-1)-2}{2(\beta-1)}$, $0<\beta<1$
Bhattacharyya	$-\log(1-y)$, $0 \le y < 1$	$-\sqrt{x}+\dfrac{x+1}{2}$
Hellinger	$y/2$, $0 \le y < 2$	$(\sqrt{x}-1)^2$

random variables [32]. Therefore, the following expression has been proposed as a possible solution to this problem [29]:

$$d_{\phi}^{\upsilon}(\boldsymbol{X},\boldsymbol{Y}) = \frac{D_{\phi}^{\upsilon}(\boldsymbol{X},\boldsymbol{Y}) + D_{\phi}^{\upsilon}(\boldsymbol{Y},\boldsymbol{X})}{2}. \tag{7.17}$$

Functions $d_{\phi}^{\upsilon} : \mathcal{A} \times \mathcal{A} \to \mathbb{R}$ are distances over $\mathcal{A}$ since, for all $\boldsymbol{X}, \boldsymbol{Y} \in \mathcal{A}$, the following properties hold:

1. $d_{\phi}^{\upsilon}(\boldsymbol{X},\boldsymbol{Y}) \ge 0$ (Nonnegativity).
2. $d_{\phi}^{\upsilon}(\boldsymbol{X},\boldsymbol{Y}) = d_{\phi}^{\upsilon}(\boldsymbol{Y},\boldsymbol{X})$ (Symmetry).
3. $d_{\phi}^{\upsilon}(\boldsymbol{X},\boldsymbol{Y}) = 0 \Leftrightarrow \boldsymbol{X} = \boldsymbol{Y}$ (Definiteness).

Table 7.2 shows the functions υ and ϕ leading to the distances considered in this work.

The hypothesis test based on stochastic distances proposed by Salicrú et al. [15] is presented below. Let p-point real vectors $\widehat{\boldsymbol{\theta}}_1 = [\widehat{\theta}_{11}, \dots, \widehat{\theta}_{1p}]^{\mathrm{T}}$ and $\widehat{\boldsymbol{\theta}}_2 = [\widehat{\theta}_{21}, \dots, \widehat{\theta}_{2p}]^{\mathrm{T}}$ be the ML estimators of the parameters $\boldsymbol{\theta}_1$ and $\boldsymbol{\theta}_2$ based on independent samples of the quantities N_1 and N_2. Under regularity conditions discussed in Salicrú et al. [[15], p. 380], the following lemma holds.

Lemma 7.2 ***Convergence in distribution of test statistics*** If the sample sizes grow as $\frac{N_1}{N_1+N_2} \xrightarrow[N_1,N_2\to\infty]{} \lambda \in (0,1)$, and $\boldsymbol{\theta}_1 = \boldsymbol{\theta}_2$, then

$$S_{\phi}^{\upsilon}(\widehat{\boldsymbol{\theta}}_1, \widehat{\boldsymbol{\theta}}_2) = \frac{2N_1N_2}{N_1+N_2}\frac{d_{\phi}^{\upsilon}(\widehat{\boldsymbol{\theta}}_1, \widehat{\boldsymbol{\theta}}_2)}{\upsilon'(0)\phi''(1)} \xrightarrow[N_1,N_2\to\infty]{\mathcal{D}} \chi_p^2, \tag{7.18}$$

where χ_p^2 represents a random variable following the chi-squared distribution with p degrees of freedom.

Based on Lemma 7.2, statistical hypothesis tests for the null hypothesis $\boldsymbol{\theta}_1 = \boldsymbol{\theta}_2$ can be derived in the form:

- Let N_1 and N_2 be large and $S^v_\phi(\widehat{\boldsymbol{\theta}}_1, \widehat{\boldsymbol{\theta}}_2) = s$, then the null hypothesis $\boldsymbol{\theta}_1 = \boldsymbol{\theta}_2$ can be rejected at level α if $\Pr(\chi^2_p > s) \leq \alpha$.

7.4 Integrated View of SAR and PolSAR Data Analysis from SIT

7.4.1 Contrast Analysis for SAR Data

The multiplicative model at position (i,j) of an intensity SAR image is defined as the result of the random variables $Z(i,j) = X(i,j)Y(i,j)$, where $X(i,j)$ and $Y(i,j)$ are independent random variables. The latter, describing the speckle, follows a gamma distribution with mean one and a shape parameter $L \geq 1$. The number of looks, L, is proportional to the signal-to-noise-ratio and is often fixed for the whole image. The unobserved quantity of interest, $X(i,j)$, is called backscatter, is positive, and contains all relevant information about the target.

In describing the backscattering with a reciprocal gamma law, Frery et al. [33] proved that the density of Z is expressed by

$$f_Z(z; \alpha, \gamma, L) = \frac{L^L \Gamma(L-\alpha)}{\gamma^\alpha \Gamma(-\alpha)\Gamma(L)} z^{L-1}(\gamma + Lz)^{\alpha - L} \mathbb{1}_{(0,\infty)}(z), \tag{7.19}$$

where $\alpha < 0$ is texture and $\gamma > 0$ is brightness. We denote this situation as $Z \sim \mathcal{G}^0(\alpha, \gamma, L)$ with parameter space $\Theta = \mathbb{R}^- \times \mathbb{R}^+ \times [1, \infty)$.

Frery et al. [33] proved the following result. Consider the sequence of random variables $Z_1, Z_2, \ldots$, in which $Z_k \sim \mathcal{G}^0_I(\alpha_k, \gamma_k, L)$. If the sequence of parameters $-\alpha_k, \gamma_k \xrightarrow[k\to\infty]{} \infty$ is such that $-\alpha_i/\gamma_i \xrightarrow[i\to\infty]{} \beta_1$, then the following convergence of the distribution holds:

$$Z_k \xrightarrow[k\to\infty]{\mathcal{D}} Z_0, \tag{7.20}$$

where Z_0 follows a gamma distribution with mean β_1 and shape parameter L. In particular, when $L = 1$, the distribution converges to an exponential law.

Let us now briefly introduce the ML estimator for the parameters α and γ from SAR intensity data. Let $\boldsymbol{Z} = [Z_1, Z_2, \ldots, Z_N]^\mathrm{T}$ be an independent and identically distributed sample obtained from $Z \sim \mathcal{G}^0_I(\alpha, \gamma, L)$. The likelihood of the observed sample $\boldsymbol{z} = [z_1, z_2, \ldots, z_N]^\mathrm{T}$ is

$$\mathcal{P}(\alpha, \gamma; \boldsymbol{z}) = \left(\frac{L^L \Gamma(L-\alpha)}{\gamma^\alpha \Gamma(-\alpha)\Gamma(L)}\right)^N \prod_{n=1}^N z_n^{L-1}(\gamma + Lz_n)^{\alpha - L}. \tag{7.21}$$

Assuming that L is fixed, the ML estimates for α and γ, i.e., $\widehat{\alpha}$ and $\widehat{\gamma}$, respectively, are the solution of the following system of nonlinear equations:

$$\psi^0(-\widehat{\alpha}) - \psi^0(L - \widehat{\alpha}) - \log\widehat{\gamma} + \frac{1}{N}\sum_{n=1}^{N}\log(\widehat{\gamma} + Lz_n) = 0, \tag{7.22}$$

$$-\frac{\widehat{\alpha}}{\widehat{\gamma}} + \frac{\widehat{\alpha} - L}{N}\sum_{n=1}^{N}(\widehat{\gamma} + Lz_n)^{-1} = 0, \tag{7.23}$$

where $\psi^0(\cdot)$ is the digamma function. There is no closed-form solution for this nonlinear system; we rely on numerical optimization methods.

Let Z_1 and Z_2 be two $\mathcal{G}_I^0$ distributed random variables with densities f_{Z_1} and f_{Z_2}. The corresponding (v, ϕ)-divergence between Z_1 and Z_2 is defined by

$$D_\phi^v(Z_1\|Z_2) = v\left(\int_0^\infty \phi\left(\frac{f_{Z_1}(z;\alpha_1,\gamma_1,L_1)}{f_{Z_2}(z;\alpha_2,\gamma_2,L_2)}\right)f_{Z_2}(z;\alpha_2,\gamma_2,L_2)dz\right). \tag{7.24}$$

A special case of (7.16) is when we are interested in identifying the contrast in the main diagonal of the PolSAR data. In this case, Kullback–Leibler, Rényi of order β, Hellinger and Bhattacharyya distances can be converted into expressions involving integral terms suitable for modern symbolic mathematical software, e.g., Mathematica [34]. The above distances are detailed and the integrals involved are given in closed formulas:

(i) *The Kullback–Leibler distance*:

$$\begin{aligned} d_{\mathrm{KL}}(\theta_1,\theta_2) &= \sum_{i=1}^{2}(-1)^i k_i d\int_0^\infty \frac{x^{a_i}}{(b_i + L_i x)^{c_i}}dx + \sum_{i=1}^{2}(-1)^i k_i d \\ &\times \int_0^\infty \frac{\log(b_i + L_i x)x^{a_i}}{(b_i + L_i x)^{c_i}}dx + \sum_{i=1}^{2}(-1)^i k_i (L_2 - L_1)\int_0^\infty \frac{\log(x)x^{a_i}}{(b_i + L_i x)^{c_i}}dx \\ &-k_1 d\int_0^\infty \frac{x^{a_1}\log(b_2 + L_2 x)}{(b_1 + L_1 x)^{c_1}}dx + k_2 d\int_0^\infty \frac{x^{a_2}\log(b_1 + L_1 x)}{(b_2 + L_2 x)^{c_2}}dx, \end{aligned} \tag{7.25}$$

where $k_i = \frac{L_i^{L_i}\Gamma(L_i - \alpha_i)}{(\frac{\gamma_i}{2})^{\alpha_i}\Gamma(\alpha_i)\Gamma(L_i)}$, $a_i = L_i - 1$, $b_i = \frac{\gamma_i}{2}$, $c_i = |\alpha_i| + L_i$ for $i = 1, 2$ and $d = \log\left(k_2/k_1\right)$. Further,

$$\int_0^\infty \frac{x^a}{(b + Lx)^c}dx = \frac{b^{-c}\left(\frac{L}{b}\right)^{-1-a}\Gamma(1+a)\Gamma(-1-a+c)}{\Gamma(c)}, \tag{7.26}$$

$$\begin{aligned} \int_0^\infty \frac{\log(b + Lx)x^a}{(b + Lx)^c}dx &= \frac{1}{\Gamma(c)b^c}\left(\frac{L}{b}\right)^{-1-a}\Gamma(1+a)\Gamma(-1-a+c) \\ &\left(\log(b) + \psi^{(0)}(c) - \psi^{(0)}(-1-a-c)\right), \end{aligned} \tag{7.27}$$

$$\int_0^\infty \frac{\log(b_1+L_1x)x^a}{(b_2+L_2x)^c}dx = \frac{b_1 b_2^{-c}\pi\csc(a\pi)}{L_1}$$
$$\left(\frac{L_2^2(b_1/L_1)^{2-a}c}{b_2^2}\frac{{}_2F_1\left(2+a,1+c;3+a;\frac{b_1L_2}{b_2L_1}\right)}{2+3a+a^2}\right.$$
$$-\frac{2+a}{2+3a+a^2}\left(\frac{b_2L_1-b_1L_2}{b_2L_1}\right)^{-c}$$
$$+\frac{(L_2/b_2)^{-a}\Gamma(-1-a+c){}_3F_2\left(1,1,-a+c;2,1-a;\frac{b_1L_2}{b_2L_1}\right)}{\Gamma(1-a)\Gamma(c)}$$
$$+\frac{L_1(b_2/L_2)^{1+a}}{b_1}\frac{\Gamma(-1-a+c)}{\Gamma(-a)\Gamma(c)}$$
$$\left.\left(\log\frac{L_2}{L_1b_2}+\psi^{(0)}(-1-a+c)-\psi^{(0)}(-a)+\pi\cot(a\pi)\right)\right), \tag{7.28}$$

where ${}_2F_1(\cdot,\cdot;\cdot;\cdot)$ and ${}_3F_2(\cdot,\cdot,\cdot;\cdot,\cdot;\cdot)$ are hypergeometric functions,

$$\int_0^\infty \frac{\log(x)x^a}{(b+Lx)^c}dx = -\frac{1}{\Gamma(c)}\left[b^{-c}(L/b)^{-1-a}\Gamma(1+a)\Gamma(-1-a+c)\right.$$
$$\left.\left(-H_{\lfloor a+1\rfloor}+H_{\lfloor -1-a+c\rfloor}+\log(L/b)\right)\right], \tag{7.29}$$

where H_L is the Lth harmonic number, and $\lfloor x\rfloor$ denotes the greatest integer less than or equal to x.

(ii) *The Rényi distance of order β:*

$$d_R^\beta(\theta_1,\theta_2) = \frac{1}{\beta-1}\left[\log\left(h_1\int_0^\infty x^{g_1}(b_1+L_1x)^{e_1}(b_2+L_2x)^{e_2}dx\right)\right.$$
$$\left.+\log\left(h_2\int_0^\infty x^{g_2}(b_1+L_1x)^{m_1}(b_2+L_2x)^{m_2}dx\right)\right], \tag{7.30}$$

where $h_1=k_2^\beta k_1^{1-\beta}$, $h_2=k_1^\beta k_2^{1-\beta}$, $g_1=(L_1-1)(1-\beta)+(L_2-1)\beta$, $g_2=(L_1-1)\beta+(L_2-1)(1-\beta)$, $e_1=(\alpha_1-L_1)(1-\beta)$, $e_2=(\alpha_2-L_2)\beta$, $m_1=(\alpha_1-L_1)\beta$

and $m_2 = (\alpha_2 - L_2)(1 - \beta)$, and $0 < \beta < 1$. Further,

$$\int_0^\infty x^a (b_1 + L_1 x)^{c_1} (b_2 + L_2 x)^{c_2} dx = b_1^{c_1} b_2^{c_2} \pi \csc((c_2 + a)\pi) \left[-\frac{\left(\frac{L_1}{b_1}\right)^{-1-c_2-a} \left(\frac{L_2}{b_2}\right)^{c_2} \Gamma(-1 - c_1 - c_2 - a)}{\Gamma(-c_1)} \frac{{}_2F_1\left(-c_2, -1 - c_1 - c_2 - a; -c_2 - a; \frac{b_2 L_1}{b_1 L_2}\right)}{\Gamma(-c_2 - a)} + \frac{\left(\frac{L_2}{b_2}\right)^{-1-a} \Gamma(1 + a)\, {}_2F_1\left(-c_1, 1 + a; 2 + c_2 + a; \frac{b_2 L_1}{b_1 L_2}\right)}{\Gamma(-c_2)\Gamma(2 + c_2 + a)} \right]. \tag{7.31}$$

(iii) *The Hellinger distance*:

$$d_{\mathrm{He}}(\boldsymbol{\theta}_1, \boldsymbol{\theta}_2) = 2\left(1 - \sqrt{k_1 k_2} \int_0^\infty x^{(a_1+a_2)/2} (b_1 + L_1 x)^{f_1} (b_2 + L_2 x)^{f_2} dx\right), \tag{7.32}$$

where $f_i = \frac{\alpha_i - L_i}{2}$, for $i = 1, 2$.

(iv) *The Bhattacharyya distance*:

$$d_B(\boldsymbol{\theta}_1, \boldsymbol{\theta}_2) = -\log\left(\sqrt{k_1 k_2} \int_0^\infty x^{(a_1+a_2)/2} (b_1 + L_1 x)^{f_1} (b_2 + L_2 x)^{f_2} dx\right). \tag{7.33}$$

Figure 7.1 shows the plot of four distances between two $\mathcal{G}_I^0$-distributed random variables. It can be seen that KL is the distance measure that is most sensitive to parameter changes. The contrasts of the Bhattacharyya and Hellinger distance measures are almost similar. Nascimento et al. [12] and Cintra et al. [35] showed more detailed studies of these distances used to quantify the contrast in SAR intensities.

It is noteworthy that, as presented in Figure 7.1, these distances are not directly comparable. They become comparable and meaningful when equipped with the statistical properties provided by (7.18).

In summary, when producing contrast, one can follow the next steps:

- Collect two intensity samples from the real image.
- Obtain the maximum likelihood estimates for each considered sample under the model assumed to describe the data.
- Plug the estimates in the stochastic distances expressions, and obtain the contrast.

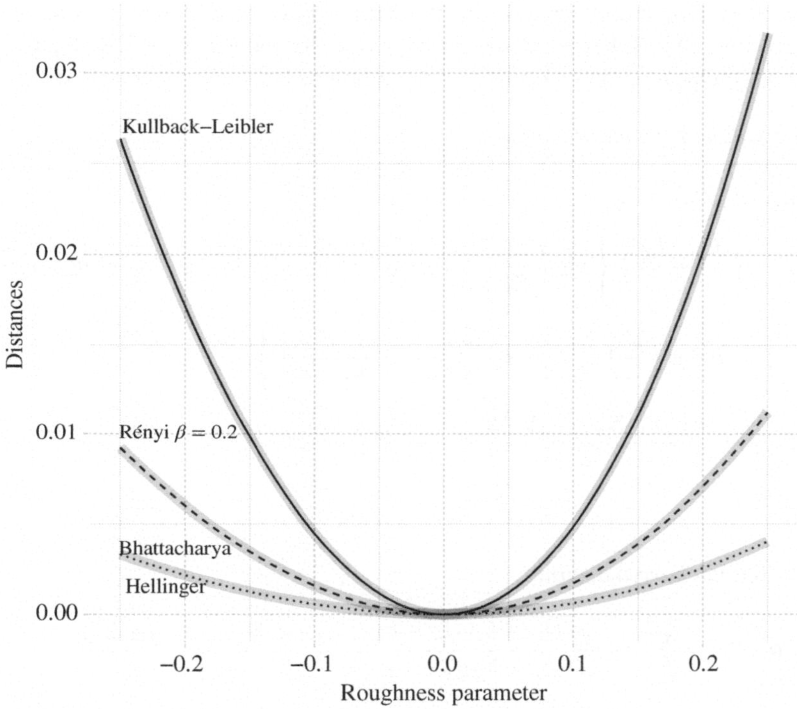

Figure 7.1 Illustration for distances between $\mathcal{G}_I^0$ distributed variables, $d_{\mathcal{M}}([-2, 3, 4], [-2 + \epsilon, 3, 4])$ for $\epsilon \in (-1, 1)$ and $\mathcal{M} \in \{\text{KL}, \text{R-}\beta, \text{B}, \text{He}\}$.

7.4.2 Contrast Analysis for PolSAR Data

In this subsection, we assume that the data follow a scaled complex Wishart law denoted by $CW(\boldsymbol{\Sigma}, L)$. This distribution has density given by

$$f_Z(\boldsymbol{Z}; \boldsymbol{\Sigma}, L) = \frac{L^{mL} |\boldsymbol{Z}|^{L-m}}{|\boldsymbol{\Sigma}|^L \Gamma_m(L)} \exp\left[-L \operatorname{tr}\left(\boldsymbol{\Sigma}^{-1} \boldsymbol{Z}\right)\right], \tag{7.34}$$

where $\boldsymbol{\Sigma}$ is the CW expected value, $\Gamma_m(L) = \pi^{m(m-1)/2} \prod_{i=1}^{m-1} \Gamma(L - i)$ is the multivariate gamma function, and $\operatorname{tr}(\cdot)$ is the trace operator.

Let $\{\boldsymbol{Z}_1, \boldsymbol{Z}_2, \dots, \boldsymbol{Z}_N\}$ be a random sample of size N obeying the $\mathcal{CW}(\boldsymbol{\Sigma}, L)$ distribution. Assuming that the parameter L is a known quantity, and considering the profile likelihood of f_Z from (7.34) with respect to $\boldsymbol{\Sigma}$, the ML estimator for $\boldsymbol{\Sigma}$ is given by [20]:

$$\widehat{\boldsymbol{\Sigma}} = \frac{1}{N} \sum_{n=1}^{N} \boldsymbol{Z}_n. \tag{7.35}$$

Frery et al. [36] found analytically the Kullback–Leibler, Rényi of order β, Hellinger, and Bhattacharyya distances between the models $\boldsymbol{Z}_A \sim \mathcal{C}W(\boldsymbol{\Sigma}_A, L_A)$ and $\boldsymbol{Z}_B \sim \mathcal{C}W(\boldsymbol{\Sigma}_B, L_B)$. These algebraic manipulations led to the expressions:

(i) *The Kullback–Leibler distance:*

$$
\begin{aligned}
&d_{\mathrm{KL}}(\boldsymbol{\theta}_A, \boldsymbol{\theta}_B) \\
&= \left(\frac{L_A - L_B}{2}\right) \left\{ \log\left(\frac{|\boldsymbol{\Sigma}_A|}{|\boldsymbol{\Sigma}_B|}\right) + m\left(\psi^{(0)}(L_A - m + 1) - \psi^{(0)}(L_B - m + 1)\right) \right. \\
&\quad \left. - m \log\left(\tfrac{L_A}{L_B}\right) + (L_B - L_A) \sum_{i=1}^{m-1} \frac{i}{(L_A - i)(L_B - i)} \right\} \\
&\quad + \frac{\operatorname{tr}(L_B \boldsymbol{\Sigma}_B^{-1} \boldsymbol{\Sigma}_A + L_A \boldsymbol{\Sigma}_A^{-1} \boldsymbol{\Sigma}_B)}{2} - \frac{m(L_A + L_B)}{2}.
\end{aligned}
\tag{7.36}
$$

(ii) *The Bhattacharyya distance:*

$$
\begin{aligned}
d_B(\boldsymbol{\theta}_A, \boldsymbol{\theta}_B) &= \sum_{k=0}^{m-1} \log \frac{\sqrt{\Gamma(L_A - k)\Gamma(L_B - k)}}{\Gamma((L_A + L_B)/2 - k)} + \frac{L_A \log|\boldsymbol{\Sigma}_A|}{2} + \frac{L_B \log|\boldsymbol{\Sigma}_B|}{2} \\
&- \frac{m}{2} \left\{ L_A \log(L_A) + L_B \log(L_B) \right\} - \left(\frac{L_A + L_B}{2}\right) \log\left| \left(\frac{L_A \boldsymbol{\Sigma}_A^{-1} + L_B \boldsymbol{\Sigma}_B^{-1}}{2}\right)^{-1} \right|.
\end{aligned}
$$

(iii) *The Hellinger distance:*

$$
\begin{aligned}
&d_{\mathrm{He}}(\boldsymbol{\theta}_B, \boldsymbol{\theta}_B) \\
&= 1 - \left\{ \prod_{k=0}^{m-1} \frac{\Gamma\left(\frac{L_A + L_B}{2} - k\right)}{\sqrt{\Gamma(L_A - k)\Gamma(L_B - k)}} \frac{\left| \left(\frac{L_A \boldsymbol{\Sigma}_A^{-1} + L_B \boldsymbol{\Sigma}_B^{-1}}{2}\right)^{-1} \right|^{\frac{L_A + L_B}{2}}}{|\boldsymbol{\Sigma}_A|^{\frac{L_A}{2}} |\boldsymbol{\Sigma}_B|^{\frac{L_B}{2}}} \sqrt{L_A^{mL_A} L_B^{mL_B}} \right\}.
\end{aligned}
\tag{7.37}
$$

(iii) *The Rényi distance:*

$$
d_{\mathrm{R}}^{\beta}(\boldsymbol{\theta}_A, \boldsymbol{\theta}_B) = \frac{1}{\beta - 1} \log\left(I(\theta_1, \theta_2)/2\right),
$$

such that

$$\begin{aligned}
I(\theta_A, \theta_B) = &\left\{ \Gamma(L_A - m + 1)^m |\mathbf{\Sigma}_A|^{L_A} L_A^{-mL_A} \prod_{i=1}^{m-1} (L_A - i)^i \right\}^{-\beta} \\
&\left\{ \Gamma(L_B - m + 1)^d |\mathbf{\Sigma}_B|^{L_B} L_B^{-mL_B} \prod_{i=1}^{m-1} (L_B - i)^i \right\}^{\beta-1} \\
&\left\{ \Gamma(E_1 - m + 1)^m |\mathbf{\Sigma}_{XY}|^{E_1} \prod_{i=1}^{m-1} (E_1 - i)^i \right\} \\
&+ \left\{ \Gamma(L_A - m + 1)^m |\mathbf{\Sigma}_A|^{L_A} L_A^{-mL_A} \prod_{i=1}^{m-1} (L_A - i)^i \right\}^{\beta-1} \\
&\left\{ \Gamma(L_B - m + 1)^m |\mathbf{\Sigma}_B|^{L_B} L_B^{-dL_B} \prod_{i=1}^{m-1} (L_B - i)^i \right\}^{-\beta} \\
&\left\{ \Gamma(E_2 - m + 1)^m |\mathbf{\Sigma}_{YX}|^{E_2} \prod_{i=1}^{m-1} (E_2 - i)^i \right\},
\end{aligned} \tag{7.38}$$

where $E_1 = \beta L_A + (1-\beta)L_B$, $E_2 = \beta L_B + (1-\beta)L_A$, $0 < \beta < 1$, and $\mathbf{\Sigma}_{ij} = |(L_i \beta \mathbf{\Sigma}_i^{-1} + L_j(1-\beta)\mathbf{\Sigma}_j^{-1})^{-1}|$ for $(i,j) = (A,B), (B,A)$.

Above, the parameter vectors $\boldsymbol{\theta}_A$ and $\boldsymbol{\theta}_B$ have replaced the symbols of the random variables $\boldsymbol{Z}_A$ and $\boldsymbol{Z}_B$ as arguments of the distance measures, since only their parameters are relevant when considering the distance between members of the same distribution [15].

Figure 7.2 shows plots of the four distances between two scaled complex Wishart distributed random matrices. For this purpose, we compute the mean covariance matrix from an urban area of the AIRSAR image of San Francisco Bay, $\mathbf{\Sigma}(0)$ as a value of Hermitian matrix-valued function:

$$\mathbf{\Sigma}(x) = \begin{bmatrix} 962892 - x & 19171 - 3579\mathrm{i} & -154638 + 191388\mathrm{i} \\ 19171 + 3579\mathrm{i} & 56707 & -5798 + 16812\mathrm{i} \\ -154638 - 191388\mathrm{i} & -5798 - 16812\mathrm{i} & 472251 \end{bmatrix}.$$

From the previous equation we draw the curve of $d_{\mathcal{M}}(\mathbf{\Sigma}(0), \mathbf{\Sigma}(\epsilon))$ for $\epsilon \in (-0.1, 0.1)$ for $\mathcal{M} \in \{\text{KL}, \text{R-}\beta, \text{B}, \text{He}\}$. It can be seen that KL and the Rényi distance measures are more sensitive to parameter changes than the others.

The same caveat made about Figure 7.1 holds for Figure 7.2: these distances are only comparable and semantic-rich after being converted into test statistics by using (7.18).

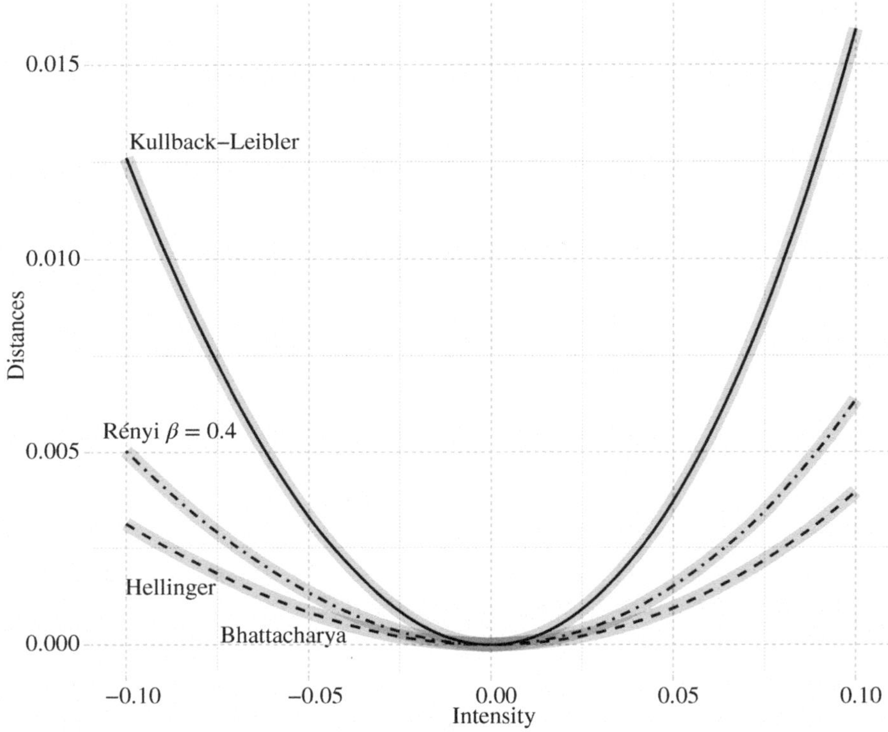

Figure 7.2 Illustration for distances between complex Wishart distributed variables, kind of $d_{\mathcal{M}}(\boldsymbol{\Sigma}(0), \boldsymbol{\Sigma}(\epsilon))$ for $\epsilon \in (-0.1, 0.1)$ and $\mathcal{M} \in \{\mathrm{KL}, \mathrm{R}\text{-}\beta, \mathrm{B}, \mathrm{He}\}$.

7.4.3 Boundary Detection for SAR Data

We will now introduce a framework for edge detection that will serve as a theoretical apparatus for all discussions of this postprocessing task.

An edge detector looks for a point on a strip of data where statistical properties change. The detection methods used in this chapter operate in three stages:

(i) Identification of the centroid of the candidate area (in an automatic, semiautomatic, or manual manner),
(ii) Identification of the transition points belonging to the edge, and
(iii) Definition of the contour using an imputation method among the transition points, such as B-splines [37].

We focus our analysis on step (ii), which is related to information theory.

Let us first assume we have a region $\mathcal{R}$ with centroid C. Rays are traced from C to control points P_i, $i = 1, 2, \ldots, S$, which lie outside $\mathcal{R}$; the S resulting rays can

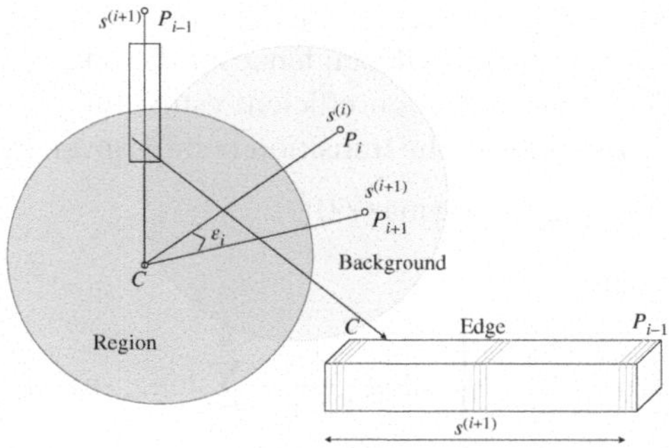

Figure 7.3 Edge detection on the polarimetric strip $s^{(i-1)}$ from the centroid C of a region to the control point P_{i-1}.

be represented by segments of the form $\boldsymbol{s}^{(i)} = \overline{CP_i}$ and the angle between successive rays is $\epsilon_i = \angle(\boldsymbol{s}^{(i)}, \boldsymbol{s}^{(i+1)})$. The rays are converted to pixels using Bresenham's Midpoint Line algorithm [38]. This representation provides the thinnest possible digital representation for a ray. This is in contrast to the 20-pixel wide strips used by Gambini et al. [37, 39], Frery et al. [40], and Giron et al. [41]. This setup is shown in Figure 7.3.

We assume that the data follow a $\mathcal{G}_I^0$ distribution, and that there are two populations: one inside the edge with $j^{(i)}$ observations, and another outside the edge with $N^{(i)} - j^{(i)}$ observations. We can then model the $N^{(i)}$ observations around the segment $\boldsymbol{s}^{(i)}$, $1 \leq i \leq S$ as

$$\begin{cases} Z_k^{(i)} \sim \mathcal{G}_I^0(\alpha_A^{(i)}, \gamma_A^{(i)}, L), & \text{for } k = 1, \dots, j^{(i)}, \\ Z_k^{(i)} \sim \mathcal{G}_I^0(\alpha_B^{(i)}, \gamma_B^{(i)}, L), & \text{for } k = j^{(i)} + 1, \dots, N^{(i)}. \end{cases} \tag{7.39}$$

In the limit case, i.e., under the conditions of (7.20), (7.39) becomes:

$$\begin{cases} Z_k^{(i)} \sim \Gamma(L_A^{(i)}, L_A^{(i)}/\beta_A^{(i)}), & \text{for } k = 1, \dots, j^{(i)}, \\ Z_k^{(i)} \sim \Gamma(L_B^{(i)}, L_B^{(i)}/\beta_B^{(i)}), & \text{for } k = j^{(i)} + 1, \dots, N^{(i)}. \end{cases} \tag{7.40}$$

Note that (7.39) collapses into (7.40) when $L_A^{(i)} = L_B^{(i)} = L$, $-\alpha_A^{(i)}/\gamma_A^{(i)} \to \beta_A^{(i)}$, and $-\alpha_B^{(i)}/\gamma_B^{(i)} \to \beta_B^{(i)}$, but these laws are not nested and therefore compete. The main idea is to find the $j^{(i)}$th edge on the segment $\boldsymbol{s}^{(i)}$ as the point that gives the best configuration with respect to a decision rule. In the following, we present three different decision rules, omitting the subscript (i) for brevity, since only one strip is considered.

7.4.3.1 Classic Edge Detector: Log-Likelihood

The log-likelihood function defined from (7.39) (or (7.40)) is given by

$$\ell(j; \mathbf{z}) = \sum_{k=1}^{j} \log f_{Z_1}(z_k; \boldsymbol{\theta}_A) + \sum_{k=j+1}^{N} \log f_{Z_2}(z_k; \boldsymbol{\theta}_B), \tag{7.41}$$

where $\boldsymbol{\theta}_A \in \{[\alpha_A, \gamma_A], [L_A, \beta_A]\}$ and $\boldsymbol{\theta}_B \in \{[\alpha_B, \gamma_B], [L_B, \beta_B]\}$ depend on j, but we have omitted this dependence for the sake of simplicity of notation. Gambini et al. [37] showed that an efficient estimator, $\widehat{J}_{\text{ML}}$, for the index on the segment that corresponds to the transition point is given by

$$\widehat{J}_{\text{ML}} = \arg\max_j \ell(j), \tag{7.42}$$

where

$$\ell(j) = \begin{cases} N \log \dfrac{L^L}{\Gamma(L)} + \displaystyle\sum_{k=1}^{N} \log z_k + j \log \dfrac{\Gamma(L - \widehat{\alpha}_A)}{\widehat{\gamma}_A^{\widehat{\alpha}_A} \Gamma(-\widehat{\alpha}_A)} + (N-j) \log \dfrac{\Gamma(L - \widehat{\alpha}_B)}{\widehat{\gamma}_B^{\widehat{\alpha}_B} \Gamma(-\widehat{\alpha}_B)} \\ \quad + (\widehat{\alpha}_A - L) \sum_{k=1}^{j} \log(\widehat{\gamma}_A + L z_k) + (\widehat{\alpha}_B - L) \sum_{k=j+1}^{N} \log(\widehat{\gamma}_B + L z_k), \\ \text{for } Z \sim \mathcal{G}_I^0, \\ j \log \dfrac{(\widehat{L}_A / \widehat{\beta}_A)^{\widehat{L}_A}}{\Gamma(\widehat{L}_A)} + (N-j) \log \dfrac{(\widehat{L}_B / \widehat{\beta}_B)^{\widehat{L}_B}}{\Gamma(\widehat{L}_B)} + (\widehat{L}_A + \widehat{L}_B - 2) \displaystyle\sum_{k=1}^{N} \log z_k, \\ \quad - \sum_{k=1}^{j} \dfrac{\widehat{L}_A}{\widehat{\beta}_A} z_k - \displaystyle\sum_{k=j+1}^{N} \dfrac{\widehat{L}_B}{\widehat{\beta}_B} z_k, \\ \text{for } Z \sim \Gamma. \end{cases} \tag{7.43}$$

However, this method is very computationally intensive because it must evaluate two likelihood functions at each search step. The main computational problem here is that the previous nonlinear system should be evaluated to find the MLEs of the unknown parameters for each position of the edge.

7.4.3.2 Information-Theoretic Edge Detectors

As discussed in Nascimento et al. [12], the distances derived in Section 7.4.1 can be scaled to be asymptotically distributed as chi-squared statistics:

$$S_D\left(\widehat{\theta}_1(j), \widehat{\theta}_2(N-j)\right) = \frac{2j(N-j)\xi_D}{N} d_D\left(\widehat{\theta}_1(j), \widehat{\theta}_2(N-j)\right), \tag{7.44}$$

where $\xi_D \in \{1, \beta^{-1}, 4, \text{and } 4\}$ for $\mathcal{D} \in \{\text{KL}, \text{RD-}\beta, \text{BA}, \text{and He}\}$, respectively, and $\widehat{\theta}_1(j) = [\widehat{\alpha}_A(j), \widehat{\gamma}_A(j)]$ and $\widehat{\theta}_2(N-j) = [\widehat{\alpha}_B(N-j), \widehat{\gamma}_B(N-j)]$ are the ML estimators for $\boldsymbol{\theta}_1 = [\alpha_A, \gamma_A]$ and $\boldsymbol{\theta}_2 = [\alpha_B, \gamma_B]$ using random samples of sizes j and $N-j$, respectively. Under mild conditions, $S_D\left(\widehat{\theta}_1(j), \widehat{\theta}_2(N-j)\right)$ is asymptotically distributed as a χ_2^2 random variable under the null hypothesis $\boldsymbol{\theta}_1 = \boldsymbol{\theta}_2$.

We, therefore, proposed new detectors to find edges on SAR intensities by looking for the point that maximizes the test statistic between the two models, i.e.,

$$\widehat{J}_D = \arg\max_j S_D\left(\widehat{\theta}_1(j), \widehat{\theta}_2(N-j)\right) = \arg\max_j S_D(j). \tag{7.45}$$

The main advantage is that it does not require the calculation of log-likelihood sums, which can be both computationally intensive and prone to numerical instabilities.

7.4.3.3 Application for SAR Data

We are now in a position to test the detectors with actual data. We present an application to a SAR image to test the proposed detectors in practice. Figure 7.4 shows a SAR image of San Francisco (USA) from the HV (horizontal–vertical) polarization channel. This image was acquired using an AIRSAR sensor with four nominal looks.

The scene has three well-defined regions: Sea, Forest, and Urban. We will use it to generate the rays that the proposed detectors work with. Figure 7.4 shows the result of detecting the edges separating the sea from the urban regions.

The $\mathcal{G}_I^0$ likelihood detectors performs consistently well. With the exception of the Hellinger distance, the detectors derived from gamma divergence did not work correctly.

7.4.4 Boundary Detection for PolSAR Data

Let us assume that the PolSAR data follow a scaled complex Wishart distribution. The strip of data collected around the ith ray $\boldsymbol{s}^{(i)}$, $i = 1, 2, \dots, S$, contains $N^{(i)}$ pixels. It is assumed that each pixel k of a given strip i is described by the return matrix $\boldsymbol{Z}_k^{(i)}$, which is distributed according to the scaled complex Wishart law [40]. If we denote the correct edge position as $j^{(i)}$, we have the following configuration:

$$\begin{cases} \boldsymbol{Z}_k^{(i)} \sim \mathcal{C}W(\boldsymbol{\Sigma}_A^{(i)}, L_A^{(i)}), & \text{for } k = 1, \dots, j^{(i)}, \\ \boldsymbol{Z}_k^{(i)} \sim \mathcal{C}W(\boldsymbol{\Sigma}_B^{(i)}, L_B^{(i)}), & \text{for } k = j^{(i)} + 1, \dots, N^{(i)}. \end{cases} \tag{7.46}$$

In other words, each strip consists of two pattern fields, and each field obeys a complex Wishart law with different parameter values. In this chapter, we assume that the number of looks is estimated beforehand and remains constant for the entire image. Thus $L_A^{(i)} = L_B^{(i)} = L$ in each strip $i = 1, 2, \dots, S$. This assumption is realistic because the number of looks is determined by the radar when it receives backscattered pulses [42].

The main idea is to estimate the edge position $j^{(i)}$ on the data strip around the ray $\boldsymbol{s}^{(i)}$ according to a decision rule. The gray lines in Figure 7.3 illustrate different configurations when the edge position is shifted. The model given in (7.46) assumes that there is at most one transition in a given beam. Problems can arise when multiple transitions occur; this is discussed in more detail in Section 7.4.3.3, Figure 7.4(a). In the following, we present three different decision rules. For brevity, we omit the superscript (i) since we focus on a single strip in our analysis.

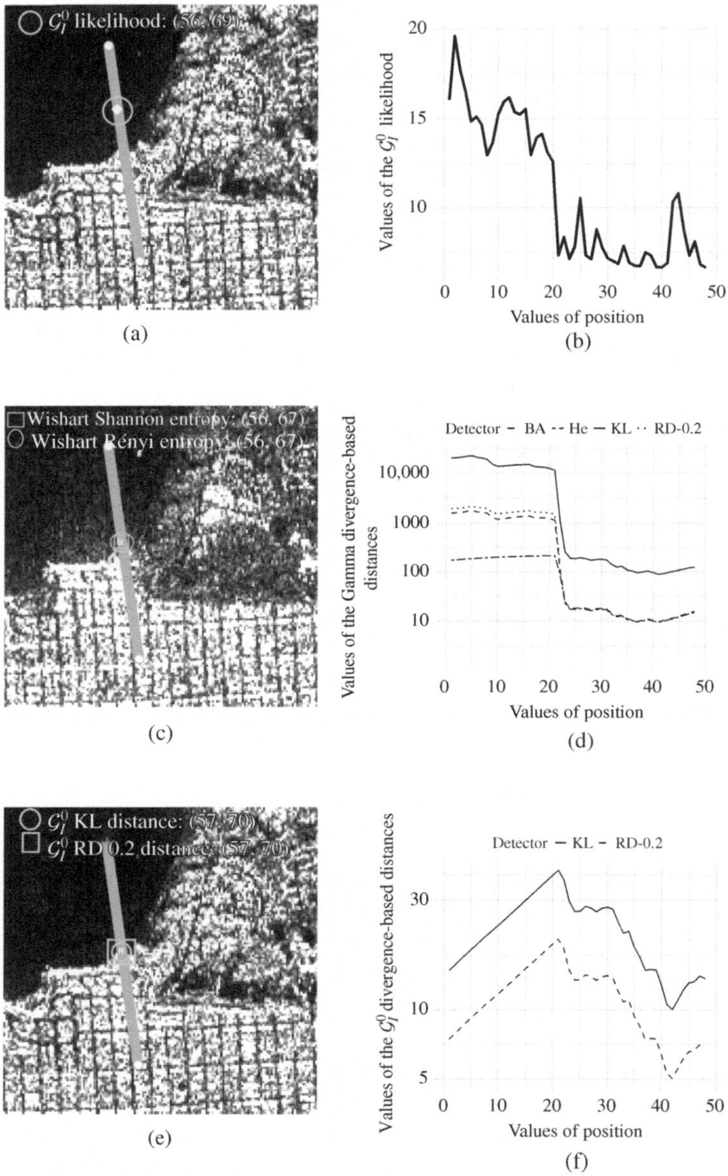

Figure 7.4 Performance of detectors based on information-theoretic measures under the gamma and $\mathcal{G}_I^0$ distribution for intensity data. (a) Likelihood-based edge estimate under the $\mathcal{G}_I^0$ law. (b) Likelihood along the ray under the $\mathcal{G}_I^0$ law. (c) Divergence-based edge estimates under the gamma law. (d) Divergences along the ray under the gamma law (logarithmic scale). (e) Divergence-based edge estimates under the $\mathcal{G}_I^0$ law. (f) Divergences along the ray under the $\mathcal{G}_I^0$ law.

7.4.4.1 Classic Edge Detector: Log-Likelihood

The likelihood function of the sample described by (7.46) is given by

$$L(j) = \prod_{k=1}^{j} f_Z(\boldsymbol{Z}_k; \boldsymbol{\Sigma}_A, L) \prod_{k=j+1}^{N} f_Z(\boldsymbol{Z}_k; \boldsymbol{\Sigma}_B, L), \tag{7.47}$$

where $\boldsymbol{Z}_k$ are possible outcomes of the random matrices described in (7.39). The log-likelihood is more suitable for maximization purposes:

$$\ell(j) = \log L(j) = \sum_{k=1}^{j} \log f_Z(\boldsymbol{Z}_k; \boldsymbol{\Sigma}_A, L) + \sum_{k=j+1}^{N} \log f_Z(\boldsymbol{Z}_k; \boldsymbol{\Sigma}_B, L). \tag{7.48}$$

The ML estimator $\hat{j}_{\mathrm{ML}}$ for the index at the point where the transition occurs is given by

$$\hat{j}_{\mathrm{ML}} = \arg\max_j \ell(j). \tag{7.49}$$

Using the model given by (7.34) in (7.49), with minor algebraic manipulation one obtains

$$\ell(j) = N\left[-mL(1-\log L) - \log\Gamma_m(L)\right] + L\left[j\log|\widehat{\boldsymbol{\Sigma}}_A(j)| + (N-j)\log|\widehat{\boldsymbol{\Sigma}}_B(j)|\right], \tag{7.50}$$

where $\widehat{\boldsymbol{\Sigma}}_A(j)$ and $\widehat{\boldsymbol{\Sigma}}_B(j)$ are the ML estimators for Σ_A and Σ_B respectively with respect to the edge position j. This estimator satisfies the following expression:

$$\widehat{\boldsymbol{\Sigma}}_I(j) = \begin{cases} j^{-1}\sum_{k=1}^{j} \boldsymbol{Z}_k, & \text{if } I = A, \\ (N-j)^{-1}\sum_{k=j+1}^{N} \boldsymbol{Z}_k, & \text{if } I = B. \end{cases} \tag{7.51}$$

7.4.4.2 Information-Theoretic Edge Detector

Let $\boldsymbol{X}$ and $\boldsymbol{Y}$ be two random matrices that obey the scaled complex Wishart distribution and whose number of looks is known. The associated probability densities are $f_{\boldsymbol{X}}(\boldsymbol{Z}; \boldsymbol{\Sigma}_A, L)$ and $f_{\boldsymbol{Y}}(\boldsymbol{Z}; \boldsymbol{\Sigma}_B, L)$. Both densities are defined over the set of Hermitian positive definite matrices $\mathcal{A}$, defined above on (7.16).

- *For stochastic distances*: Frery et al. [36] derived analytic expressions for the above distances considering Wishart distributions for the general case $\boldsymbol{\Sigma}_A \neq \boldsymbol{\Sigma}_B$ and $L_A \neq L_B$. Since we assume that the equivalent number of looks is the same, the distances can be expressed as follows:
  (i) *Kullback–Leibler*:

$$d_{\mathrm{KL}}(\boldsymbol{\theta}_1, \boldsymbol{\theta}_2) = L\left[\frac{\mathrm{tr}(\boldsymbol{\Sigma}_1^{-1}\boldsymbol{\Sigma}_2 + \boldsymbol{\Sigma}_2^{-1}\boldsymbol{\Sigma}_1)}{2} - m\right]. \tag{7.52}$$

(ii) *Rényi*:

$$d_{\text{RD-}\beta}(\boldsymbol{\theta}_1, \boldsymbol{\theta}_2) = \frac{\log 2}{1-\beta} + \frac{1}{\beta - 1} \log \left\{ \left[\frac{|(\beta \boldsymbol{\Sigma}_1^{-1} + (1-\beta)\boldsymbol{\Sigma}_2^{-1})^{-1}|}{|\boldsymbol{\Sigma}_1|^{\beta} |\boldsymbol{\Sigma}_2|^{1-\beta}} \right]^L + \left[\frac{|(\beta \boldsymbol{\Sigma}_2^{-1} + (1-\beta)\boldsymbol{\Sigma}_1^{-1})^{-1}|}{|\boldsymbol{\Sigma}_1|^{(1-\beta)} |\boldsymbol{\Sigma}_2|^{\beta}} \right]^L \right\}. \tag{7.53}$$

(iii) *Bhattacharyya*:

$$d_{\text{BA}}(\boldsymbol{\theta}_1, \boldsymbol{\theta}_2) = L \left[\frac{\log |\boldsymbol{\Sigma}_1| + \log |\boldsymbol{\Sigma}_2|}{2} - \log \left| \left(\frac{\boldsymbol{\Sigma}_1^{-1} + \boldsymbol{\Sigma}_2^{-1}}{2} \right)^{-1} \right| \right]. \tag{7.54}$$

(iv) *Hellinger*:

$$d_{\text{He}}(\boldsymbol{\theta}_1, \boldsymbol{\theta}_2) = 1 - \left[\frac{\left| 2^{-1} (\boldsymbol{\Sigma}_1^{-1} + \boldsymbol{\Sigma}_2^{-1})^{-1} \right|}{\sqrt{|\boldsymbol{\Sigma}_1||\boldsymbol{\Sigma}_2|}} \right]^L. \tag{7.55}$$

The above expressions are modified following Salicrú et al. [15] to obtain test statistics with known asymptotic properties. The test statistics based on these distances can be easily derived for the null hypothesis $\mathcal{H}_0 : \boldsymbol{\theta}_1 = \boldsymbol{\theta}_2$:

$$S_D\left(\widehat{\boldsymbol{\theta}}_1(j), \widehat{\boldsymbol{\theta}}_2(N-j)\right) = \frac{2j(N-j)\xi_D}{N} d_D\left(\widehat{\boldsymbol{\theta}}_1(j), \widehat{\boldsymbol{\theta}}_2(N-j)\right), \tag{7.56}$$

where $\mathcal{D} \in \{\text{KL}, \text{RD-}\beta, \text{BA}, \text{He}\}$, $\xi_D \in \{1, \beta^{-1}, 4, 4\}$, respectively, $\widehat{\boldsymbol{\theta}}_1(j) = \text{vec}(\widehat{\boldsymbol{\Sigma}}_A(j))$ and $\widehat{\boldsymbol{\theta}}_2(N-j) = \text{vec}(\widehat{\boldsymbol{\Sigma}}_B(N-j))$ are estimators for $\boldsymbol{\theta}_1$ and $\boldsymbol{\theta}_2$ using random samples of sizes j and $N-j$, respectively, and $\text{vec}(\cdot)$ is the vectorization operator.

Let p be the number of elements of the parameter vectors, i.e., $p = m^2$, as discussed by Anfinsen et al. [43]. As derived from Frery et al. [44], the test statistics given in (7.56) are asymptotically χ^2-distributed with p degrees of freedom. This result holds as long as the ML estimators with large samples are used [15].

Therefore, four new detectors can be proposed for searching edges on PolSAR images by looking for the point that maximizes the test statistic between the two models, i.e.,

$$\widehat{j}_D = \arg\max_j \underbrace{S_D\left(\widehat{\boldsymbol{\theta}}_1(j), \widehat{\boldsymbol{\theta}}_2(N-j)\right)}_{\equiv S_D(j)} = \arg\max_j S_D(j), \tag{7.57}$$

where $\mathcal{D} = \{\text{KL}, \text{RD-}\beta, \text{BA}, \text{He}\}$.

De Borba et al. [45] measured the amount of information contained in components of the fully PolSAR matrix.

- *For stochastic entropies*: As suggested by Pardo et al. [28], we consider the Rényi of order $0 < \beta < 1$ and the Shannon entropies, denoted $H_{\text{RE-}\beta}$ and H_S, respectively. A discrimination statistic based on stochastic entropies can be defined as follows:

$$S_{\mathcal{M}}\left(\widehat{\theta}_1(j), \widehat{\theta}_2(N-j)\right) = \frac{j\left(H_{\mathcal{M}}(\widehat{\theta}_1(j)) - \overline{o}\right)^2}{\sigma^2_{\mathcal{M}}(\widehat{\theta}_1(j))} + \frac{(N-j)\left(H_{\mathcal{M}}(\widehat{\theta}_2(N-j)) - \overline{o}\right)^2}{\sigma^2_{\mathcal{M}}(\widehat{\theta}_2(N-j))}, \tag{7.58}$$

where

$$\overline{o} = \left[\frac{j}{\sigma^2_{\mathcal{M}}(\widehat{\theta}_1(j))} + \frac{(N-j)}{\sigma^2_{\mathcal{M}}(\widehat{\theta}_2(N-j))}\right]^{-1} \times \left[\frac{jH_{\mathcal{M}}(\widehat{\theta}_1(j))}{\sigma^2_{\mathcal{M}}(\widehat{\theta}_1(j))} + \frac{(N-j)H_{\mathcal{M}}(\widehat{\theta}_2(N-j))}{\sigma^2_{\mathcal{M}}(\widehat{\theta}_2(N-j))}\right]. \tag{7.59}$$

Therefore, two additional detectors can be defined by the points that maximize the test statistics based on the entropies between the two distributions:

$$\widehat{j}_{\mathcal{M}} = \arg\max_j \underbrace{S_{\mathcal{M}}(\widehat{\theta}_1(j), \widehat{\theta}_2(N-j))}_{\equiv S_{\mathcal{M}}(j)} = \arg\max_j S_{\mathcal{M}}(j), \tag{7.60}$$

where $\mathcal{M} = \{S, \text{RE-}\beta\}$.

7.4.4.3 Edge Detector Based on Information-Theoretic Measures

Now we apply the detectors based on the joint likelihood ($\widehat{j}_{\text{ML}}$), distances (Rényi $\widehat{j}_{\text{RD-}\beta}$, Hellinger $\widehat{j}_{\text{He}}$, Bhattacharyya $\widehat{j}_{\text{BA}}$, and Kullback–Leibler $\widehat{j}_{\text{KL}}$), and entropies (attributed by Rényi $\widehat{j}_{\text{RE-}\beta}$, and Shannon $\widehat{j}_S$). This application was performed on an image of San Francisco Bay taken by the AIRSAR sensor, as discussed previously.

Figure 7.5 shows the detection results after identifying the transition points associated with the edge. In general, all the used detectors perform well except the one derived from the Kullback–Leibler distance. From Figure 7.5(a) and (b), it is evident that the Wishart-likelihood detector provides good detection and even has a bimodal detection curve. For the divergence-based detectors in Figure 7.5(c) and (d), the regularity of the Hellinger divergence detection curve is remarkable. Entropy-based detectors perform equally well in Figure 7.5(e) and (f). It is worth noting that the maximum value of the detection curve is better defined than that resulting from the divergence-based criteria.

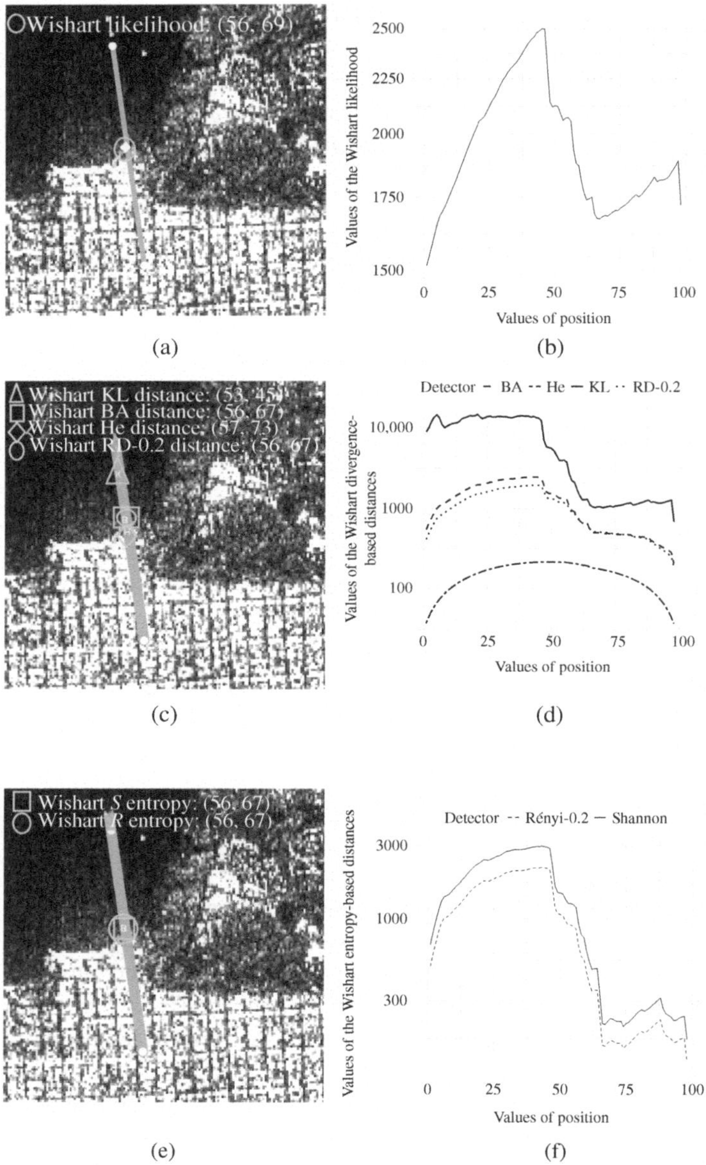

Figure 7.5 Performance of detectors based on information-theoretic measures under the scaled complex Wishart distribution. (a) Likelihood-based edge estimate. (b) Likelihood along the ray. (c) Divergence-based edge estimates. (d) Divergences along the ray (logarithmic scale). (e) Entropy-based edge estimates. (f) Entropies along the ray (logarithmic scale).

7.4.5 Speckle Reduction with Nonlocal Means

It is well known that the preservation of fine structures and textures requires local adjustment of the quantities used in noise reduction. The nonlocal mean is one of the most successful approaches in this field. It was proposed by Buades et al. [46] and has been adapted and used in countless applications since its introduction. The nonlocal mean approach filters each pixel by convolving it with a typically large mask whose weights reflect the similarity of the patches to the central value. This is a flexible approach that produces excellent results when set appropriately. Torres et al. [47] used it for fully polarimetric data. Deledalle et al. [48] extend the use of nonlocal mean filters for interferometric SAR data. The reader is referred to the work of Gomez et al. [49] for a detailed explanation, literature review, and examples.

A general structure of a nonlocal means filter is shown in Figure 7.6. The right grid shows the upper part of the desired $n \times n$, with $n = 9$ convolutional mask. This mask is created by comparing a computed, in our case estimated, feature for the central observation. This estimation is done with nine observations: $z_{0,1}, z_{0,2}, \dots, z_{0,9}$, and the estimate is $\widehat{\theta}_0$. Suppose we want to obtain the weight w_1 in the convolutional mask. First, we estimate the feature $\widehat{\theta}_1$ in the corresponding pixel using 25 observations around it: $z_{1,1}, z_{1,2}, \dots, z_{1,25}$. Note that the size of the estimation windows may vary. After this estimation step, we compare $\widehat{\theta}_0$ and $\widehat{\theta}_1$ with a statistical test $S(\widehat{\theta}_0, \widehat{\theta}_1)$ that yields the p value p_1 for the null hypothesis that the features were generated by the same model. Using p-values as mask weights is possible but not advisable, as they would assign very little influence to pixel values for which one has no evidence to reject. For this reason, one usually

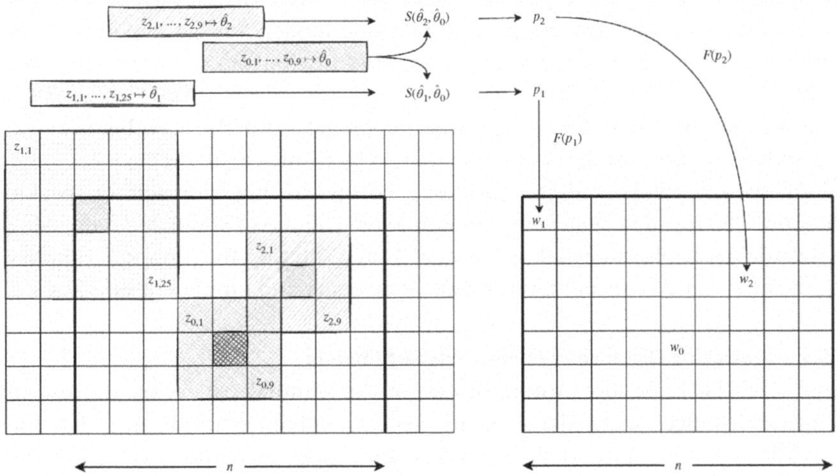

Figure 7.6 Overview of a general nonlocal means filter.

transforms these p-values by means of a nondecreasing function F. Figure 7.6 also shows that the features can be computed with partially overlapping samples, i.e., $\widehat{\theta}_0$ and $\widehat{\theta}_2$ are computed with a common observation.

In the following subsections, we summarize the use of nonlocal mean filters for monopolarized (Section 7.4.5.1) and PolSAR (Section 7.4.5.2), which incorporate concepts from information theory.

7.4.5.1 Entropy and Intensity Data

Chan et al. [50] worked with the noisiest case of intensity-speckled data, namely the single-look situation, i.e., $L = 1$ in (7.19). Instead of working directly with the data whose signal-to-noise-ratio is unfavorable, the authors extracted Shannon and Rényi entropy using local operations. To compare the sampling entropy of each patch with that of the central pixel, the asymptotic variance of the estimator is determined. The following results were obtained:

$$\sigma^2_{H_S} = \frac{(\alpha-1)^2\left[(-2+4\alpha^3-\gamma^2-\alpha^2(10+\gamma^2)+\alpha(5+2\gamma^2))\right]}{\alpha^2\left[\alpha(5+2\gamma^2)-2-\gamma^2-\alpha^2(2+\gamma^2)\right]}, \tag{7.61}$$

and

$$\begin{aligned}\sigma^2_{H_{\text{RE-}\beta}} &= \frac{(\alpha-1)^2\left[-2+4\beta+4\alpha^3\beta^2-\beta^2(4+\gamma^2)\right]}{[1+(\alpha-1)\beta]^2\left[\alpha(5+2\gamma^2)-\alpha^2(2+\gamma^2)-\gamma^2-2\right]} \\ &\quad+\frac{\alpha\left[1-10\beta+2\beta^2\left(7+\gamma^2-\alpha^2\beta(-4+\beta(14+\gamma^2))\right)\right]}{(1+(\alpha-1)\beta)^2\left(\alpha(5+2\gamma^2)-\alpha^2(2+\gamma^2)-\gamma^2-2\right)}.\end{aligned} \tag{7.62}$$

The parameters α and γ should be estimated here.

With this information, it is possible to develop tests for the null hypothesis that two observed entropies are samples of the same model. Such a hypothesis test yields a p value that is transformed by a smooth function and then converted to a weight.

The authors show that the entropy-based approach outperforms other state-of-the-art filters in terms of the $\mathcal{M}$ metric defined by Gomez et al. [51].

More recently, Vitale et al. [52] proposed a deep learning approach for speckle reduction. The training stage employs a massive volume of simulated data using the statistical models discussed by Yue et al. [18]. With this, the proposal complies with the often hard to meet requirements of diversity and number of samples.

7.4.5.2 Stochastic Distances Between Wishart Models

Torres et al. [47] are the authors of one of the seminal papers on the use of stochastic distances in nonlocal means speckle reduction. Assuming that the observations follow Wishart distributions, they calculated the p values of the hypothesis test for comparing two samples using the Hellinger distance (given in (7.37), unconstrained, and (7.55), for the number of looks) between them. Then, they converted this p-value into a weight by a piecewise linear function and obtained the convolution mask. Their results outperform the Boxcar, Refined Lee, and Intensity-Driven Adaptive Neighborhood (IDAN) filters in terms of equivalent number of looks, structural similarity measure (SSM) index, and visual quality of edges computed after noise reduction.

7.4.6 Classification and Segmentation

The general statistical classification problem can be formulated as follows: assume that the data were generated by one of K possible models $\mathcal{D}_1, \mathcal{D}_2, \dots, \mathcal{D}_K$, where $K \geq 2$. Given an observation z, the problem is "guessing" which model generated it. The segmentation problem is similar, but it refers to "clusters" or "groups" rather than to a number of models. Given $z_1, z_2, \dots, z_n$ observations, the task is splitting them into groups such that those belonging to the same group are simultaneously "similar" and "different" from those assigned to a different group.

Classification and segmentation may be supervised (if a decision rule can be learned from examples of observations that have already been classified or grouped, and in particular the number of classes or groups is known), otherwise the problems are unsupervised. The reader is referred to the book by Devroye et al. [53] for a comprehensive and thorough treatment of these and related problems.

Classification and segmentation can work with the original observations or with features extracted from them. The following sections summarize approaches that use different types of entropy and techniques to segment and classify speckled data into intensity (Section 7.4.6.1) and polarimetric (Section 7.4.6.2) data.

7.4.6.1 Entropy and Intensity Data

Assuming that the $\mathcal{G}^0$ is characterized by the density given in (7.19), Cassetti et al. [54] computed locally the ML estimators of its parameters, i.e., $\widehat{\theta} = [\widehat{\alpha}, \widehat{\gamma}]$ for a known number of looks L. They used $\widehat{\theta}$ to index the expression of Shannon entropy under this model and obtained the ML estimator of the Shannon entropy. Knowing that $\widehat{\theta}$ is prone to numerical problems, the authors computed seven nonparametric estimators based on differences of order statistics. All these estimators converge in probability, for large samples, to the Shannon entropy.

The authors used these Shannon entropy estimators as input for two classification algorithms: K-means and a support vector machine, and used the observed measures of quality for tuning the parameters (mostly, the spacing between observations of the nonparametric estimators). The authors concluded that the use of nonparametric estimators of the Shannon entropy yields better results than using the same feature based on ML estimation.

7.4.6.2 Entropy and PolSAR Data

Nascimento et al. [55] proposed an unsupervised machine learning approach to classify PolSAR data based on entropy measures. To this end, they theoretically derived the Shannon entropy measure for the $\mathcal{G}_m^0$ law and used the complex Wishart–Shannon entropy proposed by Frery et al. [44] as a special case. By considering these measures as elements of the υ–ϕ entropy class, the authors formulated four segmentation methods using asymptotic results from the estimation of the parameters used in different settings. First, their contributions provided rich maps of features that are a kind of projection of the full PolSAR data onto the real set. Second, new segmentation methods for polarimetric data were proposed.

Numerical results provided information superior to Wishart entropy-based segmentation for objects with scattered clusters, while $\mathcal{G}_m^0$ entropy-based segmentation is better suited to deal with more complex images. In all the studies performed, the new proposals outperformed those considered state-of-the-art.

7.4.7 Computational Aspects of Information-Theoretic Tools

In general, the use of information-theoretic measures provides advantages in computational complexity compared to Monte Carlo and likelihood-based techniques. The latter ones are based on finite summation, while information-theoretic techniques tend to generate functions in terms of the parameter space in question (capable of quantifying disorder in the case of entropy and discrepancies in the case of divergence) whose computation depends only on the estimates of the parameters.

In particular, stochastic distances between $\mathcal{G}_I^0$-distributed variables, as proposed by Nascimento et al. [12], have been computed numerically, but they can be used in the form of special functions. In contrast, the expressions for stochastic distances between complex Wishart matrices are simple and essentially involve only two operations: the determinant and inverse of the complex covariance matrix. Recently, Coelho et al. [56] proposed a method to perform these two operations with less computational effort and discuss an implementation using general-purpose graphics processing units.

7.5 Conclusions and Future Work

Divergence and entropy have been used by the remote sensing community for a long time, but they are most useful within the context of Statistics. We have seen that seemingly disparate image processing and analysis tasks can be solved by using those Information Theory concepts along with statistical inference. In particular, the previous sections have presented:

Contrast analysis: The question "Are two regions significantly different?" can be formulated as "How similar or different are these samples?"

Boundary detection: Finding an edge between two regions is locating the point at which their statistical properties differ the most.

Noise reduction: The convolution mask is formed by coefficients related to the p-value of tests that verify if small patches of data belong to the same distribution.

Classification: Classes are characterized by models, and observations can be assigned to the one with closest statistical properties.

Segmentation: Clusters can be formed by minimizing the difference among statistical properties of their samples and by maximizing the difference to other clusters.

The solutions provided by Statistical Information Theory to these problems have remarkable properties, including excellent performance even with small samples and less computational burden than several alternative approaches.

Among the questions that remain open today are the following:

- The asymptotic properties of these tests require the use of ML estimators. How do they behave when other inference techniques, e.g., robust estimators, estimators by analogy, are used?
- How do these techniques perform when there is no explicit model and this is derived from the data, e.g., with kernel estimators?
- How do these test statistics perform with different types of contamination?

The approach based on deterministic distances is relatively new and has already shown excellent performance [57–62]. This type of data representation in terms of its dissimilarity to prototypical backscatterers is flexible and extensible, as it accepts an arbitrary number of reference points as input. The statistical properties of such features are largely unknown to date.

We conclude this chapter by reflecting upon the differences between the statistical and artificial intelligence approaches. The progress made with neural networks in image processing and analysis is remarkable, signifying a significant leap forward in the field. Nevertheless, these techniques typically require massive annotated samples for training and operate as black boxes. As discussed in this chapter, the statistical approach does not suffer from these shortcomings. Recent works aim at lessening the impact of these limitations. For instance, Vitale et al. [52] trained a neural network with simulated samples obtained from models that arise from the ones discussed in this chapter and produced a remarkably efficient despeckling technique. Datcu et al. [63] considered ways of incorporating the physical properties of the imaging process with SAR in an explainable and trustworthy artificial intelligence pipeline.

We, thus, conclude this chapter with an additional open question:

- How to effectively integrate the knowledge obtained from the statistical approach with techniques based on machine learning and artificial intelligence?

Acknowledgment

The authors would like to thank the National Council for Scientific and Technological Development, Grant/Award Number: CNPq 200819/2022-4.

References

1 Jafarzadeh, H., Mahdianpari, M., Homayouni, S. et al. (2021). Oil spill detection from synthetic aperture radar earth observations: a meta-analysis and comprehensive review. *GIScience & Remote Sensing* 58: 1022–1051.

2 Carta, F., Zidda, C., Putzu, M. et al. (2023). Advancements in forest fire prevention: a comprehensive survey. *Sensors* 23(14): 6635.

3 Hartley, R.V.L. (1928). Transmission of information. *Bell System Technical Journal* 7: 535–563.

4 Shannon, C.E. (1948). A mathematical theory of communication. *Bell System Technical Journal* 27: 379–423.

5 Nadarajah, S. and Zografos, K. (2003). Formulas for Rényi information and related measures for univariate distributions. *Information Sciences* 155: 119–138.

6 Nadarajah, S. and Zografos, K. (2005). Expressions for Rényi and Shannon entropies for bivariate distributions. *Information Sciences* 170(2–4): 173–189.

7 Morio, J., Réfrégier, P., Goudail, F. et al. (2009). A characterization of Shannon entropy and Bhattacharyya measure of contrast in polarimetric and interferometric SAR image. *Proceedings of the IEEE* 97(6): 1097–1108.

8 Cloude, S.R. and Pottier, E. (1997). An entropy based classification scheme for land application of polarimetric SAR. *IEEE Transactions on Geoscience and Remote Sensing* 35(1): 68–78.

9 Yan, W., Yang, W., Liu, Y., and Sun, H. (2010). Unsupervised classification of PolinSAR image based on Shannon entropy characterization. *IEEE International Conference on Signal Processing (ICSP'2010)*, 2192–2195.

10 Eguchi, S. and Copas, J. (2006). Interpreting Kullback–Leibler divergence with the Neyman–Pearson lemma. *Journal of Multivariate Analysis* 97(9): 2034–2040.

11 Basseville, M. (1989). Distance measures for signal processing and pattern recognition. *Signal Processing* 18: 349–369.

12 Nascimento, A.D.C., Cintra, R.J., and Frery, A.C. (2010). Hypothesis testing in speckled data with stochastic distances. *IEEE Transactions on Geoscience and Remote Sensing* 48(1): 373–385.

13 Ali, S.M. and Silvey, S.D. (1996). A general class of coefficients of divergence of one distribution from another. *Journal of the Royal Statistical Society, Series B: Statistical Methodology* 26: 131–142.

14 Csiszár, I. (1967). Information type measures of difference of probability distributions and indirect observations. *Studia Scientiarum Mathematicarum Hungarica* 2: 299–318.

15 Salicrú, M., Menéndez, M.L., Pardo, L., and Morales, D. (1994). On the applications of divergence type measures in testing statistical hypothesis. *Journal of Multivariate Analysis* 51(2): 372–391.

16 Ulaby, F.T. and Long, D.G. (2014). *Microwave Radar and Radiometric Remote Sensing*. Norwood, OH: The University of Michigan Press.

17 Frery, A.C., Wu, J., and Gomez, L. (2022). *SAR Image Analysis –A Computational Statistics Approach: With R Code, Data, and Applications.* Wiley-IEEE Press.

18 Yue, D.-X., Xu, F., Frery, A.C., and Jin, Y.-Q. (2021). SAR image statistical modeling Part I: single-pixel statistical models. *IEEE Geoscience and Remote Sensing Magazine* 9(1): 82–114.

19 Delignon, Y. and Pieczynski, W. (2002). Modeling non-Rayleigh speckle distribution in SAR images. *IEEE Transactions on Geoscience and Remote Sensing* 40(6): 1430–1435.

20 Goodman, N.R. (1963). Statistical analysis based on a certain complex Gaussian distribution (an introduction). *The Annals of Mathematical Statistics* 34(1): 152–177.

21 Hagedorn, M., Smith, P.J., Bones, P.J. et al. (2006). A trivariate chi-squared distribution derived from the complex Wishart distribution. *Journal of Multivariate Analysis* 97(3): 655–674.

22 Nascimento, A.D.C., Frery, A.C., and Cintra, R.J. (2014). Bias correction and modified profile likelihood under the Wishart complex distribution. *IEEE Transactions on Geoscience and Remote Sensing* 52(8): 4932–4941.

23 Anfinsen, S.N. and Eltoft, T. (2011). Application of the matrix-variate Mellin transform to analysis of polarimetric radar images. *IEEE Transactions on Geoscience and Remote Sensing* 49(6): 2281–2295.

24 Khan, S. and Guida, R. (2012). On single-look multivariate distribution for PolSAR data. *IEEE Journal of Selected Topics in Applied Earth Observations and Remote Sensing* 5: 1149–1163.

25 Salicrú, M., Mendéndez, M.L., and Pardo, L. (1993). Asymptotic distribution of (h, ϕ)*-entropy*. *Communications in Statistics - Theory and Methods* 22(7): 2015–2031.

26 Havrda, J. and Chalvat, F. (1967). Concept of structural α-entropy. *Kybernetika* 3: 30–35.

27 Rényi, A. (1961). On measures of entropy and information. *4th Berkeley Symposium on Mathematical Statistics and Probability*, volume 1, 547–561.

28 Pardo, L., Morales, D., Salicrú, M., and Menéndez, M.L. (1997). Large sample behavior of entropy measures when parameters are estimated. *Communications in Statistics - Theory and Methods* 26(2): 483–501.

29 Seghouane, A.K. and Amari, S.I. (2007). The AIC criterion and symmetrizing the Kullback–Leibler divergence. *IEEE Transactions on Neural Networks* 18(1): 97–106.

30 Kailath, T. (1967). The divergence and Bhattacharyya distance measures in signal selection. *IEEE Transactions on Communication Technology* 15: 52–60.

31 Burbea, J. and Rao, C. (1982). Entropy differential metric, distance and divergence measures in probability spaces: a unified approach. *Journal of Multivariate Analysis* 12: 575–596.

32 Jager, L. and Wellner, J.A. (2007). Goodness-of-fit tests via phi-divergences. *The Annals of Statistics* 35(5): 2018–2053.

33 Frery, A.C., Muller, H.J., Yanasse, C.C.F., and Sant'Anna, S.J.S. (1997). A model for extremely heterogeneous clutter. *IEEE Transactions on Geoscience and Remote Sensing* 35(3): 648–659.

34 Wolfram Research, Inc. (2005). *Mathematica*. Champaign, IL, Version 5.2.

35 Cintra, R.J., Frery, A.C., and Nascimento, A.D.C. (2011). Parametric and nonparametric tests for speckled imagery. *Pattern Analysis and Applications* 16: 1–21.

36 Frery, A.C., Nascimento, A.D.C., and Cintra, R.J. (2014). Analytic expressions for stochastic distances between relaxed complex Wishart distributions. *IEEE Transactions on Geoscience and Remote Sensing* 52(2): 1213–1226.

37 Gambini, J., Mejail, M., Jacobo-Berlles, J., and Frery, A.C. (2006). Feature extraction in speckled imagery using dynamic B-spline deformable contours under the G0 model. *International Journal of Remote Sensing* 27(22): 5037–5059.

38 McConnell, J.J. (2005). *Computer Graphics: Theory Into Practice*. USA: Jones and Bartlett Publishers, Inc.

39 Gambini, J., Mejail, M., Jacobo-Berlles, J., and Frery, A.C. (2008). Accuracy of edge detection methods with local information in speckled imagery. *Statistics and Computing* 18(1): 15–26.

40 Frery, A.C., Jacobo-Berlles, J., Gambini, J., and Mejail, M. (2010). Polarimetric SAR image segmentation with B-Splines and a new statistical model. *Multidimensional Systems and Signal Processing* 21(4): 319–342.

41 Giron, E., Frery, A.C., and Cribari-Neto, F. (2012). Nonparametric edge detection in speckled imagery. *Mathematics and Computers in Simulation* 82: 2182–2198.

42 Freitas, C.C., Frery, A.C., and Correia, A.H. (2005). The polarimetric $\mathcal{G}$distribution for SAR data analysis. *Environmetrics* 16(1): 13–31.

43 Anfinsen, S.N., Doulgeris, A.P., and Eltoft, T. (2009). Estimation of the equivalent number of looks in polarimetric synthetic aperture radar imagery. *IEEE Transactions on Geoscience and Remote Sensing* 47(11): 3795–3809.

44 Frery, A.C., Cintra, R.J., and Nascimento, A.D.C. (2013). Entropy-based statistical analysis of PolSAR data. *IEEE Transactions on Geoscience and Remote Sensing* 51(6): 3733–3743.

45 De Borba, A.A., Muhuri, A., Marengoni, M., and Frery, A.C. (2023). Feature selection for edge detection in PolSAR images. *Remote Sensing* 15(9): 2479.

46 Buades, A., Coll, B., and Morel, J.M. (2005). A review of image denoising algorithms, with a new one. *Multiscale Modeling and Simulation* 4(2): 490–530.

47 Torres, L., Sant'Anna, S.J.S., Freitas, C.C., and Frery, A.C. (2014). Speckle reduction in polarimetric SAR imagery with stochastic distances and nonlocal means. *Pattern Recognition* 47(1): 141–157.

48 Deledalle, C.-A., Denis, L., Tupin, F. et al. (2013). NL-SAR: a unified non-local framework for resolution-preserving (Pol)(In)SAR denoising. *IEEE Transactions on Geoscience and Remote Sensing* 53(4): 2021–2038.

49 Gomez, L., Wu, J., and Frery, A.C. (2022). Non-local means filters for full polarimetric synthetic aperture radar images with stochastic distances. *Image Processing On Line* 12: 142–172.

50 Chan, D., Gambini, J., and Frery, A.C. (2022). Entropy-based non-local means filter for single-look SAR speckle reduction. *Remote Sensing* 14(3): 509.

51 Gomez, L., Ospina, R., and Frery, A.C. (2017). Unassisted quantitative evaluation of despeckling filters. *Remote Sensing* 9(9): 389.

52 Vitale, S., Ferraioli, G., Frery, A.C. et al. (2023). SAR despeckling using multi-objective neural network trained with generic statistical samples. *IEEE Transactions on Geoscience and Remote Sensing* 61: 1–12.

53 Devroye, L., Györfi, L., and Lugosi, G. (1996). *A Probabilistic Theory of Pattern Recognition*, Number 31 in Applications of Mathematics. New York: Springer-Verlag.

54 Cassetti, J., Delgadino, D., Rey, A.A., and Frery, A.C. (2022). Entropy estimators in SAR image classification. *Entropy* 24(4): 509.

55 Nascimento, A.D.C., Ferreira, J.A., and Frery, A.C. (2023). Unsupervised segmentation of PolSAR data with Complex Wishart and $\mathcal{G}_m^0$ distributions and Shannon entropy. *Statistics and Computing* 33(6).

56 Coelho, D., Cintra, R., Frery, A.C., and Dimitrov, V. (2018). Fast matrix inversion and determinant computation for polarimetric synthetic aperture radar. *Computers & Geosciences* 119: 109–114.

57 Dey, S., Bhattacharya, A., Ratha, D. et al. (2021). Target characterization and scattering power decomposition for full and compact polarimetric SAR data. *IEEE Transactions on Geoscience and Remote Sensing* 59(5): 3981–3998.

58 Mandal, D., Ratha, D., Bhattacharya, A. et al. (2020). A radar vegetation index for crop monitoring using compact polarimetric SAR data. *IEEE Transactions on Geoscience and Remote Sensing* 58(9): 6321–6335.

59 Ratha, D., Mandal, D., Kumar, V. et al. (2019). A generalized volume scattering model based vegetation index from polarimetric SAR data. *IEEE Geoscience and Remote Sensing Letters* 16(11): 1791–1795.

60 Ratha, D., Gamba, P., Bhattacharya, A., and Frery, A.C. (2020). Novel techniques for built-up area extraction from polarimetric SAR images. *IEEE Geoscience and Remote Sensing Letters* 17(1): 177–181.

61 Ratha, D., Pottier, E., Bhattacharya, A., and Frery, A.C. (2020). A PolSAR scattering power factorization framework and novel roll-invariant parameters based unsupervised classification scheme using a geodesic distance. *IEEE Transactions on Geoscience and Remote Sensing* 58(5): 3509–3525.

62 Yue, D.-X., Xu, F., Frery, A.C., and Jin, Y.-Q. (2020). A generalized Gaussian coherent scatterer model for correlated SAR texture. *IEEE Transactions on Geoscience and Remote Sensing* 58(4): 2947–2964.

63 Datcu, M., Huang, Z., Anghel, A. et al. (2023). Explainable, physics-aware, trustworthy artificial intelligence: a paradigm shift for synthetic aperture radar. *IEEE Geoscience and Remote Sensing Magazine* 11(1): 8–25.

8

Information Fusion and Target Tracking: Information-Theoretic Sensor Selection

Nianxia Cao[1], *Pramod K. Varshney*[2], *Engin Masazade*[3], *and Sora Haley*[2]

[1] *Advanced Safety and User Experience, Aptiv, Agoura Hills, CA, USA*
[2] *College of Engineering and Computer Science, Syracuse University, Syracuse, NY, USA*
[3] *Department of Electrical and Electronics Engineering, Faculty of Engineering, Marmara University, Istanbul, Türkiye*

8.1 Introduction

In numerous scenarios involving distributed sensing, such as system monitoring, surveillance, and the Internet of Things (IoT), vast number of sensors are densely scattered throughout the designated area. This chapter delves into the challenges of target tracking within a sensor network boasting a substantial sensor count. In such networks, where resources like energy and bandwidth are constrained, utilizing all sensors continuously proves to be inefficient. This is due to the presence of uninformative sensors, which consume resources without significantly contributing to the task at hand. Hence, sensor selection emerges as a critical approach, aiming to identify the optimal subset of sensors within the region during each observation period to ensure effective tracking performance within the given resource limitations [1–8], has been investigated in the literature. These inquiries entail the selection of sensor sets with the aim of accomplishing particular objectives, such as attaining good information gain or minimizing the error in estimation during state estimation of the target. For the problems considered in [2–4, 9] sensor selection metrics are derived based on involve mutual information (MI) and entropy. Conversely, in [5, 6], the focus of sensor selection lies in pinpointing sensors which achieve the minimum posterior Cramér–Rao lower bound (PCRLB), where PCRLB is inversely related to Fisher information (FI). For quantized sensor measurements, the work in [10], compares these two selection criteria, i.e., MI and PCRLB, noting that the PCRLB-driven approach yields comparable mean square error (MSE) and outcomes MI with substantially reduced computational demands. In a different context, the sensor selection is defined as an integer programming

Information-Theoretic Radar Signal Processing, First Edition. Edited by Yujie Gu and Yimin D. Zhang.

problem [7]. The integer-valued decision variables are first replaced with their continuous counterparts, and then solved using convex optimization techniques. The authors in [8] achieved a sensor selection strategy for multiple time steps ahead, through reformulating the Kalman filter and proposing an approach adaptable to diverse performance metrics and resource constraints.

Prior research on sensor selection has proceeded under the assumption of sensors operating reliably throughout target tracking, uninterrupted. However, real-world sensor observations are susceptible to uncertainty [11–15]. This uncertainty stems from factors like temporary sensor failures, environmental changes [13, 14], or external interference such as traffic or wildlife, causing fluctuations in sensor-received power. Communication channels may encounter random interruptions, and malicious entities may attempt jamming attacks with various tactics [15]. These uncertainties lead to incomplete and unreliable sets of sensor observations reaching the fusion center. Essentially, in an uncertain environment, sensor observations may carry valuable information about the target with a certain probability, which may diminish the network's reliability. Thus, it is imperative to work on the problem of sensor selection within the context of observation uncertainty. This study specifically examines uncertainty arising from occlusions, where obstacles obstruct sensors' view of the target. Initially proposed by Nahi [11] and Hadidi and Schwartz [12], sensor observations are described by a stochastic model. Then, the work presented in [13] and [14] expanded this model to include multiple sensors, providing a more realistic perspective by considering varying uncertainties over time. While past research explored Kalman filters for target tracking under uncertainty [12, 16–18], and target localization is addressed for nonideal channels [19, 20], the specific challenge of sensor selection with uncertain observations remains unexplored. This chapter aims to bridge this gap by focusing on this challenge.

The literature mentioned earlier revolves around sensor selection strategies where the prior knowledge of the number of sensors to be selected, A, is available. Efficient algorithms are then developed to identify the optimal A sensors capable of maximizing system performance. However, practical scenarios often lack foreknowledge of the exact number of sensors required per tracking at each time step. Hence, exploring sensor selection strategies capable of optimally and dynamically determining both the number and identity of sensors becomes crucial. As will be illustrated in this chapter, the FI-based selection scheme (FI-SS) tends to favor sensors which are in close proximity of the target. Conversely, the MI-based selection scheme (MI-SS) selects sensors which have higher sensing probabilities, resulting in significantly improved estimation performance. Notably, the enhanced performance of MI-SS is counterbalanced by its high computational complexity. To address this, we adopt a new metric based on the mutual information upper bound (MIUB) for sensor selection. The computational complexity of MIUB is similar to that of FI, and significantly lower than MI. Simulation experiments confirm that the MIUB-based selection scheme (MIUB-SS) achieves good tracking performance with less computational load. Furthermore, we extend our study by formulating a multiobjective optimization problem (MOP). We employ the Nondominating Sorting Genetic Algorithm-II (NSGA-II) to dynamically determine an optimal set of sensors to be selected at each time step of tracking. Numerical findings reveal that MIUB-SS tends to select more sensors than FI-SS. Comparisons are drawn with an alternative sensor selection method which maximize the weighted sum of objectives and selecting fixed number of sensors by using convex optimization. Experimental results demonstrate the adaptability of our proposed framework to determine the optimal number of sensors selected for each time step of tracking while ensuring satisfactory estimation performance.

We organize the subsequent sections of the chapter as follows: In Section 8.2, we present the system model. In Section 8.3, we present the target tracking framework utilizing a particle filter. Then, in Section 8.4, we introduce FI and MI as criteria for sensor selection and conduct comparative analyses through numerical experiments. Transitioning to Section 8.5, we first state the fundamentals of MOP, and then utilize NSGA-II to define the proposed MOP. This section also includes a comprehensive examination of the MOP framework's performance through simulations. Finally, Section 8.6 is dedicated to our conclusions and outlooks for future research directions.

8.2 Target Tracking Model

In this chapter, we consider a mobile target, in a given region of interest, emitting a signal. The target movement is monitored by a wireless sensor network (WSN) which involves a total of N sensors. The state of the target is defined by a vector of size four, $\boldsymbol{x}_t = [x_t, y_t, \dot{x}_t, \dot{y}_t]^{\mathrm{T}}$, where x_t and y_t represent the horizontal and the vertical positions of the target at time step t, while $\dot{x}_t$ and $\dot{y}_t$ denote the target velocities in the respective directions. The notation $[\cdot]^{\mathrm{T}}$ signifies the transpose operation on the vector. Although the methodologies discussed can accommodate more complex dynamic models for target motion, we opt for a linear constant velocity for the target's motion, outlined as follows:

$$\boldsymbol{x}_{t+1} = \boldsymbol{F}\boldsymbol{x}_t + \boldsymbol{v}_t, \tag{8.1}$$

where the state transition matrix $\boldsymbol{F}$ is defined as

$$\boldsymbol{F} = \begin{bmatrix} 1 & 0 & \Delta & 0 \\ 0 & 1 & 0 & \Delta \\ 0 & 0 & 1 & 0 \\ 0 & 0 & 0 & 1 \end{bmatrix}, \tag{8.2}$$

and Gaussian process noise $\boldsymbol{v}_t$ is assumed to be zero mean and has covariance matrix $\boldsymbol{Q}$,

$$\boldsymbol{Q} = \tau \begin{bmatrix} \frac{\Delta^3}{3} & 0 & \frac{\Delta^2}{2} & 0 \\ 0 & \frac{\Delta^3}{3} & 0 & \frac{\Delta^2}{2} \\ \frac{\Delta^2}{2} & 0 & \Delta & 0 \\ 0 & \frac{\Delta^2}{2} & 0 & \Delta \end{bmatrix}. \tag{8.3}$$

In equations (8.2) and (8.3), Δ represents the sampling interval, while τ denotes the parameter for process noise.

The received power by sensor n obeys the following attenuation model [19]:

$$P_{n,t}(\boldsymbol{x}_t) = \frac{P_0}{1 + \zeta d_{n,t}^{\gamma}}, \tag{8.4}$$

Here, P_0 denotes the signal power emitted by the target, γ signifies the decay exponent, ζ is a scaling parameter. At time step t, $d_{n,t} = \sqrt{(x_t - x_n^s)^2 + (y_t - y_n^s)^2}$ is the distance between the target and the location of the nth sensor (x_n^s, y_n^s).

8.2.1 Sensor Observation Model

As previously noted, sensor observations can have uncertainties originating from sensor failures, environmental interference, and/or random disruptions [21]. In this work, we narrow our focus to a specific scenario wherein the uncertainties in sensor observations arise due to obstacles. We adopt a probabilistic measurement model, initially proposed in [11] and subsequently generalized in works such as [13] and [14]. According to this model, sensor observations are assumed to consist solely of noise if the sensor's perception of the target is obstructed by obstacles. Considering that this uncertainty may arise unpredictably at any given time and for any sensor, it is crucial to recognize that the sensing probability of sensors may not be the same. Then, the measurement at the nth sensor is expressed as

$$z_{n,t} = \begin{cases} g_{n,t}(\boldsymbol{x}_t) + \mathrm{v}_{n,t}, & \text{with probability } p_s^{(n)}, \\ \mathrm{v}_{n,t}, & \text{with probability } 1 - p_s^{(n)}, \end{cases} \tag{8.5}$$

where we denote $p_s^{(n)}$ as the *sensing probability* associated with sensor n. Here, $g_{n,t}(\boldsymbol{x}_t) = \sqrt{P_{n,t}(\boldsymbol{x}_t)}$ denotes the received signal amplitude at the sensor, and $\mathrm{v}_{n,t}$ represents the measurement noise. This noise is assumed to be independent across time steps and sensors, assumed to be Gaussian distribution with mean 0 and variance σ^2 respectively.

When conditioned on $\boldsymbol{x}_t$, $z_{n,t}$ follows Gaussian distribution with mean $g_{n,t}(\boldsymbol{x}_t)$ and variance σ^2 with probability of $p_s^{(n)}$. Conversely, when conditioned on the same $\boldsymbol{x}_t$, $z_{n,t}$ follows Gaussian distribution with mean 0 and variance σ^2 with a probability $1 - p_s^{(n)}$, i.e.,

$$p(z_{n,t}|\boldsymbol{x}_t) = \left(1 - p_s^{(n)}\right) \mathcal{N}\left(0, \sigma^2\right) + p_s^{(n)} \mathcal{N}\left(g_{n,t}(\boldsymbol{x}_t), \sigma^2\right). \tag{8.6}$$

We can then define the sensor measurement vector at time step t as $\boldsymbol{z}_t = [z_{1,t}, \dots, z_{N,t}]^{\mathrm{T}}$, and the likelihood function of $\boldsymbol{z}_t$ given the target state $\boldsymbol{x}_t$, is essentially the product of the individual likelihood functions for each sensor n.

$$p(\boldsymbol{z}_t|\boldsymbol{x}_t) = \prod_{n=1}^{N} p(z_{n,t}|\boldsymbol{x}_t). \tag{8.7}$$

Regarding the communication between sensors and the fusion center, we consider the following scenarios:

1. Sensors transmit their raw (analog) measurements $\boldsymbol{z}_t$ directly to the fusion center.
2. Sensors first quantize their raw measurements, then M-bits quantized sensor measurements are sent to fusion center.

The use of raw sensor measurements ensures the availability of comprehensive observation data, while incurring high communication costs. Conversely, employing quantized measurements alleviates the communication burden but results in a loss of information regarding the target.

We define $\lambda_{n,t}$ as the M-bit quantized measurement of sensor n at time step t as:

$$\lambda_{n,t} = \begin{cases} 0, & -\infty < z_{n,t} < \eta_1 \\ 1, & \eta_1 \le z_{n,t} < \eta_2 \\ \vdots & \\ L-1 & \eta_{L-1} \le z_{n,t} < \infty \end{cases}, \tag{8.8}$$

where $\{\eta_1, \dots, \eta_{L-1}\}$ is the set of quantization decision thresholds and $L = 2^M$ is the total number of quantization levels. The probability that $\lambda_{n,t}$ takes the value l ($l \in \{1, \dots, L-2\}$) is

$$\begin{aligned} p(\lambda_{n,t} = l|\boldsymbol{x}_t) &= \Pr\left(\eta_l \le z_{n,t} < \eta_{l+1}|\boldsymbol{x}_t\right) \\ &= \left(1 - p_s^{(n)}\right) \Pr\left(\eta_l \le z_{n,t} < \eta_{l+1}|z_{n,t} \sim \mathcal{N}(0, \sigma^2)\right) \\ &\quad + p_s^{(n)} \Pr\left(\eta_l \le z_{n,t} < \eta_{l+1}|z_{n,t} \sim \mathcal{N}(g_{n,t}(\boldsymbol{x}_t), \sigma^2)\right) \\ &= \left(1 - p_s^{(n)}\right) \left[Q\left(\frac{\eta_l}{\sigma}\right) - Q\left(\frac{\eta_{l+1}}{\sigma}\right)\right] \\ &\quad + p_s^{(n)} \left[Q\left(\frac{\eta_l - g_{n,t}(\boldsymbol{x}_t)}{\sigma}\right) - Q\left(\frac{\eta_{l+1} - g_{n,t}(\boldsymbol{x}_t)}{\sigma}\right)\right] \end{aligned}, \tag{8.9}$$

where $Q(\cdot)$ is the complementary cumulative distribution of Gaussian distribution with zero mean and unit variance as,

$$Q(x) = \int_x^\infty \frac{1}{\sqrt{2\pi}} \exp\left\{-\frac{y^2}{2}\right\} dy. \tag{8.10}$$

Note that $p(\lambda_{n,t} = 0|\boldsymbol{x}_t) = \Pr(-\infty < z_{n,t} < \eta_1|\boldsymbol{x}_t)$ and $p(\lambda_{n,t} = L-1|\boldsymbol{x}_t) = \Pr(\eta_{L-1} < z_{n,t} < \infty|\boldsymbol{x}_t)$ can be calculated similarly with (8.9). Given the target state $\boldsymbol{x}_t$, the sensor measurements are conditionally independent, the likelihood function for $\lambda_t = [\lambda_{1,t}, \lambda_{2,t}, \dots, \lambda_{N,t}]^T$ can be expressed as the product of individual likelihood functions for each sensor n as in (8.7).

8.3 Particle Filtering for Target Tracking

In WSNs, we define tracking as the state estimation based on a vector of sensor measurements at each time step of tracking. If the process and/or measurement models are nonlinear, the extended Kalman filter (EKF) may provide suboptimal solutions. On the other hand, the performance of the EKF may deteriorate significantly, particularly when sensor measurements are quantized, especially with a small number of quantization levels [22]. Since the target tracking problem considered in this chapter has nonlinear measurement model involving both analog or quantized sensor measurements, we employ sequential importance resampling (SIR) particle filter [23, 24].

The fundamental idea of particle filter is to construct a discrete representation of the posterior probability distribution of the target state $p(\boldsymbol{x}_t|\boldsymbol{z}_{1:t})$ for analog sensor measurements, or $p(\boldsymbol{x}_t|\boldsymbol{\lambda}_{1:t})$ for quantized sensor measurements, through a set of particles $\boldsymbol{x}_t^p$ accompanied by corresponding weights w_t^p as,

$$p(\boldsymbol{x}_t|\boldsymbol{z}_{1:t}) \approx \sum_{p=1}^{N_p} w_t^p \delta(\boldsymbol{x}_t - \boldsymbol{x}_t^p), \tag{8.11}$$

Above, $\delta(\cdot)$ is the Dirac delta function, N_p is the number of particles. Algorithm 8.1 summarizes the SIR particle filtering with analog data $\boldsymbol{z}_t$, where the notation can transition to $\boldsymbol{\lambda}_t$ easily if quantized data is considered. T_s denotes the total number of time steps. The particle weights are determined based on the likelihoods given in (8.6) or (8.9) for the analog ($\boldsymbol{z}_t$) or quantized ($\boldsymbol{\lambda}_t$) measurements, respectively. Upon normalizing the weights, $\hat{\boldsymbol{x}}_t$ represents the state estimate $\boldsymbol{x}_t$. Finally, the step of resampling prevents the occurrence of the degeneracy phenomenon, where most important weights tend toward zero after a few iterations.

Algorithm 8.1: SIR particle filter for target tracking. Reproduced from [25]/with permission of IEEE.

Set $t = 1$. Produce initial set of particles $\boldsymbol{x}_0^p \sim p(\boldsymbol{x}_0)$ with $\forall p$, $w_0^p = N_p^{-1}$.
while $t \leq T_s$ **do**
  $\boldsymbol{x}_t^p = \boldsymbol{F}\boldsymbol{x}_{t-1}^p + \boldsymbol{v}_t$ (Propagating particles)
  $p(\boldsymbol{x}_t|\boldsymbol{z}_{1:t}) = \frac{1}{N_p}\sum_{p=1}^{N_p}\delta(\boldsymbol{x}_t - \boldsymbol{x}_t^p)$
  Obtain sensor measurements $\boldsymbol{z}_t$
  $w_t^p \propto p(\boldsymbol{z}_t|\boldsymbol{x}_t^p)$ (Update weights)
  $w_t^p = \frac{w_t^p}{\sum_{p=1}^{N_p} w_t^p}$ (Normalizing weights)
  $\hat{\boldsymbol{x}}_t = \sum_{p=1}^{N_p} w_t^p \boldsymbol{x}_t^p$
  $\{\boldsymbol{x}_t^p, N_p^{-1}\} = \text{Resampling}(\boldsymbol{x}_t^p, w_t^p)$
  $t = t + 1$
end while

8.4 Information-Theoretic Sensor Selection

In this section, our focus is directed toward the introduction and exploration of three distinct information-theoretic performance metrics for sensor selection: FI, MI, and MIUB. These metrics serve as analytical tools in addressing the sensor selection problem under uncertain conditions. Our subsequent efforts encompass the formulation of these metrics for both the analog data and quantized data scenarios as well as a comprehensive comparison based on the resultant tracking performance.

8.4.1 Fisher Information

Within the context of Bayesian estimation, the PCRLB stands as a theoretical performance bound [26]. Let $p(\boldsymbol{z}_t, \boldsymbol{x}_t)$ denote the joint probability density function of sensor measurements and the target state. The PCRLB, characterizing the lower bound of estimation error, is then expressed as [26]:

$$\mathbb{E}\left\{[\hat{\boldsymbol{x}}_t - \boldsymbol{x}_t][\hat{\boldsymbol{x}}_t - \boldsymbol{x}_t]^{\mathrm{T}}\right\} \geq \boldsymbol{J}_t^{-1}, \tag{8.12}$$

Here $\boldsymbol{J}_t$ is the FI matrix. An important observation highlighted in [27] is the decomposition of the FI matrix into two distinct components: the FI matrix of sensor measurements, and the FI matrix of the *a priori* information. Additionally, since sensor measurements as conditionally independent given the target state $\boldsymbol{x}_t$, the FI matrix of the measurements can be represented as the sum of each sensor's individual FI matrix. Then, the FI matrix is written as,

$$\boldsymbol{J}_t \triangleq \sum_{n=1}^{N} \int_{\boldsymbol{x}_t} \boldsymbol{J}^P_{n,t}(\boldsymbol{x}_t) p(\boldsymbol{x}_t) d\boldsymbol{x}_t + \boldsymbol{J}^P_t, \tag{8.13}$$

where $p(\boldsymbol{x}_t)$ is the approximated probability distribution of the target $\boldsymbol{x}_t$ as in (8.11), $\boldsymbol{J}^P_t$ denotes the FI matrix of the *a priori* information, and

$$\boldsymbol{J}^P_{n,t}(\boldsymbol{x}_t) = \int_{z_{n,t}} \frac{1}{p(z_{n,t}|\boldsymbol{x}_t)} \left[\frac{\partial p(z_{n,t}|\boldsymbol{x}_t)}{\partial \boldsymbol{x}_t}\right] \left[\frac{\partial p(z_{n,t}|\boldsymbol{x}_t)}{\partial \boldsymbol{x}_t}\right]^{\mathrm{T}} dz_{n,t} \tag{8.14}$$

is the FI matrix of sensor n with $\boldsymbol{x}_t$ as its target state.

8.4.2 Fisher Information Derivation for Analog Sensor Measurements

For analog data, the FI can be derived by substituting the likelihood function $p(z_{n,t}|\boldsymbol{x}_t)$, as presented in (8.6), into the expression (8.14). The derivative of $p(z_{n,t}|\boldsymbol{x}_t)$ is then computed as follows:

$$\frac{\partial p(z_{n,t}|\boldsymbol{x}_t)}{\partial \boldsymbol{x}_t} = p_s^{(n)} \frac{z_{n,t} - g_{n,t}(\boldsymbol{x}_t)}{\sigma^3\sqrt{2\pi}} \exp\left\{-\frac{(z_{n,t} - g_{n,t}(\boldsymbol{x}_t))^2}{2\sigma^2}\right\} \frac{\partial g_{n,t}(\boldsymbol{x}_t)}{\partial \boldsymbol{x}_t}, \tag{8.15}$$

where

$$\frac{\partial g_{n,t}(\boldsymbol{x}_t)}{\partial \boldsymbol{x}_t} = -\frac{\zeta\gamma}{2} \frac{g_{n,t}(\boldsymbol{x}_t) d_{n,t}^{\gamma-2}}{1+\zeta d_{n,t}^{\gamma}} \begin{bmatrix} x_n^s - x_t \\ y_n^s - y_t \\ 0 \\ 0 \end{bmatrix}. \tag{8.16}$$

By incorporating (8.6), (8.15), and (8.16) into (8.14), we can define the standard FI matrix for analog data as follows:

$$\begin{aligned}
\boldsymbol{J}^S_{n,t}(\boldsymbol{x}_t) &= \left(p_s^{(n)}\right)^2 \kappa_{n,t}(\boldsymbol{x}_t) \left[\frac{\partial g_{n,t}(\boldsymbol{x}_t)}{\partial \boldsymbol{x}_t}\right] \left[\frac{\partial g_{n,t}(\boldsymbol{x}_t)}{\partial \boldsymbol{x}_t}\right]^{\mathrm{T}} \\
&= \kappa_{n,t}(\boldsymbol{x}_t) \frac{\left(p_s^{(n)}\right)^2 \zeta^2\gamma^2 g_{n,t}^2(\boldsymbol{x}_t) d_{n,t}^{2\gamma-4}}{4(1+\zeta d_{n,t}^{\gamma})^2} \\
&\times \begin{bmatrix} (x_n^s - x_t)^2 & (x_n^s - x_t)(y_n^s - y_t) & 0 & 0 \\ (x_n^s - x_t)(y_n^s - y_t) & (y_n^s - y_t)^2 & 0 & 0 \\ 0 & 0 & 0 & 0 \\ 0 & 0 & 0 & 0 \end{bmatrix},
\end{aligned} \tag{8.17}$$

where

$$\kappa_{n,t}(\boldsymbol{x}_t) = \int_{z_{n,t}} \frac{1}{p(z_{n,t}|\boldsymbol{x}_t)} \left[\frac{z_{n,t} - g_{n,t}(\boldsymbol{x}_t)}{\sigma^3\sqrt{2\pi}} \exp\left\{ -\frac{\left(z_{n,t} - g_{n,t}(\boldsymbol{x}_t)\right)^2}{2\sigma^2} \right\} \right]^2 dz_{n,t}. \tag{8.18}$$

This leads us to the derivation of the FI for the analog observation model.

8.4.3 Fisher Information Derivation for Quantized Sensor Measurements

FI calculation for quantized data entails substituting the likelihood function $p(z_{n,t}|\boldsymbol{x}_t)$ from (8.14) with $p(\lambda_{n,t}|\boldsymbol{x}_t)$ as defined in (8.9). Given that the derivative of the likelihood function for quantized observations is:

$$\begin{aligned}\frac{\partial p(\lambda_{n,t} = l|\boldsymbol{x}_t)}{\partial \boldsymbol{x}_t} &= \frac{p_s^{(n)}}{\sigma\sqrt{2\pi}} \left[\exp\left\{ -\frac{(\eta_l - g_{n,t}(\boldsymbol{x}_t))^2}{2\sigma^2} \right\} \right. \\ &\quad \left. - \exp\left\{ -\frac{(\eta_{l+1} - g_{n,t}(\boldsymbol{x}_t))^2}{2\sigma^2} \right\} \right] \frac{\partial g_{n,t}(\boldsymbol{x}_t)}{\partial \boldsymbol{x}_t},\end{aligned} \tag{8.19}$$

the FI for quantized data by substituting (8.16) into (8.19) is derived as follows:

$$\begin{aligned}\boldsymbol{J}_{n,t}^{SQ}(\boldsymbol{x}_t) &= \sum_{l=0}^{L-1} \frac{1}{p(\lambda_{n,t} = l|\boldsymbol{x}_t)} \left[\frac{\partial p(\lambda_{n,t} = l|\boldsymbol{x}_t)}{\partial \boldsymbol{x}_t} \right]^2 \\ &= \left(p_s^{(n)}\right)^2 \kappa_{n,t}^Q(\boldsymbol{x}_t) \left[\frac{\partial g_{n,t}(\boldsymbol{x}_t)}{\partial \boldsymbol{x}_t} \right] \left[\frac{\partial g_{n,t}(\boldsymbol{x}_t)}{\partial \boldsymbol{x}_t} \right]^{\mathrm{T}} \\ &= \kappa_{n,t}^Q(\boldsymbol{x}_t) \frac{\left(p_s^{(n)}\right)^2 \zeta^2\gamma^2 g_{n,t}^2(\boldsymbol{x}_t) d_{n,t}^{2\gamma-4}}{4(1 + \zeta d_{n,t}^{\gamma})^2} \\ &\quad \times \begin{bmatrix} (x_n^s - x_t)^2 & (x_n^s - x_t)(y_n^s - y_t) & 0 & 0 \\ (x_n^s - x_t)(y_n^s - y_t) & (y_n^s - y_t)^2 & 0 & 0 \\ 0 & 0 & 0 & 0 \\ 0 & 0 & 0 & 0 \end{bmatrix},\end{aligned} \tag{8.20}$$

where

$$\begin{aligned}\kappa_{n,t}^Q(\boldsymbol{x}_t) &= \sum_{l=0}^{L-1} \frac{1}{2\pi\sigma^2 p(\lambda_{n,t} = l|\boldsymbol{x}_t)} \left[\exp\left\{ -\frac{(\eta_l - g_{n,t}(\boldsymbol{x}_t))^2}{2\sigma^2} \right\} \right. \\ &\quad \left. - \exp\left\{ -\frac{(\eta_{l+1} - g_{n,t}(\boldsymbol{x}_t))^2}{2\sigma^2} \right\} \right]^2.\end{aligned} \tag{8.21}$$

This leads us to the derivation of the FI for the quantized observation model.

8.4.4 Mutual Information

Given the probability distribution of the target state and the likelihood function associated with sensor measurements, the MI between the target state $\boldsymbol{x}_t$ and analog sensor measurements $\boldsymbol{z}_t$ is expressed as [3, 10]

$$
\begin{aligned}
I(\boldsymbol{x}_t;\boldsymbol{z}_t) &= h(\boldsymbol{z}_t) - h(\boldsymbol{z}_t|\boldsymbol{x}_t) \\
&= -\int_{\boldsymbol{z}_t}\left\{\int_{\boldsymbol{x}_t} p(\boldsymbol{z}_t|\boldsymbol{x}_t)p(\boldsymbol{x}_t)d\boldsymbol{x}_t\right\}\left\{\log_2\left[\int_{\boldsymbol{x}_t} p(\boldsymbol{z}_t|\boldsymbol{x}_t)p(\boldsymbol{x}_t)d\boldsymbol{x}_t\right]\right\} d\boldsymbol{z}_t \\
&\quad + \sum_{n=1}^{N}\int_{\boldsymbol{x}_t}\left[\int_{z_{n,t}} p(z_{n,t}|\boldsymbol{x}_t)\log_2 p(z_{n,t}|\boldsymbol{x}_t)dz_{n,t}\right] p(\boldsymbol{x}_t)d\boldsymbol{x}_t,
\end{aligned}
\tag{8.22}
$$

Here, $h(\boldsymbol{z}_t)$ denotes the differential entropy of the analog sensor measurements $\boldsymbol{z}_t$, while $h(\boldsymbol{z}_t|\boldsymbol{x}_t)$ indicates its conditional differential entropy given the target state $\boldsymbol{x}_t$. Then, the MI for quantized sensor measurements can be formulated in a similar fashion as:

$$
\begin{aligned}
I(\boldsymbol{x}_t;\lambda_t) &= H(\lambda_t) - H(\lambda_t|\boldsymbol{x}_t) \\
&= -\sum_{l_1=0}^{L-1}\cdots\sum_{l_N=0}^{L-1}\left\{\int_{\boldsymbol{x}_t}\left(\prod_{n=1}^{N} p(\lambda_{n,t}=l_n|\boldsymbol{x}_t)\right)p(\boldsymbol{x}_t)d\boldsymbol{x}_t\right. \\
&\quad \left.\times\log_2\left[\int_{\boldsymbol{x}_t}\left(\prod_{n=1}^{N} p(\lambda_{n,t}=l_n|\boldsymbol{x}_t)\right)p(\boldsymbol{x}_t)d\boldsymbol{x}_t\right]\right\} \\
&\quad + \sum_{n=1}^{N}\int_{\boldsymbol{x}_t}\left[\sum_{l=0}^{L-1} p(\lambda_{n,t}=l|\boldsymbol{x}_t)\log_2 p(\lambda_{n,t}|\boldsymbol{x}_t)\right] p(\boldsymbol{x}_t)d\boldsymbol{x}_t,
\end{aligned}
\tag{8.23}
$$

where $H(\lambda_t)$ is the entropy of the quantized sensor measurements λ_t, and $H(\lambda_t|\boldsymbol{x}_t)$ represents its conditional entropy. Note that $H(\lambda_t)$ requires a computationally intense N-fold summation over all possible sensors and over all possible quantization levels.

8.4.5 Mutual Information Upper Bound

The computational load associated with computing the MI for a set of A sensors escalates exponentially with the number of selected sensors A [10]. As a result, assessing the MI in equations (8.22) and (8.23) becomes unfeasible when dealing with a large number of selected sensors. This challenge has been underscored in [3] and [28]. Using the chain rule [29], the MI can be writted as,

$$
I(\boldsymbol{z}_t;\boldsymbol{x}_t) = \sum_{n=1}^{N} I(z_{n,t};\boldsymbol{x}_t|z_{n-1,t},\ldots,z_{1,t}). \tag{8.24}
$$

leading to,

$$I(z_{n,t}; \boldsymbol{x}_t | z_{n-1,t}, \dots, z_{1,t}) \leq I(z_{n,t}; \boldsymbol{x}_t). \tag{8.25}$$

Hence, the summation $\sum_{n=1}^{N} I(z_{n,t}; \boldsymbol{x}_t)$ serves as an upper bound for MI in (8.22). Note that same principles apply to quantized sensor measurements as well. Note that the computational complexity associated with evaluating MIUB for selecting A out of N sensors now increases linearly with A, which is comparable to that of computing FI. We use the MIUB as the performance metric for our sensor selection problem.

8.4.6 Performance Analysis of Sensor Selection Methods by Numerical Experiments

In this part, we carry out numerical experiments to assess the effectiveness of the three sensor selection metrics based on FI, MI, and MIUB in terms of estimation error in tracking and number of sensors selected at each time step of tracking.

8.4.6.1 Simulation Parameters

In our simulations, we consider the WSN as depicted in Figure 8.1, featuring $N = 6 \times 6 = 36$ sensors distributed uniformly across an area measuring

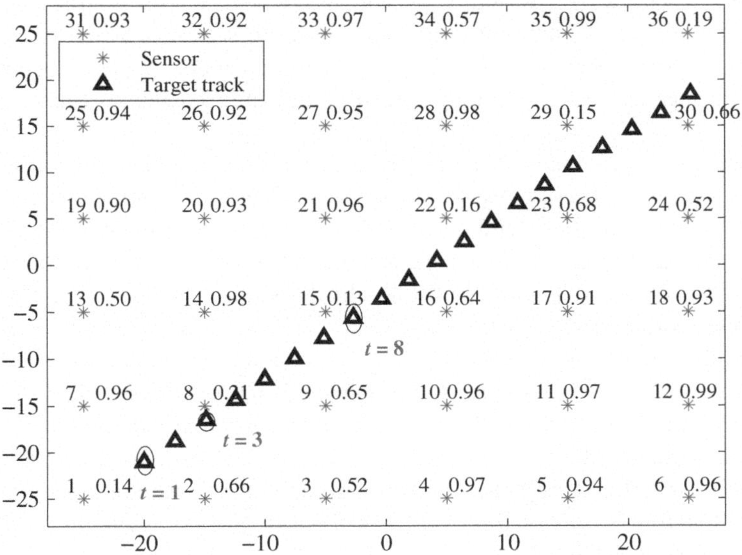

Figure 8.1 A WSN with $N = 36$ unreliable sensors. The numbers above the asterisks denote the sensor index (left) and its corresponding sensing probability (right). x-axis and y-axis represent the horizontal and vertical dimensions in meters in a two-dimensional plane. Source: Reproduced from [25]/with permission of IEEE.

$b^2 = 50 \times 50\,\text{m}^2$. In this study, we presume that the fusion center is already acquainted with each sensors sensing probability. Prior literature, such as [30, 31], studies detection (or sensing) probability estimation of each sensor. Moreover, as cited in [11], sensing probabilities can also be inferred from historical data. For a detector with a false alarm rate p_f, sensing probabilities are modeled as a function of p_f, contingent upon the detector's decision regarding signal presence or absence. Hence, for a target tracking scenario, one can either utilize past data for learning the sensing probabilities by employing methods like logistic regression [31], or directly assign each sensors sensing probability as functions of the false alarm probability. In real-world scenarios, sensing probability p_s is influenced by various reasons such as sensor characteristics, attributes or positions of the obstacles, weather conditions, and other environmental factors [30, 31]. Given the context-specific nature of these probabilities, we do not specifically address the sensing probability estimation in this study, leaving it as a subject for future research. Typically, when sensors positioned near the trajectory of the target exhibit higher sensing probabilities compared to others, the algorithm tends to prioritize selecting these sensors, leveraging their dual advantages of heightened signal power and sensing likelihood. However, to assess the robustness of our algorithm, we deliberately examine more challenging scenarios. In particular, we assume that sensors situated close to the target trajectory possess lower sensing probabilities, as depicted in Figure 8.1. Additionally, we concentrate on scenarios where each sensor boasts different sensing probabilities. This approach enables us to delve into a wider range of sensor selection outcomes and their effects on tracking performance.

In the linear dynamical model described by (8.1), we select the time interval as $\Delta = 1.25$ seconds, the process noise parameter as $\tau = 2.5 \times 10^{-3}$, emitted signal power as $P_0 = 1000$, and the standard deviation of the measurement noise as $\sigma = 0.2$. For M-bit quantized sensor measurements, quantization thresholds $[\eta_1, \dots, \eta_{L-1}]$ are obtained by uniformly dividing the interval $[-\sigma, \sigma + \sqrt{P_0}]$. The prior distribution regarding the target state, $p(\boldsymbol{x}_0)$, is also assumed to be Gaussian, with a mean of $\boldsymbol{\mu}_0 = [-23, -24, 2, 2]^{\mathrm{T}}$ and a covariance matrix of $\boldsymbol{\Sigma}_0 = \text{diag}[6^2, 6^2, 0.1^2, 0.1^2]$, represented as a diagonal matrix. We initialize the particle filter with $N_p = 5000$ particles drawn from the initial distribution $p(\boldsymbol{x}_0)$. To gauge the accuracy of our estimates, MSE between the estimated position values and the actual ground truth is used. Throughout our experiments, we conduct multiple trials to ensure the robustness of our results, averaging the MSE over T_{total} trials, as depicted below:

$$\text{MSE}_t = \frac{1}{T_{\text{total}}} \sum_{tr=1}^{T_{\text{total}}} \left(\hat{\boldsymbol{x}}_t^{\text{tr}}(1) - \boldsymbol{x}_t^{\text{tr}}(1)\right)^2 + \left(\hat{\boldsymbol{x}}_t^{\text{tr}}(2) - \boldsymbol{x}_t^{\text{tr}}(2)\right)^2, \tag{8.26}$$

where $\hat{x}_t^{tr}(1)$ and $\hat{x}_t^{tr}(2)$ represent the first two elements of the estimated target location corresponding to the estimated target location on x-axis and y-axis at time step t of trial tr.

8.4.6.2 Sensors with Highest MI or FI at Different Time Steps

In our analysis, we start with comparing analog sensor measurements and two different quantized sensor measurements ($M = 5$ and $M = 2$) transmitted to the fusion center for a sample target trajectory. The sensors with the highest MI or FI values are illustrated in Table 8.1. Several critical assumptions are stated follows:

- Given the matrix nature of FI in our context, we evaluate the FI matrix determinant, corresponding to the area of the uncertainty ellipsoid [32].
- Our primary focus lies in understanding how the distance between sensors and the target location, and the associated sensing probabilities impact performance metrics. Therefore, we assess the performance metric for each sensor individually, rather than examining various sets of multiple sensors.
- It is also noteworthy that for individual sensors, MI and MIUB coincide.

In general, quantization introduces information loss as compared to raw sensor measurements. To commence our discussion, let's first delve into the outcomes concerning analog data and 5-bit quantized data. Table 8.1 shows that sensors with the highest MI or FI values keep unchanged for both analog and 5-bit quantized data. This shows that 5-bit quantization keeps a substantial amount of information regarding sensor selection. Furthermore, we explore a comparison of results across three specific time steps:

- At time step 1, the target is in close proximity to both sensors 2 and 7, with nearly identical distances from the target. Consequently, sensors 2 and 7 have

Table 8.1 Sensors exhibiting the highest MI or FI across various time steps.

Metric	Time step	Analog	Quantized $M = 5$	Quantized $M = 2$
MI	$t = 1$	2,7	2,7	2,7
FI	$t = 1$	2,7	2,7	2,7
MI	$t = 3$	8	8	2,7,14
FI	$t = 3$	8	8	2,7,9
MI	$t = 8$	16	16	10,16,21
FI	$t = 8$	15	15	10,16

Source: Adapted from [25].

the highest MI and FI values. Despite sensors 1 and 8 being similarly close to the target as sensors 2 and 7, their reduced sensing probabilities result in lower MI and FI. This scenario shows the significance of both distance and sensing probability in influencing MI and FI values.
- At time step 3, the target is very close to sensor 8 is notably closer compared to the other sensors. Despite sensor 8 having a lower sensing probability, its close proximity to the target leads to the highest MI and FI values among all sensors.
- At time step 8, sensor 15 emerges as the closest one to the target, despite having a notably low sensing probability. Following closely is sensor 16, which has a high sensing probability but is still slightly farther from the target. This close proximity of sensor 15 results in highest FI, despite its low sensing probability. On the other hand, sensor 16 demonstrates the highest MI, indicating the combined impact of its relatively greater distance from the target and better sensing probability.

When examining the case of 2-bit quantized data, it becomes apparent that the level of information conveyed about the target is notably lower than that of analog data or 5-bit quantized data. Consequently, the impact of sensing probabilities on both FI and MI becomes more significant. Consequently, sensors with relatively higher sensing probabilities exhibit elevated FI and MI values, as illustrated in Table 8.1. In summary, our analysis yields the following conclusions:

- *Impact of sensing probabilities*: In scenarios involving analog or quantized sensor measurements with a considerable number of quantization levels, MI tends to be more influenced by the sensing probabilities compared to FI. However, when dealing with quantized data featuring only a small number of quantization levels, both MI and FI demonstrate significant sensitivity to sensing probabilities.
- *Selection patterns of FI and MI*: The sensor selection method based on FI tends to prioritize sensors that are physically closer to the target. This tendency can be explained by Equations (8.17) and (8.20), where the distance between the target and the sensors plays a predominant role in determining the value of FI. However, a similarly straightforward explanation does not exist for MI. In essence, both the sensor's distance to the target and its sensing probability, along with the number of quantization levels employed, collectively contribute to the overall tracking performance.

8.4.6.3 Tracking Performance

In Figures 8.2 and 8.3, we show the MSE performance and the average percentage of reliable sensors, respectively, for the WSN configuration illustrated in

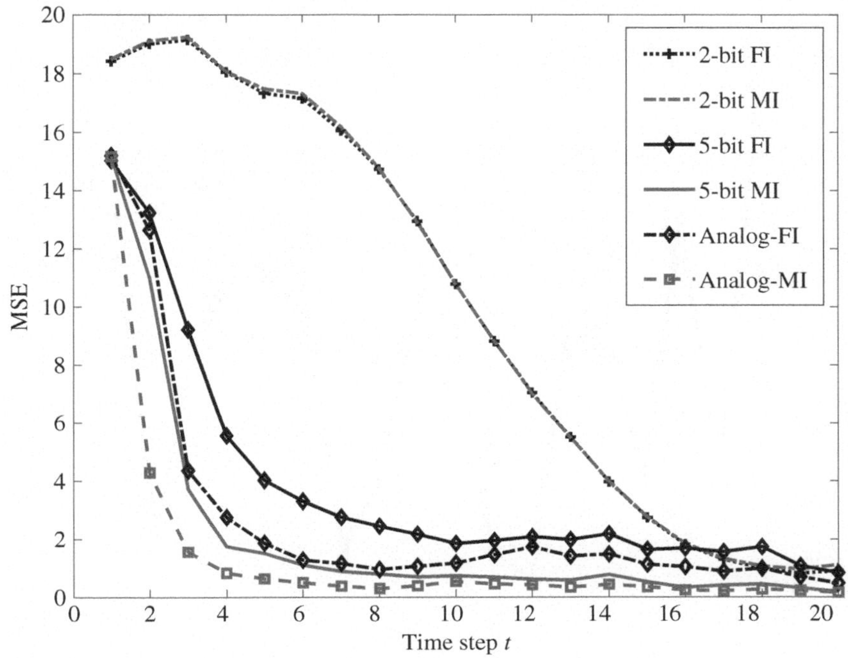

Figure 8.2 MSE of target tracking performance with analog, 5-bit quantized, and 2-bit quantized sensor measurements. (MSE has unit m^2). Source: Reproduced from [25]/with permission of IEEE.

Figure 8.1, while selecting only one sensor at each time step across $T_{\text{total}} = 500$ Monte Carlo runs. Figure 8.2 illustrates that the MI-SS consistently outperforms the FI-SS concerning MSE for both analog and 5-bit quantized measurements. This observed outcome can be attributed to the proportion of *reliable* sensors selected, as depicted in Figure 8.3. It is worth noting that a sensor is deemed *unreliable* if its signal amplitude closely matches noise.[1] From our observations, it is evident that approximately 60% of the sensors chosen by the MI-SS approach are reliable, whereas only about 40% of the sensors chosen by the FI-SS approach meet this criterion. This discrepancy explains why MI-SS achieves superior MSE compared to FI-SS. Additionally, our findings indicate that the sensor selection scheme operating with 2-bit quantized measurements yields an even higher proportion of reliable sensors. However, this enhancement does not translate into improved MSE performance due to the significant information loss during the quantization process. Figure 8.2 indicates that the sensor selection scheme

1 In our experiments, we determine whether the signal amplitude falls within the interval of $[-3\sigma, 3\sigma]$.

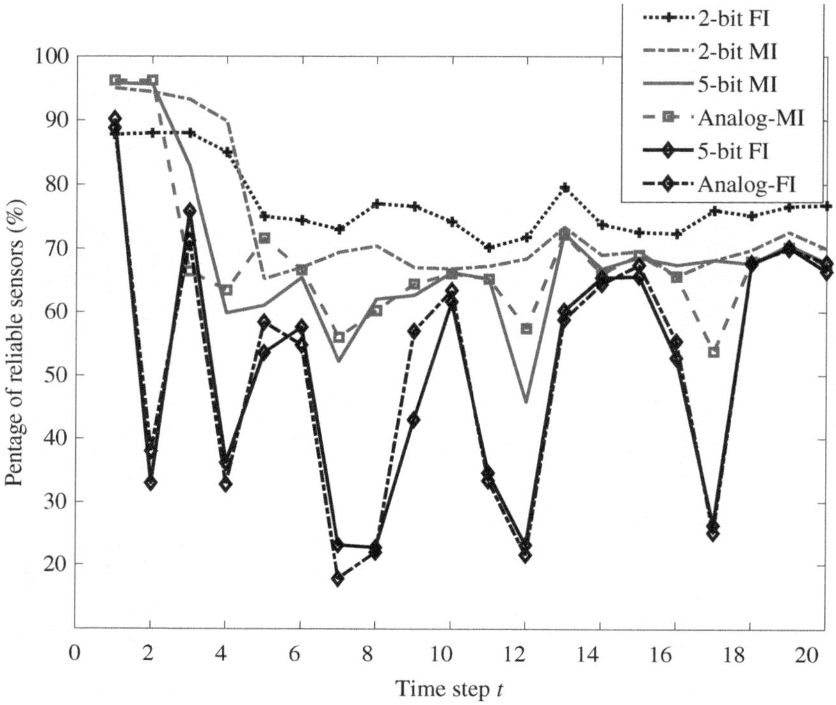

Figure 8.3 Average percentage of selected reliable sensors with analog, 5-bit quantized, and 2-bit quantized sensor measurements. Source: Reproduced from [25]/with permission of IEEE.

utilizing analog data displays the most favorable tracking performance. The performance of the sensor selection scheme using 5-bit quantized measurements closely approximates that of analog data. Conversely, the sensor selection scheme operating with 2-bit quantized measurements exhibits considerably poorer tracking performance. Consequently, we focus on presenting simulation results solely for the 5-bit quantized measurements in the subsequent experiments.

8.4.6.4 Performance of MIUB-SS

The computational complexity associated with calculating MIUB when a subset of A sensors are selected out of a total of N sensors is equivalent to that of computing FI. While the computational complexity of Both FI and MIUB increase linearly with the number of selected sensors A, the computational complexity of MI grows exponentially with the number of selected sensors. In Figure 8.4, we present the results obtained from the MI-SS and MIUB-SS algorithms when

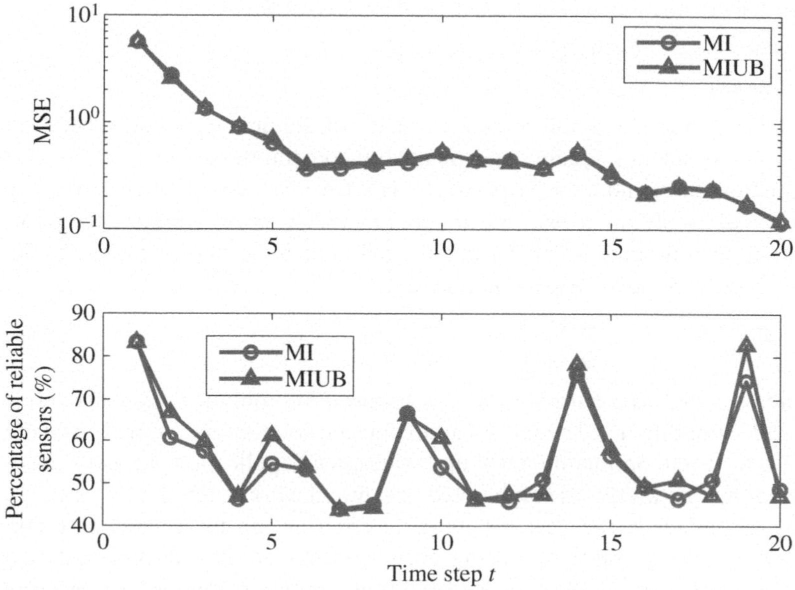

Figure 8.4 Target tracking performance for MI and MIUB where MSE has unit m^2, $A = 2$. Source: Reproduced from [25]/with permission of IEEE.

$A = 2$ sensors are selected at each time step of tracking. The performance of MI-SS and MIUB-SS is observed to be similar, not only in terms of the proportion of reliable sensors selected but also in terms of MSE as well. Therefore, MIUB-SS achieves comparable performance to MI-SS while requiring significantly less computational complexity. Consequently, in Section 8.5, we employ MIUB-SS instead of MI-SS within the multiobjective optimization framework, and then compare it with FI-SS.

8.5 Sensor Selection Using Multiobjective Optimization

This section studies the use of the MOP to determine the optimal sensor selection strategy. Our MOP is composed of two conflicting objective functions: estimation error in tracking and the total number of sensors selected at each time step of tracking. Between these two objective functions various sensor selection strategies depict diverse tradeoffs, both of which rely on the binary sensor selection variables.

An M-objective optimization problem can be described by:

$$\begin{array}{ll}\min\limits_{\boldsymbol{\alpha}} & \{f_1(\boldsymbol{\alpha}), f_2(\boldsymbol{\alpha}), \dots, f_M(\boldsymbol{\alpha})\} \\ \text{subject to} & a \leq \alpha_n \leq b, c_1(\boldsymbol{\alpha}) = 0, c_2(\boldsymbol{\alpha}) \leq 0,\end{array} \tag{8.27}$$

Here $\boldsymbol{\alpha}$ is the vector of decision variables with its nth element α_n bounded between a and b. The problem's equality and inequality constraints are denoted by arbitrary functions $c_1(\cdot)$ and $c_2(\cdot)$, respectively. The feasible set C of MOP comprises solutions that satisfy the constraints defined by (8.27). In an optimization problem aiming to minimize all objectives, the solution $\boldsymbol{\alpha}^1$ is said to dominate the another solution $\boldsymbol{\alpha}^2$ ($\boldsymbol{\alpha}^1 \succ \boldsymbol{\alpha}^2$) if and only if

$$\begin{array}{ll} f_u(\boldsymbol{\alpha}^1) \leq f_u(\boldsymbol{\alpha}^2) & \forall u \in \{1, 2, \dots, M\} \\ f_v(\boldsymbol{\alpha}^1) < f_v(\boldsymbol{\alpha}^2) & \exists v \in \{1, 2, \dots, M\}, \end{array} \tag{8.28}$$

Then, $\boldsymbol{\alpha}^\star$ is stated as a Pareto optimal solution if and only if there exists no $\boldsymbol{\alpha}$ within C dominating $\boldsymbol{\alpha}^\star$. The set of Pareto optimal solutions is referred to as the Pareto front. Pareto optimality has found widespread applications in methodologies for addressing MOPs, as emphasized in seminal works such as [33] and [34]. Strategies founded on the Pareto optimality criterion are formulated to retain only Pareto optimal solutions during each iteration of the solution updating process. The authors of [35] have asserted that a solution initially not meeting Pareto optimality criteria may improve subsequently, potentially surpassing one or more Pareto optimal solutions across all objectives. In this chapter, we adhere to the Pareto optimality criterion at every stage of the solution update procedure. An alternative, simple, and widely recognized strategy for solving a MOP is to minimize a weighted sum of all objectives, yielding a single solution corresponding to the set of weights chosen for each objective function. On the other hand, uniform weight selection to each objective may seldom results in a uniformly distribution of solutions on the Pareto front. Hence, certain optimal solutions may become closely clustered, diminishing the diversity of feasible design alternatives, as highlighted in [36].

8.5.1 FI- and MIUB-Based Objective Functions

a) *FI-based objective function*: Consider $\boldsymbol{\alpha}_t = [\alpha_{1,t}, \dots, \alpha_{N,t}]$ as the representation of the sensor selection strategy at time t, where $\alpha_{n,t}$ is binary, and $\alpha_{n,t} = 1$ means that sensor n is selected and $\alpha_{n,t} = 0$ otherwise. Hence, the aggregate number of sensors chosen at time t is $A = \sum_{n=1}^{N} \alpha_{n,t}$. With $\boldsymbol{\alpha}_t$ provided, the FI matrix at time step t is stated as follows:

$$\boldsymbol{J}_t(\boldsymbol{\alpha}_t) = \sum_{n=1}^{N} \alpha_{n,t} \boldsymbol{J}_{n,t}^D + \boldsymbol{J}_t^P, \tag{8.29}$$

Here, $\boldsymbol{J}_{n,t}^D \triangleq \int \boldsymbol{x}_t \boldsymbol{J}_{n,t}^P(\boldsymbol{x}_t) p(\boldsymbol{x}_t) d\boldsymbol{x}_t$ is derived from (8.13).

For the first objective function, we determine the sensor selection strategy which minimize the normalized FI gap, which is defined as the FI obtained from all sensor measurements and the FI computed using the sensors selected by $\boldsymbol{\alpha}_t$.

$$f_1(\boldsymbol{\alpha}_t) = \frac{\log \det\left(\sum_{n=1}^{N} \boldsymbol{J}_{n,t}^{D} + \boldsymbol{J}_t^{P}\right) - \log \det\left(\sum_{n=1}^{N} \alpha_{n,t}\boldsymbol{J}_{n,t}^{D} + \boldsymbol{J}_t^{P}\right)}{\log \det\left(\sum_{n=1}^{N} \boldsymbol{J}_{n,t}^{D} + \boldsymbol{J}_t^{P}\right)}. \tag{8.30}$$

Here, $\det(\cdot)$ denotes the determination of a matrix.

Then, the second objective function minimizes the percentage of selected sensors,

$$f_2(\boldsymbol{\alpha}_t) = \frac{1}{N}\sum_{n=1}^{N} \alpha_{n,t}. \tag{8.31}$$

b) *MIUB-based objective function*: Similar to FI, the objective function to be minimized using MIUB is defined as the information gap between the MIUB if all the sensors are selected and the MIUB computed using the sensors selected by $\boldsymbol{\alpha}_t$,

$$f_1(\boldsymbol{\alpha}_t) = \frac{\sum_{n=1}^{N} I^{(n)} - \sum_{n=1}^{N} \alpha_{n,t} I^{(n)}}{\sum_{n=1}^{N} I^{(n)}}, \tag{8.32}$$

where $I^{(n)}$ represents $I(z_{n,t}; \boldsymbol{x}_t)$ (or $I(\lambda_{n,t}; \boldsymbol{x}_t)$), and minimizing the percentage of selected sensors (the same as (8.31)).

8.5.2 NSGA-II

In this section, we solve the previously stated MOP featuring binary decision variables using the multiobjective evolutionary algorithm, NSGA-II, as introduced in [37]. This algorithm is engineered to yield all solutions on the Pareto front, showcasing the various potential tradeoffs between the two conflicting objectives.

NSGA-II [37] commences its execution by generating an initial population of a predefined size P, wherein each individual solution is one of the feasible solutions of the MOP. In the context of our specific problem, each solution in the population is represented as a binary vector of size of N indicating whether a particular sensor is selected or not. NSGA-II operates as an elitist algorithm, ensuring that high-quality solutions remain within the population. Each solution is assigned a fitness value by assessing the corresponding values of the objective functions. Subsequently, all solutions in the population are sorted ascending based on their rank of nondomination. For example, solutions with Rank 1 comprise all nondominated solutions, indicating they are not dominated by any other solution in the population. Solutions with Rank 2 include solutions that are dominated by only one solution from the population. This ranking scheme continues, with higher ranks indicating solutions that are increasingly dominated by others. In cases where two or more solutions possess identical fitness values, they are ordered based on their crowding distance. The crowding distance serves as a metric quantifying the amount of space surrounding a solution in the objective space.

NSGA-II utilizes solution ranks to form the mating population. The generation of offspring entails binary tournament selection, as outlined in [37]. In cases where both selected solutions possess identical fitness values, preference is given to the solution with the greater crowding distance. In our scenario, we utilize uniform crossover (UX) which produces binary offspring solutions s_1^c and s_2^c from binary parent solutions s_1^p and s_2^p based on,

$$\begin{aligned} s_1^c &= \xi_c s_1^p + (1-\xi_c)s_2^p, \\ s_2^c &= (1-\xi_c)s_1^p + \xi_c s_2^p, \end{aligned} \tag{8.33}$$

where ξ_c is defined based on a random number q_c between $[0, 1]$ [38]

$$\begin{aligned} \xi_c &= 1 \qquad q_c \leq 0.5 \\ \xi_c &= 0 \qquad q_c > 0.5. \end{aligned} \tag{8.34}$$

In addition to UX, we also employ uniform mutation. Uniform mutation generates a binary offspring solution s_l^c from the binary parent solution s_l^p according to

$$s_l^c = \xi_m(1 - s_l^p) + (1 - \xi_m)s_l^p, \tag{8.35}$$

where ξ_m is determined by (8.34) as well. The population including both parent and offspring solutions, undergoes another round of sorting based on their non-dominance. Following this, the population size is returned back to its original value P by discarding the lower-rank solutions. The remaining solutions are then subjected to a new iteration of binary tournament selection, crossover, and mutation operations. This iterative cycle persists for multiple generations G, progressively maintaining solutions that lie around or on the Pareto optimal front.

8.5.3 Pareto Optimal Solution Decision for Sensor Selection

With NSGA-II generating P nondominated tradeoff solutions on the Pareto front, the challenge lies in selecting a specific solution that effectively embodies the desired tradeoff among the conflicting objectives. As discussed in [33–35], the decision-making process regarding such a selection is influenced by the preferences of the decision maker. For instance, the decision maker can contribute to solving a MOP by suggesting a reference direction, reference points, or other guidelines. These inputs facilitate the single solution selection or a preferred subset of solution selections from the Pareto optimal front. Our selection process prioritizes the knee point solution or the compromise solution of the Pareto optimal front. As outlined in [39], the knee point represents the solution where a slight reduction in one objective results in a significant increase in the other. On the Pareto optimal front, consider two neighboring solutions denoted as $\boldsymbol{\alpha}^a$ and $\boldsymbol{\alpha}^b$ where $f_1(\boldsymbol{\alpha}^a) > f_1(\boldsymbol{\alpha}^b)$ and $f_2(\boldsymbol{\alpha}^a) < f_2(\boldsymbol{\alpha}^b)$. The slope between solutions $\boldsymbol{\alpha}^a$ and $\boldsymbol{\alpha}^b$ is calculated as:

$$\text{slope}\{\boldsymbol{\alpha}^b\} = 180 - \left[\arctan\left(\frac{f_1(\boldsymbol{\alpha}^a) - f_1(\boldsymbol{\alpha}^b)}{f_2(\boldsymbol{\alpha}^a) - f_2(\boldsymbol{\alpha}^b)}\right)\frac{180}{\pi}\right]. \tag{8.36}$$

Let $\boldsymbol{\alpha}^1$ be the solution where no sensors are selected, thus $f_1(\boldsymbol{\alpha}^1) = 1$ and $f_2(\boldsymbol{\alpha}^1) = 0$. Also, let $\boldsymbol{\alpha}^P$ be the solution where all the sensors are selected, so that $f_1(\boldsymbol{\alpha}^P) = 0$ and $f_2(\boldsymbol{\alpha}^P) = 1$. The Pareto optimal solution maximizing (8.36) is then called the knee point solution as,

$$\boldsymbol{\alpha}_t^\star = \arg\max_{\boldsymbol{\alpha}^2,\ldots,\boldsymbol{\alpha}^P} \text{slope}\{\boldsymbol{\alpha}^\rho\}, \tag{8.37}$$

where $\boldsymbol{\alpha}^\rho$ ($\rho \in \{2, 3, \ldots P\}$) is the solution on the Pareto optimal front.

Alternatively, the utopia point $\boldsymbol{f}^\star$ is defined as the individual minimized values of all objectives under consideration as [36],

$$\boldsymbol{f}^\star = [f_1^\star, \ldots, f_M^\star]^{\mathrm{T}}, \tag{8.38}$$

where $f_m^\star$ represents the individual minimum of objective f_m ($m \in \{1, \ldots, M\}$),

$$f_m^\star = \min_{\boldsymbol{\alpha}} \{f_m(\boldsymbol{\alpha}) \mid \boldsymbol{\alpha} \in C\}, \tag{8.39}$$

Here, we define $\boldsymbol{f}(\boldsymbol{\alpha}^\rho) = [f_1(\boldsymbol{\alpha}^\rho), \ldots, f_M(\boldsymbol{\alpha}^\rho)]^{\mathrm{T}}$ where $\rho \in 1, 2, \ldots, P$. As described in [36], the Pareto optimal solution nearest to the utopia point is named as the compromise solution. Then, the Euclidean distance metric is employed to determine the compromise solution as,

$$\boldsymbol{\alpha}_t = \arg\min_{\boldsymbol{\alpha}^1,\ldots,\boldsymbol{\alpha}^P} \sqrt{\sum_{m=1}^{M} (f_m^\star - f_m(\boldsymbol{\alpha}^\rho))^2}. \tag{8.40}$$

Numerical results will be presented in Section 8.5.4.

8.5.4 Numerical Experiments for the MOP Framework

Within this subsection, we assess the effectiveness of our MOP framework by conducting a series of simulation experiments. This subsection considers the same sensor network as Figure 8.1 in Section 8.4.6, and also uses the same system parameters as in Section 8.4.6. In our experiment, the population size $P = 100$ has been selected for NSGA-II. The determination of the number of generations relies on a diversity metric proposed in [37]. This metric assesses the degree of extent (dispersion) among the acquired solutions. The extent is defined as follows:

$$\Gamma = \frac{d_f^{(E)} + d_l^{(E)} + \sum_{n=1}^{N-1} |d_n^{(E)} - \overline{d}^{(E)}|}{d_f^{(E)} + d_l^{(E)} + (N-1)\overline{d}^{(E)}}, \tag{8.41}$$

Here, we define $d_f^{(E)}$ and $d_l^{(E)}$ the Euclidean distances connecting the extreme solutions and the boundary solutions respectively within the obtained nondominated set. Furthermore, we define $d_n^{(E)}$ as the Euclidean distance between two consecutive solutions in the nondominated solution set. We then define $\overline{d}^{(E)}$ as the average of $d_n^{(E)}$ for $n \in 1, \ldots, N$. Throughout our observations, both the FI-SS and MIUB-SS diversity metrics converge within first 100 generations across all 20 time steps. Hence, in our simulation experiments, we set the total number of NSGA-II generations as $G = 100$. Furthermore, before initiating the NSGA-II process, we enrich the initial population by incorporating the two extreme solutions: the all-zero solution and the all-one solution. It is noteworthy that all simulation results presented in this subsection employ 5-bit quantized sensor measurements.

8.5.4.1 Pareto Optimal Solutions

In Figures 8.5 and 8.6, we show the Pareto optimal solutions obtained with NSGA-II for FI-SS, and MIUB-SS respectively. An intriguing observation is that NSGA-II yields $N + 1$ distinct solutions on the Pareto optimal front, each representing the optimal selection of A out of N sensors, where $A \in 0, 1, \ldots, N$. Referring to (8.17), (8.20), and Table 8.1, we infer that in the case of FI-SS, the proximity between the sensor and the target holds greater significance than the sensing probability. For instance, at time step $t = 3$, significant FI can be attained with just sensor 8 due to its proximity to the target. However, at time step $t = 6$, there is considerable uncertainty regarding the target's location as it is not relatively close to any sensor. Consequently, the Pareto front for FI-SS at $t = 3$ exhibits a steeper slope compared to that at $t = 6$. Conversely, MIUB-SS prioritizes reliable sensors (with relatively high sensing probability) and tends to select more sensors. Therefore, compared to FI-SS, the Pareto front of MIUB-SS at $t = 3$ or $t = 6$ is less steep. Furthermore, our observation suggests that the compromise and knee point solutions align when the Pareto front exhibits a steep slope.

8.5.4.2 Effect of Solution Selection on Estimation Performance and Number of Selected Sensors

In Figure 8.7, we show the average number of selected sensors at each time step of tracking. Additionally, we illustrate the tracking performance using MSE obtained from MIUB-SS and FI-SS with both the knee point solution (8.37) and the compromise solution (8.40). Similar outcomes are observed for both MIUB-SS and

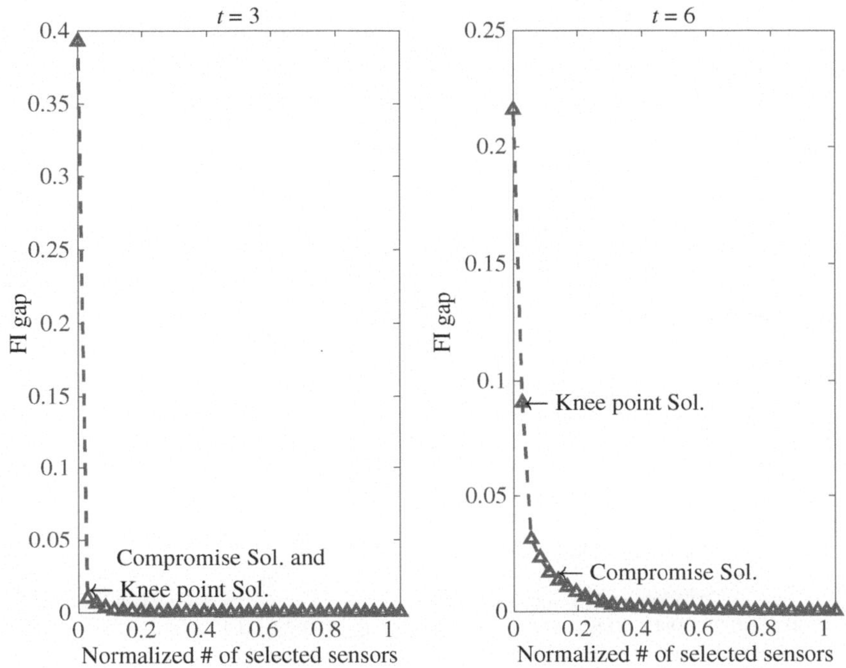

Figure 8.5 Pareto optimal solutions for FI-based objective function (in (8.30)) at time steps $t = 3$ and $t = 6$. Source: Reproduced from [25]/with permission of IEEE.

FI-SS, indicating that the knee point solution, which selects only one sensor at every step, yields worse performance. However, the compromise solution in (8.40) strikes a balance between the total number of activated sensors and the tracking performance. Hence, we employ the compromise solution to further investigate the sensor selection strategy in the subsequent simulations.

Note that Figure 8.7 revisits the outcomes previously presented in Figure 8.2–8.4, where MIUB-SS prioritizes reliable sensors considering the predetermined number of sensors to be selected. Additionally, MIUB-SS chooses a larger number of sensors than FI-SS when the exact number of selected sensors is uncertain, consequently leading to lower MSE compared to FI-SS.

8.5.4.3 NSGA-II, Convex Optimization, and Weighted Sum Methods

In Figures 8.8 and 8.9, we conduct a comparative analysis of tracking performance using NSGA-II and (8.40) against a sensor selection scheme based on convex relaxation [7, 27], where at each time step of tracking A sensors are chosen out of N. It is worth noting that the sensor selection scheme, based on MOP, dynamically selects varying number of sensors at each tracking step, as depicted in Figure 8.7.

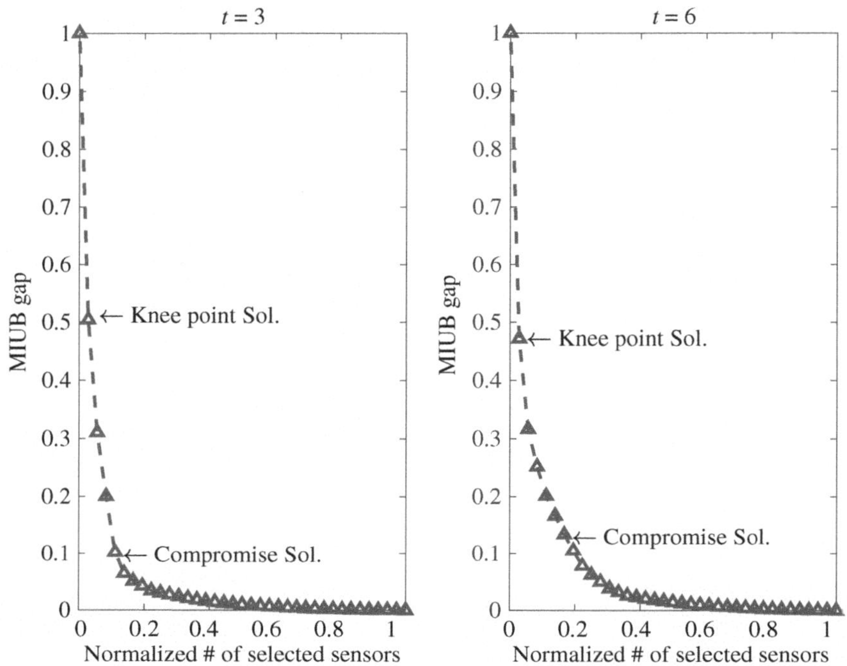

Figure 8.6 Pareto optimal solutions for MIUB-based objective function (in (8.32)) at time steps $t = 3$ and $t = 6$. Source: Reproduced from [25]/with permission of IEEE.

In contrast, the convex optimization method requires a predetermined number of sensors to be selected. We proceed to employ the convex relaxation method to select both the minimum and maximum numbers of sensors, as determined by NSGA-II using the compromise solution in Figure 8.7. Opting for the minimum number of sensors results in notably suboptimal tracking performance derived from the convex relaxation-based sensor selection method. Conversely, selecting the maximum number of sensors or utilizing the convex relaxation method to include all available sensors yields marginal enhancements in terms of MSE compared to the MOP approach. Thus, unlike the convex relaxation method, the multiobjective optimization approach offers significantly improved tracking performance while effectively reducing the number of sensors selected for both MIUB-SS and FI-SS when few sensors have most information about the target. Furthermore, we compare the MSE performance between the MOP framework and the weighted sum approach. In the weighted sum approach, the sensor selection scheme combines both objectives as, $w_1 f_1(\boldsymbol{\alpha}_t) + (1 - w_1) f_2(\boldsymbol{\alpha}_t)$ with equal weights, i.e., $w_1 = 0.5$. The simulation findings indicate that for MIUB-SS (as illustrated in Figure 8.8), the NSGA-II method yields MSE performance similar

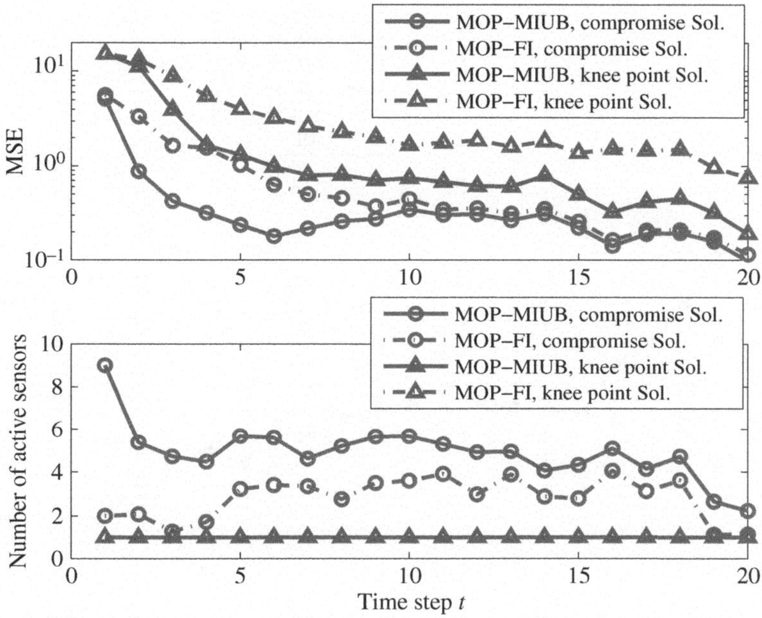

Figure 8.7 MSE and total number of selected sensors averaged over 100 Monte Carlo trials (MSE has unit m^2). Source: Reproduced from [25]/with permission of IEEE.

to that of the weighted sum method. However, concerning FI-SS (Figure 8.9), the weighted sum method results in significantly worse MSE performance. Notably, the weighted sum method requires specifying the weight w_1. Additionally, despite adopting an evenly distributed range of weights, the weighted sum method often struggles to produce a well-distributed array of points on the Pareto front. This circumstance leads to certain optimal solutions clustering closely together, ultimately limiting the available design alternatives [36].

8.5.4.4 A Naive Strategy

We investigate a straightforward sensor selection approach where the fusion center deactivates the sensors with lower sensing probabilities before initiating the sensor selection process. Figure 8.10 illustrates the outcomes when sensors whose sensing probabilities fall below a certain threshold p_s^{th} are disabled. In this context, we examine two threshold values: $p_s^{\text{th}} = 0.5$ and $p_s^{\text{th}} = 0.15$. It is noteworthy that for the sensor network depicted in Figure 8.1, sensors positioned near the target tend to have lower sensing probabilities and are thus considered candidates for deactivation. As observed in previous results, MIUB-SS tends to favor the selection of more reliable sensors, while FI-SS is biased toward choosing sensors closer to the

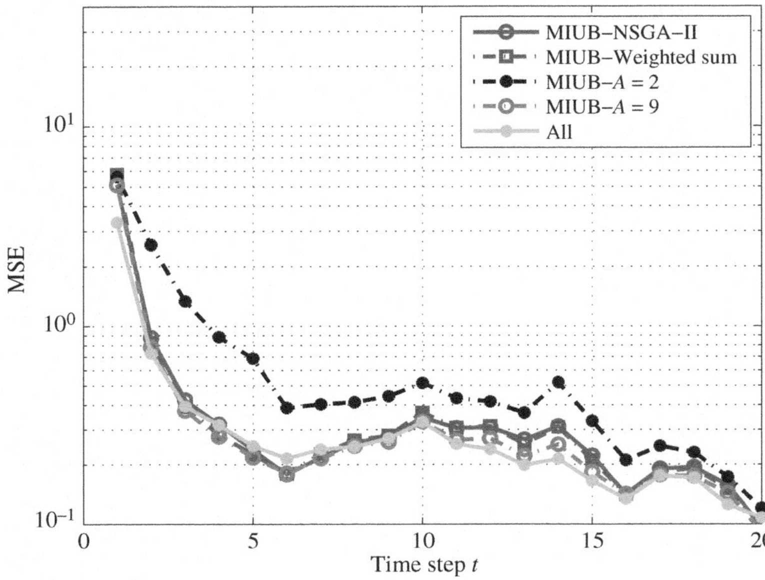

Figure 8.8 MSE comparison of NSGA-II, convex relaxation, and weighted sum methods with MIUB as the objective (MSE has unit m^2). Source: Reproduced from [25]/with permission of IEEE.

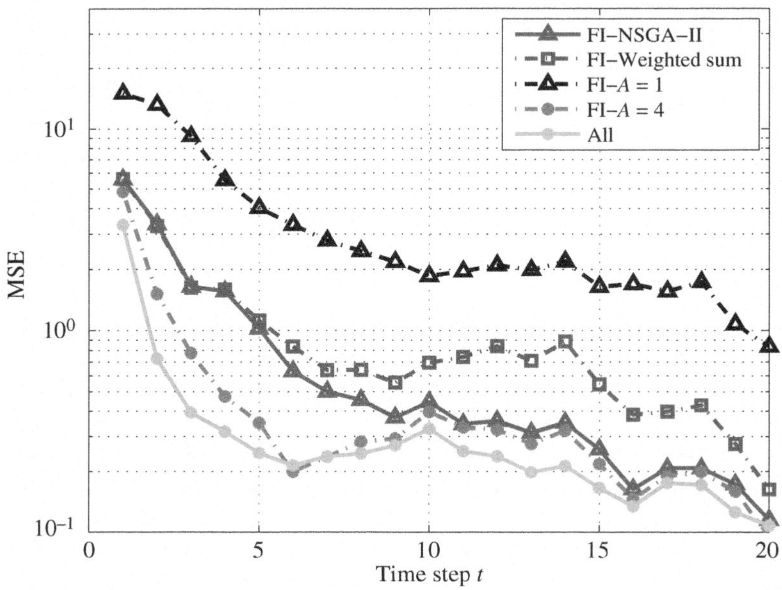

Figure 8.9 MSE comparison of NSGA-II, convex relaxation, and weighted sum methods with FI as the objective (MSE has unit m^2). Source: Reproduced from [25]/with permission of IEEE.

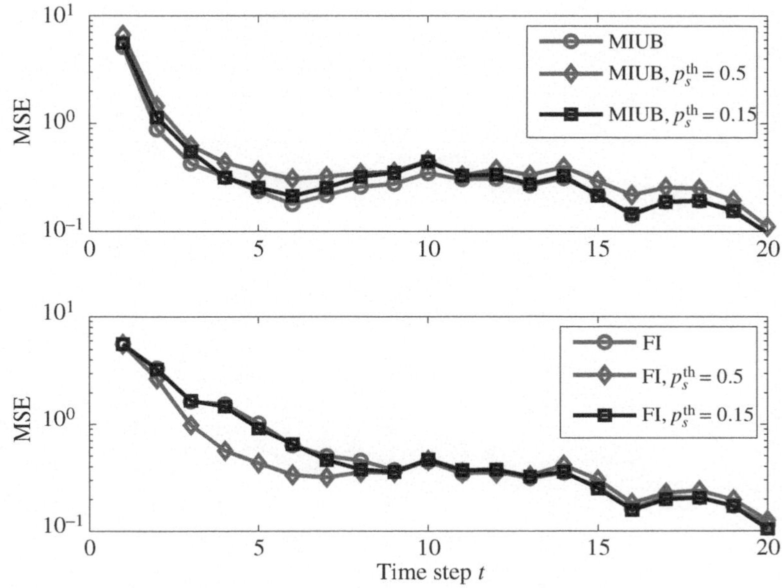

Figure 8.10 MSE performance of MIUB and FI when sensors with low sensing probabilities are turned off (MSE has unit m^2). Source: Reproduced from [25]/with permission of IEEE.

target. The impact of deactivating sensors before the selection process yields mixed results: for MIUB-SS, the approach leads to inferior results due to the reduced pool of selection alternatives. Conversely, for FI-SS, MSE is reduced as it allows the selection of more reliable sensors when the closest sensors to the estimated target location with low sensing probabilities are unavailable.

8.5.4.5 Performance Comparison When There Is No Uncertainty

In Figure 8.11, we show the MSE of the target tracking under the scenario where all sensors are assumed to be reliable, i.e., $p_s = 1$ for every sensor in the network, and then conduct a comparison with the outcomes obtained when uncertainties are present in observations. An observation is that when sensor observations have uncertainties, both FI-SS and MIUB-SS prefer to select a greater number of sensors; however, they still achieve poor MSE performance. Furthermore, it becomes evident that in this context, MIUB-SS significantly outperforms FI-SS since MIUB-SS selects a much larger number of sensors compared to FI-SS.

8.5.4.6 Sensor Networks with Different Densities

For MIUB-SS and FI-SS, Figure 8.12 provides an insight into the MSE and average number of sensors selected at each time step of tracking averaged over 100 trials,

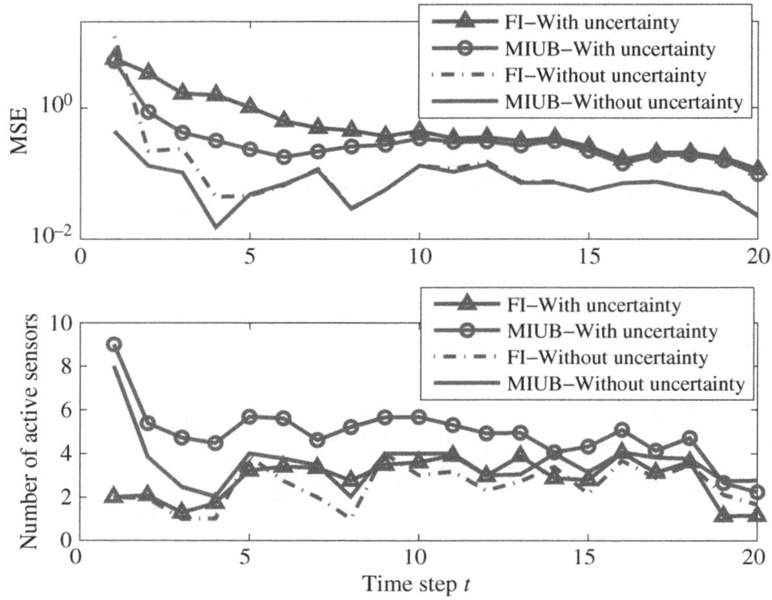

Figure 8.11 Comparison of the performance for networks with and without uncertainty where MSE has unit m^2. Source: Reproduced from [25]/with permission of IEEE.

considering $N = 9, 16, 25$, and 49 sensors in the WSN. It is crucial to emphasize that the sensing probability for each sensor is randomly generated from a standard uniform distribution $p_s \in [0, 1]$. Notably, the findings from Figure 8.12 corroborate earlier observations, wherein MIUB-SS exhibits a preference for selecting more reliable sensors, and surpasses FI-SS in terms of MSE performance.

8.5.4.7 Further Examples of Sensing Probabilities

In Figure 8.13, each sensor's sensing probability is reversed as compared to Figure 8.1. In other words, sensors closer to the target now exhibit higher sensing probabilities. Under this scenario, both MIUB-SS and FI-SS converge toward selecting an equivalent number of sensors, consequently achieving comparable MSE. This convergence stems from the fact that, in this setup, both MIUB-SS and FI-SS prioritize selecting sensors positioned near the target, which inherently possess higher sensing probabilities.

8.5.5 Impact of Modeling Uncertainties

In this chapter, we considered a particle filter-based target tracking-algorithm, where particle filters require system and measurement models in a probabilistic

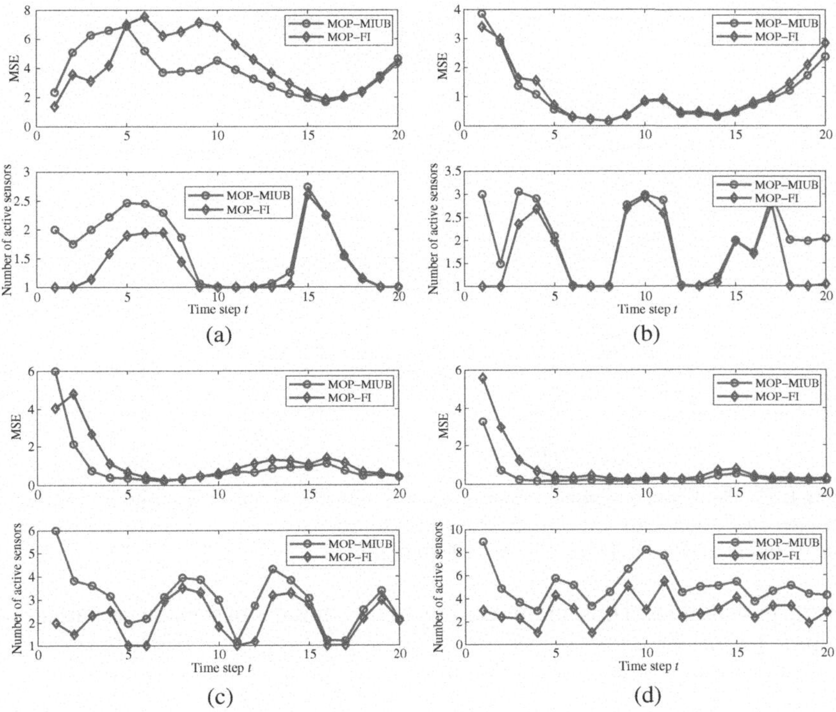

Figure 8.12 MSE and number of sensors selected at each time step of tracking averaged over 100 Monte Carlo trials (MSE has unit m^2) (a) $N = 9$, (b) $N = 16$, (c) $N = 25$, (d) $N = 49$. Source: Reproduced from [25]/with permission of IEEE.

form which provides a broader framework for dynamic state estimation as compared to Kalman filter. Model accuracy naturally depends on the number of particles employed to represent probability distributions. Since larger number of particles will require increased computational effort, there is a tradeoff between model accuracy and computational complexity.

Furthermore, in an uncertain environment, we assumed that the sensing probabilities of sensors are known in advance where sensing probabilities may depend on many factors such as the number and locations of the obstacles in the region of interest, and the distance between the sensors and the fusion center. When FI-based sensor selection metrics were used, the sensors were selected according to their distances to the estimated target location. In this case, accurate estimates of sensing probabilities have less importance. On the other hand, accurate estimates of sensing probabilities have great importance when MI-based sensor selection metrics were employed, since the sensors were selected based on both the sensors' distances to the estimated target location, and their respective

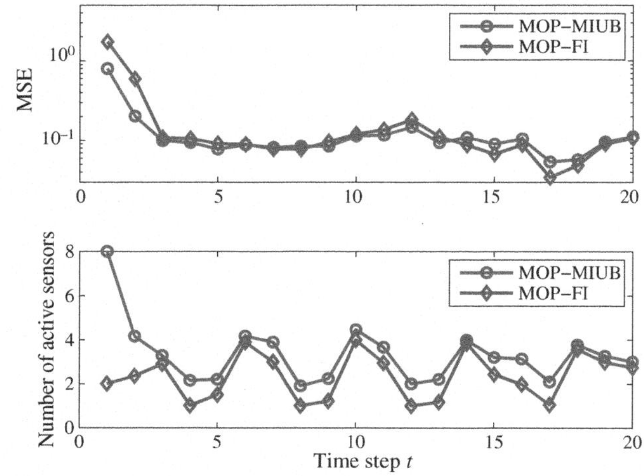

Figure 8.13 MSE and average number of sensors selected at each time step of tracking averaged over 100 Monte Carlo trials when each sensor's sensing probability is reversed. Source: Reproduced from [25]/with permission of IEEE.

sensing probabilities. Thus, in this case accurate estimates of sensing probabilities are quite important.

8.6 Conclusion

In this chapter, we have outlined a multiobjective optimization approach for addressing the sensor selection problem in resource-constrained WSN dedicated to target tracking. Here, sensors operate under conditions of unreliability, meaning each sensor may observe signals with a probability less than one. At each time step of the tracking process, the sensor selection strategy has been obtained from a set of Pareto optimal solutions, which encapsulate various tradeoffs between the accuracy of estimation and the total number of selected sensors. We have employed three performance metrics FI, MI, and MIUB as objective functions to characterize the estimation performance within the MOP framework. Numerical results demonstrate that MIUB-SS tends to select more reliable sensors as compared to FI-SS while maintaining computational resources compared to MI-SS. Moreover, within the MOP framework, when compromise solution is selected on the Pareto front, it yields satisfactory MSE performance while also saves the number of sensors selected. Looking ahead, promising future research is to investigate the MOP framework under sensing uncertainty for multiple target tracking scenarios.

References

1 Rowaihy, H., Eswaran, S., Johnson, M. et al. (2007). A survey of sensor selection schemes in wireless sensor networks. *Defense and Security Symposium*, 65621A. International Society for Optics and Photonics.
2 Williams, J.L., Fisher, J.W., and Willsky, A.S. (2007). Approximate dynamic programming for communication-constrained sensor network management. *IEEE Transactions on Signal Processing* 55(8): 4300–4311.
3 Hoffmann, G.M. and Tomlin, C.J. (2010). Mobile sensor network control using mutual information methods and particle filters. *IEEE Transactions on Automatic Control* 55(1): 32–47.
4 Zhao, F., Shin, J., and Reich, J. (2002). Information-driven dynamic sensor collaboration. *IEEE Transactions on Signal Processing* 19(1): 61–72.
5 Zuo, L., Niu, R., and Varshney, P.K. (2007). Posterior CRLB based sensor selection for target tracking in sensor networks. *IEEE International Conference on Acoustics, Speech, and Signal Processing (ICASSP)*, volume 2, 1041–1044.
6 Zuo, L., Niu, R., and Varshney, P.K. (2008). A sensor selection approach for target tracking in sensor networks with quantized measurements. *IEEE International Conference on Acoustics, Speech, and Signal Processing (ICASSP)*, 2521–2524.
7 Joshi, S. and Boyd, S. (2009). Sensor selection via convex optimization. *IEEE Transactions on Signal Processing* 57(2): 451–462.
8 Mo, Y., Ambrosino, R., and Sinopoli, B. (2011). Sensor selection strategies for state estimation in energy constrained wireless sensor networks. *Automatica* 47: 1330–1338.
9 Wang, H., Yao, K., Pottie, G., and Estrin, D. (2004). Entropy-based sensor selection heuristic for target localization. *Proceedings of the 3rd International Symposium on Information Processing in Sensor Networks*, 36–45. ACM.
10 Masazade, E., Niu, R., Varshney, P.K., and Keskinoz, M. (2010). Energy aware iterative source localization for wireless sensor networks. *IEEE Transactions on Signal Processing* 58(9): 4824–4835.
11 Nahi, N.E. (1969). Optimal recursive estimation with uncertain observation. *IEEE Transactions on Information Theory* 15(4): 457–462.
12 Hadidi, M. and Schwartz, S. (1979). Linear recursive state estimators under uncertain observations. *IEEE Transactions on Automatic Control* 24(6): 944–948.
13 Hounkpevi, F.O. and Yaz, E.E. (2007). Robust minimum variance linear state estimators for multiple sensors with different failure rates. *Automatica* 43(7): 1274–1280.

14 Zhang, H., Shi, Y., and Mehr, A.S. (2011). Robust weighted H_∞ filtering for networked systems with intermittent measurements of multiple sensors. *International Journal of Adaptive Control and Signal Processing* 25(4): 313–330.

15 Xu, W., Ma, K., Trappe, W., and Zhang, Y. (2006). Jamming sensor networks: attack and defense strategies. *IEEE Network* 20(3): 41–47.

16 Mariton, M. (1990). *Jump Linear Systems in Automatic Control.* CRC Press.

17 Costa, O.L.V. and Guerra, S. (2002). Stationary filter for linear minimum mean square error estimator of discrete-time Markovian jump systems. *IEEE Transactions on Automatic Control* 47(8): 1351–1356.

18 Sinopoli, B., Schenato, L., Franceschetti, M. et al. (2004). Kalman filtering with intermittent observations. *IEEE Transactions on Automatic Control* 49(9): 1453–1464.

19 Ozdemir, O., Niu, R., and Varshney, P.K. (2009). Channel aware target localization with quantized data in wireless sensor networks. *IEEE Transactions on Signal Processing* 57(3): 1190–1202.

20 Masazade, E., Niu, R., Varshney, P.K., and Keskinoz, M. (2010). Channel aware iterative source localization for wireless sensor networks. *Proceedings IEEE 13th International Conference on Information Fusion (FUSION)*, 1–7. IEEE.

21 Lee, W.-C. (2010). Uncertainty in wireless sensor networks. *Workshop on AFRL.*

22 Ruan, Y., Willett, P., Marrs, A. et al. (2008). Practical fusion of quantized measurements via particle filtering. *IEEE Transactions on Aerospace and Electronic Systems* 44(1): 15–29.

23 Gordon, N.J., Salmond, D.J., and Smith, A.F. (1993). Novel approach to nonlinear/non-Gaussian Bayesian state estimation. *IEEE Proceedings F (Radar and Signal Processing)* 140(2): 107–113.

24 Arulampalam, M.S., Maskell, S., Gordon, N., and Tim, C. (2002). A tutorial on particle filters for online nonlinear/ non-Gaussian Bayesian tracking. *IEEE Transactions on Signal Processing* 50(2): 174–188.

25 Cao, N., Choi, S., Masazade, E., and Varshney, P.K. (2016). Sensor selection for target tracking in wireless sensor networks with uncertainty. *IEEE Transactions on Signal Processing* 64(20): 5191–5204.

26 Trees, H.V. (2001). *Detection, Estimation, and Linear Modulation Theory, Part I.* Wiley Interscience.

27 Masazade, E., Niu, R., and Varshney, P.K. (2012). Dynamic bit allocation for object tracking in wireless sensor networks. *IEEE Transactions on Signal Processing* 60(10): 5048–5063.

28 Zhang, Y. and Ji, Q. (2010). Efficient sensor selection for active information fusion. *IEEE Transactions on Systems, Man, and Cybernetics, Part B (Cybernetics)* 40(3): 719–728.

29 Cover, T.M. and Thomas, J.A. (2012). *Elements of Information Theory.* Wiley.

30 Osman, E.A., El-Gazar, M.I., Shaat, M.K. et al. (2009). An estimation of a passive infra-red sensor probability of detection. *Proceedings of the 17th Nuclear and Particle Physics Conference (NUPPAC)*, 662.

31 Oymak, O. (2007). Sample size determination for estimation of sensor detection probabilities based on a test variable. PhD dissertation. Monterey, California: Naval Postgraduate School.

32 Bar-Shalom, Y., Willett, P.K., and Tian, X. (2011). *Tracking and Data Fusion: A Handbook of Algorithms*. Storrs, CT: YBS Publishing.

33 Deb, K., Sinha, A., and Kukkonen, S. (2006). Multi-objective test problems, linkages, and evolutionary methodologies. *Proceedings of the 8th Annual Conference on Genetic and Evolutionary Computation*, 1141–1148. ACM.

34 Miettinen, K. (2008). Introduction to multiobjective optimization: noninteractive approaches. *Multiobjective Optimization*, 1–26. Springer.

35 Carreno Jara, E. (2014). Multi-objective optimization by using evolutionary algorithms: the p-optimality criteria. *IEEE Transactions on Evolutionary Computation* 18(2): 167–179.

36 Marler, R.T. and Arora, J.S. (2004). Survey of multi-objective optimization methods for engineering. *Structural and Multidisciplinary Optimization* 26: 369–395.

37 Deb, K., Pratap, A., Agarwal, S., and Meyarivan, T. (2002). A fast and elitist multiobjective genetic algorithm: NSGA-II. *IEEE Transactions on Evolutionary Computation* 6(2): 182–197.

38 Issimakis, N. and Adam, M. (1998). A genetic algorithm for the multidimensional knapsack problem. *Journal of Heuristics* 4(4): 63–86.

39 Boyd, S. and Vandenberghe, L. (2004). *Convex Optimization*. Cambridge: Cambridge University Press.

9

Robust Filtering Under Minimum Error Entropy Criterion

Siyuan Peng[1], Lujuan Dang[2], Badong Chen[2], and Jose C. Principe[3]

[1] *School of Information Engineering, Guangdong University of Technology, Guangzhou, China*
[2] *Institute of Artificial Intelligence and Robotics, Xi'an Jiaotong University, Xi'an, China*
[3] *Department of Electrical and Computer Engineering, University of Florida, Gainesville, FL, USA*

9.1 Introduction

Adaptive filtering (AF) plays an indispensable role in radar signal processing by significantly enhancing the detection and tracking capabilities of radar systems [1–5]. It effectively mitigates the adverse effects of noise, interference, and clutter, thereby improving target detection and estimation accuracy. AF techniques are particularly adept at addressing nonstationary and time-varying environments frequently encountered in radar applications, such as background clutter suppression [6, 7], interference rejection [8, 9], moving target detection and Doppler processing [10, 11], adaptive beamforming [12, 13], and tracking and estimation [14, 15]. Among various AF approaches, the least mean square (LMS) algorithm and the recursive least square (RLS) algorithm emerge as the most prevalent techniques in radar signal processing [16]. These algorithms have demonstrated their efficacy in adapting to changing signal conditions and optimizing the filter coefficients to enhance the performance of radar systems.

In general, the majority of existing AF algorithms in radar signal processing are developed based on the commonly used mean square error (MSE) criterion, due to its desirable characteristics, including mathematical tractability, computational simplicity, and optimality under the assumption of Gaussian-distributed noise. However, it is important to note that the Gaussian assumption is not always valid in real-world radar system environments. In scenarios where radar signals are affected by various non-Gaussian disturbances, such as background clutter and interference from multiple sources, the MSE-based LMS and RLS algorithms often exhibit poor performance or encounter instability issues. An important reason behind these limitations lies in the fact that, from a statistical perspective,

Information-Theoretic Radar Signal Processing, First Edition. Edited by Yujie Gu and Yimin D. Zhang.

the MSE criterion only considers the second-order statistics, which fails to capture all the relevant information present in radar signals [17]. As a result, alternative criteria are required to effectively handle the non-Gaussian nature of disturbances in adaptive filters.

To enhance the robustness of conventional AF algorithms, several novel AF algorithms have been proposed that utilize robust optimization criteria. In recent years, information-theoretic learning (ITL) has gained significant attention due to its applications in adaptive signal processing [18]. ITL leverages information-theoretic descriptors, such as entropy, divergence, and mutual information, to directly estimate crucial statistical properties from the data. The adoption of ITL enables the utilization of robust optimization criteria. Unlike the traditional MSE criterion, information-theoretic criteria can capture higher-order statistics of signals, providing distinct advantages in addressing non-Gaussian noise challenges. Consequently, algorithms based on information-theoretic criteria hold promise for improving learning performance in practical scenarios, particularly when data is corrupted by non-Gaussian noises or outliers [19].

Thus far, ITL optimization criteria have found success in diverse applications, including adaptive filter, system identification, linear regression, and classification [18]. Notably, robust adaptive filters have been developed in recent years to handle various types of non-Gaussian noise, utilizing two popular information-theoretic criteria: the maximum correntropy criterion (MCC) [20] and the minimum error entropy (MEE) criterion [21]. While MCC enhances the robustness of AF algorithms, particularly in heavy-tailed noise environments, it encounters challenges when dealing with complex non-Gaussian noises. However, MEE can overcome this limitation of MCC, and has demonstrated greater robustness in AF algorithms, albeit at the expense of higher computational complexity [22].

In this chapter, we present two robust MEE-based AF algorithms, including the sparse AF algorithm [23, 24] and the constrained AF algorithm [25]. These algorithms utilize the MEE criterion as the optimization objective, to enhance the robustness of existing AF algorithms. Additionally, in order to reduce the computational complexity of MEE-based AF algorithms, we further propose a novel criterion called quantized minimum error entropy (QMEE), and accordingly develop the QMEE-based AF algorithm [26]. Finally, the effectiveness and robustness of these MEE- and QMEE-based AF algorithms are verified through computer simulations.

This chapter is structured as follows. In Section 9.2, we provide an overview of the MEE criterion and the original MEE-based AF algorithm. Section 9.3 focuses on the MEE-based sparse AF algorithm, which incorporates a sparsity penalty term (SPT) into the MEE-based AF algorithm. Furthermore, in Section 9.4, we introduce the constrained AF algorithm based on MEE and provide a convergence analysis for this algorithm. Section 9.5 is dedicated to the introduction of the QMEE criterion and its properties. We also present the development of QMEE-based AF algorithm in this section. To evaluate the performance of these introduced AF algorithms, simulation results are provided in Section 9.6. Finally, in Section 9.7, we give the conclusion.

9.2 Minimum Error Entropy Criterion

In information theory, entropy is a fundamental measure that quantifies the average uncertainty associated with predicting the value of a random variable. Minimizing entropy leads to a concentration of errors, typically resulting in higher peaks [18]. Figure 9.1 illustrates a general scheme of the AF model under the MEE criterion. The MEE criterion seeks to minimize the entropy of the training error, thereby enabling the AF model to closely approximate the unknown system. Consider a linear AF model in which the desired signal is generated by

$$d_n = \overline{\boldsymbol{w}}^{\mathrm{T}} \boldsymbol{x}_n + v_n, \tag{9.1}$$

where $\overline{\boldsymbol{w}} = [\overline{w}_1, \overline{w}_2, \dots, \overline{w}_M]^{\mathrm{T}}$ is the weight vector of the unknown system (M being the memory size), $\boldsymbol{x}_n = [x_{n,1}, x_{n,2}, \dots, x_{n,M}]^{\mathrm{T}}$ denotes the input vector at instant n, v_n is the interference noise, and $[\cdot]^{\mathrm{T}}$ stands for the transpose. Assume that the adaptive filter system is a finite impulse response (FIR) filter with the weight vector $\boldsymbol{w}_n = [w_{n,1}, w_{n,2}, \dots, w_{n,M}]^{\mathrm{T}}$. Then the AF error is

$$e_n = d_n - y_n = d_n - \boldsymbol{w}_n^{\mathrm{T}} \boldsymbol{x}_n, \tag{9.2}$$

where $y_n = \boldsymbol{w}_n^{\mathrm{T}} \boldsymbol{x}_n$ is the output signal of the adaptive filter. Under the MEE criterion, the weight vector $\boldsymbol{w}$ is solved by minimizing the error entropy. With Rényi entropy of order α ($\alpha > 0$, but $\alpha \neq 1$), the MEE solution is

$$\boldsymbol{w}_{\mathrm{MEE}} = \arg\min_{\boldsymbol{w}} H_\alpha(e) = \arg\min_{\boldsymbol{w}} \frac{1}{1-\alpha} \log \int (p_e(x))^\alpha dx, \tag{9.3}$$

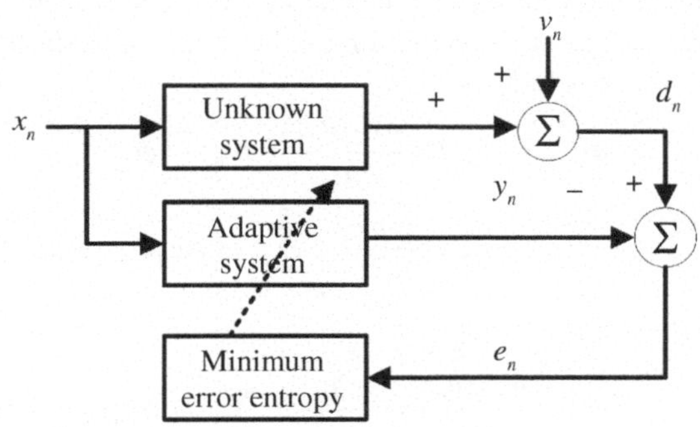

Figure 9.1 Adaptive filtering under minimum error entropy criterion.

where $H_\alpha(e)$ denotes Rényi entropy of the estimated error e, and $p_e(x)$ is the error probability density function (PDF). Obviously, when $\alpha \to 1$, Rényi entropy approaches the well-known Shannon entropy. In AF applications, Rényi entropy of order $\alpha = 2$ has been widely used in ITL, due to its desirable features and simple computation. Thus, Rényi entropy is considered only for the case of $\alpha = 2$ in this chapter. Clearly, minimizing $H_2(e)$ is equivalent to maximizing the following quadratic information potential (QIP)

$$\boldsymbol{w}_{\text{MEE}} = \arg\max_{\boldsymbol{w}} V_2(e) = \arg\max_{\boldsymbol{w}} \int (p_e(x))^2 dx, \tag{9.4}$$

where $V_2(e) = \int (p_e(x))^2 dx$ denotes the QIP cost. In general, the error entropy is a function of the error distribution. However, in practical applications, the error distribution is often unknown, and we only have access to a finite number (N) of error samples. In such cases, it becomes necessary to estimate the error entropy from these error samples. One approach to estimate the error PDF is by using the Parzen windowing method

$$\hat{p}_e(x) = \frac{1}{N}\sum_{n=1}^{N} \kappa_\sigma(e - e_n), \tag{9.5}$$

where $\kappa_\sigma(\cdot)$ is the Gaussian kernel function with width σ. Accordingly, an empirical version of the QIP cost can easily be obtained [17]

$$\hat{V}_2(\boldsymbol{e}) \approx \frac{1}{N^2}\sum_{i,j=1}^{N} \kappa_{\sqrt{2}\sigma}(e_i - e_j), \tag{9.6}$$

where $\boldsymbol{e}$ denotes the error vector, and $\sum_{i,j=1}^{N} = \sum_{i=1}^{N}\sum_{j=1}^{N}$. Therefore, the optimal weight vector under the MEE criterion can be formulated as

$$\boldsymbol{w}_{\text{MEE}} = \arg\max_{\boldsymbol{w}_n} \hat{V}_2(\boldsymbol{e}_n), \tag{9.7}$$

where $\boldsymbol{e}_n = [e_n, e_{n-1}, \dots, e_{n-L+1}]^{\text{T}}$, and L is the sliding window length. When the MEE criterion is employed as the optimization criterion in AF algorithms, the stochastic information gradient algorithm is commonly used to solve the optimization problems. Building upon the optimization problem presented in (9.7), the stochastic information gradient approach for the MEE-based AF algorithm is derived as follows:

$$\begin{aligned} \boldsymbol{w}_{n+1} &= \boldsymbol{w}_n + \eta\frac{\partial \hat{V}_2(\boldsymbol{e}_n)}{\partial \boldsymbol{w}_n} = \boldsymbol{w}_n + \eta\boldsymbol{\phi}_n \\ &= \boldsymbol{w}_n + \frac{\eta}{2\sigma^2 L^2}\sum_{i,j=n-L+1}^{n}\left(\Delta e_{ij}\kappa_{\sqrt{2}\sigma}(\Delta e_{ij})\left(\boldsymbol{x}_i - \boldsymbol{x}_j\right)\right), \end{aligned} \tag{9.8}$$

where η is the step size,

$$\boldsymbol{\phi}_n = \frac{\partial \hat{V}_2(\boldsymbol{e}_n)}{\partial \boldsymbol{w}_n} = \frac{1}{2\sigma^2 L^2}\sum_{i,j=n-L+1}^{n}\left(\Delta e_{ij}\kappa_{\sqrt{2}\sigma}\left(\Delta e_{ij}\right)\left(\boldsymbol{x}_i - \boldsymbol{x}_j\right)\right)$$

is the stochastic information gradient, and $\Delta e_{ij} = e_i - e_j$.

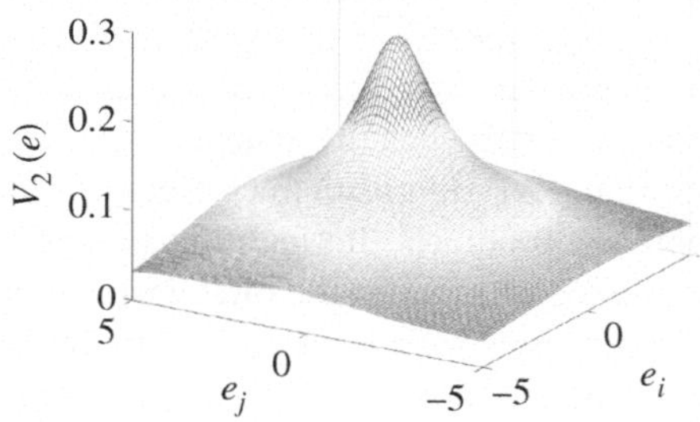

Figure 9.2 Performance surface of Rényi entropy ($\alpha = 2$) in the joint space. Source: [18]/with permission of Springer Nature.

Unlike the traditional MSE criterion, which relies solely on second-order moments, Rényi entropy is a global descriptor of the underlying distribution that can capture higher-order moments of the prediction error. Figure 9.2 illustrates the performance surface of Rényi entropy in the joint space. When an appropriate kernel width is selected, the performance surface of Rényi entropy exhibits not only a fixed curvature but also high smoothness. As a result, the MEE criterion performs well even in the presence of non-Gaussian data or significant outliers. The kernel width σ is a crucial parameter that governs the smoothness of the performance surface of Rényi entropy. In practical applications, the kernel width can be determined in advance using Silverman's rule or manually set based on domain knowledge. To converge to the global solution in MEE-based AF algorithms, a useful strategy is to initiate the algorithms with a large kernel width and then gradually decrease it during the course of adaptation [17].

In addition, several important variants of MEE have been proposed for various applications, such as quantized MEE [27], smoothed MEE [28], minimum total error entropy [29], quaternion MEE [30], complex MEE [31], minimum fuzzy error entropy [32], and generalized MEE [33]. In recent years, alternative techniques such as the fixed-point algorithm and recursive algorithm have also been developed to address the optimization challenges encountered in MEE-based AF algorithms [34, 35].

9.3 Sparse Adaptive Filter Under Minimum Error Entropy Criterion

Sparse characteristics are prevalent in many real-world radar applications, particularly in scenarios where the radar signal encounters a limited number of active targets or reflections. The utilization of sparse characteristics in radar applications enables efficient target detection, accurate localization, improved imaging resolution, and enhanced signal processing capabilities [36]. Therefore, how to identify the sparse characteristics effectively and quickly in radar applications is a valuable problem, especially in challenging environments with limited resources and complicated scenarios. Sparse AF algorithms have been successfully employed in sparse system identification [37]. Among these sparse AF algorithms, one of the most popular algorithms is the sparse AF algorithm with a sparsity penalty constraint, which mainly incorporates a SPT into a traditional AF algorithm to encourage sparsity in the estimated system [38]. This approach promotes the identification of sparse systems by explicitly enforcing sparsity constraints on the filter coefficients. However, most of the existing sparse AF algorithms have been derived under the MSE criterion, which easily lead to poor performance in the presence of impulsive noises and/or abrupt changes. To enhance the robustness of traditional MSE-based sparse AF algorithms, we proposed the MEE algorithm with correntropy-induced metric (MEECIM) [23], which was designed for sparse system identification in non-Gaussian noise environments, providing stronger robustness compared to the MSE-based AF algorithms.

9.3.1 Correntropy-Induced Metric

In most existing sparse AF algorithms, the used SPTs are usually the l_0-norm, the l_1-norm, and the reweighted l_1-norm. However, minimizing the l_0-norm is a computationally challenging problem that is NP-hard and difficult to solve in practice. Although the l_1-norm can achieve similar performance to the l_0-norm, it is not as smooth, which can limit its effectiveness in certain scenarios [18]. To address the limitations of previous SPTs, a new SPT called correntropy-induced metric (CIM) has been developed in [39]. The CIM is an approximation of the l_0-norm and offers improved smoothness compared to the l_1-norm.

Correntropy is a new nonlinear similarity measure that quantifies the similarity between two random variables within a neighborhood of the joint space. Given two vectors of samples: $\boldsymbol{a} = [a_1, a_2, \dots, a_N]^{\mathrm{T}}$ and $\boldsymbol{b} = [b_1, b_2, \dots, b_N]^{\mathrm{T}}$, a sample mean estimator of the correntropy between $\boldsymbol{a}$ and $\boldsymbol{b}$ can be computed as follow:

$$\mathrm{Cor}(\boldsymbol{a}, \boldsymbol{b}) = \frac{1}{N}\sum_{n=1}^{N} \kappa_\sigma(a_n - b_n). \tag{9.9}$$

In order to find the sparsest vector (minimum l_0-norm), CIM is used as an approximation of the l_0-norm. Based on the definition of correntropy, CIM is defined as

$$\mathrm{CIM}(\boldsymbol{a}, \boldsymbol{b}) = (\kappa_\sigma(0) - \mathrm{Cor}(\boldsymbol{a}, \boldsymbol{b}))^{1/2}. \tag{9.10}$$

The l_0-norm of $\boldsymbol{a}$ can be approximated by

$$\|\boldsymbol{a}\|_0 \approx \mathrm{CIM}^2(\boldsymbol{a}, \boldsymbol{0}) = \frac{1}{\sqrt{2\pi}\sigma N}\sum_{n=1}^{N}\left(1 - \exp\left(-\frac{a_n^2}{2\sigma^2}\right)\right). \tag{9.11}$$

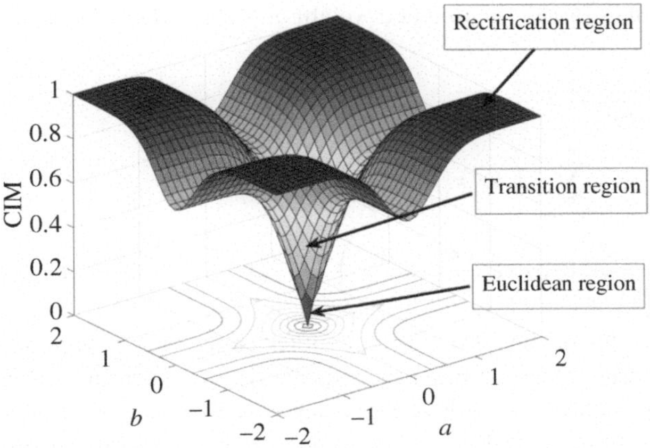

Figure 9.3 Contour plot of CIM in a 3-D space ($\sigma = 0.4$). Source: [18]/with permission of Springer Nature.

The contour plot of CIM is depicted in Figure 9.3. The plot reveals that CIM divides the space into three distinct regions: the Euclidean region, the transition region, and the rectification region. In the Euclidean region, CIM behaves similar to an l_2-norm. In the transition region, it exhibits characteristics akin to an l_1-norm. Finally, in the rectification region, CIM emulates an l_0-norm behavior. If $a_n > \delta$ with δ being a small positive number, then as the kernel width σ approaches zero, the solution obtained using CIM can closely approximate the solution obtained using the l_0-norm. This emphasizes that with an appropriate choice of kernel width, CIM promotes sparsity and can serve as an effective penalty term in sparse adaptive filters.

9.3.2 MEE Algorithm with CIM

In this context, a robust and sparse AF algorithm, namely MEECIM, is introduced. The MEECIM algorithm leverages the MEE criterion and the CIM as the SPT to achieve robust and sparse system identification. By incorporating the CIM into the optimization framework, the MEECIM algorithm provides an effective approach to overcome the drawbacks of traditional MSE-based sparse AF algorithms. To develop the MEECIM algorithm, the CIM penalty term is integrated into the MEE algorithm. By combining (9.6) and (9.11), the optimization problem of MEECIM can be formulated as follows:

$$\arg\min_{\boldsymbol{w}} J_{\text{MEECIM}} = -\hat{V}_2(\boldsymbol{e}_n) + \lambda \text{CIM}^2(\boldsymbol{w}_n, 0), \tag{9.12}$$

where J_{MEECIM} denotes the cost function of the MEECIM algorithm, and λ is the regularization parameter. Using the stochastic information gradient algorithm for the optimization problem in (9.12), the updated weight of MEECIM can be easily derived as

$$\begin{aligned} \boldsymbol{w}_{n+1} &= \boldsymbol{w}_n - \eta \frac{\partial J_{\text{MEECIM}}}{\partial \boldsymbol{w}_n} \\ &= \boldsymbol{w}_n + \frac{\eta}{2\sigma^2 L^2} \sum_{i,j=n-L+1}^{n} \left(\Delta e_{ij} \kappa_{\sqrt{2}\sigma}(\Delta e_{ij}) \left(\boldsymbol{x}_i - \boldsymbol{x}_j\right) \right) \\ &\quad - \frac{\eta\lambda}{\sqrt{2\pi\sigma_1^3}M} \boldsymbol{w}_n \odot \exp\left(-\frac{\boldsymbol{w}_n \odot \boldsymbol{w}_n}{2\sigma_1^2} \right), \end{aligned} \tag{9.13}$$

where σ_1 is the kernel width in the CIM, and $\odot$ denotes element-wise product. Therefore, the MEECIM algorithm is summarized in Algorithm 9.1. Additionally, several fixed-point MEE algorithms with different sparsity penalty constraints have been proposed in [40], including the zero-attracting algorithm (using the l_1-norm), the reweighted zero-attracting algorithm (using the reweighted l_1-norm), and the CIM-based algorithm. These algorithms have been developed to improve the performance for addressing the problem of sparse system identification.

Algorithm 9.1: MEECIM algorithm.

Input: samples $\{\boldsymbol{x}_n, d_n\}_{n=1}^{N}$

Output: weight vector $\boldsymbol{w}$

1: Parameters setting: step size η, kernel widths σ and σ_1, regularization parameter λ, and sliding window length L
2: Initialization: Set $\boldsymbol{w}_0 = \boldsymbol{0}$
3: **for** $n = 1, 2, \ldots, N$ **do**
4: Compute the error samples: $e_n = d_n - \boldsymbol{w}_{n-1}^{\mathrm{T}} \boldsymbol{x}_n$;
5: Update the weight vector by (9.13);
6: **end for**

9.4 Constrained Adaptive Filter Under MEE Criterion

Constrained adaptive filter is a technique commonly used in radar signal processing for enhancing the detection and estimation of radar signals in the presence of noise and interference [16]. By incorporating the constraints into the adaptation process, constrained adaptive filtering (CAF) algorithms can effectively enhance the radar signal processing performance, such as improving the signal-to-interference-plus-noise ratio (SINR), the target detection capabilities, and the overall radar system's robustness. Particularly, the constraints in CAF can be related to various aspects, such as the desired signal's characteristics, interference sources, or system requirements. For example, in radar applications, constraints can be applied to maintain the desired signal's amplitude, phase, or shape, while suppressing interference from other radar systems or clutter from the environment. However, most of the existing CAF approaches are developed by using the MSE criterion as the optimization criterion, whose performance tends to degrade significantly when the signals are corrupted by non-Gaussian noises. To address this limitation, a robust CAF algorithm based on the MEE criterion, called CAFMEE, was proposed in [25], to enhance the robustness of CAF algorithms in the presence of non-Gaussian noise. In Sections 9.4.1 and 9.4.2, we will introduce the CAFMEE algorithm and provide an analysis of its convergence properties.

9.4.1 CAFMEE Algorithm

In general, for most of constrained adaptive filters, a linear constraint will be imposed upon the filter weight vector $\boldsymbol{w}_n$ such that

$$\boldsymbol{C}^{\mathrm{T}}\boldsymbol{w}_n = \boldsymbol{f}, \tag{9.14}$$

where $\boldsymbol{C}$ is an $M \times K$ constraint matrix of full rank, and $\boldsymbol{f}$ is a vector containing K constraint values. Combining (9.7) and (9.14), the CAFMEE algorithm can then be derived by solving the following optimization problem:

$$\arg\max_{\boldsymbol{w}} \hat{V}_2(\boldsymbol{e}_n) \quad \text{subject to} \quad \boldsymbol{C}^{\mathrm{T}}\boldsymbol{w}_n = \boldsymbol{f}. \tag{9.15}$$

Accordingly, the cost function of CAFMEE is

$$J_{\text{CAFMEE}} = \hat{V}_2(\boldsymbol{e}_n) + \xi_n^{\mathrm{T}}(\boldsymbol{C}^{\mathrm{T}}\boldsymbol{w}_n - \boldsymbol{f}), \tag{9.16}$$

where ξ_n is a $K \times 1$ vector containing the undetermined Lagrange multipliers. The stochastic gradient of J_{CAFMEE} with respect to $\boldsymbol{w}_n$ is given by

$$\frac{\partial J_{\text{CAFMEE}}}{\partial \boldsymbol{w}_n} = \boldsymbol{\phi}_n + \boldsymbol{C}\xi_n, \tag{9.17}$$

Then a gradient-based CAFMEE algorithm can be obtained as

$$\boldsymbol{w}_{n+1} = \boldsymbol{w}_n + \eta\frac{\partial J_{\text{CAFMEE}}}{\partial \boldsymbol{w}_n} = \boldsymbol{w}_n + \eta\boldsymbol{C}\xi_n + \eta\boldsymbol{\phi}_n. \tag{9.18}$$

Due to $\boldsymbol{f} = \boldsymbol{C}^{\mathrm{T}}(\boldsymbol{w}_n + \eta\boldsymbol{C}\xi_n + \eta\boldsymbol{\phi}_n)$, we have

$$\xi_n = \frac{(\boldsymbol{C}^{\mathrm{T}}\boldsymbol{C})^{-1}}{\eta}\left(\boldsymbol{f} - \boldsymbol{C}^{\mathrm{T}}\boldsymbol{w}_n - \eta\boldsymbol{C}^{\mathrm{T}}\boldsymbol{\phi}_n\right). \tag{9.19}$$

Substituting (9.19) into (9.18), and after some algebraic manipulations, the weight update equation of CAFMEE can be obtained

$$\begin{aligned}\boldsymbol{w}_{n+1} &= \boldsymbol{w}_n + \boldsymbol{C}(\boldsymbol{C}^{\mathrm{T}}\boldsymbol{C})^{-1}\left(\boldsymbol{f} - \boldsymbol{C}^{\mathrm{T}}\boldsymbol{w}_n - \eta\boldsymbol{C}^{\mathrm{T}}\boldsymbol{\phi}_n\right) + \eta\boldsymbol{\phi}_n \\ &= \boldsymbol{P}\left(\boldsymbol{w}_n + \eta\boldsymbol{\phi}_n\right) + \boldsymbol{q},\end{aligned} \tag{9.20}$$

where $\boldsymbol{P} = \boldsymbol{I}_M - \boldsymbol{C}(\boldsymbol{C}^{\mathrm{T}}\boldsymbol{C})^{-1}\boldsymbol{C}^{\mathrm{T}}$, $\boldsymbol{q} = \boldsymbol{C}(\boldsymbol{C}^{\mathrm{T}}\boldsymbol{C})^{-1}\boldsymbol{f}$, and $\boldsymbol{I}_M$ is an $M \times M$ identity matrix. The above algorithm is referred to as the CAFMEE algorithm.

9.4.2 Convergence Analysis

9.4.2.1 Energy Conservation Relation

Here, we will present the mean square convergence analysis for the CAFMEE algorithm by applying the energy conservation relation to analyze its convergence behavior [16]. First, we rewrite (9.2) in a matrix–vector form as

$$\boldsymbol{e}_n = \boldsymbol{d}_n - \boldsymbol{X}_n \boldsymbol{w}_n = \boldsymbol{X}_n \tilde{\boldsymbol{w}}_n + \boldsymbol{v}_n, \tag{9.21}$$

where $\boldsymbol{d}_n = [d_n, d_{n-1}, \ldots, d_{n-L+1}]^{\mathrm{T}}$ is the $L \times 1$ desired signal vector, $\boldsymbol{X}_n = [\boldsymbol{x}_n, \boldsymbol{x}_{n-1}, \ldots, \boldsymbol{x}_{n-L+1}]^{\mathrm{T}}$ is the $L \times M$ input matrix, $\boldsymbol{v}_n = [v_n, v_{n-1}, \ldots, v_{n-L+1}]^{\mathrm{T}}$ is the $L \times 1$ noise vector, and $\tilde{\boldsymbol{w}}_n = \overline{\boldsymbol{w}} - \boldsymbol{w}_n$ denotes the weight error vector. Additionally, the *a priori* and *a posteriori* error vectors are defined as:

$$\begin{cases} \mathbf{e}_n^a = \mathbf{X}_n \tilde{\mathbf{w}}_n, \\ \mathbf{e}_n^p = \mathbf{X}_n \tilde{\mathbf{w}}_{n+1}, \end{cases} \tag{9.22}$$

where $\boldsymbol{e}_n^a = [e_n^a, e_{n-1}^a, \ldots, e_{n-L+1}^a]^{\mathrm{T}}$, and $\boldsymbol{e}_n^p = [e_n^p, e_{n-1}^p, \ldots, e_{n-L+1}^p]^{\mathrm{T}}$. Obviously, we obtain

$$\boldsymbol{e}_n^p = \boldsymbol{e}_n^a + \boldsymbol{X}_n \left(\tilde{\boldsymbol{w}}_{n+1} - \tilde{\boldsymbol{w}}_n \right). \tag{9.23}$$

Due to

$$\frac{\partial \hat{V}_2(\boldsymbol{e}_n)}{\partial \boldsymbol{w}_n} = \frac{\partial \boldsymbol{e}_n}{\partial \boldsymbol{w}_n} \frac{\partial \hat{V}_2(\boldsymbol{e}_n)}{\partial \boldsymbol{e}_n} = \boldsymbol{X}_n^{\mathrm{T}} \boldsymbol{z}(\boldsymbol{e}_n), \tag{9.24}$$

where $\boldsymbol{z}(\boldsymbol{e}_n) = [z_L(\boldsymbol{e}_n), z_{L-1}(\boldsymbol{e}_n), \ldots, z_1(\boldsymbol{e}_n)]^{\mathrm{T}}$ with

$$z_l(\boldsymbol{e}_n) = -\frac{\partial \hat{V}_2(\boldsymbol{e}_n)}{\partial e_{n-L+l}}, \tag{9.25}$$

By using $\tilde{\boldsymbol{w}}_n = \overline{\boldsymbol{w}} - \boldsymbol{w}_n$, (9.20) can be rewritten as:

$$\tilde{\boldsymbol{w}}_{n+1} = \boldsymbol{P} \tilde{\boldsymbol{w}}_n - \eta \boldsymbol{P} \boldsymbol{X}_n^{\mathrm{T}} \boldsymbol{z}(\boldsymbol{e}_n). \tag{9.26}$$

It is observed that the matrix $\boldsymbol{P}$ is idempotent since $\boldsymbol{P}^2 = \boldsymbol{P}$. Multiplying both sides of (9.26) by $\boldsymbol{P}$, we have

$$\boldsymbol{P} \tilde{\boldsymbol{w}}_{n+1} = \tilde{\boldsymbol{w}}_{n+1}. \tag{9.27}$$

Hence we rewrite (9.26) as

$$\tilde{\boldsymbol{w}}_{n+1} = \tilde{\boldsymbol{w}}_n - \eta \boldsymbol{P} \boldsymbol{X}_n^{\mathrm{T}} \boldsymbol{z}(\boldsymbol{e}_n). \tag{9.28}$$

By substituting (9.23) into (9.28), we have

$$\boldsymbol{e}_n^p - \boldsymbol{e}_n^a = -\eta \boldsymbol{R}_n^{\boldsymbol{P}} \boldsymbol{z}(\boldsymbol{e}_n), \tag{9.29}$$

where $\boldsymbol{R}_n^{\boldsymbol{P}} = \boldsymbol{X}_n\boldsymbol{P}\boldsymbol{X}_n^{\mathrm{T}}$ is an $L \times L$-dimensional symmetric matrix with the ith row jth column element $\boldsymbol{R}_n^{\boldsymbol{P}}(i,j) = \boldsymbol{x}_{n-L+i}^{\mathrm{T}}\boldsymbol{P}\boldsymbol{x}_{n-L+j}$. Assume that $\boldsymbol{R}_n^{\boldsymbol{P}}$ is invertible. Incorporating (9.29) into (9.28) yields

$$\tilde{\boldsymbol{w}}_{n+1} = \tilde{\boldsymbol{w}}_n + \boldsymbol{P}\boldsymbol{X}_n^{\mathrm{T}}\left[\boldsymbol{R}_n^{\boldsymbol{P}}\right]^{-1}\left(\boldsymbol{e}_n^p - \boldsymbol{e}_n^a\right). \tag{9.30}$$

Squaring and taking expectations of both sides of (9.30), after some straightforward manipulations, the following weighted-energy conservation relation can be obtained:

$$\mathbb{E}\left[\|\tilde{\boldsymbol{w}}_{n+1}\|^2\right] + \mathbb{E}\left[\|\boldsymbol{e}_n^a\|^2_{[\boldsymbol{R}_n^{\boldsymbol{P}}]^{-1}}\right] = \mathbb{E}\left[\|\tilde{\boldsymbol{w}}_n\|^2\right] + \mathbb{E}\left[\|\boldsymbol{e}_n^p\|^2_{[\boldsymbol{R}_n^{\boldsymbol{P}}]^{-1}}\right], \tag{9.31}$$

where $\mathbb{E}[\cdot]$ denotes the expectation operator, $\mathbb{E}\left[\|\tilde{\boldsymbol{w}}_n\|^2\right]$ is the weight error power (WEP), and $\|\boldsymbol{e}_n^p\|^2_{[\boldsymbol{R}_n^{\boldsymbol{P}}]^{-1}}$ denotes $[\boldsymbol{e}_n^p]^{\mathrm{T}}[\boldsymbol{R}_n^{\boldsymbol{P}}]^{-1}\boldsymbol{e}_n^p$.

9.4.2.2 Steady-State Mean Square Deviation

Based on the energy conservation relation in (9.31), and $\boldsymbol{e}_n^p - \boldsymbol{e}_n^a = -\eta\boldsymbol{R}_n^{\boldsymbol{P}}\boldsymbol{z}(\boldsymbol{e}_n)$, we have

$$\begin{aligned}\mathbb{E}\left[\|\tilde{\boldsymbol{w}}_{n+1}\|^2\right] = \mathbb{E}\left[\|\tilde{\boldsymbol{w}}_n\|^2\right] - 2\eta\mathbb{E}\left[\left[\boldsymbol{e}_n^a\right]^{\mathrm{T}}\boldsymbol{z}(\boldsymbol{e}_n)\right] \\ + \eta^2\mathbb{E}\left[\boldsymbol{z}^{\mathrm{T}}(\boldsymbol{e}_n)\boldsymbol{R}_n^{\boldsymbol{P}}\boldsymbol{z}(\boldsymbol{e}_n)\right].\end{aligned} \tag{9.32}$$

In order to calculate the expectations $\mathbb{E}\left[\left[\boldsymbol{e}_n^a\right]^{\mathrm{T}}\boldsymbol{z}(\boldsymbol{e}_n)\right]$ and $\mathbb{E}\left[\boldsymbol{z}^{\mathrm{T}}(\boldsymbol{e}_n)\boldsymbol{R}_n^{\boldsymbol{P}}\boldsymbol{z}(\boldsymbol{e}_n)\right]$, some useful assumptions are given as follows:

A1 The noise $\{v_n\}$ is zero-mean, independent and identically distributed (i.i.d.) with variance σ_v^2, and is independent of any other signals in the system.
A2 The *a priori* error vector $\boldsymbol{e}_n^a$ is jointly Gaussian distributed.
A3 The input vectors $\{\boldsymbol{x}_n\}$ are i.i.d. with zero-mean.
A4 $\boldsymbol{R}_n^{\boldsymbol{P}}(i,j)$ is independent of $\{e_i, e_j\}$, $\forall i,j \in \{n, \dots, n-L+1\}$.

Remark 9.1 Assumption A1 has been widely applied in the convergence analysis of most AF algorithms. Assumption A2 can be justified in many practical scenarios, especially for long adaptive filters, through the central limit theorem [16]. Assumption A3 restricts the input vectors to white regression data, which simplifies the computation of expectations. Under Assumptions A1 and A3, Assumption A4 is reasonable, especially as the filter length increases. The main reason behind this is that when the CAFMEE algorithm achieves perfect identification, we can assume that $\boldsymbol{e}_n^a = \boldsymbol{0}$, and $\boldsymbol{e}_n \approx \boldsymbol{v}_n$. Therefore, this independent condition will approximately hold.

Due to the independence between $\boldsymbol{X}_n$ and $\tilde{\boldsymbol{w}}_n$, under Assumptions A2 and A3, *a priori* error vector $\boldsymbol{e}_n^a$ can be regarded as zero-mean i.i.d. Gaussian distributed.

Therefore, the expectation $\mathbb{E}\left[\left[\boldsymbol{e}_n^a\right]^{\mathrm{T}}\boldsymbol{z}(\boldsymbol{e}_n)\right]$ only relies on $\boldsymbol{e}_n^a$ through the second moment $\sigma_{\boldsymbol{e}_n^a}^2$, where $\sigma_{\boldsymbol{e}_n^a}^2 = \mathbb{E}\left[(e_i^a)^2\right]$ denotes the variance of e_i^a. Assuming that $\hbar_G$ denotes a function of $\sigma_{\boldsymbol{e}_n^a}^2$, the expectation $\mathbb{E}\left[\left[\boldsymbol{e}_n^a\right]^{\mathrm{T}}\boldsymbol{z}(\boldsymbol{e}_n)\right]$ can be rewritten as:

$$\mathbb{E}\left[\left[\boldsymbol{e}_n^a\right]^{\mathrm{T}}\boldsymbol{z}(\boldsymbol{e}_n)\right] = \sigma_{\boldsymbol{e}_n^a}^2 \hbar_G(\sigma_{\boldsymbol{e}_n^a}^2), \tag{9.33}$$

where the subscript G in $\hbar_G$ demonstrates that Gaussian Assumption A2 plays a key role in defining (9.33). Meanwhile, under Assumptions A1, A3, and A4, the expectation $\mathbb{E}\left[\boldsymbol{z}^{\mathrm{T}}(\boldsymbol{e}_n)\boldsymbol{R}_n^{\boldsymbol{P}}\boldsymbol{z}(\boldsymbol{e}_n)\right]$ can be rewritten as

$$\mathbb{E}\left[\boldsymbol{z}^{\mathrm{T}}(\boldsymbol{e}_n)\boldsymbol{R}_n^{\boldsymbol{P}}\boldsymbol{z}(\boldsymbol{e}_n)\right] = \hbar_I(\sigma_{\boldsymbol{e}_n^a}^2)\mathbb{E}\left[\boldsymbol{R}_n^{\boldsymbol{P}}(i,i)\right], \tag{9.34}$$

in which

$$\hbar_I(\sigma_{\boldsymbol{e}_n^a}^2) = \sum_{l=1}^{L}\mathbb{E}\left[z_l^2(\boldsymbol{e}_n)\right]. \tag{9.35}$$

Here, the subscript I in $\hbar_I$ reveals that independent Assumption A4 is the main assumption in evaluating the expectation. Since $\boldsymbol{P} = \boldsymbol{P}^2$, we have

$$\mathbb{E}\left[\boldsymbol{R}_n^{\boldsymbol{P}}(l,l)\right] = \mathbb{E}\left[\boldsymbol{x}_{n-L+l}^{\mathrm{T}}\boldsymbol{P}\boldsymbol{P}\boldsymbol{x}_{n-L+l}\right] = \mathrm{Tr}\left\{\boldsymbol{PRP}\right\}, \tag{9.36}$$

where $\mathrm{Tr}\left\{\cdot\right\}$ is the trace operator and $\boldsymbol{R} = \mathbb{E}\left[\boldsymbol{x}_n\boldsymbol{x}_n^{\mathrm{T}}\right]$ denotes the covariance matrix of the input vector.

In general, the steady-state mean square deviation (MSD)

$$S_{\mathrm{MSD}} = \lim_{n\to\infty}\mathbb{E}\left[\|\tilde{\boldsymbol{w}}_n\|^2\right] \tag{9.37}$$

is commonly used as a key performance measure in adaptive filters. Under Assumption A3, we have $\boldsymbol{R} = \sigma_x^2\boldsymbol{I}_M$, where σ_x^2 denotes the variance of the white input regression data. Therefore, using $\boldsymbol{P} = \boldsymbol{P}^2$, we derive

$$\mathbb{E}\left[\boldsymbol{R}_n^{\boldsymbol{P}}(l,l)\right] = \mathrm{Tr}\left\{\boldsymbol{PRP}\right\} = \sigma_x^2\mathrm{Tr}\left\{\boldsymbol{P}\right\}. \tag{9.38}$$

Substituting (9.33), (9.34) with (9.38) into (9.31), we have

$$\mathbb{E}\left[\left\|\tilde{\boldsymbol{w}}_{n+1}\right\|^2\right] = \mathbb{E}\left[\left\|\tilde{\boldsymbol{w}}_n\right\|^2\right] - 2\eta\sigma_{\boldsymbol{e}_n^a}^2\hbar_G(\sigma_{\boldsymbol{e}_n^a}^2) + \eta^2\sigma_x^2\mathrm{Tr}\left\{\boldsymbol{P}\right\}\hbar_I(\sigma_{\boldsymbol{e}_n^a}^2). \tag{9.39}$$

Considering that the WEP achieves a steady-state value, i.e.,

$$\lim_{n\to\infty}\mathbb{E}\left[\left\|\tilde{\boldsymbol{w}}_{n+1}\right\|^2\right] = \lim_{n\to\infty}\mathbb{E}\left[\left\|\tilde{\boldsymbol{w}}_n\right\|^2\right], \tag{9.40}$$

after some simple manipulations, (9.39) is reduced to the following relation

$$\lim_{n\to\infty}\mathbb{E}\left[\left\|\tilde{\boldsymbol{w}}_n\right\|^2\right]\hbar_G\left(\sigma_x^2\lim_{n\to\infty}\mathbb{E}\left[\left\|\tilde{\boldsymbol{w}}_n\right\|^2\right]\right) = \frac{\eta\mathrm{Tr}\left\{\boldsymbol{P}\right\}}{2}\hbar_I\left(\sigma_x^2\lim_{n\to\infty}\mathbb{E}\left[\left\|\tilde{\boldsymbol{w}}_n\right\|^2\right]\right). \tag{9.41}$$

Furthermore, combining (9.37) and (9.41) yields

$$S_{\mathrm{MSD}} = \frac{\eta \mathrm{Tr}\left\{\boldsymbol{P}\right\} \hbar_I(\sigma_x^2 S_{\mathrm{MSD}})}{2\hbar_G\left(\sigma_x^2 S_{\mathrm{MSD}}\right)}. \tag{9.42}$$

Remark 9.2 If the CAFMEE algorithm converges, the value of steady-state MSD, S_{MSD}, will be equal to a positive solution of (9.42). In other words, S_{MSD} will be a fixed point of the function $\eta \mathrm{Tr}\left\{\boldsymbol{P}\right\} \hbar_I(\sigma_x^2 S_{\mathrm{MSD}})/2\hbar_G\left(\sigma_x^2 S_{\mathrm{MSD}}\right)$. In addition, (9.42) mainly relies on two functions (i.e., $\hbar_G(\cdot)$ and $\hbar_I(\cdot)$), whose calculation process is usually rather complicated and tedious in practical problems.

9.5 Adaptive Filter Under Quantized Minimum Error Entropy Criterion

In adaptive filters, the MEE cost is typically measured using the information potential (IP), which is directly estimated from the data. However, the IP computation can be computationally expensive due to the double summation involved, leading to potential bottlenecks when dealing with large-scale datasets. To address this problem, the QMEE criterion was proposed in [27] to reduce the computational complexity of the IP from $\mathcal{O}(N^2)$ to $\mathcal{O}(PN)$, where N represents the number of samples and P denotes the number of real-valued code words in the codebook of a quantization operator. The fundamental idea behind the QMEE is to simplify the inner summation by quantizing the error samples using a straightforward quantization method. This quantization helps reduce the computational burden by decreasing the number of pairwise computations required. In addition to introducing the QMEE criterion, this section also presents some properties of the QMEE and derives the adaptive filter under the QMEE criterion. These advancements provide a computationally efficient approach to MEE-based AF algorithms, making them more feasible for large-scale datasets.

9.5.1 Quantized Minimum Error Entropy Criterion

To reduce the computational burden of estimating the error's PDF in the classical MEE cost, the QMEE cost aims to estimate the PDF of error using fewer bits. This is achieved by representing the N error samples $\left\{e_1, e_2, \dots, e_N\right\}$ with a quantization method. The quantization method involves a quantization operator or quantizer denoted as $Q[\cdot]$, with a codebook C containing P real-valued code words (typically $P \ll N$), i.e., $C = \left\{c_1, c_2, \dots, c_P \in \mathbb{R}\right\}$. The quantization operator $Q[\cdot]$ maps the error sample e_j to one of the P code words in C, i.e., $Q[e_j] \in C$. It is assumed that each error sample is quantized to the nearest code word. With the quantizer, the empirical IP for the QMEE can be simplified to:

$$\begin{aligned}\hat{V}_2(\boldsymbol{e}) \approx \hat{V}_2^Q(\boldsymbol{e}) &= \frac{1}{N^2}\sum_{i,j=1}^{N}\kappa_\sigma\left(e_i - Q[e_j]\right)\\ &= \frac{1}{N^2}\sum_{n=1}^{N}\left(\sum_{m=1}^{M}\zeta_m\kappa_\sigma\left(e_n - c_m\right)\right)\\ &= \frac{1}{N}\sum_{n=1}^{N}\hat{p}_Q\left(e_n\right),\end{aligned} \tag{9.43}$$

where ζ_m is the number of error samples that are quantized to the code word c_m, and $\hat{p}_Q(x) = \frac{1}{N}\sum_{m=1}^{M}\zeta_m\kappa_\sigma\left(x - c_m\right)$ is the PDF estimator based on the quantized error samples. Clearly, we have $\sum_{m=1}^{M}\zeta_m = N$ and $\int \hat{p}_Q(x)\,dx = 1$. The computational complexity of the QMEE cost $\hat{V}_2^Q(\boldsymbol{e})$ is $\mathcal{O}(PN)$, which is much simpler than the original MEE cost especially for large-scale datasets ($N \gg P$).

9.5.2 Properties of QMEE

In the QMEE cost framework, several fundamental properties are presented to provide a theoretical foundation. These properties provide insights into the behavior and characteristics of the QMEE cost, supporting its use as a computationally efficient approximation of the MEE cost. For detailed proofs and a deeper understanding, the readers can refer to [27].

Property 9.1 When the codebook $C = \left\{e_1, e_2, \dots, e_N\right\}$, we have $\hat{V}_2^Q(\boldsymbol{e}) = \hat{V}_2(\boldsymbol{e})$.

Property 9.2 The QMEE cost $\hat{V}_2^Q(\boldsymbol{e})$ is bounded, i.e., $\hat{V}_2^Q(\boldsymbol{e}) \leq \frac{1}{\sqrt{2\pi}\sigma}$, with equality if and only if $e_1 = e_2 = \cdots = e_N = c$, where c is an element of C.

Property 9.3 It holds that $\hat{V}_2^Q(\boldsymbol{e}) = \frac{1}{M}\sum_{m=1}^{M}\beta_m\hat{p}\left(c_m\right)$, where $\beta_m = \frac{\zeta_m}{N}$, satisfying $\sum_{m=1}^{M}\beta_m = 1$.

Remark 9.3 Property 3 states that the QMEE cost $\hat{V}_2^Q(\boldsymbol{e})$ can be expressed as a weighted average of Parzen's PDF estimator evaluated at the code words in the codebook C. Specifically, when there is only one code word in C, i.e., $C = \{c\}$, we have $\hat{V}_2^Q(\boldsymbol{e}) = \hat{p}(c)$. In particular, when $C = \{0\}$, we have $\hat{V}_2^Q(\boldsymbol{e}) = \hat{V}_2(\boldsymbol{e}) = \frac{1}{N}\sum_{n=1}^{N}\kappa_\sigma(e_n) = \hat{p}(0)$. Here, $\hat{V}_2(\boldsymbol{e})$ represents the empirical correntropy [20], which is a well-known local similarity measure in ITL. In this context, the correntropy measures the local similarity around zero. On the other hand, the QMEE cost $\hat{V}_2^Q(\boldsymbol{e})$ measures the average similarity around each code word in C. Therefore, the correntropy can be regarded as a special case of the QMEE cost, where it focuses on the similarity to zero, while the QMEE cost considers the average similarity to each code word in the codebook. This observation indicates that the QMEE cost provides a more generalized and flexible measure of similarity by considering multiple code words in the codebook.

Property 9.4 When σ is large enough, we have $\hat{V}_2^Q(\boldsymbol{e}) \approx \frac{1}{\sqrt{2\pi}\sigma} - \frac{1}{2\sqrt{2\pi}\sigma^3} \sum_{m=1}^{M} \beta_m \mu_m$, where $\mu_m = \frac{1}{N} \sum_{n=1}^{N} \left(e_n - c_m\right)^2$ is the second-order moment of error about the code word c_m.

Remark 9.4 Based on Property 4, as the width parameter σ tends to infinity, the second-order moments become the dominant factor in the QMEE cost $\hat{V}_2^Q(\boldsymbol{e})$. In this scenario, maximizing the QMEE cost is equivalent to minimizing a weighted average of the second-order moments around the code words.

Property 9.5 If $\forall j$, $\left|e_j - Q[e_j]\right| \leq \varepsilon$ with ε being a positive number, then $\left|\hat{V}_2^Q(\boldsymbol{e}) - \hat{V}_2(\boldsymbol{e})\right| \leq \frac{\varepsilon \exp(-1/2)}{\sigma}$.

Remark 9.5 From Property 5, when ε is very small or σ is very large, the difference between the values of $\hat{V}_2^Q(\boldsymbol{e})$ and $\hat{V}_2(\boldsymbol{e})$ will be very small.

For the QMEE, designing a simple and efficient quantizer is a crucial aspect, encompassing the construction of the codebook and the assignment of code words to the data. To address this, we can employ a straightforward online vector quantization (VQ) method, which has previously been proposed in kernel adaptive filters to control network growth in [41]. The online VQ method offers simplicity and online adaptability as its primary advantages. The online VQ algorithm is summarized in Algorithm 9.2.

Remark 9.6 The online VQ method described in Algorithm 9.2 constructs the codebook sequentially from the samples, resulting in a computationally efficient process. Its computational complexity is linear in the number of samples.

9.5.3 QMEE-Based AF Algorithm

In the context of linear AF, the optimal weight vector of the filter can be obtained by maximizing the following function using the QMEE criterion:

$$J_{\text{QMEE}} = \max_{\boldsymbol{w}_n} \hat{V}_2^Q(\boldsymbol{e}_n). \tag{9.44}$$

Algorithm 9.2: Online vector quantization.

Input: error samples $\{e_n\}_{n=1}^{N}$
Output: quantized errors $\{Q[e_n]\}_{n=1}^{N}$
Parameters setting: quantization threshold ε
Initialization: Set $C_1 = \{e_1\}$
for $n = 2, \ldots, N$ **do**
  Compute the distance between e_n and C_{n-1}:$|e_n - C_{n-1}(j)|$ where $j = \underset{1 \leq j \leq |C_{n-1}|}{\text{argmin}} |e_n - C_{n-1}(j)|$, $C_{n-1}(j)$ denotes the jth element of C_{n-1}, and $|C_{n-1}|$ stands for the cardinality of C_{n-1}.
  if $|e_n - C_{n-1}(j)| \leq \varepsilon$ **then**
    Keep the codebook unchanged: $C_n = C_{n-1}$ and quantize e_n to the closest code word $Q[e_n] = C_{n-1}(j)$;
  else
    Update the codebook: $C_n = \{C_{n-1}, e_n\}$ and quantize e_n to itself: $Q[e_n] = e_n$;
  end if
end for

According to the cost function in (9.44), a gradient-based adaptive algorithm is obtained, whose update equation can be described by:

$$\begin{aligned}
\boldsymbol{w}_{n+1} &= \boldsymbol{w}_n + \eta \frac{\partial \hat{V}_2^Q(\boldsymbol{e}_n)}{\partial \boldsymbol{w}_n} \\
&= \boldsymbol{w}_n + \eta \frac{1}{L^2} \sum_{n=N-L+1}^{N} \sum_{m=1}^{M} \zeta_m \frac{\partial}{\partial \boldsymbol{w}_n} \kappa_\sigma (e_n - c_m) \\
&= \boldsymbol{w}_n + \eta \frac{1}{\sigma^2 L^2} \sum_{n=N-L+1}^{N} \sum_{m=1}^{M} \zeta_m (e_n - c_m) \kappa_\sigma (e_n - c_m) \boldsymbol{x}_n,
\end{aligned} \tag{9.45}$$

where L is the sliding window length. The algorithm described above is commonly referred to as the QMEE algorithm. From (9.45), one can observe that the weight update in the QMEE algorithm follows a gradient descent direction. This means that the weight vector is adjusted iteratively in the direction opposite to the gradient of the IP with respect to the weight vector. By updating the weights in this manner, the algorithm aims to minimize the quantized error and maximize the similarity between the desired response and the filter output. The QMEE algorithm provides an effective approach for adapting the weight vector in linear adaptive filter. By leveraging the QMEE criterion and its gradient, it enables the filter to learn and adapt its weights to optimize the performance objective in terms of quantized error and similarity measures. Moreover, one can see that

(1) As quantization threshold $\varepsilon \to 0$, we have $M \to L$. In this case, the QMEE algorithm becomes the original MEE algorithm.
(2) As quantization threshold $\varepsilon \to \infty$, we have $M \to 1$. In this case, the QMEE algorithm becomes the traditional MCC algorithm

$$\boldsymbol{w}_{n+1} = \boldsymbol{w}_n + \frac{\eta}{L\sigma^2} \sum_{l=1}^{L} e_l \kappa_\sigma \left(e_l\right) \boldsymbol{x}_l. \tag{9.46}$$

(3) The computational complexity of the QMEE algorithm is approximately $\mathcal{O}(LP)$, which is significantly lower that of the MEE algorithm (i.e., $\mathcal{O}(L^2)$), particularly when $P \ll L$.

The QMEE-based AF algorithm is summarized in Algorithm 9.3.

Algorithm 9.3: QMEE-based adaptive filtering algorithm.

Input: samples $\{\boldsymbol{x}_n, d_n\}_{n=1}^{N}$
Output: weight vector $\boldsymbol{w}$
1: Parameters setting: iteration number T, step size η, kernel width σ, sliding window length L, and quantization threshold ε
2: Initialization: Set $\boldsymbol{w}_0 = \boldsymbol{0}$
3: **for** $t = 1, \dots, T$ **do**
4: Compute the error samples based on $\boldsymbol{w}_{t-1}$: $e_n = d_n - \boldsymbol{w}_{t-1}^{\mathrm{T}} \boldsymbol{x}_n$, $n = 1, \dots, N$;
5: Create the quantization codebook C and quantize the N error samples by Algorithm 9.2;
6: Update the weight vector by (9.45);
7: **end for**

9.6 Simulation Results

In this section, we present several simulation results to showcase the performance of the introduced MEE-based AF algorithms in terms of accuracy, robustness, convergence behavior, and other relevant metrics. These simulations aim to provide empirical evidence of the effectiveness and advantages of the MEE-based AF algorithms.

9.6.1 Sparse Adaptive Filter Under MEE

In this example, we conduct simulations on time-varying channel estimation to evaluate the performance of the MEECIM algorithm. We compare its performance with other two AF algorithms, including the MSE-based zero-attracting LMS

(ZALMS) algorithm [42], and the original MEE-based AF algorithm [18]. To assess the performance of these algorithms, we adopt the MSD as the performance measure, whose definition is:

$$\text{MSD} = \mathbb{E}\left[\|\overline{\boldsymbol{w}} - \boldsymbol{w}_n\|^2\right]. \tag{9.47}$$

To identify the sparsity of the system, we use a filter of $M = 20$ coefficients in the time-varying system. The parameter vector of the unknown channel is

$$\overline{\boldsymbol{w}} = \begin{cases} [0,0,0,0,1,0,0,0,0,0,0,0,0,0,0,0,0,0,0,0], & N \leq 2000, \\ [1,0,1,0,1,0,1,0,1,0,1,0,1,0,1,0,1,0,1,0], & 2000 < N \leq 3000, \\ [1,-1,1,-1,1,-1,1,-1,1,-1,1,-1,1,-1,1,-1,1,-1,1,-1], & N > 3000. \end{cases} \tag{9.48}$$

In the simulation, the input signal $\boldsymbol{x}_n$ is modeled as a white Gaussian random sequence with zero mean and unit variance. The impulsive noise is modeled using the alpha-stable distribution. Four main parameters are adopted in the characteristic function of the alpha-stable distribution which are the characteristic parameter, the symmetry parameter, the location parameter, and the dispersion parameter. We set these parameters to 1.2, 0, 0.2, and 0, respectively. The specific details of these parameters and their interpretations in the alpha-stable distribution can be found in [23].

We conducted simulations to investigate the convergence behaviors of the MEECIM algorithm in an impulsive alpha-stable noise environment. The MEECIM algorithm utilizes a sliding data length of $L = 20$ and a step size of 0.03. The kernel widths used in MEE and CIM are 2.0 and 0.04, respectively. In Figure 9.4, we present the convergence curves in terms of MSD for all algorithms. The simulation results indicate that when the channel system is highly sparse (before the 2000th iteration), MEECIM exhibits a faster convergence rate and

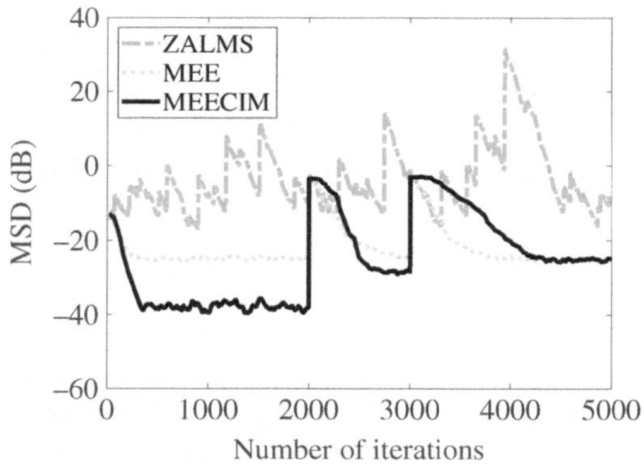

Figure 9.4 Tracking and steady-state behaviors of 20-order adaptive filters. Source: Adapted from [23].

superior steady-state performance compared to the other algorithms. After the 2000th iteration, as the number of nonzero taps increases to 10, the performance of MEECIM continues to outperform other sparse-aware filters. Even after 3000 iterations, when the system becomes completely nonsparse, MEECIM still performs comparably to the traditional MEE algorithm. These results highlight the effectiveness of the MEECIM algorithm in sparse signal recovery and its ability to adapt to changes in sparsity levels under impulsive alpha-stable noise environments.

9.6.2 Constrained Adaptive Filter Under MEE

In the first example, we consider a linear system identification problem with $M = 7$, where the parameters $\boldsymbol{C}$ and $\boldsymbol{f}$ are available in [25]. We use a white Gaussian process with unit power as the input signal. The simulation investigates the steady-state MSD with different step sizes in the presence of Gaussian and binary noises. The Gaussian noise has a zero mean and unit variance, while the binary noise has probabilities $\Pr\{v = 0.1\} = \Pr\{v = -0.1\} = 0.5$. Figure 9.5 presents the simulated and theoretical steady-state MSDs for the different noise scenarios and step sizes, where the step-sizes are in the range of $[0.01, 0.08]$. The simulation results demonstrate that the steady-state MSDs obtained from the simulations reasonably match the theoretical results. Furthermore, the results indicate that increasing the sliding data length L can improve the steady-state performance, particularly when the step size is large.

In the second example, we conduct a comparison between the CAFMEE algorithm and other typical algorithms, including the constrained least mean square (CLMS) algorithm [43], the constrained affine projection (CAP) algorithm [44],

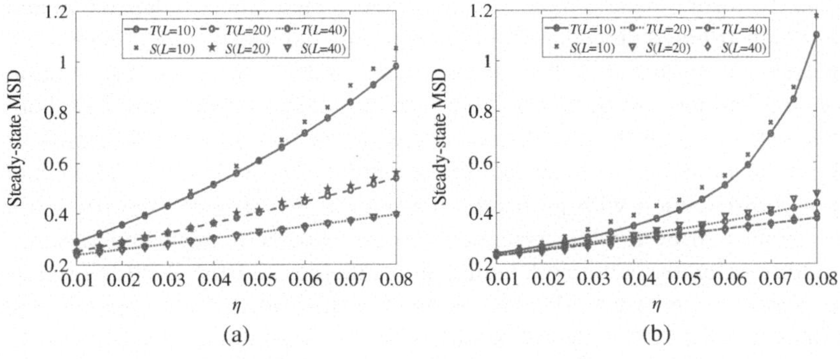

Figure 9.5 Simulated and theoretical steady-state MSDs (S and T in short respectively) versus different step-sizes. Source: Adapted from [25]. (a) Gaussian noise and (b) binary noise.

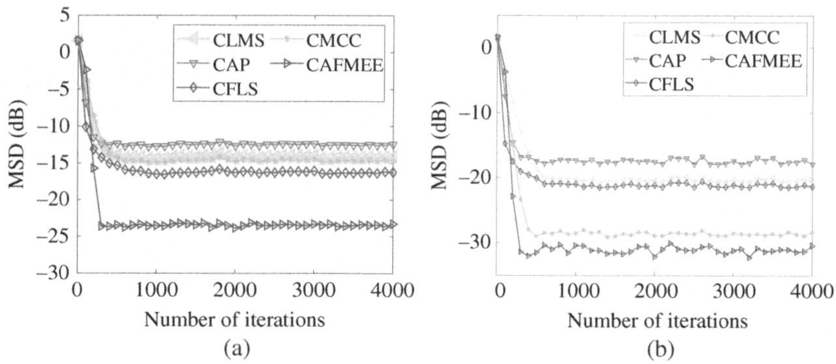

Figure 9.6 Convergence comparison of CLMS, CAP, CFLS, CMCC, and CAFMEE. Source: [25]/with permission of IEEE. (a) Binary noise and (b) mixed Gaussian noise.

the constrained fast least square (CFLS) algorithm [45], and the constrained maximum correntropy criterion (CMCC) algorithm [46]. The convergence curves of these algorithms are illustrated in Figure 9.6, considering two different types of noise: binary noise and mixed Gaussian noise. From the simulation results, it is evident that CAFMEE exhibits significant performance advantages over the MSE-based constrained AF algorithms, as well as outperforms CMCC in non-Gaussian noises characterized by a light-tailed or multi-peak distribution.

9.6.3 Adaptive Filter Under QMEE

In the first example, we consider an unknown system with a weight vector assumed to be $\overline{\boldsymbol{w}} = [0.1, 0.2, 0.3, 0.4, 0.5, 0.4, 0.3, 0.2, 0.1]^{\mathrm{T}}$. The input signal is a zero-mean white Gaussian process with a variance of 1.0. We introduce a noise model of the form $v_n = (1 - a_n)A_n + a_n B_n$, where a_n is a binary independent and identically distributed process with $\Pr\{a_n = 0\} = 1 - \rho$, and $0 \leq \rho \leq 1$ represents the occurrence probability. The processes A_n and B_n represent background noises and outliers, respectively. They are mutually independent and both independent of a_n. In the simulations, we consider two different distributions for A_n: (i) asymmetric Gaussian mixture density: $\frac{2}{3}\mathcal{N}(-10, 1) + \frac{1}{3}\mathcal{N}(12, 1)$; and (ii) binary distribution with probability mass $\Pr\{x = -10\} = \Pr\{x = 10\} = 0.5$. B_n is assumed to be a white Gaussian process with zero-mean and variance 100. Additionally, the window length L is set to 100, and the WEP is used as the performance measure, defined as $\mathrm{WEP} = ||\overline{\boldsymbol{w}} - \boldsymbol{w}_n||^2$. We compare the performance of the QMEE algorithm with these AF algorithms under MSE, MCC, and MEE. The simulation results are depicted in Figure 9.7. It is evident from the results that both the MEE and QMEE algorithms outperform the other compared algorithms significantly. This superior performance can be attributed to the

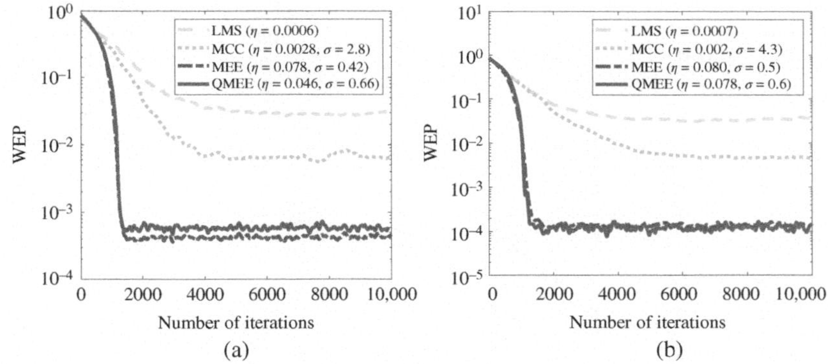

Figure 9.7 Convergence comparison of LMS, MCC, MEE, and QMEE. Source: [26]/with permission of Elsevier. (a) Asymmetric Gaussian mixture density and (b) binary distribution.

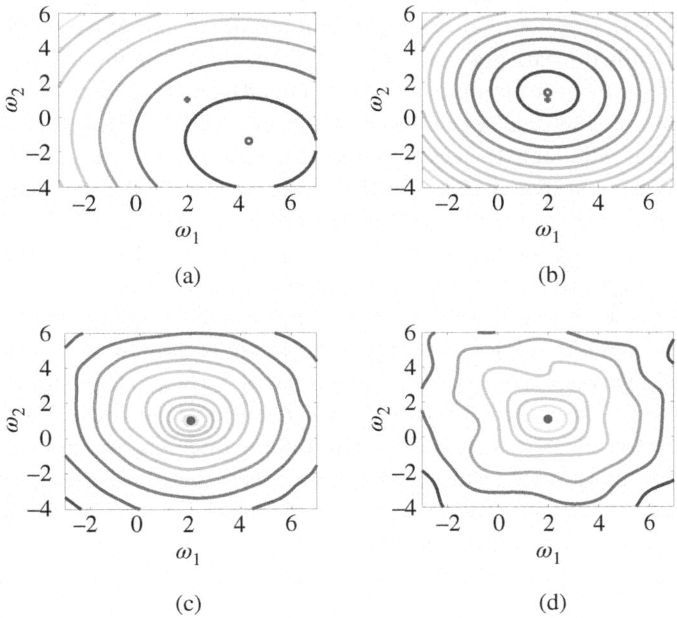

Figure 9.8 Contour plots of the performance surfaces. Source: [27]/with permission of IEEE. (a) MSE, (b) MCC, (c) MEE and (d) QMEE.

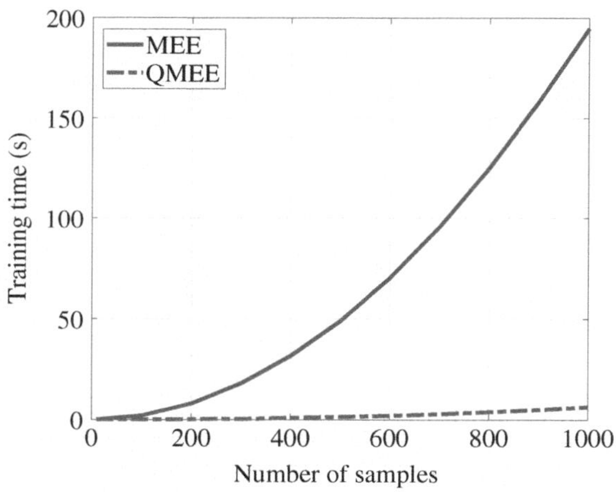

Figure 9.9 Computing time of MEE and QMEE versus the number of samples. Source: [26]/with permission of Elsevier.

contour plots of the performance surfaces for MSE, MCC, MEE, and QMEE, as shown in Figure 9.8. In this figure, the target weight vector is represented by the light gray crosses, while the dark gray circles indicate the optimal solutions. From Figure 9.8, one can see that the optimal solutions obtained using MEE and QMEE closely resemble the target weight vector. This suggests that MEE and QMEE yield solutions that are more similar to the desired target value compared to the optimal solutions obtained using MSE and MCC. Furthermore, the QMEE algorithm can achieve almost the same performance as the original MEE algorithm.

In the second example, we investigate the running time of the QMEE and MEE algorithms, considering varying the number of samples ranging from 1 to 1000. Figure 9.9 illustrates the average computation time of both algorithms as the number of samples increases. From Figure 9.9, it is evident that the computation time of the MEE algorithm experiences a significant increase as the increasing training sample size. In contrast, the computation time of the QMEE algorithm increases at a slower rate due to the effect of quantization. This finding underscores the superiority of the QMEE algorithm in terms of computational efficiency, especially for larger sample sizes, rendering it a more viable and efficient option in situations where processing time is of utmost importance.

9.7 Conclusion

This chapter outlines the fundamental principle of the MEE criterion and presents derivations of adaptive filters based on this criterion. The MEE criterion is utilized as an optimality metric in adaptive filters to improve the resilience of traditional MSE-based filtering algorithms in the presence of non-Gaussian noise and interference. Apart from the original MEE-based AF algorithm, this chapter also introduces advanced techniques such as MEE-based sparse and constrained AF algorithms, which further enhance the performance and adaptability of the original algorithm. Furthermore, to address the computational complexity of the MEE criterion, the QMEE criterion is developed, and the QMEE-based AF algorithm is presented as a computationally efficient alternative that retains the advantages of the MEE criterion. Extensive simulations have demonstrated the effectiveness and robustness of these MEE-based AF algorithms in linear system identification applications.

Acknowledgments

This work was partially supported by the National Natural Science Foundation of China (Grants U21A20485, 62306080, and 61976175) and the Science and Technology Project of Guangdong Province (Grant 2023A1515011185).

References

1 Li, J. and Stoica, P. (1996). An adaptive filtering approach to spectral estimation and SAR imaging. *IEEE Transactions on Signal Processing* 44 (6): 1469–1484.

2 Roman, J.R., Rangaswamy, M., Davis, D.W. et al. (2000). Parametric adaptive matched filter for airborne radar applications. *IEEE Transactions on Aerospace and Electronic Systems* 36 (2): 677–692.

3 Rangaswamy, M., Lin, F.C., and Gerlach, K.R. (2004). Robust adaptive signal processing methods for heterogeneous radar clutter scenarios. *Signal Processing* 84 (9): 1653–1665.

4 Patole, S.M., Torlak, M., Wang, D., and Ali, M. (2017). Automotive radars: a review of signal processing techniques. *IEEE Signal Processing Magazine* 34 (2): 22–35.

5 Hakobyan, G. and Yang, B. (2019). High-performance automotive radar: a review of signal processing algorithms and modulation schemes. *IEEE Signal Processing Magazine* 36 (5): 32–44.

6 Song, P., Manduca, A., Trzasko, J.D., and Chen, S. (2016). Ultrasound small vessel imaging with block-wise adaptive local clutter filtering. *IEEE Transactions on Medical Imaging* 36 (1): 251–262.

7 Liu, S., Cao, Y., Yeo, T.S. et al. (2020). Adaptive clutter suppression in randomized stepped-frequency radar. *IEEE Transactions on Aerospace and Electronic Systems* 57 (2): 1317–1333.

8 Gebhard, A., Lang, O., Lunglmayr, M. et al. (2019). A robust nonlinear RLS type adaptive filter for second-order-intermodulation distortion cancellation in FDD LTE and 5G direct conversion transceivers. *IEEE Transactions on Microwave Theory and Techniques* 67 (5): 1946–1961.

9 Qin, W., Gamba, M.T., Falletti, E., and Dovis, F. (2020). An assessment of impact of adaptive notch filters for interference removal on the signal processing stages of a GNSS receiver. *IEEE Transactions on Aerospace and Electronic Systems* 56 (5): 4067–4082.

10 Shi, S.N., Liang, X., Shui, P.L. et al. (2019). Low-velocity small target detection with Doppler-guided retrospective filter in high-resolution radar at fast scan mode. *IEEE Transactions on Geoscience and Remote Sensing* 57 (11): 8937–8953.

11 Chen, P., Qi, C., Wu, L., and Wang, X. (2018). Waveform design for Kalman filter-based target scattering coefficient estimation in adaptive radar system. *IEEE Transactions on Vehicular Technology* 67 (12): 11805–11817.

12 Abualhayja'a, M. and Hussein, M. (2021). Comparative study of adaptive beamforming algorithms for smart antenna applications. *2020 International Conference on Communications, Signal Processing, and their Applications (ICC-SPA)*, 1–5. IEEE.

13 Gogineni, V.C. and Mula, S. (2018). Logarithmic cost based constrained adaptive filtering algorithms for sensor array beamforming. *IEEE Sensors Journal* 18 (14): 5897–5905.

14 Claser, R. and Nascimento, V.H. (2021). On the tracking performance of adaptive filters and their combinations. *IEEE Transactions on Signal Processing* 69: 3104–3116.

15 Zarai, K. and Cherif, A. (2021). Adaptive filter based on Monte Carlo method to improve the non-linear target tracking in the radar system. *Aerospace Systems* 4 (1): 67–74.

16 Sayed, A.H. (2003). *Fundamentals of Adaptive Filtering*. Wiley.

17 Chen, B., Zhu, Y., Hu, J., and Príncipe, J.C. (2013). *System Parameter Identification: Information Criteria and Algorithms*. Newnes.

18 Príncipe, J.C. (2010). *Information Theoretic Learning: Renyi's Entropy and Kernel Perspectives*. Springer Science & Business Media.

19 Chen, B., Dang, L., Zheng, N., and Príncipe, J.C. (2023). *Kalman Filtering Under Information Theoretic Criteria*, 89–126. Springer.

20 Liu, W., Pokharel, P.P., and Príncipe, J.C. (2007). Correntropy: properties and applications in non-Gaussian signal processing. *IEEE Transactions on Signal Processing* 55 (11): 5286–5298.

21 Chen, B., Xing, L., Xu, B. et al. (2016). Insights into the robustness of minimum error entropy estimation. *IEEE Transactions on Neural Networks and Learning Systems* 29 (3): 731–737.

22 Chen, B., Zhu, P., and Príncipe, J.C. (2011). Survival information potential: a new criterion for adaptive system training. *IEEE Transactions on Signal Processing* 60 (3): 1184–1194.

23 Wu, Z., Peng, S., Ma, W. et al. (2015). Minimum error entropy algorithms with sparsity penalty constraints. *Entropy* 17 (5): 3419–3437.

24 Wu, Z., Peng, S., Chen, B. et al. (2015). Proportionate minimum error entropy algorithm for sparse system identification. *Entropy* 17 (9): 5995–6006.

25 Peng, S., Ser, W., Chen, B. et al. (2018). Robust constrained adaptive filtering under minimum error entropy criterion. *IEEE Transactions on Circuits and Systems II: Express Briefs* 65 (8): 1119–1123.

26 Li, Z., Xing, L., and Chen, B. (2020). Adaptive filtering with quantized minimum error entropy criterion. *Signal Processing* 172: 107534.

27 Chen, B., Xing, L., Zheng, N., and Príncipe, J.C. (2018). Quantized minimum error entropy criterion. *IEEE Transactions on Neural Networks and Learning Systems* 30 (5): 1370–1380.

28 Chen, B. and Príncipe, J.C. (2012). On the smoothed minimum error entropy criterion. *Entropy* 14 (11): 2311–2323.

29 Shen, P. and Li, C. (2015). Minimum total error entropy method for parameter estimation. *IEEE Transactions on Signal Processing* 63 (15): 4079–4090.

30 Ogunfunmi, T. and Safarian, C. (2019). The quaternion stochastic information gradient algorithm for nonlinear adaptive systems. *IEEE Transactions on Signal Processing* 67 (23): 5909–5921.

31 Zhao, H., Nefabas, G.L., and Wang, Z. (2021). Augmented complex minimum error entropy for adaptive frequency estimation of power system. *IEEE Transactions on Circuits and Systems II: Express Briefs* 69 (3): 1972–1976.

32 Li, L.Q. and Yy, C. (2022). Constrained minimum fuzzy error entropy filtering for target tracking. *Digital Signal Processing* 132: 103796.

33 He, J., Wang, G., Cao, K. et al. (2023). Generalized minimum error entropy for robust learning. *Pattern Recognition* 135: 109188.

34 Zhang, Y., Chen, B., Liu, X. et al. (2015). Convergence of a fixed-point minimum error entropy algorithm. *Entropy* 17 (8): 5549–5560.

35 Wang, G., Peng, B., Feng, Z. et al. (2021). Adaptive filtering based on recursive minimum error entropy criterion. *Signal Processing* 179: 107836.

36 Gogineni, S. and Nehorai, A. (2011). Target estimation using sparse modeling for distributed MIMO radar. *IEEE Transactions on Signal Processing* 59 (11): 5315–5325.

37 Kalouptsidis, N., Mileounis, G., Babadi, B., and Tarokh, V. (2011). Adaptive algorithms for sparse system identification. *Signal Processing* 91 (8): 1910–1919.

38 Das, B.K. and Chakraborty, M. (2014). Sparse adaptive filtering by an adaptive convex combination of the LMS and the ZA-LMS algorithms. *IEEE Transactions on Circuits and Systems I: Regular Papers* 61 (5): 1499–1507.

39 Seth, S. and Príncipe, J.C. (2008). Compressed signal reconstruction using the correntropy induced metric. *2008 IEEE International Conference on Acoustics, Speech and Signal Processing*, 3845–3848. IEEE.

40 Li, J., Xie, Y., Dang, L. et al. (2021). Fixed-point minimum error entropy with sparsity penalty constraints. *IEEE Transactions on Circuits and Systems II: Express Briefs* 68 (8): 2997–3001.

41 Liu, W., Principe, J.C., and Haykin, S. (2011). *Kernel Adaptive Filtering: A Comprehensive Introduction*. New Jersey: Wiley.

42 Chen, Y., Gu, Y., and Hero, A.O. (2009). Sparse LMS for system identification. *2009 IEEE International Conference on Acoustics, Speech and Signal Processing*, 3125–3128. IEEE.

43 Frost, O.L. (1972). An algorithm for linearly constrained adaptive array processing. *Proceedings of the IEEE* 60 (8): 926–935.

44 Werner, S., Apolinario, J.A., de Campos, M.L., and Diniz, P.S. (2005). Low-complexity constrained affine-projection algorithms. *IEEE Transactions on Signal Processing* 53 (12): 4545–4555.

45 Resende, L.S., Romano, J.M.T., and Bellanger, M.G. (1996). A fast least-squares algorithm for linearly constrained adaptive filtering. *IEEE Transactions on Signal Processing* 44 (5): 1168–1174.

46 Peng, S., Chen, B., Sun, L. et al. (2017). Constrained maximum correntropy adaptive filtering. *Signal Processing* 140: 116–126.

10

Dynamic Control of Radar Systems Using Information Rate

Bryan Paul[1] and Daniel W. Bliss[2]

[1] *General Dynamics Mission Systems, Fairfax, VA USA*
[2] *Center for Wireless Information Systems and Computational Architectures (WISCA), Arizona State University, Tempe, AZ USA*

10.1 Introduction

While – depending on one's definition – radar technology is at least several decades old [1, 2], because of the combination of recent advances in processing and increased use of the spectrum, radar operation is becoming more complicated [3]. Sophisticated radar systems have long had to deal with the resource management challenges of multiple simultaneous goals, such as search, track, and identification [4, 5]. The complexities have compounded in recent years as contention for spectrum has increased to keep pace with a growing demand for wireless commercial technologies [1, 6]. Multiple, potentially disparate, functions (radar, communications, positioning, navigation, timing, etc.) need to be performed simultaneously within a given region of spectrum [7].

Here we present a method of dynamic radar control by using a metric of performance and signal processing approach that provides a formalism for investigating resource management issues. As a metric of performance, we consider measuring how much we are learning from the radar operation from an information theory perspective [8]. We can use this metric in a variety of ways to improve radar performance. Specifically, we consider the data required to describe what we have learned through radar measurement given ideal information source compression. We then control the radar system operating parameters to achieve a particular information rate, or information gain per radar spectral access.

We might use this estimation information rate metric for each target or even each of a target's estimated parameters, separately. Each type of application likely leads to different uses. In general, we expect that this metric is useful for dynamic system optimization. Historically, a complicated set of heuristic rules

Information-Theoretic Radar Signal Processing, First Edition. Edited by Yujie Gu and Yimin D. Zhang.

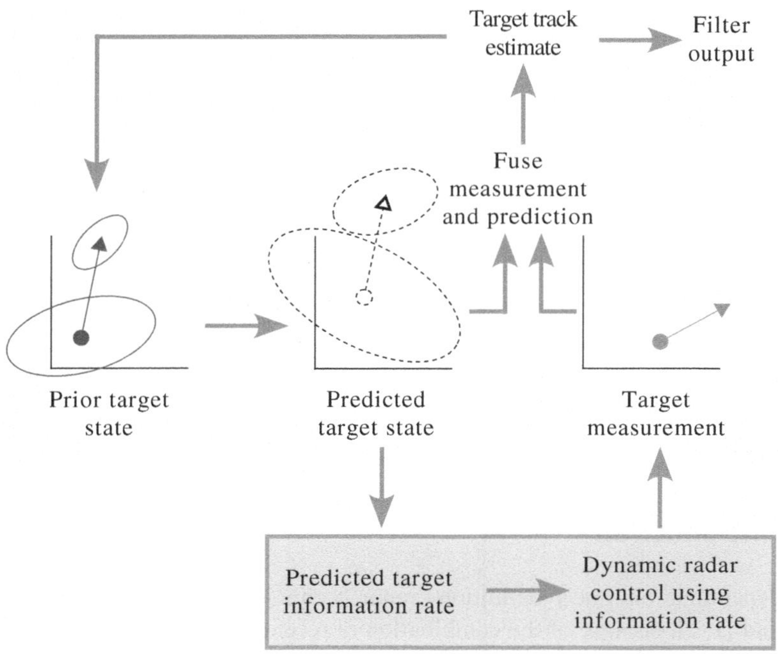

Figure 10.1 Top-level view of the information rate Bayesian/Kalman filter used to dynamically control a radar system. The darkened square is added to generalize the classical method of radar track filtering.

have been employed to control the radar. However, heuristic-based approaches are potentially fragile when exposed to novel scenarios. We offer information rate as a tool that may be used to increase robustness in dynamic radar system optimization.

In Figure 10.1, we depict a top-level view of the approach derived in this chapter. The darkened square is our extension and generalization of the Bayesian tracker often used in radar systems. By using the same measurements and parameters available to classical tracking methods, we also predict target information and use this to control the radar prior to measurement to optimize available radar resources and spectrum usage.

Radar systems are often constrained by their use of precious resources: spectrum, array access and size, transmit power, etc. Classical systems may have these parameters statically defined, leading to inefficient use of the allocated resources. For example, a radar may use maximum power and all available bandwidth regardless of the target conditions or priority, even if a given mission or objective could be completed with less resources. Not long after the introduction of radar, dynamic configurations started to emerge, with early versions manifesting as operators changing physical configuration of the radar in response to radar returns [2]. Even as the digital era ushered in electronic control, ad hoc rules-based approaches or heuristics were often used to set increasingly complex radar configurations [9, 10]. As more advances were made to radar technology, such as the advent of the electronically steered array (ESA), more heuristics were employed to adapt control of radar systems [11–14]. For example, some radar systems proposed allocating beam power based on predicted multitarget mean square error (MSE) [15] or scheduling in a multitarget scenario based on tracking prediction error [16]. Rules-based methods rooted in heuristics can get increasingly complex to deal with new scenarios, corner cases, and edge conditions as they arise in operation. Managing radar behavior for every conceivable scenario creates a convoluted approach that often becomes expensive to sustain. Unexpected target tracking scenarios may be handled poorly, and new rules are created, further complicating the radar system.

Information-based approaches offer an attractive alternative to ad hoc methods by providing a mathematically sound way of dynamically changing radar parameters while incorporating all of the complexities of the radar tracking scenario. Information-theoretic measures distill all of the target tracking parameters into a single output that has an intrinsic value. This is because information is at the heart of a radar system. Radars transmit, using spectrum, to illuminate and measure targets to learn information. Otherwise, it would be a fruitless endeavor and waste of resources. Modeling a complex target scenario as a target tracking estimation distribution allows incorporation of radar system parameters and target dynamics jointly. Once we have a distribution, mutual information can be used to provide insight into the achievable compression data rate of a target in a radar tracking scenario: radar estimation information rate. This key metric captures the predicted bits per spectral access for a given radar configuration and target scenario.

The use of information rate to dynamically control radar systems was first conceptually introduced by Bliss [8] and with a specific approach developed by Paul and Bliss [17, 18]. For example, using the extended framework shown in Figure 10.1, we can choose to optimize the radar configuration to hold the information rate to a desired setpoint. By predicting the information rate which quantifies what the radar stands to learn in terms of information through spectral access, the radar is modulated so that every illumination is set to gain a desired threshold of information, or commensurately reduce target uncertainty by a set amount. The initial application of this "constant information radar" was developed to address the emerging challenge of radio frequency (RF) convergence of radar and communications [19]. In addition to spectral efficiency and radar resource allocation, information rate can be used to prioritize targets in a multitarget tracking scenario. Static approaches may revisit targets in a "round-robin" fashion. Ad hoc approaches may prioritize targets based on a set of rules, for example, using residual magnitude in the measurement update step to revisit more dynamic targets more frequently [9, 16]. Using information rate provides an information-theoretic criterion for target prioritization to replace a rules-based approach that may fail when extrapolating to new scenarios.

This approach is very powerful in its simplicity of control for complex scenarios. However, depending on the application, this may not always be the right approach. For example, critical military applications most often prioritize protecting the warfighter, and so radar resource efficiency is weighted lower in the overall priorities within mission execution. Illuminating targets as fast and high power as possible is often the best solution in these cases. In addition, while the information-driven approach is elegant in its reduction of parameters, there are drawbacks to controlling based on a singular metric. While this approach provides a single measure of information gain, *what* information is gained is not discernible. For example, reducing uncertainty in target range may be more important in some applications than range-rate. Two different radar configurations may produce the same information rate for the same target in the same scenario, but result in vastly different *posterior* tracking distributions. In other words, not all information bits are created equal.

A simplified example we derive and demonstrate in this chapter is given in Figure 10.2 as a special case of the more general approach. In this case, we show a simplified two-dimensional (2D) radar tracking scenario where the target starts well behaved, but soon starts to accelerate in a serpentine pattern. In this example radar system, we assumed a linear, nonaccelerating motion model. Therefore, during well-modeled segments, the predicted information rate from measuring is small. Using information rate to dynamically control the radar, the target revisit is delayed until the predicted information rate triggers the threshold of usefulness. At a time in the future when the predicted information rate exceeds the set threshold, the radar makes a measurement. During turns, the target is naturally revisited more often. During straight, well-modeled segments, the radar revisits the target less frequently. This is all done using information rate, not rules about acceleration, model mismatch, or any other heuristic. Using information rate to dynamically control the radar results in intuitive tracking behavior.

In Section 10.2, we go through the background of information-based control of radar systems. This section builds up the general framework for estimation information rate, Bayesian radar tracking, physical target modeling, and radar measurements.

In Section 10.3, we develop the general approach to dynamic control of radar systems using information. We start by generalizing the Bayesian framework from the Section 10.2 to include predicted mutual information between the predicted target state and the predicted innovation, or measurement residual. This is the estimation of the information the radar stands to gain through physical measurement and spectrum use assuming a given radar configuration. This

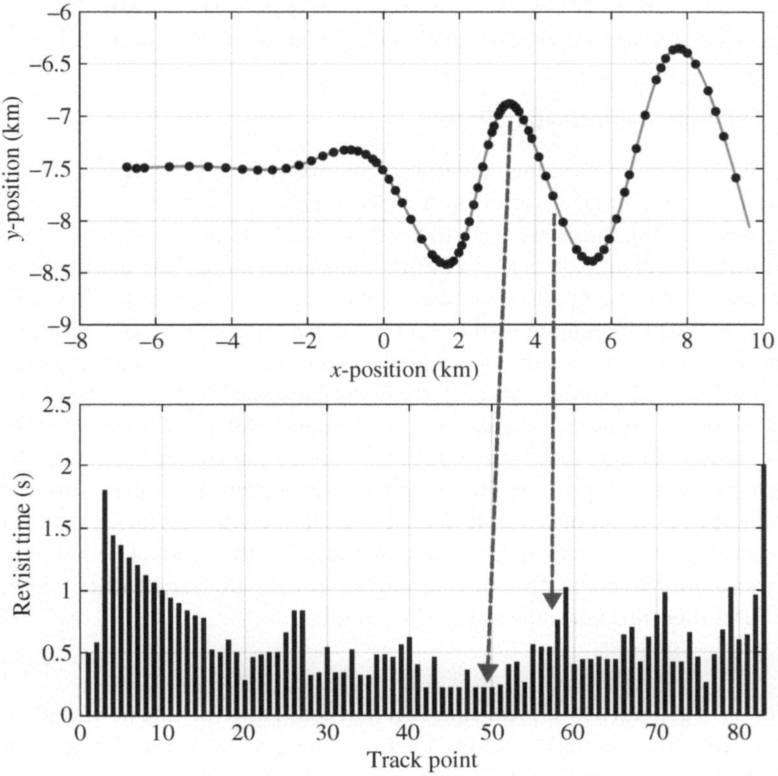

Figure 10.2 Example using information to control revisit time for a radar tracking scenario. After acquisition, the revisit time is intuitively adjusted based on target behavior, sampling the target more frequently when the target is exhibiting unexpected behavior.

allows the radar designer to optimize the radar configuration to achieve a desired predicted information gain.

Finally, in Section 10.4, we derive a special, simplified case for our approach in a 2D tracking scenario using the extended Kalman filter (EKF) and a nonaccelerating motion model. This allows us to derive specific distributions to illustrate the concept more clearly and show example scenarios that motivate this approach with metrics of performance to compare with static methods.

10.2 Signal Model Framework

In this section, we motivate and derive estimation information rate for radar systems. We provide a general signal model framework for using information rate for dynamic control of radar systems, including a Bayesian tracking filtering approach, a general target physical model, and a radar measurement model.

10.2.1 Estimation Information Rate

For our application of dynamic radar control, we derive general estimation information rate for radar as follows. Assume a general target state vector: $\boldsymbol{s}$. This represents all the things we wish to know about the target. In general, since it is unknown, it is characterized by a multidimensional probability distribution. This distribution can occupy any number of dimensions, in any desired coordinate space, and can include any state vector information about a given target. For example, it can simply be the range of the target to the radar, occupying only a single dimension. It could also be a three-dimensional (3D) target position, 3D target velocity, 3D target acceleration, and target radar cross section (RCS). We make no assumptions in this general derivation about the distribution; this could have been a target prior in a tracking scenario, a propagated prior through a motion model, a fusion of multiple data sets. Regardless, it represents the best estimate we have of the target at this time step, modeling the residual uncertainty as a distribution. We generally are not able to sample the target distribution directly, but rather through some noisy measurement

$$\boldsymbol{z} = f(\boldsymbol{s}), \tag{10.1}$$

where $f(\boldsymbol{s})$ is in general a nonlinear function that could also include new random sources of error like thermal noise on a sensor. Our goal is to use the radar to measure the target and reduce the uncertainty of the target state vector $\boldsymbol{s}$. How much we reduce that uncertainty can be computed via the mutual information between the target state vector we wish to estimate, and the noisy measurement we actually observe. This quantifies what information we learn about the target by making the measurement [20]:

$$I(\boldsymbol{s};\boldsymbol{z}) = \int_{\boldsymbol{z}}\int_{\boldsymbol{s}} p_{\boldsymbol{s},\boldsymbol{z}}(\boldsymbol{s},\boldsymbol{z}) \log\left[\frac{p_{\boldsymbol{s},\boldsymbol{z}}(\boldsymbol{s},\boldsymbol{z})}{p_{\boldsymbol{s}}(\boldsymbol{s})\, p_{\boldsymbol{z}}(\boldsymbol{z})}\right] d\boldsymbol{s}\, d\boldsymbol{z} = I(\boldsymbol{s}; f(\boldsymbol{s})), \tag{10.2}$$

where $I(\boldsymbol{s};\boldsymbol{z})$ denotes the mutual information between $\boldsymbol{s}$ and $\boldsymbol{z}$, $p_{\boldsymbol{s},\boldsymbol{z}}(\boldsymbol{s},\boldsymbol{z})$ is the joint probability distribution of $\boldsymbol{s}$ and $\boldsymbol{z}$, and $p_{\boldsymbol{s}}(\boldsymbol{s})$ and $p_{\boldsymbol{z}}(\boldsymbol{z})$ are the respective marginals. If we use a base 2 logarithm, then we have an expression in bits after computing this value. Since this occurs at some periodic rate, such as the revisit rate of a target, we can now intuitively derive how many bits over a time period we learn from measuring a given target, the radar estimation information rate [8]:

$$R_{\text{est}} = \frac{I(\boldsymbol{s};\boldsymbol{z})}{T}, \tag{10.3}$$

where T represents a time period between radar measurements or spectral access.

This is the radar estimation information rate for a given target scenario. There are many analogies we can draw from information theory to describe this more intuitively. If we consider the radar system in the context of a communications channel, this can be thought of as the information rate the target is uncooperatively communicating to the radar about its state. Estimation information rate also quantifies the minimum number of bits needed to describe the target track over time, or the compressibility of the target track for a given scenario. A well-behaved target will track easily, and measurements will often confirm track predictions, making deviations from the target physical model small. These deviations are more easily compressed if they are distributed close to the prediction mean (low covariance) or if the measurement noise is relatively high, where it becomes more difficult to unambiguously resolve two closely spaced locations and more bits would only encode undesirable noise. Note we have made no assumptions about the prior distribution, the target dynamics (modeled or reality), nor how the radar measurement is made. In a shared spectrum scenario with both radar and communications systems operating in the same band, we can now compare the target channel capacity to the communications user on the same scale of bits-per-second. This metric can also be used to control the radar dynamically, since it quantifies the value of radar measurement in a given scenario and radar configuration.

10.2.2 Bayesian Tracking Estimation

Estimation information rate is a flow of target information over time. For targets with static parameters, information gained will decay rapidly with successful measurements. A more interesting case is to consider dynamic target tracking, where information is gained with every radar measurement. Therefore, as we build up toward a practical approach to leveraging this new metric for dynamic radar control, we start by considering the tracking distribution framework. In a tracking scenario, we have estimates for target states at previous time steps. A typical assumption to make radar tracking filtering mathematically tractable is that the probability distribution of the current target state at time step k, given the previous target states, is conditionally independent of all the earlier target states except the prior state at time step $k-1$:

$$p(\boldsymbol{s}_k|\boldsymbol{s}_{k-1}, \boldsymbol{s}_{k-2}, \dots, \boldsymbol{s}_0) = p(\boldsymbol{s}_k|\boldsymbol{s}_{k-1}). \tag{10.4}$$

That is, we only need to consider the previous target state to optimally compute the current state distribution. This is known as the Markov property [21]. We assert this is a very reasonable assumption for nearly all radar tracking systems and scenarios. This means determining the information gain for a radar track is dramatically simplified, so we do not have to consider all prior states, but rather update the information based on the previous distribution. How we derive this distribution depends on how we want to model the target, which we cover in Section 10.2.3.

The same argument can be made for the current measurement, where we assume conditional independence from all but the current state:

$$p(\boldsymbol{z}_k|\boldsymbol{s}_k, \boldsymbol{s}_{k-1}, \ldots, \boldsymbol{s}_0) = p(\boldsymbol{z}_k|\boldsymbol{s}_k). \tag{10.5}$$

To predict the next target state, considering all of the prior measurements (all of the observed evidence seen so far), we can solve for the state prediction distribution [22]:

$$p(\boldsymbol{s}_k|\boldsymbol{z}_{k-1}, \boldsymbol{z}_{k-2}, \ldots, \boldsymbol{z}_0) = \int p(\boldsymbol{s}_k|\boldsymbol{s}_{k-1})\, p(\boldsymbol{s}_{k-1}|\boldsymbol{z}_{k-1}, \boldsymbol{z}_{k-2}, \ldots, \boldsymbol{z}_0)\, d\boldsymbol{s}_{k-1}. \tag{10.6}$$

This provides a rigorous, statistical way of deriving the distribution of the future target state, or the probability of it being in a given location based on all the measurements taken so far and our knowledge of the physical target model. Using these distributions, we can also recursively predict the estimation information rate and use it to dynamically control the radar system.

10.2.3 Target Physical Modeling

To fully and accurately quantify the information we learn through radar measurement, and therefore the information rate for a given target, we need to remove from consideration any prior knowledge we have of the target. A target physical model is used to capture existing knowledge of the prior state, prior measurements, and any physical constraints of the target in the radar context. As a radar tracking design parameter, ultimately the system designer must specify Equation (10.4). This distribution defines for a target the probability of the next state given the prior states. Ultimately, Equation (10.4) defines the physical motion model of a radar target. If we know approximately where a target is located at the prior time step $k-1$, the target is traveling at a constant velocity and we assume it is not accelerating, what is the probability the target is at some new location $\boldsymbol{s}_k = \boldsymbol{\alpha}$ at the next track point? The state prediction distribution we derive in the Bayesian framework inherently makes assumptions about the target. Any higher-order motion moments omitted from the model are implicitly assumed to be 0. For example, a constant velocity model assumes the target is not accelerating, since acceleration is not tracked as a dynamic parameter.

The more accurate this model, the better the radar predicts the target state, and the more redundant measurement becomes. If we perfectly model the target, and these parameters do not vary, then measurement stands to gain little information and the estimation information rate decays to zero with every successive measurement. However, if the target is dynamic or if the model is mismatched to reality, measuring the target shows significant deviation from our predictions. Therefore, significant information is gained through radar illumination and the estimation information rate is higher.

This physical model is a distribution, as there is always inherent modeling error or uncertainty in the target dynamics. This distribution quantifies how confident we are in our assumptions. For the constant velocity model example, we know realistically a target will eventually stop, so acceleration is likely at some point. We may also make this assumption knowing it is flawed, but do so anyway to make the math tractable or the radar implementable on a low-power processor, understanding the target may accelerate frequently. Therefore we may assume this model process variance is very high. The dynamics that are modeled for a given target are specified by the radar designer. The more complex the model, in general the more complex the distribution, and ultimately, the more difficult it may be to predict information rate to control the radar dynamically, so there is a tradeoff similar to existing radar tracking design.

10.2.4 Radar Measurements

We are rarely afforded the luxury of directly measuring the target state parameters. For RF-based radar systems, this is a known reality. Rather, we assume a hidden Markov model where the target parameters of interest are not necessarily observed directly (i.e., they are "hidden"), but rather through some other, related observable processes. These measurements are done with imperfect instruments, and so noise is introduced, as given in Equation (10.1). In general, this measurement transform could be highly nonlinear and with new random sources of error introduced, additive or otherwise.

Recall we started with the prior distribution and then predicted the next state using all of the prior measurements and the target physical model. Once we measure the target, we have learned new information we can introduce into our tracking Bayesian framework. We use this new information to update the target state distribution and ultimately the target state estimate [22]:

$$p(\mathbf{s}_k \,|\, \mathbf{z}_k, \mathbf{z}_{k-1}, \dots, \mathbf{z}_0) = \frac{p(\mathbf{z}_k \,|\, \mathbf{s}_k)\, p(\mathbf{s}_k \,|\, \mathbf{z}_{k-1}, \mathbf{z}_{k-2}, \dots, \mathbf{z}_0)}{p(\mathbf{z}_k \,|\, \mathbf{z}_{k-1}, \mathbf{z}_{k-2}, \dots, \mathbf{z}_0)}. \tag{10.7}$$

We are updating the prediction distribution to account for the new measurement distribution. If we could predict this information gain, we could use it to configure the radar *before* measurement to optimize the amount of information gained. We now have the necessary general framework to formally derive the information rate controlled radar.

10.3 Information Rate Controlled Radar

With the basic theoretical foundation in place, we can extend this general tracking framework to compute the predicted estimation information rate and demonstrate how it can be used to dynamically control the radar system.

10.3.1 Generalized Bayesian Framework

We now extend the canonical Bayesian framework to add radar tracking information gain. We want to quantify what new information we learn through this new measurement in the update step, or the mutual information between the measurement $\boldsymbol{z}_k$ and the predicted target state $\boldsymbol{s}_{k|k-1}$. We start by denoting the predicted target state probability distribution from Equation (10.6) as

$$p_{\boldsymbol{s}_{k|k-1}}\left(\boldsymbol{s}_{k|k-1}\right) = p(\boldsymbol{s}_k | \boldsymbol{z}_{k-1}, \boldsymbol{z}_{k-2}, \dots, \boldsymbol{z}_0). \tag{10.8}$$

Then the information gained during the Bayesian update step in Equation (10.7) is given as

$$I\left(\boldsymbol{s}_{k|k-1}; \boldsymbol{z}_k\right) = \iint p_{\boldsymbol{s}_{k|k-1},\boldsymbol{z}_k}\left(\boldsymbol{s}_{k|k-1}, \boldsymbol{z}_k\right) \log\left[\frac{p_{\boldsymbol{s}_{k|k-1},\boldsymbol{z}_k}\left(\boldsymbol{s}_{k|k-1}, \boldsymbol{z}_k\right)}{p_{\boldsymbol{s}_{k|k-1}}\left(\boldsymbol{s}_{k|k-1}\right) p_{\boldsymbol{z}}\left(\boldsymbol{z}_k\right)}\right] d\boldsymbol{s}_{k|k-1}\, d\boldsymbol{z}_k. \tag{10.9}$$

This is the information gained from measurement. This is the heart of *why* the radar is measuring. A perfectly modeled target with static parameters would need only a few measurements, then could be tracked without additional measurements. This is not practical, as no model is perfect in practice. Further, targets may adhere to the model or break from the model throughout the life of a track. Finally, parameters of the model are often unknown but dynamic instead of static, and not able to be fully characterized by a few initial measurements. Therefore, there is almost always information to be gained through measurement, and Equation (10.9) assigns that information a value.

This is a general framework that can be used to calculate information rate for any target tracking scenario in a Bayesian framework with the Markov property. This can include few or many dimensions, single or multiple targets, assuming linearized Kalman filters or particle filters. So how can we use this generalized framework to modulate our radar system? We start by generalizing the prediction step of the Bayesian filter to predict this mutual information, then use this as an input to a general optimization routine for our radar system.

10.3.2 Information Rate Prediction Algorithm

To further extend the Bayesian filter to accommodate information rate, we define a similar mutual information to Equation (10.9), but between the predicted state and the *predicted* measurement. To start, we revisit the target state prediction distribution from Equation (10.6).

This is the predicted target state considering the prior measurements, which depends on the prior state as well. This prediction inherently has modeling error from model mismatch which is partially reconciled through measurement. Target model mismatch is observed through the update step, where the difference between the prediction and the measurement (called the residual or innovation) is computed. Therefore, the mutual information between the target state prediction and the target measurement prediction allow us to predict the information gain.

We predict the target measurement by taking the target state prediction through the measurement function $f(\boldsymbol{s})$:

$$\boldsymbol{z}_{k|k-1} = f\left(\boldsymbol{s}_{k|k-1}\right). \tag{10.10}$$

Note the distribution of $\boldsymbol{z}_{k|k-1}$ varies greatly depending on if there is a 1-for-1 mapping, the dimensionality of the transformation, and other properties like monotonicity [21]. We can now predict the mutual information using Equation (10.9), substituting $\boldsymbol{z}_{k|k-1}$ for $\boldsymbol{z}_k$. Without more constraints on the measurement model (for example, independent additive noise), we cannot simplify the predicted mutual information in general. Simplifying assumptions can reduce this to very easily defined or computed closed-forms. We show this in Section 10.4 for a complete system to demonstrate this approach more concretely. Alternatively, with the input distributions specified, Monte Carlo or other numerical methods may be used to estimate predicted information rate in practical systems.

Finally, we can finish extending our Bayesian filter to modulate radar parameters based on predicted information gain. Table 10.1 provides the summary of the information rate tracking filter algorithm under the general Bayesian framework. Section 10.3.2.1 includes things that need to be determined by the radar designer in advance. Sections 10.3.2.2–10.3.2.4 follow directly from standard literature on Bayesian filtering but are generalized for using estimation information rate to control the radar.

As we will demonstrate in Section 10.4.5, there are multiple factors that modulate this predicted information. If we want to achieve some minimum, fixed, or otherwise optimum information gain, the radar configuration and target dynamics affect the outcome. Ultimately, what radar configuration changes that can be made, system constraints, desired paretos, cost functions, etc., are all up to the designer. In the next few paragraphs, we describe each of these elements for the algorithm defined in Table 10.1 in detail to help the radar designer understand the general formulation.

10.3.2.1 Design

This section covers the elements of the algorithm that are defined in advance of the tracking solution.

Table 10.1 Information rate tracking filter algorithm.

Design	
State vector	$\boldsymbol{s}$
Motion model	$p\left(\boldsymbol{s}_k \mid \boldsymbol{s}_{k-1}\right)$
Measurement model	$\boldsymbol{z} = f(\boldsymbol{s})$
Optimizer and radar configurations	$\boldsymbol{\Omega}$
Initialize	$k = 0$, Repeat for $k = 1, 2, \ldots,$ end
Predict	
Predicted state	$\boldsymbol{s}_{k\|k-1} \sim p(\boldsymbol{s}_k \mid \boldsymbol{z}_{k-1}, \boldsymbol{z}_{k-2}, \ldots, \boldsymbol{z}_0) = \int p(\boldsymbol{s}_k \mid \boldsymbol{s}_{k-1})\, p(\boldsymbol{s}_{k-1} \mid \boldsymbol{z}_{k-1}, \ldots, \boldsymbol{z}_0)\, d\boldsymbol{s}_{k-1}$
Predicted measurement	$\boldsymbol{z}_{k\|k-1}$
Predicted information rate (new)	$\hat{I}\left(\boldsymbol{s}_{k\|k-1}; \boldsymbol{z}_{k\|k-1}\right)$ [refer to Equation (10.9)]
Optimize radar configuration (new)	Select "best" $\Omega^{\star} \in \boldsymbol{\Omega}$
Update	
Measurement	$\boldsymbol{z}_k = f\left(\boldsymbol{s}_k\right)$
Updated state distribution	$p(\boldsymbol{s}_k \mid \boldsymbol{z}_k, \ldots, \boldsymbol{z}_0) = \dfrac{p(\boldsymbol{z}_k \mid \boldsymbol{s}_k)\, p(\boldsymbol{s}_k \mid \boldsymbol{z}_{k-1}, \ldots, \boldsymbol{z}_0)}{p(\boldsymbol{z}_k \mid \boldsymbol{z}_{k-1}, \ldots, \boldsymbol{z}_0)}$
Advance to next time step	$k = k + 1$

State Vector This is the target parameter vector. The radar designer needs to determine what elements of a target to estimate in the radar target track. These may be constrained by what is measurable. For example, if the radar aperture has no discernible resolution for the elevation angle, this may limit the coordinate space in which the target is tracked.

Motion Model This is where the constraints and assumptions about the target dynamics are imposed. Recall these can be physical constraints matching typical target behavior or assumptions we make, knowing they are not true for simplicity, where model noise is used to capture this discrepancy.

Measurement Model This should represent how the state is observed, including any new injection of random processes from sensors or other sources. Often we may simplify this as well to make things mathematically tractable. For example, we may choose to linearize the more general nonlinear function about the predicted state vector.

Optimizer and Radar Configurations The optimization method, constraints, and cost functions are ultimately up to the radar designer. If radar power is fixed, then it cannot be used to modulate information rate. If information gain is weighted with some other cost function associated with a communications user for a joint RF system, this needs to be included. Since optimization depends on the available parameters, we leave this as a design parameter.

These design choices are made once but are important for setting up the rest of the algorithm. The rest of Table 10.1 shows the continuous, recursive process once the target track is established.

10.3.2.2 Initialize

The target track is initialized. Target track initiation (as well as termination) is thoroughly covered in standard radar literature and omitted here for brevity.

10.3.2.3 Predict

At every measurement time step, we predict the target state in the parameter and measurement domain. This is the vanilla Bayesian framework. However we extend this by adding the new items shown italicized. Using the same distributions available, we predict the target information gain, and use this to optimize the radar configuration.

Predicted State Using the target motion model and the Markov property, we take the prior distribution and propagate it forward in time. This is needed to incorporate as much *a priori* knowledge as possible so we accurately quantify the estimation information rate, removing constraints and redundancy from consideration.

Predicted Measurement With the measurement model defined, we propagate the target state prediction distribution through the measurement function and account for measurement uncertainty due to noise and other distorting factors.

Predicted Information Rate (New) With the predicted measurement and state distributions, we have all the knowledge we need to predict the target information rate by taking the mutual information between the two predictions. Again, this may be challenging to solve for in closed-form or need to be computed numerically if the distributions are complicated.

Optimize Radar Configuration (New) Since all three predicted elements preceding this step depend on the radar configuration, we can use any optimization routine within the radar designer constraints to select the "best" radar configuration $\Omega^\star \in \mathbf{\Omega}$ to drive to the desired information rate, where $\mathbf{\Omega}$ is the valid set of radar configurations available and $\Omega^\star$ is the optimal candidate configuration within that set.

10.3.2.4 Update

After we optimize the radar configuration using the predicted information rate, we make the measurement and update the target state prediction which becomes the prior we use to predict the next state and the recursion continues.

Measurement This is the physical measurement made using the radar configuration selected in the optimization routine.

Updated State Distribution and Time Step Advance Recursively, the measurement and prediction are combined in the Bayesian update distribution. This distribution becomes the prior in the next time step $k + 1$ and the process repeats starting with the prediction step again until the track is terminated.

Selecting a given configuration and the optimization approach employed depends on the goal of the system and is not covered here. A minimum information gain may try to use as little of resources as possible from the radar such as power and bandwidth to conserve resources. It may delay target revisit until the minimum threshold of usefulness is achieved, some minimum information gain. It may allocate resources to maximize information gain, though this has predictable outcomes. It may decide which target to measure in a multitarget tracking scenario based on which target has the most information to be gained. We leave this up to the designer since it is central to the goal of a given radar.

What we have presented so far represents nearly the most general framework with a few select constraints that cover most radar systems (like the Markov assumption). While this is a powerful framework for dynamically controlling radar systems using information rate, it is too high level to discuss some of the system sensitivities, illustrate with concrete examples, and demonstrate usefulness. To help cement understanding of this approach, the Section 10.4 provides an extensive special case of the framework derived here.

10.4 Example: Simplified 2D Target Tracking Kalman Filter

In this section we develop the predicted information rate algorithm using a simplified set of assumptions to demonstrate using information rate in a radar control scheme for a specific application. While we started with a more general Bayesian framework, we use a number of simplifications here to aid the reader in designing a radar using information rate to dynamically control the radar system and explore some intuitive relationships between predicted information and typical radar design parameters. The key assumptions made are as follows:

- 2D target tracking scenario in Cartesian space.
- Linear, nonaccelerating motion model for the target.

- Gaussian prior with Gaussian noise, linear models (remains Gaussian).
- Linearized EKF instead of the more general nonlinear filtering.
- Hard detection decision and constant false alarm threshold.
- System is narrowband (models Doppler shift instead of Doppler time scaling).
- Only consider local errors in estimation, no global errors.

All of these assumptions can be relaxed or tightened. They could be removed, or more could be added. All that is important is that the mutual information between the predicted state and the predicted measurement can be solved for in some fashion as a function of the radar parameters the radar designer intends to use to control the radar system. For example, one could generalize a radar tracking particle filter to predict information using Monte Carlo methods and relax many of the assumptions above, provided the radar had the computational power to leverage such methods.

10.4.1 Simplified Target Physical Model

To show a closed-form solution for the predicted estimation information rate, we make some simplifying assumptions about the target physical model. Here we assume a constant velocity, linear 2D motion model for the target. To capture uncertainty with this model, we also assume a Gaussian perturbative acceleration distribution [23]:

$$\boldsymbol{s}_k = \begin{bmatrix} x_k \\ \dot{x}_k \\ y_k \\ \dot{y}_k \end{bmatrix} = \boldsymbol{A}\boldsymbol{s}_{k-1} + \boldsymbol{w}_k = \begin{bmatrix} 1 & T & 0 & 0 \\ 0 & 1 & 0 & 0 \\ 0 & 0 & 1 & T \\ 0 & 0 & 0 & 1 \end{bmatrix} \begin{bmatrix} x_{k-1} \\ \dot{x}_{k-1} \\ y_{k-1} \\ \dot{y}_{k-1} \end{bmatrix} + \boldsymbol{w}_k, \tag{10.11}$$

where x_k and y_k are, respectively, the target position along the x-axis and the y-axis, $\dot{x}_k$ and $\dot{y}_k$ are, respectively, the target velocity along the x-axis and the y-axis, $\boldsymbol{A}$ is the linear prediction matrix used to predict the next target state, and T is the period of time between time steps k and $k-1$ (also between target measurements, known as the revisit time).

The variable $\boldsymbol{w}_k$ is the model (or process) noise vector with covariance defined as

$$\boldsymbol{Q}_k = \begin{bmatrix} q_{x,k}\dfrac{T^3}{3} & q_{x,k}\dfrac{T^2}{2} & 0 & 0 \\ q_{x,k}\dfrac{T^2}{2} & q_{x,k}T & 0 & 0 \\ 0 & 0 & q_{y,k}\dfrac{T^3}{3} & q_{y,k}\dfrac{T^2}{2} \\ 0 & 0 & q_{y,k}\dfrac{T^2}{2} & q_{y,k}T \end{bmatrix}, \tag{10.12}$$

where $q_{x,k}$ and $q_{y,k}$ are, respectively, the model error intensity for the x-axis and the y-axis, assumed independent from one another. This captures our uncertainty in the model as a distribution around the prediction. We know radar targets, in general, will accelerate. However, we are assuming a constant velocity to make this derivation and tracking algorithm simpler. This model noise distribution provides a means within our framework to keep this simplification, but incorporate our doubt about that model into the solution. The intensity parameters are estimated at each track step using maximum likelihood (ML) estimation to capture the model mismatch dynamically. For example, a highly accelerating target is estimated as having a strong position and velocity model noise. This naturally weights the measurement more in the Kalman fusion, meaning more information is gained through measurement, and thus, a higher radar information rate. If the target stops accelerating, the model intensity terms are estimated to be smaller and the algorithm adapts accordingly.

10.4.2 Simplified Radar Measurements

Similar to the more general case, here we observe the target state through a measurement, using observations gleaned from the radar target reflection. The optimum radar processor extracts range r_k and range-rate $\dot{r}_k$ through a matched filter that measures a target's reflected waveform delay τ and Doppler shift f_D: the cross-ambiguity function [5, 24]. Once again, to focus on the information rate based control, we adopt a simple set of assumptions that could easily be extended into more complicated scenarios. We assume the narrowband assumption for our system, or that $T_p B \ll \frac{c}{2\dot{r}_k}$, where T_p is the radar pulse duration, B is the bandwidth of the radar waveform, and c is the speed of the transmitted radar wave. An independent measurement is made for the bearing angle using a phased antenna array with half-wavelength spacing on a steerable manifold.

The mapping from the target state space defined in Section 10.4.1 to the measurement space defined here is highly nonlinear. Since we assume we are propagating a Gaussian prior state with Gaussian model noise, the nonlinear transformation of the measurement results in a more complicated set of distributions. To simplify both the Kalman tracking fusion and the information models we use to control the radar, we linearize the measurement about the predicted state using principles from the EKF [22]. Though this simplifies the assumptions and subsequent mathematics, it results in a time-varying measurement model that depends on the tracking state. However, it means we can continue to propagate a multivariate Gaussian distribution into our information model. This makes predicting mutual information a simple special case with an intuitive closed-form that is instructive for demonstrating this technique in practice.

To keep things relatively simple and maintain focus on information rate, we use the Cramér-Rao lower bound (CRLB) for the measurement variance focusing on the performance with signal-to-noise-ratio (SNR) governed by the standard monostatic radar link budget and receiver thermal noise. This derivation ignores phase noise, calibration biases, global error terms, and other complicating factors. The time delay and Doppler shift estimates under these assumptions are perturbed by correlated Gaussian noise. Successive measurements are assumed independent. Ultimately, after performing the narrowband cross-ambiguity matched filter, we measure the location of the peak which is offset by the target range r and range-rate $\dot{r}$. These relate to the position and velocity state as follows [24, 25]:

$$r = \sqrt{x^2 + y^2}, \tag{10.13}$$

and

$$\dot{r} = \frac{x\dot{x} + y\dot{y}}{\sqrt{x^2 + y^2}}. \tag{10.14}$$

We assume the bearing angle measurement is made through the phased array processing, perturbed by additive white Gaussian noise (AWGN) that is independent of the delay and Doppler estimation errors, and also independent in time [1]. We assume we have an ESA with N_A elements to measure the target bearing, or rotational angle relative to the radar [1]:

$$\theta = \text{atan2}\,(y, x)\,, \tag{10.15}$$

where atan2(·) is the 2-argument arctangent function.

We linearize the measurement model and noise distribution using the EKF so the resulting distributions are Gaussian after the time-varying but linear transform. This is accomplished by solving for the Jacobian of the measurement transformation [22, 26]:

$$\boldsymbol{C} = \begin{bmatrix} \frac{2}{c}\frac{\partial R}{\partial x} & \frac{2}{c}\frac{\partial R}{\partial \dot{x}} & \frac{2}{c}\frac{\partial R}{\partial y} & \frac{2}{c}\frac{\partial R}{\partial \dot{y}} \\ \frac{2f_c}{c}\frac{\partial \dot{R}}{\partial x} & \frac{2f_c}{c}\frac{\partial \dot{R}}{\partial \dot{x}} & \frac{2f_c}{c}\frac{\partial \dot{R}}{\partial y} & \frac{2f_c}{c}\frac{\partial \dot{R}}{\partial \dot{y}} \\ \frac{\partial \Theta}{\partial x} & \frac{\partial \Theta}{\partial \dot{x}} & \frac{\partial \Theta}{\partial y} & \frac{\partial \Theta}{\partial \dot{y}} \end{bmatrix}, \tag{10.16}$$

where R, $\dot{R}$, and Θ are the measurement functions for the range, range-rate, and bearing, respectively, capitalized to distinguish from the parameter values. All the partial derivatives shown in the matrix are evaluated at the predicted target state at each time step k. For the range, we get the following:

$$\frac{\partial R}{\partial x} = \frac{\partial}{\partial x}\left[\sqrt{x^2+y^2}\right]\Bigg|_{\mathbf{s}=\mathbf{s}_{k|k-1}} = \frac{x_{k|k-1}}{\sqrt{x^2_{k|k-1}+y^2_{k|k-1}}},$$

$$\frac{\partial R}{\partial \dot{x}} = \frac{\partial}{\partial \dot{x}}\left[\sqrt{x^2+y^2}\right]\Bigg|_{\mathbf{s}=\mathbf{s}_{k|k-1}} = 0,$$

where the same applies for $\frac{\partial R}{\partial y}$ and $\frac{\partial R}{\partial \dot{y}}$ by swapping x and y in the equations above. Repeating for the range-rate, we get the following:

$$\frac{\partial \dot{R}}{\partial x} = \frac{\dot{x}_{k|k-1}}{\sqrt{x^2_{k|k-1}+y^2_{k|k-1}}} - \frac{x^2_{k|k-1}\,\dot{x}_{k|k-1} + y_{k|k-1}\,\dot{y}_{k|k-1}\,x_{k|k-1}}{\left(\sqrt{x^2_{k|k-1}+y^2_{k|k-1}}\right)^3},$$

$$\frac{\partial \dot{R}}{\partial \dot{x}} = \frac{x_{k|k-1}}{\sqrt{x^2_{k|k-1}+y^2_{k|k-1}}},$$

where the same applies for $\frac{\partial \dot{R}}{\partial y}$ and $\frac{\partial \dot{R}}{\partial \dot{y}}$ by swapping x and y in the equations above. Finally, for the bearing, we get the following:

$$\frac{\partial \Theta}{\partial x} = \frac{-y_{k|k-1}}{x^2_{k|k-1}+y^2_{k|k-1}},$$

$$\frac{\partial \Theta}{\partial y} = \frac{x_{k|k-1}}{x^2_{k|k-1}+y^2_{k|k-1}},$$

$$\frac{\partial \Theta}{\partial \dot{x}} = \frac{\partial \Theta}{\partial \dot{y}} = 0.$$

What, in general, would be very complicated distributions most likely with no closed-form, has now been reduced to a simple linear Gaussian case. It can be extended further to include higher-order physics like acceleration and also to higher tracking dimensions like 3D space. If the model is accurate (a big assumption), including more prior knowledge and dimensions further improves prediction performance. This means the target residual or innovation which affects the information rate will be smaller. Therefore, less information is learned through measurement. For more complicated, nonlinear extensions, mutual information can be bounded using Gaussian mixture models and other approximations [27].

We now must define the estimation noise of the target measurements in our simplified scenario. We start by modeling the radar return signal at the receiver

$$s_{\mathrm{RX}}(t) = \sqrt{2}\,\Re\left[\sqrt{E_t}\,s_{\mathrm{rad}}(t-\tau)e^{j2\pi(f_c+f_D)(t-\tau)}\right] + n(t), \tag{10.17}$$

where E_t is the transmitted energy, $s_{\mathrm{rad}}(t)$ is the complex baseband radar transmit waveform, f_c is the carrier frequency in Hz, and $n(t)$ is the complex baseband receiver noise with variance σ^2_{noise}. Here, $\Re[\cdot]$ denotes the real part of a complex number. As noted before, we assume only local errors dominate performance of the receiver, and the CRLB is achievable. The Fisher information matrix (FIM) elements are then given by [24]

$$J_{11} = 2\,\frac{\text{ISNR}}{2\pi}\int_{-\infty}^{\infty}\omega^2\,|S_{\text{rad}}(j\omega)|^2\,d\omega = 2\,T_p\,s^2_{\text{rms}}\,\text{ISNR}\,\left(2\pi\,B_{\text{rms}}\right)^2, \tag{10.18}$$

$$J_{22} = 2\,\text{ISNR}\int_{-\infty}^{\infty}t^2\,|s_{\text{rad}}(t)|^2\,dt = 2\,T_p\,s^2_{\text{rms}}\,\text{ISNR}\,\left(T_{\text{rms}}\right)^2, \tag{10.19}$$

$$J_{12} = J_{21} = 2\,\Im\left\{\text{ISNR}\int_{-\infty}^{\infty}t\,s_{\text{rad}}(t)\,\frac{\partial s^*_{\text{rad}}(t)}{\partial t}\,dt\right\}, \tag{10.20}$$

for $S_{\text{rad}}(j\omega)$ the Fourier transform of the baseband radar signal, B_{rms} the root mean square (RMS) bandwidth, and T_{rms} the RMS pulse duration. $\Im[\cdot]$ denotes the imaginary part of a complex number. Here,

$$s_{\text{rms}} = \sqrt{\frac{1}{T_p}\int_{-T_p/2}^{T_p/2}\left|s_{\text{rad}}(t)\right|^2 dt} \tag{10.21}$$

is the waveform finite RMS [24], and

$$\text{ISNR} = \frac{N_p\,T_p\,B\,P_{\text{tx}}\,G^2\,\gamma\,c^2}{(4\pi)^3 r^4 f_c^2\,k_{\text{B}}\,T_{\text{temp}}\,B} \tag{10.22}$$

is the integrated SNR [1, 5, 24], where N_p is the number of pulses over the coherent processing interval (CPI), P_{tx} is the radar transmit power, G is the radar antenna gain relative to an isotropic antenna, γ is the target RCS in square meters, k_{B} is the Boltzmann constant in Joules/Kelvin, and T_{temp} is the absolute temperature of the radar receiver in Kelvin. To simplify the calculations, we can assume the carrier is chosen such that the mean frequency is 0 [24].

For our simplified scenario, we assume a Gaussian windowed linear frequency modulated (FM) chirp [24] with chirp rate B/T_p, and the covariance matrix for (τ, f_D) is given by [24, 28]:

$$\boldsymbol{J}^{-1} = \frac{1}{\text{ISNR}}\begin{bmatrix} T_p^2 & -2\,T_p\,B \\ -2\,T_p\,B & \left(2\pi\,T_p\right)^{-2} + (2\,B)^2 \end{bmatrix}. \tag{10.23}$$

Finally, the CRLB for the independent bearing estimate is given by [1]:

$$\sigma^2_\theta = \frac{1}{\text{ISNR}}\,\frac{1}{2k^2\,N_A\,\sigma^2_A}, \tag{10.24}$$

where N_A is the number of antennas, k is the wavenumber, and σ_A^2 is the RMS element position. If we assume a uniform linear array steered to the predicted target position, then [28]:

$$\sigma_\theta^2 = \frac{1}{\text{ISNR}} \frac{6}{\pi^2 \cos^2(\Delta\theta) N_A \left(N_A^2 - 1\right)}, \tag{10.25}$$

where $\Delta\theta$ is the difference between the predicted bearing and the true target bearing. Finally, we can define the full measurement covariance matrix as

$$\boldsymbol{\Sigma} = \frac{1}{\text{ISNR}} \begin{bmatrix} \frac{T_p c^2}{4} & -\frac{T_p B c^2}{2 f_c} & 0 \\ -\frac{T_p B c^2}{2 f_c} & \left(\frac{c}{T_p 4\pi f_c}\right)^2 + \left(\frac{B c}{f_c}\right)^2 & 0 \\ 0 & 0 & \frac{6}{\pi^2 \cos^2(\Delta\theta) N_A (N_A^2 - 1)} \end{bmatrix}, \tag{10.26}$$

where we have augmented the cross-ambiguity function measurement covariance matrix to include the independent bearing measurement variance. Note the scalar terms that present a difference between this final covariance matrix and Equation (10.23) is due to a conversion from delay and Doppler to range and range-rate. With the simplified source distribution, measurement model, waveform, and measurement noise model defined, we have enough definition to derive a Kalman tracking framework and extend for predicted information to dynamically control the simplified radar system.

10.4.3 Simplified Estimation Information Rate

Here we extend the EKF framework to include a predicted information rate. Once we derive a predicted state covariance and predicted measurement covariance, we can predict the mutual information between the predicted state and prediction measurement as a prediction for the information we stand to gain through measurement. We start with the target prior $\boldsymbol{s}_{k-1}$ which we model as a multivariate Gaussian distribution $\mathcal{N}\left(\hat{\boldsymbol{s}}_{k-1}, \boldsymbol{P}_{k-1|k-1}\right)$, where $\hat{\boldsymbol{s}}_{k-1}$ is the prior state *estimate* and also the mean of the target prior distribution, and $\boldsymbol{P}_{k-1|k-1}$ is the prior state covariance matrix. This models the uncertainty that exists in the prior target state estimate that we would like to reduce via learning information.

We can now predict the next target state. This is injecting *a priori* knowledge of the target with some modeling uncertainty. This is an important step to remove this model knowledge from consideration to avoid overestimating the target information gain. Based on the model physics of the target, the next state can be predicted as

$$\boldsymbol{s}_{k|k-1} = \boldsymbol{A}\, \boldsymbol{s}_{k-1}, \tag{10.27}$$

which is distributed as a multivariate Gaussian since our prediction operator $\boldsymbol{A}$ is linear: $\mathcal{N}\left(\hat{\boldsymbol{s}}_{k|k-1}, \boldsymbol{A}\,\boldsymbol{P}_{k-1|k-1}\,\boldsymbol{A}^{\mathrm{T}} + \boldsymbol{Q}_k\right)$. Note the model noise covariance $\boldsymbol{Q}_k$ is added to our prediction covariance to capture the model noise in our framework.

To characterize the information we stand to gain, we have one more step in the process. We do not observe $\boldsymbol{s}$ directly, but rather some measurement $\boldsymbol{z}$ often in a different domain. We can predict the target measurement by applying the linearized measurement model $\boldsymbol{C}_k$ to the target prediction:

$$\boldsymbol{z}_{k|k-1} = \boldsymbol{C}_k\,\boldsymbol{s}_{k|k-1}, \tag{10.28}$$

which is also distributed as a multivariate Gaussian since we have linearized our measurement operator: $\mathcal{N}\left(\hat{\boldsymbol{z}}_{k|k-1}, \boldsymbol{C}_k\,\boldsymbol{P}_{k|k-1}\,\boldsymbol{C}_k^{\mathrm{T}} + \boldsymbol{\Sigma}_{k|k-1}\right)$, where

$$\boldsymbol{P}_{k|k-1} = \boldsymbol{A}\,\boldsymbol{P}_{k-1|k-1}\,\boldsymbol{A}^{\mathrm{T}} + \boldsymbol{Q}_k, \tag{10.29}$$

and $\boldsymbol{\Sigma}_{k|k-1}$ is the predicted measurement covariance matrix. We need to predict this covariance that depends on the radar parameters like pulse duration and bandwidth, but also the integrated SNR. Recall from Equation (10.22), this term includes not only many radar parameters like transmit power but also the range of the target. Substituting in the predicted range from the predicted measurement vector in Equation (10.28) and our radar control parameters as a part of the optimization routine, we can solve for the predicted measurement covariance matrix $\boldsymbol{\Sigma}_{k|k-1}$. Finally, we observe an actual measurement:

$$\boldsymbol{z}_k = g\left(\boldsymbol{s}_k\right) + \boldsymbol{n}_k, \tag{10.30}$$

where $g(\cdot)$ is the (in general) nonlinear measurement function, and $\boldsymbol{n}_k$ is the measurement noise. In our case, this is a transformation from Cartesian coordinates to polar coordinates via range, range-rate, and bearing measurements.

After measurement, we can solve for the innovation by subtracting the target prediction from the actual measurement:

$$\tilde{\boldsymbol{z}}_k = \boldsymbol{z}_k - \boldsymbol{z}_{k|k-1}, \tag{10.31}$$

which is distributed as (assuming the linear measurement transform as true): $\mathcal{N}\left(\boldsymbol{0}, \boldsymbol{C}_k\,\boldsymbol{P}_{k|k-1}\,\boldsymbol{C}_k^{\mathrm{T}} + \boldsymbol{\Sigma}_k\right)$. The mean of the innovation is $\boldsymbol{0}$, as we assume the prediction is unbiased, i.e., the mean of the prediction is equal to the mean of the measurement. Therefore, we can assume this innovation distribution is identical to the predicted measurement as it pertains to the mutual information, substituting $\boldsymbol{\Sigma}_{k|k-1}$ for $\boldsymbol{\Sigma}_k$.

It is well known that the differential entropy of an N-dimensional multivariate Gaussian random variable $\boldsymbol{x} \sim \mathcal{N}(\boldsymbol{\mu}, \boldsymbol{\Sigma})$ is [20]:

$$h(\boldsymbol{x}) = \frac{1}{2}\log_2\left[(2\pi e)^N |\boldsymbol{\Sigma}|\right], \tag{10.32}$$

where $|\cdot|$ is the matrix determinant operator. The mutual information between two random vectors $\boldsymbol{x}$ and $\boldsymbol{y}$ is given by [20]:

$$I(\boldsymbol{x};\boldsymbol{y}) = h(\boldsymbol{y}) - h(\boldsymbol{y}|\boldsymbol{x}). \tag{10.33}$$

In our case, $\tilde{\boldsymbol{z}}_k$ is a linear transformation of $\boldsymbol{s}_{k|k-1}$ with independent AWGN. Therefore, if we let $\hat{\boldsymbol{R}}_k = \boldsymbol{C}_k\boldsymbol{P}_{k|k-1}\boldsymbol{C}_k^{\mathrm{T}} + \boldsymbol{\Sigma}_{k|k-1}$, then

$$\begin{aligned} I(\boldsymbol{s}_{k|k-1};\tilde{\boldsymbol{z}}_k) &= h(\tilde{\boldsymbol{z}}_k) - h(\tilde{\boldsymbol{z}}_k|\boldsymbol{s}_{k|k-1}) \\ &= \frac{1}{2}\log_2\left[(2\pi e)^N|\hat{\boldsymbol{R}}_k|\right] - \frac{1}{2}\log_2\left[(2\pi e)^N|\boldsymbol{\Sigma}_{k|k-1}|\right], \end{aligned} \tag{10.34}$$

and therefore

$$I(\boldsymbol{s}_{k|k-1};\tilde{\boldsymbol{z}}_k) = \frac{1}{2}\log_2\left[\frac{|\hat{\boldsymbol{R}}_k|}{|\boldsymbol{\Sigma}_{k|k-1}|}\right] = \frac{1}{2}\log_2\left[\frac{|\boldsymbol{C}_k\boldsymbol{P}_{k|k-1}\boldsymbol{C}_k^{\mathrm{T}} + \boldsymbol{\Sigma}_{k|k-1}|}{|\boldsymbol{\Sigma}_{k|k-1}|}\right]. \tag{10.35}$$

We finally have the predicted information rate for this simplified framework. Note it depends on the prior covariance, the linearized motion model, the process or model covariance, the linearized measurement transform, and the measurement noise covariance. We can now vary the radar parameters and predict what information can be gained by using Equation (10.35) in order to control the radar. The new information rate tracking filter algorithm for our simplified EKF scenario is given in Table 10.2. Note, we have omitted the design parameters from the more general algorithm given in Table 10.1 because in our simplified scenarios we have included specified designs for those parameters. We cover the optimization routine in more detail in Section 10.4.4.

10.4.4 Simplified Radar Optimization

Now, we are ready to add the optimization for this simplified scenario in order to complete the development of a system that dynamically controls the radar in response to a target tracking scenario to maintain a minimum viable information gain per radar access.

In Algorithm 10.1, we see the optimization algorithm that completes the simplified information rate controlled radar defined generally in Sections 10.4.1–10.4.3. Here, we use the predicted information rate for a given radar track scenario to dynamically configure these radar parameters to be optimal for the current target tracking situation. This allows for better performance or more efficient radar resource utilization than static methods designed to envelope all scenarios, and is more simple and extensible than heuristics-based approaches. Note again the optimization cost function and other constraints are something the radar designer should select for their application.

Table 10.2 Simplified EKF information rate tracking filter algorithm.

Predict	
Predicted state estimate (mean)	$\hat{\boldsymbol{s}}_{k\|k-1} = \boldsymbol{A}\,\hat{\boldsymbol{s}}_{k-1\|k-1}$
Predicted state covariance	$\boldsymbol{P}_{k\|k-1} = \boldsymbol{A}\,\boldsymbol{P}_{k-1\|k-1}\,\boldsymbol{A}^{\mathrm{T}} + \boldsymbol{Q}_k$
Predicted measurement (mean)	$\hat{\boldsymbol{z}}_{k\|k-1} = \boldsymbol{C}_k\,\hat{\boldsymbol{s}}_{k\|k-1}$
Predicted measurement covariance	$\hat{\boldsymbol{R}}_k = \boldsymbol{C}_k\,\boldsymbol{P}_{k\|k-1}\,\boldsymbol{C}_k^{\mathrm{T}} + \boldsymbol{\Sigma}_{k\|k-1}$
Predicted information rate (new)	$I\left(\boldsymbol{s}_{k\|k-1}; \tilde{\boldsymbol{z}}_k\right) = \frac{1}{2}\log_2\left(\frac{\|\hat{\boldsymbol{R}}_k\|}{\|\boldsymbol{\Sigma}_{k\|k-1}\|}\right)$
Optimize radar configuration (new)	Select "best" $\Omega^\star \in \boldsymbol{\Omega}$
Update	
Innovation/residual	$\tilde{\boldsymbol{z}}_k = \boldsymbol{z}_k - \boldsymbol{z}_{k\|k-1}$
Innovation covariance	$\boldsymbol{R}_k = \boldsymbol{C}_k\,\boldsymbol{P}_{k\|k-1}\,\boldsymbol{C}_k^{\mathrm{T}} + \boldsymbol{\Sigma}_k$
Kalman gain	$\boldsymbol{K}_k = \boldsymbol{P}_{k\|k-1}\,\boldsymbol{C}_k^{\mathrm{T}}\,\boldsymbol{R}_k^{-1}$
Updated state estimate	$\hat{\boldsymbol{s}}_{k\|k} = \hat{\boldsymbol{s}}_{k\|k-1} + \boldsymbol{K}_k\,\tilde{\boldsymbol{z}}_k$
Updated estimate covariance	$\boldsymbol{P}_{k\|k} = \left(\boldsymbol{I} - \boldsymbol{K}_k\,\boldsymbol{C}_k\right)\boldsymbol{P}_{k\|k-1}$

Algorithm 10.1: Example radar optimization algorithm.

$\boldsymbol{\Omega} = \{\Omega_1, \Omega_2, \ldots, \Omega_M\}$ (Radar Parameter Table)
for $i = 1$ to M **do**
  Pick $\Omega_i \in \boldsymbol{\Omega}$ (Hypothesis Radar Configuration)
  $\hat{\boldsymbol{s}}_{k|k-1} = \boldsymbol{A}\,\hat{\boldsymbol{s}}_{k-1|k-1}$ (State Mean Prediction)
  $\boldsymbol{P}_{k|k-1} = \boldsymbol{A}\,\boldsymbol{P}_{k-1|k-1}\,\boldsymbol{A}^{\mathrm{T}} + \boldsymbol{Q}_k$ (Source Covariance Prediction)
  $\hat{r}_{k|k-1} = \sqrt{\hat{x}^2_{k|k-1} + \hat{y}^2_{k|k-1}}$ (Range Prediction)
  $\mathrm{ISNR}_{k|k-1} = \frac{N_p\,T_p\,B\,P_{\mathrm{tx}}\,G^2\,\gamma\,c^2}{(4\pi)^3\,\hat{r}^4_{k|k-1}\,f_c^2\,k_{\mathrm{B}}\,T_{\mathrm{temp}}\,B}$ (Integrated SNR Prediction)
  $\boldsymbol{\Sigma}_{k|k-1} = \boldsymbol{J}^{-1}/\mathrm{ISNR}_{k|k-1}$ (Scaled Inverse Fisher Information Prediction)
  $\hat{\boldsymbol{R}}_k = \boldsymbol{C}(\boldsymbol{s}_{k|k-1})\,\boldsymbol{P}_{k|k-1}\,\boldsymbol{C}^{\mathrm{T}}(\boldsymbol{s}_{k|k-1}) + \boldsymbol{\Sigma}_{k|k-1}$ (Meas. Covariance Prediction)
  $\hat{I}\,(i) = \frac{1}{2}\,\log_2\left[\frac{|\hat{\boldsymbol{R}}_k|}{|\boldsymbol{\Sigma}_{k|k-1}|}\right]$ (Predicted Information)
end for
Select $\Omega^\star = \Omega_i$ using cost function with optimal $\hat{I}\,(i)$

For each potential radar configuration, the predicted state and measurement distributions are estimated. The predicted information rate can then be computed. The best configuration out of the allowable radar configurations is selected using a cost function. For example, the radar configurations could include only varying the revisit time T over some set of permissible values $\boldsymbol{T}$. The value

$T = T^{\star} \in \boldsymbol{T}$ that yields the predicted information $\hat{I}(i)$ closest to some setpoint I_{const} is selected. This optimization routine yields the constant information radar algorithm demonstrated in [17, 18]. We show an example of this algorithm in Section 10.4.6.

10.4.5 Information Rate Influencing Factors

Now that we have a more concrete, special case to examine, we can look mathematically how the various radar configuration parameters and target state dynamics affect the predicted information rate used to control the radar shown in Algorithm 10.1. Just about every term in the radar tracking formulation affects the information rate of the system and a target. As radar information rate quantifies uncertainty of the target from the perspective of the radar, both the target characteristics and the radar characteristics affect the information rate in a given tracking scenario. The target characteristics are not controllable by the system, but we describe them here because they factor into the optimization. The radar characteristics are controllable by the system and can be used to offset the uncontrollable target characteristics. For example, the uncertainty of a target is increased if the target's movement and dynamics are outside of the tracking filter's modeled assumptions (e.g., accelerating target in a constant velocity tracking model). To control the information gain, the target can be revisited quicker than the previous track update.

10.4.5.1 Target Characteristics

We start by breaking down the target characteristics that affect information rate. These are generally not controllable by the radar. However, they affect information rate. As a result, the radar needs to dynamically control something in the radar system in response to target behavior.

Model Mismatch When the target exhibits dynamics that are inconsistent with the tracking filter's modeled assumptions (e.g., accelerating in nonaccelerating model), model mismatch occurs. In this case, uncertainty about the target from the radar's perspective is increased, and more information is gained through measurement.

Target Reflected Power The target's return signal strength affects information rate. More information can be gained for closer targets or for larger RCS targets. Obstructions or nonreturns result in null information gain.

Measurement Ambiguity Information rate is generally reduced when considering the effect of the radar measurement transformation. This can be characterized by the mutual information between the predicted state and the noiseless predicted measurement after application of the Jacobian transformation. The main influencing factor is ultimately the hypervolume of predicted measurement covariance, which characterizes the measurement distortion of the prior distribution. As an example, if the radar is tracking a target which is accelerating to a vector where its velocity will be tangential to the radar, the Doppler measurement will not reduce uncertainty in the target velocity, reducing the amount of information that can be gained through measurement.

It is important to note that changing basis has no effect on information. Tracking in the polar coordinate domain would result in no loss of information, as it represents an invertible transform [20]. However, radar designers often make design choices that are suboptimal for performance in order to make the solution mathematically tractable or for computational limitations [28]. For example, linearization of the Bayesian physical model using the EKF reduces the information rate in general.

10.4.5.2 Radar Characteristics

The following factors are radar parameters typically controllable by the radar system designer or operator. They can be used either statically in the definition of the radar for a given application, or dynamically controlled using information rate.

Revisit Time The longer the period between target revisits, the larger the model error grows. This means we gain more information the longer we delay revisiting a given target.

Receiver SNR Key terms driving the integrated SNR directly affect the information rate, e.g., CPI, antenna gain, and amplifier design parameters like noise figure. Similarly, antenna resources can be dynamically allocated such as elements dedicated to the radar instead of other RF purposes which impacts the SNR of the reflected waveform. The receiver SNR is also increased by increasing the radar transmit power and the aperture size.

Radar Waveform The radar waveform performance, largely driven by the RMS bandwidth and envelope, and ultimately Fisher information affect the information gain. The larger the determinant of the FIM, the more information is gained about a target.

Estimation Process As previously discussed, we can estimate the radar state optimally, but often simplifications like linearizing are done to make the problem tractable. In general, applying noninvertible functions to random variables results in loss of information rate.

Table 10.3 Summary of information rate-influencing factors.

Parameter (increased)	Description	Information rate change
SNR	Signal-to-noise-ratio	↑
CPI	Coherent processing interval	↑
q_k	Process/model noise power	↑
r	Target range	↓
$\|\boldsymbol{CP}_{k\|k-1}\boldsymbol{C}^{\mathrm{T}}\|$	Hypervolume of predicted measurement covariance	↑
γ	Radar cross section	↑
$\|\boldsymbol{J}\|$	Fisher information matrix determinant	↑
G	Antenna gain	↑
P_{tx}	Transmit power	↑
B	Bandwidth	↑
N_{A}	Number of antenna elements	↑
T	Revisit time	↑
N_{P}	Number of pulses	↑
B_{rms}	RMS bandwidth	↑
T_{rms}	RMS pulse duration	↑

Table 10.3 summarizes all of these factors and how the information rate responds to those parameters increasing. In the special case of modulating revisit time to maintain a constant information rate as in [17], for any parameter that increased information rate, the revisit period would be shortened and the target would be scheduled for a tracking update sooner than the previous update.

Imagine building heuristics-based controls on all these factors! All of the radar characteristics can be dynamically modulated to achieve a fixed information, and the resulting systems and optimization routines can get quite complex. More general optimization approaches are out of the scope of this chapter, as they quickly become application specific. For example, the same information rate may be achieved with one configuration using low bandwidth and high power, and another using high bandwidth and low transmit power. These have very different system-level implications, and so the optimization routine needs to consider this, paretos, and other factors depending on the application.

In the more general sense, other parameters that are used in a given system may contribute to the information rate. For example, in a system model that assumes the presence of an interfering signal in the same band, the interference signal power would be an influencing factor.

In Section 10.4.6, we provide some example scenarios. To simplify the approaches used in these examples, we optimize only one parameter at a time, or reconfigure the system at a fixed interval to show an example of re-prioritization.

10.4.6 Simplified Radar Example Scenarios

In this section, we use our simplified derivation of dynamic radar control using information rate as a testbed for various scenarios to understand the advantages and limitations of this approach. To help illustrate the usefulness of this approach, we explore a few metrics of performance such as spectral efficiency, multitarget priority track MSE, and radar resource utilization. In general, performance metrics are a significant system design parameter that would influence the cost function of the dynamic radar control optimization routine. It is expected that these metrics would vary dramatically from application to application in practice.

10.4.6.1 Example Scenarios and Results

To start, we set some of the basic radar parameters in Table 10.4. At each simulated measurement, a Bernoulli random variable is drawn to determine if a target is observed. The radar performs validation gating of detects and excludes detection of targets if they are outside of the 3 dB beamwidth of the antenna. The antenna assumes a shaping factor of 0.89. For the model noise power estimation, the last innovation is used to compute an ML estimation within a lookup for q_x and q_y separately.

Scenario 1: Configuration Change For the first scenario, shown in Figure 10.3, we see the track and revisit time for a radar that has resources reallocated in the middle of the track. In this case we reduce the bandwidth by a factor of 4 (5 MHz

Table 10.4 Example scenario radar configuration and target parameters.

Parameter	Value	Parameter	Value
Bandwidth	5 MHz	Absolute temperature	1000 K
Pulse duration	5 μs	Probability of false alarm	10^{-6}
Radar antenna gain	30 dBi	Radar transmit power	100 kW
Target cross section	10 m^2	Chirp rate	B/T_p
Window variance	T_p	Wave speed	3×10^8 m/s
Number of pulses	10	Number of array elements	10

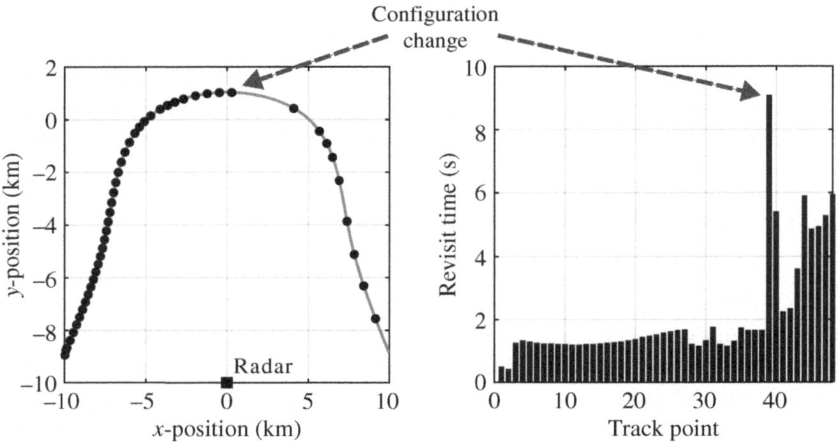

Figure 10.3 Radar track with a configuration change. On the left is the radar target track, where the radar is indicated by the square. The radar resources are reallocated in the middle of the track when the target *x*-position crosses 0. On the right figure, we see the revisit time in seconds, modulated using predicted target information rate.

to 1.25 MHz) and the power by a factor of 100 (100 kW to 1 kW). We also reduced the number of array elements from 10 to 5. For multifunction RF systems, this is a typical scenario when a shared radio allocates power, bandwidth, and antenna elements to another function such as communications. In these track plots, the true target track is indicated by the solid line on the plot. The dots are radar track samples when measurements are taken, and represent the filtered output, post fusion.

In this example, we had the radar modulate target revisit time to hold the information rate at 15 bits. When the resources the radar devoted to this track were cut, the information the radar stood to gain through measurement dropped for an equivalent revisit time. Therefore the revisit time was delayed longer in order to maintain a 15 bits per spectral access efficiency. We see on the right side of Figure 10.3, the revisit time on average goes from one to two seconds to around five to six seconds after the first correction.

In Figure 10.4, we compare the spectral efficiency of this dynamic approach to a fixed revisit period approach. We took roughly the average revisit period of the dynamic radar before the configuration change (around 1.4 seconds) and kept it constant even through the radar reconfiguring. The dynamic radar maintains a fixed information gain by delaying revisit, and therefore, the spectral efficiency over time is maintained by this optimization. The fixed revisit approach, however, sees a drop in spectral efficiency per radar access as the reduction in performance means less information is gained for each sample. Note with this fixed revisit time,

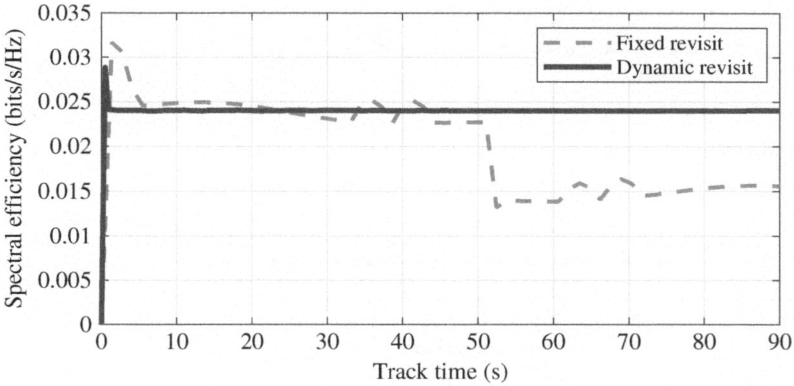

Figure 10.4 Comparison of spectral efficiency between fixed revisit rate and dynamic revisit rate using information. The dynamically controlled radar responds to the radar configuration change to maintain the set spectral efficiency.

the tracking MSE is roughly equivalent to the dynamic radar (the dynamic radar was negligibly better), but the spectral efficiency is 40% worse for the fixed radar in the second half of the track.

It may seem counterintuitive that the radar samples less frequently when the radar performance is reduced. However, the explanation is much like the sphere packing argument for channel capacity. The resolvable target positions can be thought of as symbols in 2D space much like communications symbols. Each symbol has a distribution around it dominated by thermal and estimation noise. Pushing the spheres closer together and encoding would only waste bits. Therefore, the adaptive radar delays illumination until the innovation results in the minimum viable information gain.

Scenario 2: Multitarget Priority In Figure 10.5, we show two simultaneous targets where information rate is used to dynamically select which target to revisit. The radar minimum period between measurements was set to 1.5 seconds to force a challenging priority call between the targets, where both exhibited high information gains. This is not an unrealistic expectation of some radars that may have average power limits, thermal management challenges, or mechanical rotation mechanisms that slow slewing between targets.

At the start, both targets are well behaved and sampled evenly. That is, the radar alternates between the two targets, interleaving revisits in a round-robin fashion. However, once the dynamic target Track 1, indicated by the hollow points in Figure 10.5, begins to behave more dynamically, the radar starts to revisit it more frequently. That is, more revisits in a row on Track 1 before revisiting Track 2, the

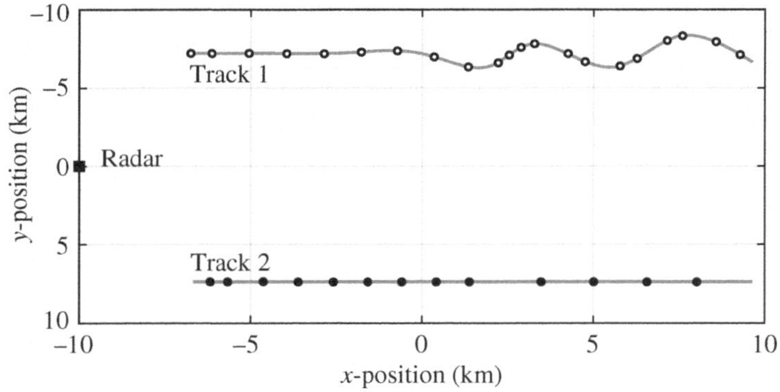

Figure 10.5 Multiple target track scenario with dynamical radar using information. When Track 1, indicated by the hollow points, starts to behave more dynamically, the radar naturally revisits it more often than the well-behaved track, Track 2 (dark points).

linear track indicated by the dark sample points. This can be seen by the large spaces between samples late in the simulation for Track 2.

This is better illustrated in Figure 10.6 where we show predicted information as a function of revisit time for each track against the overall track time. We set the information threshold to 16 bits, indicating we would like to revisit a target if at least 16 bits is available upon measurement. This is an intentionally challenging limit to force a prioritization. As revisit is delayed, we can see the logarithmic rise of predicted information. In this zoomed in segment, we can see around time 25 seconds, Track 1 is behaving so dynamically by the accelerating turn that the

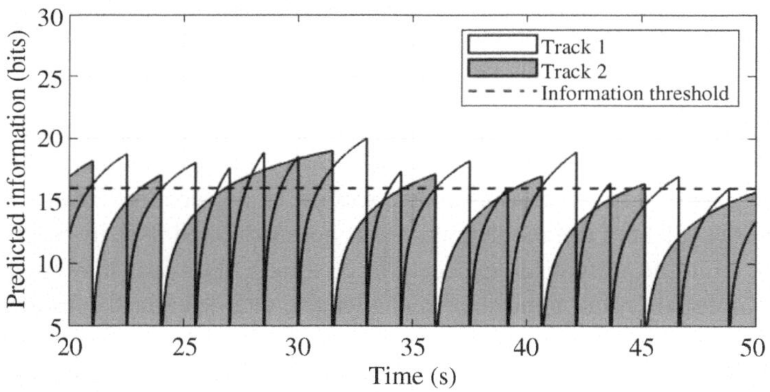

Figure 10.6 Predicted information for a two-target tracking scenario. The more dynamic track, Track 1, shows a more rapid increase in predicted information. The well-behaved track, Track 2, has a much slower predicted information rise and is prioritized less often as expected.

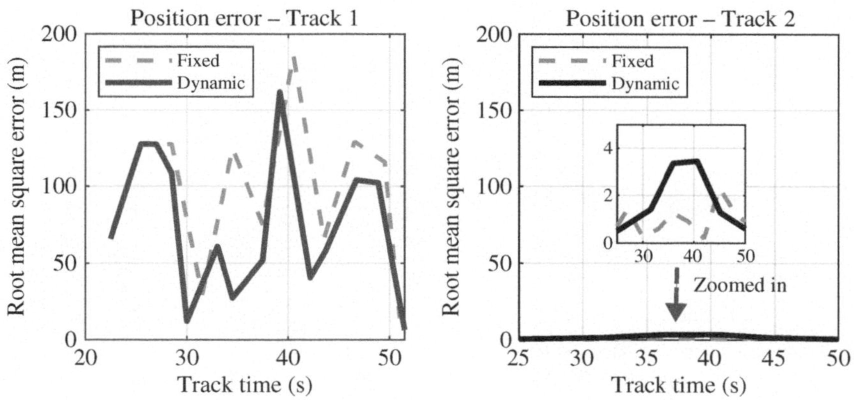

Figure 10.7 Comparison of track position MSE for information-based multitarget priority and fixed round-robin revisiting.

predicted information quickly shoots up past the threshold in the 1.5 seconds minimum down time between measurements. Even though at this point Track 2 is above the threshold, the information gain is higher on Track 1 so it is prioritized. In fact, the radar revisits Track 1 four times before going back to sampling Track 2!

So how does this compare to a round-robin approach where the targets are switched every 1.5 seconds, regardless of predicted information gain? In Figure 10.7, we see the dynamic, information-driven priority approach outperforms the fixed round-robin revisit under the same constraints for Track 1 (the serpentine target) when comparing track position MSE. For the well-behaved target, the fixed revisit performed better which is not surprising considering how much it was neglected in favor of the dynamic target in the information controlled radar. However, this position error is on the order of meters compared to the hundreds of meters for the dynamic target. This tradeoff is heavily dependent on the radar application and purpose in the broader mission context.

Note that the position of the tracks relative to the radar affects the priority of target revisits using this algorithm. As we saw previously, predicted SNR and measurement ambiguity affect the predicted information rates. Therefore if the radar was not centered between both tracks, we would expect to see the relative target ranges vary the priority of targets in addition to the targets' dynamics.

In the next two examples, we highlight how elegantly this information-driven control framework handles unexpected behavior.

Scenario 3: Radar Blockage In the track shown in Figure 10.8, we see how dynamically modulating the radar in response to information rate affects a number of

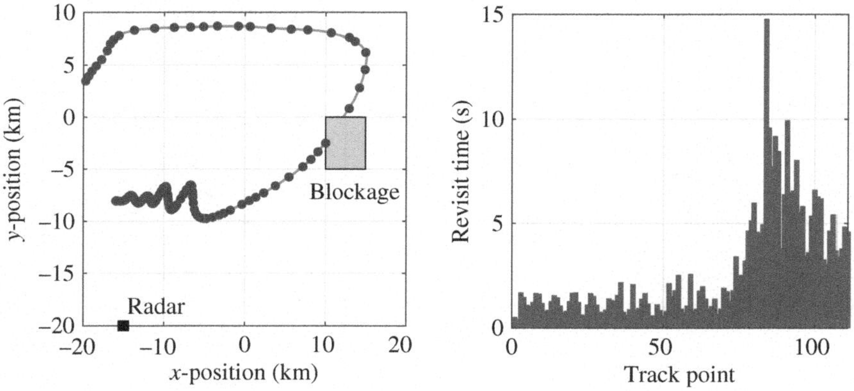

Figure 10.8 Radar track with a target blockage (shown as the darkened square). On the left is the radar target track, where the radar is indicated by the square. The dynamic radar modulates revisit time based on proximity, target dynamics, and blockages.

target-track dynamics, including one surprise. In general, the information rate modulates the revisit time intuitively. As the target turns, accelerates, or otherwise breaks the constant velocity assumption, the revisit time is reduced as shown on the right side of the plot. Overall, as the target is further away and SNR is reduced, revisit times are increased. The superposition of these two influencing factors create global and local trends based on the target dynamics with respect to the radar.

In this scenario we also inserted a known blockage, shown as the darkened square on the track plot. This blockage is known to the radar, and so in a tracking loop it can predict when the target is expected to be blocked and the radar illumination goes unreturned. Ad hoc methods may implement rules-based approaches to deal with this blockage. However, the predicted SNR in the information prediction step naturally modulates the information rate-driven radar to delay until the target is expected to be on the other side of the blockage. This demonstrates in practice the earlier point that information is at the heart of the radar measurement process.

Scenario 4: Unexpected Jammer In the next example, we show this adaption in another unexpected event. It is typical for radars to measure background noise power between target illuminations. As a result, noise and other unexpected sources can be measured in real time to understand the target SNR. In this example, additional energy is measured in the middle of the track, as a high-power in-band interference "jammer" source is suddenly turned on, hitting the receiver at −90 dBm. This naturally decreases the predicted SNR, and subsequently, the predicted information gain for a commensurate revisit time drops. The target revisit period is delayed as shown in Figure 10.9.

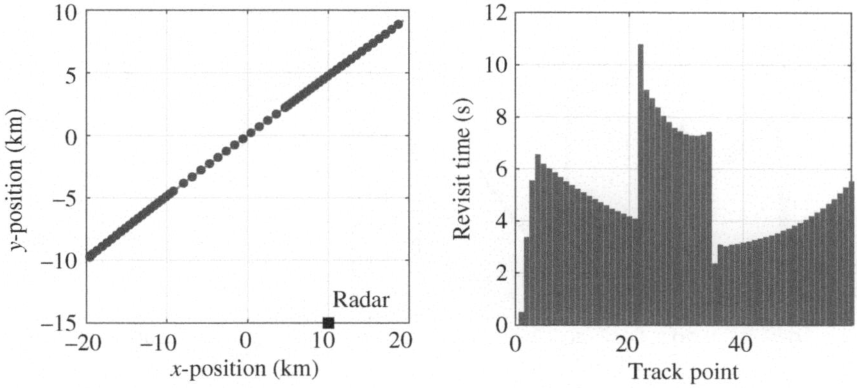

Figure 10.9 Radar track with unexpected in-band interference in the middle of the track. While the jammer is active, the revisit rates are increased to maintain a constant information rate.

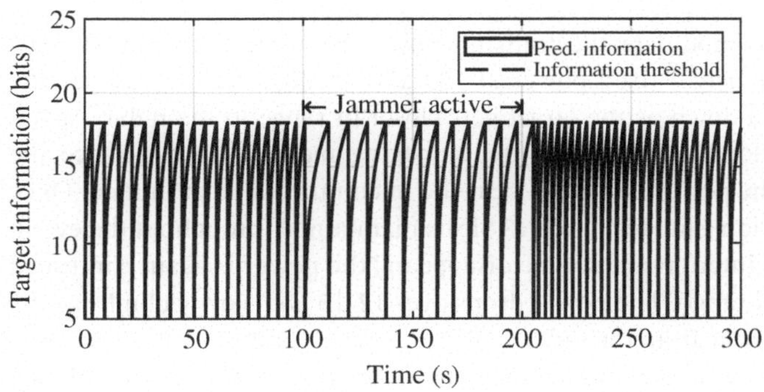

Figure 10.10 Radar information as a function of radar track time with a jamming event. When the jammer is active, the predicted information takes longer to trip the information threshold, so revisit of the target is delayed.

Before the jamming event, the radar functions as expected with the target being revisited right when the predicted information reaches the setpoint. In this scenario, the setpoint is 18 bits. However, when the jammer degrades the SNR, the threshold of information cannot be reached within the same revisit period so the period is delayed. Once the jammer ceases, the radar recovers normal operation and the target information rate is maintained by increasing the revisit rate. This is illustrated in Figure 10.10.

We can use this scenario to show another example of control. If we fix the revisit time but instead modulate the radar transmit power, we can see the predicted information optimize a different configuration using the same metric of control. Now we fix the revisit time to 5 seconds, but allow the radar to use predicted

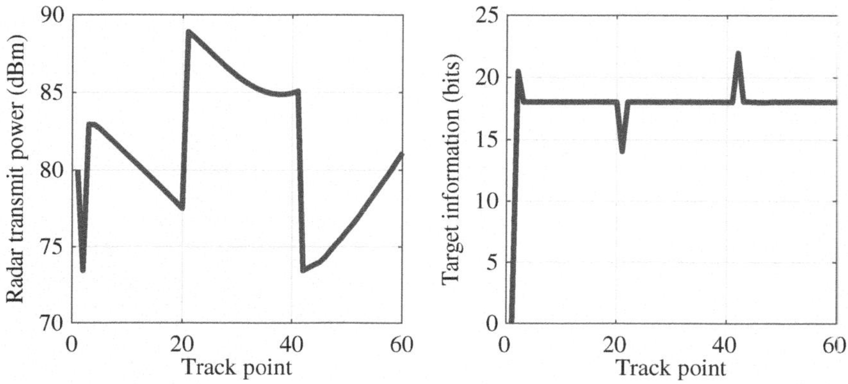

Figure 10.11 Radar track with unexpected in-band interference in the middle of the track, but now we modulate radar transmit power while keeping revisit time fixed. Once again, the control parameter is modulated during the jamming event to maintain the desired information gain.

information rate to modulate the transmit power between 1 kW and 1 MW. Recall it was previously a fixed power at 100 kW. In Figure 10.11, we see the power is modulated down as the target gets closer in range to the radar. Again, an intuitive behavior that is desired for managing radar resources. Once the jammer is active, the power shoots up near the maximum value permitted. Once again, the information gain desired of 18 bits is maintained throughout the event. Once the jamming interference source disappears, the power is again controlled within the nominal range. The power drops over 12 dB and slowly climbs as the target gets further away from the radar.

Again, this may not be the optimal thing to do in all scenarios. As mentioned previously, not all information bits are equal and depending on the radar scenario modulating power alone may result in an information-controlled radar that performs poorly. For example, for targets with varying dynamics exhibiting changes in modeling error like examples shown previously in this chapter, a power-only information rate control scheme would reduce transmit power when the target modeling error increases, resulting in poor target tracking MSE.

Nevertheless, these last two examples highlight the power of the singular information rate metric in handling unexpected or unintended scenarios. Since information truly encapsulates the goal of the radar measurement instrument, it naturally subsumes all target, radar, and environmental variables into the framework. As long as these can be modeled, information responds intuitively to changes in these parameters and can drive the radar configuration to more efficiently manage radar or spectral resources. Unexpected corner cases do not need to be added, as information changes in response to new scenarios and the framework holds as expected.

10.5 Conclusion

Using information rate to dynamically control a radar configuration is an elegant, mathematically sound way to optimize radar resources in complex scenarios. Information is at the heart of radar, and so using it to modulate radar behavior leads to a radar system that takes into consideration multiple factors about the radar and target state without resorting to complicated sets of ad hoc rules. Predicting information requires no elements beyond those found in a typical Kalman filter formulation, making it very easy to extend the framework to include information as a control measure. The approach outlined in this chapter does this in a way that is familiar to the radar tracking community as demonstrated in the simplified example using the EKF. While we implemented very simple optimization and control routines, the more general Bayesian framework leveraging all of the configurable radar resources can be costed and optimized using this framework in an application-specific way.

References

1 Bliss, D.W. and Govindasamy, S. (2013). *Adaptive Wireless Communications: MIMO Channels and Networks*. Cambridge University Press.

2 James, R.J. (1989). A history of radar. *IEE Review* 35 (9): 343–349.

3 Griffiths, H., Cohen, L., Watts, S. et al. (2015). Radar spectrum engineering and management: technical and regulatory issues. *Proceedings of the IEEE* 103 (1): 85–102.

4 Skolnik, M.I. (1990). *Radar Handbook*, 2e. McGraw-Hill Professional.

5 Richards, M.A., Sheer, J.A., and Holm, W.A. (ed.) (2010). *Principles of Modern Radar: Basic Principles*. Scitech Publishing.

6 Bliss, D.W. (2021). *Modern Communications: A Systematic Introduction*. Cambridge University Press.

7 Herschfelt, A., Chiriyath, A.R., Srinivas, S., and Bliss, D.W. (2021). An introduction to spectral convergence: challenges and paths to solutions. *2021 1st IEEE International Online Symposium on Joint Communications & Sensing (JC&S)*.

8 Bliss, D.W. (2014). Cooperative radar and communications signaling: the estimation and information theory odd couple. *IEEE Radar Conference*.

9 Ding, Z. (2008). A survey of radar resource management algorithms. *Canadian Conference on Electrical and Computer Engineering*.

10 Yan, J., Dai, J., Pu, W. et al. (2021). Target capacity based resource optimization for multiple target tracking in radar network. *IEEE Transactions on Signal Processing* 69: 2410–2421.

11 Gillespie, B., Hughes, E., and Lewis, M. (2005). Scan scheduling of multi-function phased array radars using heuristic techniques. *IEEE International Radar Conference.*

12 Stoffel, A.P. (1994). Heuristic energy management for active array multifunction radars. *Proceedings of IEEE National Telesystems Conference.*

13 Orman, A.J., Potts, C.N., Shahani, A.K., and Moore, A.R. (1996). Scheduling for a multifunction phased array radar system. *European Journal of Operational Research* 90 (1): 13–25.

14 Noyes, S.P. (1998). Calculation of next time for track update in the MESAR phased array radar. *IEE Colloquium on Target Tracking and Data Fusion (Digest No 1998/282).*

15 Yan, J., Liu, H., Pu, W. et al. (2016). Joint beam selection and power allocation for multiple target tracking in netted colocated MIMO radar system. *IEEE Transactions on Signal Processing* 64 (24): 6417–6427.

16 Baek, S.H., Seok, H., Park, K.H., and Chun, J. (2010). An adaptive update-rate control of a phased array radar for efficient usage of tracking tasks. *IEEE Radar Conference.*

17 Paul, B. and Bliss, D.W. (2015). Constant information radar for dynamic shared spectrum access. 49th Asilomar Conference on Signals, Systems and Computers.

18 Paul, B. and Bliss, D.W. (2016). The constant information radar. *Entropy* 18 (9): 338.

19 Paul, B., Chiriyath, A.R., and Bliss, D.W. (2016). Survey of RF communications and sensing convergence research. *IEEE Access* 5: 252–270.

20 Cover, T.M. and Thomas, J.A. (2006). *Elements of Information Theory, 2e.* Hoboken, NJ: Wiley-Interscience.

21 Kay, S.M. (2005). *Intuitive Probability and Random Processes Using MATLAB.* Springer.

22 Candy, J. (2009). *Bayesian Signal Processing: Classical, Modern, and Particle Filtering Methods.* New Jersey: Wiley.

23 Li, X.R. and Jilkov, V.P. (2003). Survey of maneuvering target tracking. Part I: dynamic models. *IEEE Transactions on Aerospace and Electronic Systems* 39 (4): 1333–1364.

24 Trees, H.L.V. (1992). *Detection, Estimation, and Modulation Theory: Radar-Sonar Signal Processing and Gaussian Signals in Noise.* New Jersey: Wiley.

25 Bourgeois, B.S. (2007). Using range and range rate for relative navigation. Naval Research Laboratory, Mapping, Charting, Geodesy Branch, Marine Geosciences Division. *Tech. Rep. NRL/FR/7440--07-10,153.*

26 Paul, B. and Bliss, D.W. (2015). Extending joint radar-communications bounds for FMCW radar with Doppler estimation. *IEEE Radar Conference.*

27 Paul, B. and Bliss, D.W. (2016). Estimation information bounds using the I-MMSE formula and Gaussian mixture models. *50th Annual Conference on Information Sciences and Systems.*

28 Richards, M.A. (2013). *Fundamentals of Radar Signal Processing.* McGraw Hill.

11

Power Allocation Strategies for Localization in Distributed Multiple-Radar Architectures

Hana Godrich[1], Athina P. Petropulu[1], and H. Vincent Poor[2]

[1] *Department of Electrical and Computer Engineering, Rutgers University, Piscataway, NJ, USA*
[2] *Department of Electrical and Computer Engineering, Princeton University, Princeton, NJ, USA*

11.1 Introduction

Traditional radar systems, referred to as monostatic radars [1], use a single antenna or colocated antennas to transmit a signal and receive the reflection from a target. The measured round trip time, or time-of-arrival (TOA), provides a range estimate that marks a circle around the radar where a target might be located. When using multiple monostatic radars, the individual TOAs can be used to determine a target's location through multilateration [2]. Lateration refers to computing the position of an object by measuring its distance from multiple reference points/antennas [3–6]. Calculating a target's position in two dimensions requires at least trilateration (see Figure 11.1).

To improve on the performance of monostatic radars, architectures with multiple, widely distributed antennas have been introduced. Such architectures include multiple-input multiple-output (MIMO) radars [7–10], multisite/multistatic radars [11], and bistatic radars [12]. These systems benefit from their inherent spatial diversity. The use of multiple, geographically spread, transmit and receive antennas results in multiple propagation paths that, when processed jointly, offer significant advantages when it comes to target detection, localization, tracking, and more. In MIMO radar, localization is based on measuring multiple propagation times of signals transmitted by various transmit antennas, reflected by the target, and received at multiple receive antennas (see Figure 11.2) [10].

The relationship between the geometric spread of antennas and target location estimation mean square error (MSE) has been evaluated for multistatic and bistatic radar systems in [11, 12]. A gain in localization accuracy has been shown

Information-Theoretic Radar Signal Processing, First Edition. Edited by Yujie Gu and Yimin D. Zhang.

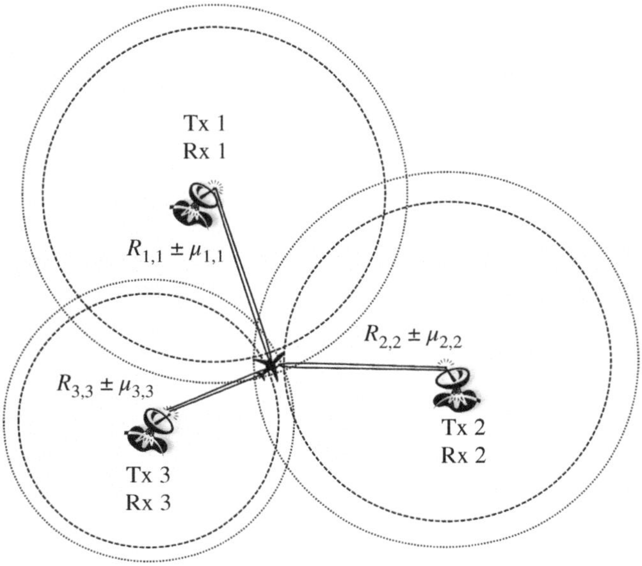

Figure 11.1 Localization using trilateration: three monostatic radars are used, and the round trip time represents a circle of potential locations. The intersection of these circles marks the area where the target is located.

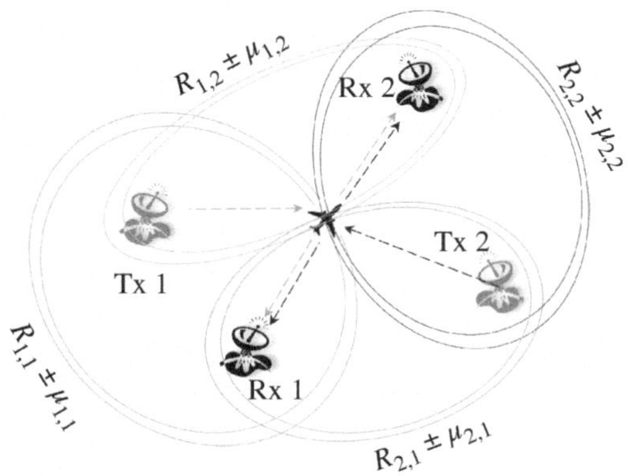

Figure 11.2 Localization with multiple radar systems.

to result from a wider spatial spread in multistatic systems [11]. The relationship between antenna distribution and estimation capabilities was quantified for MIMO radar systems in [13]. This work demonstrated that with appropriate, optimal, antenna placement with respect to a given search area, target location estimation MSE can be reduced by a factor proportional to the product of the number of transmit antennas and receive antennas. In addition to its dependence on antenna locations, the estimation MSE is inversely proportional to the effective signal bandwidth and the signal-to-noise-ratio (SNR). Thus, one can improve system localization accuracy by increasing the number of antennas, changing their location, or modifying power at the transmit antennas, among other techniques [14–18].

The advantages of multiple radar systems come with operational challenges. An increase in the number of antennas means higher communication load, increased power needs, and greater computational complexity. In practical applications, missions have predetermined performance goals when it comes to localization accuracy, overall radiated power, etc. Antennas have their own operational limits, among which are bandwidth and maximum radiated power [1, 2]. Consequently, full system utilization may result in inefficient use of resources, such as the transmit power, antennas, and communication links between radars and a central fusion center. Therefore, the notion of resource-aware operation is of critical importance whether it is mobile deployment of stations or systems operating over prolonged time periods, in which the cost of operation becomes significant. For example, in the case of ground surveillance radars (GSRs) for border patrol and security, the notion of power-aware operation becomes critical. Similar constraints are relevant in various tactical battlefield applications, such as antimissile defense radar systems which are powered off-grid by diesel generators. For those, power-aware design is crucial in extending the ability to operate before refueling. Power management is an essential part of military operations in hostile environments, where a low probability of intercept (LPI) may be needed [1], reinforcing the need for low-power operation schemes. In these scenarios, the ability of operators to manage all available resources may be key to prolonging system operation and availability to address mission's needs.

In this chapter, power-aware operation schemes are detailed that support task oriented resource management. In Section 11.2, mathematical modeling fundamentals for multiple radar systems are detailed. In Section 11.3, two power allocation schemes are proposed. In Section 11.3.1, the first optimizes localization estimation MSE given a total transmit power budget, while in Section 11.3.2, the second minimizes total power use for a given localization accuracy objective. Insights are provided for both optimization problems and proposed algorithms in Sections 11.3.1.3 and 11.3.2.3. In Section 11.4, a deeper understanding of the proposed optimization algorithms is gained through the use of numerical analysis. Finally, Section 11.5 extends the discussion of resource allocation in multiple radar systems to subset selection and additional opportunities for resource savings.

11.2 Mathematical Modeling

We consider a distributed multiple radar system, with M transmit antennas and N receive antennas, forming an $M \times N$ radar system. In general, it is assumed that transmit and receive antennas are spread across a wide geographic area in which the target is located (see Figure 11.3). These antennas are not necessarily colocated and the number of transmit antennas can differ from that of receive antennas.

11.2.1 Fundamentals

In this subsection, we detail the basic system assumptions. The transmit and receive radars are assumed to be located in a two-dimensional plane, where the M transmit antennas are arbitrarily located at coordinates (x_{tx_m}, y_{tx_m}), $m = 1, \dots, M$, and the N receive antennas are arbitrarily located at coordinates (x_{rx_n}, y_{rx_n}), $n = 1, \dots, N$. A transmit antenna set is defined as: $S_{tx} := \{(x_{tx_1}, y_{tx_1}), (x_{tx_2}, y_{tx_2}), \dots, (x_{tx_M}, y_{tx_M})\}$ and a receive antenna set is defined as: $S_{rx} := \{(x_{rx_1}, y_{rx_1}), (x_{rx_2}, y_{rx_2}), \dots, (x_{rx_N}, y_{rx_N})\}$. An extended target is assumed with a center of mass located at position (x, y) [13]. The variation in the target center of mass is assumed to be small with respect to the system resolution capabilities. As localization happens while tracking the target, an estimate of the target's unknown parameters from previous cycles is available. A set of orthogonal waveform is transmitted, each with a baseband equivalent $S_m(t)$, where $\int_{T_m} |S_m(t)|^2 dt = 1$, and T_m is the duration of the mth transmitted signal. The mth waveform's effective bandwidth is denoted by β_m and the individual transmit antenna power by p_m. The search cell is confined to $\left(x_c \pm \frac{kc}{\beta}, y_c \pm \frac{kc}{\beta}\right)$, where k is an integer, c is the speed of light, and β is the waveforms' averaged effective bandwidth, defined as $\beta^2 = \frac{1}{M} \sum_{m=1}^{M} \beta_m^2$. Next, the mathematical model for target localization in multiple radar systems is presented.

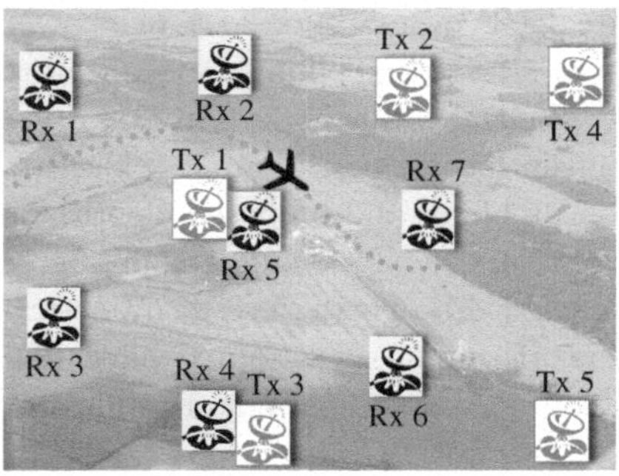

Figure 11.3 Topology of multiple radar systems.

11.2.2 Localization

As mentioned earlier, in multiple radar systems, localization is based on measuring multiple propagation times of signals transmitted by the multiple transmit antennas, reflected by the target, and received at multiple receive antennas (as depicted in Figure 11.4). Let $\tau_{m,n}(x, y)$ denote the propagation time of a signal transmitted by radar m, reflected by the target, and received by radar n:

$$\tau_{m,n}(x, y) = \frac{R_{tx_m} + R_{rx_n}}{c}, \tag{11.1}$$

where R_{tx_m} is the range from transmitter m to the target, i.e.,

$$R_{tx_m} = \sqrt{(x_{tx_m} - x)^2 + (y_{tx_m} - y)^2}, \tag{11.2}$$

and the range from the target to receiver n is denoted by R_{rx_n}, calculated as

$$R_{rx_n} = \sqrt{(x_{rx_n} - x)^2 + (y_{rx_n} - y)^2}. \tag{11.3}$$

The signal transmitted from radar m, reflected from the target, and received at radar n (baseband representation) is

$$r_{m,n}(t) = \sqrt{\alpha_{m,n} p_m} h_{m,n} s_m(t - \tau_{m,n}) + \omega_{m,n}(t), \tag{11.4}$$

where $\alpha_{m,n} \propto \frac{1}{R^2_{tx_m} R^2_{rx_n}}$ represents the variation in the signal strength due to path loss effects. The term $h_{m,n}$ incorporates the phase effect of multipath fading and the target impact on the phase and amplitude. The unknown target radar cross section (RCS) is modeled as deterministic and complex. The noise expression, $w_{m,n}(t)$, is

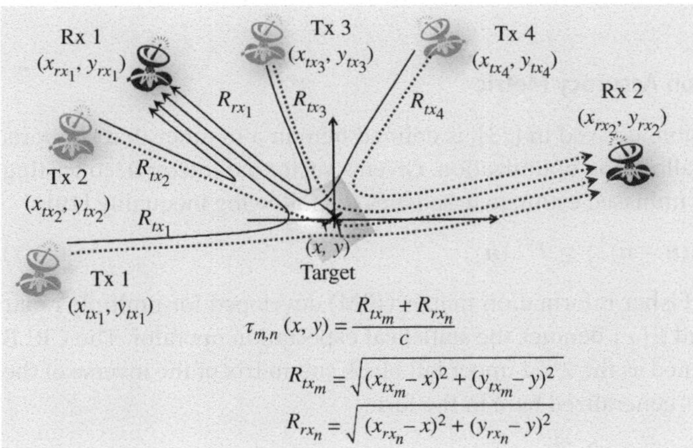

Figure 11.4 TOAs measurements in multiple radar systems.

a circularly symmetric, zero-mean, complex Gaussian noise, spatially and temporally white with autocorrelation function $\sigma_w^2\delta(\tau)$.

The SNR for the propagation path from transmit antenna m to the target and from the target to receive antenna n, defined as channel (m, n), is $\mathrm{SNR}_{m,n} = \frac{\alpha_{m,n}|h_{m,n}|^2 p_m}{\sigma_w^2}$.

We define a vector of transmit power $\boldsymbol{p} = [p_1, p_2, \ldots, p_M]^\mathrm{T}$ and a vector of unknown parameters:

$$\boldsymbol{u} = \left[x, y, \boldsymbol{h}^\mathrm{T}\right]^\mathrm{T}, \tag{11.5}$$

where $\boldsymbol{h} = [h_{1,1}, h_{1,2}, \ldots, h_{M,N}]^\mathrm{T}$. In addition, we also defined additional vectors $\boldsymbol{\beta} = [\beta_1, \beta_2, \ldots, \beta_M]^\mathrm{T}$, $\boldsymbol{\alpha} = [\alpha_{1,1}, \alpha_{1,2}, \ldots, \alpha_{M,N}]^\mathrm{T}$, and $\boldsymbol{\tau} = [\tau_{1,1}, \tau_{1,2}, \ldots, \tau_{M,N}]^\mathrm{T}$ for later use.

The maximum likelihood estimator (MLE) [19] of the unknown parameter vector $\boldsymbol{u}$, given the observation vector $\boldsymbol{r} = [r_{1,1}, \ldots, r_{m,n}, \ldots, r_{M,N}]^\mathrm{T}$, where $r_{m,n}$ is given in (11.4), is

$$\hat{\boldsymbol{u}}_{\mathrm{ML}} = \arg\max_{\boldsymbol{u}} \left[\log p(\boldsymbol{r}|\boldsymbol{u})\right], \tag{11.6}$$

where $p(\boldsymbol{r}|\boldsymbol{u})$ is the conditional, joint probability density function (PDF) of $\boldsymbol{r}$ given $\boldsymbol{u}$.

The Cramér–Rao lower bound (CRLB) is known to be asymptotically tight to the mean square error of the MLE at high SNR [19]. In [13] it was demonstrated that the CRLB for target localization in MIMO radar systems is asymptotically unbiased, and the MLE is asymptotically tight to the CRLB at high SNR (over 10 dB). Thus, the CRLB, detailed in Section 11.2.3, is used here to represent the localization MSE.

11.2.3 Estimation Accuracy Metric

The CRLB expression, derived in [13], is defined here in a manner that supports its use in resource allocation optimization. Given a vector parameter $\boldsymbol{u}$, consisting of elements u_i, the unbiased estimate $\hat{u}_i$ satisfies the following inequality [19]:

$$\mathbb{E}_{\boldsymbol{u}}\left\{(\hat{\boldsymbol{u}} - \boldsymbol{u})(\hat{\boldsymbol{u}} - \boldsymbol{u})^\mathrm{T}\right\} \geq \boldsymbol{J}^{-1}(\boldsymbol{u}), \tag{11.7}$$

where $\boldsymbol{J}(\boldsymbol{u})$ is the Fisher information matrix (FIM) developed for multiple radar systems in [20], and $\mathbb{E}\{\cdot\}$ denotes the statistical expectation operator. The CRLB matrix, $\boldsymbol{C}_{x,y}$, is defined as the 2×2 upper left block submatrix of the inverse of the FIM, $\boldsymbol{J}^{-1}(\boldsymbol{u})$, and is generalized here in the form

$$\boldsymbol{C}_{x,y} = \left\{\sum_{m=1}^{M}\sum_{n=1}^{N} \boldsymbol{J}_{m,n}\right\}^{-1}, \tag{11.8}$$

where the submatrix $\boldsymbol{J}_{m,n}$ is defined as

$$\boldsymbol{J}_{m,n} = \begin{bmatrix} g_{a_{m,n}} & g_{c_{m,n}} \\ g_{c_{m,n}} & g_{b_{m,n}} \end{bmatrix}, \tag{11.9}$$

and the elements $g_{a_{m,n}}$, $g_{b_{m,n}}$, and $g_{c_{m,n}}$ are defined as

$$g_{a_{m,n}} = \alpha_{m,n} p_m \xi_m \left|h_{m,n}\right|^2 \left(\frac{x_{tx_m} - x}{R_{tx_m}} + \frac{x_{rx_n} - x}{R_{rx_n}} \right)^2, \tag{11.10}$$

$$g_{b_{m,n}} = \alpha_{m,n} p_m \xi_m \left|h_{m,n}\right|^2 \left(\frac{y_{tx_m} - y}{R_{tx_m}} + \frac{y_{rx_n} - y}{R_{rx_n}} \right)^2, \tag{11.11}$$

and

$$g_{c_{m,n}} = \alpha_{m,n} p_m \xi_m \left|h_{m,n}\right|^2 \left(\frac{x_{tx_m} - x}{R_{tx_m}} + \frac{x_{rx_n} - x}{R_{rx_n}} \right) \times \left(\frac{y_{tx_m} - y}{R_{tx_m}} + \frac{y_{rx_n} - y}{R_{rx_n}} \right), \tag{11.12}$$

where $\xi_m = \frac{8\pi^2 \beta_m^2}{\sigma_w^2 c^2}$. The elements in the above equations are expressed in a vector form as $\mathbf{g}_a = \left[g_{a_{1,1}}, g_{a_{1,2}}, \dots, g_{a_{M,N}}\right]^{\mathrm{T}}$, $\mathbf{g}_b = \left[g_{b_{1,1}}, g_{b_{1,2}}, \dots, g_{b_{M,N}}\right]^{\mathrm{T}}$, and $\mathbf{g}_c = \left[g_{c_{1,1}}, g_{c_{1,2}}, \dots, g_{c_{M,N}}\right]^{\mathrm{T}}$.

The expression for the CRLB offers a metric that may be used to represent the MLE MSE in any power allocation schemes. The trace of the matrix $\boldsymbol{C}_{x,y}$ represents a lower bound on the sum of the MSEs for the target location estimation, i.e., $\mathrm{tr}\left(\boldsymbol{C}_{x,y}\right) \leq \sigma_x^2 + \sigma_y^2$, where σ_x^2 and σ_y^2 are the target's x and y location estimation MSEs, respectively. From (11.8) and (11.9), the metric can be expressed in terms of transmit power [20] as

$$\mathrm{tr}\left(\boldsymbol{C}_{x,y}\right) = \sigma_{x,y}^2\left(\boldsymbol{p}\right) = \frac{\boldsymbol{b}^{\mathrm{T}}\boldsymbol{p}}{\boldsymbol{p}^{\mathrm{T}}\boldsymbol{A}\boldsymbol{p}}, \tag{11.13}$$

where $\boldsymbol{b} = \mathbf{g}_a + \mathbf{g}_b$ and $\boldsymbol{A} = \mathbf{g}_a\mathbf{g}_b^{\mathrm{T}} - \mathbf{g}_c\mathbf{g}_c^{\mathrm{T}}$. Here, $\mathrm{tr}\left(\cdot\right)$ refers to the trace of a matrix.

11.2.4 Extended Discussion on Multilateration and Localization MSE

As previously mentioned, localization is essentially based on multilateration. Each transmit/receiver pair, (m, n), measures a combined range, $R_{m,n} = R_{tx_m} + R_{rx_n}$, through the measurement of time delays, $\tau_{m,n}(x, y)$. This range traces out an imaginary ellipse, which represents all possible target locations, as illustrated in Figure 11.2. An intersection of three or more of these ellipses determines the target location. In practice, the time delay estimation, and therefore, the range sum,

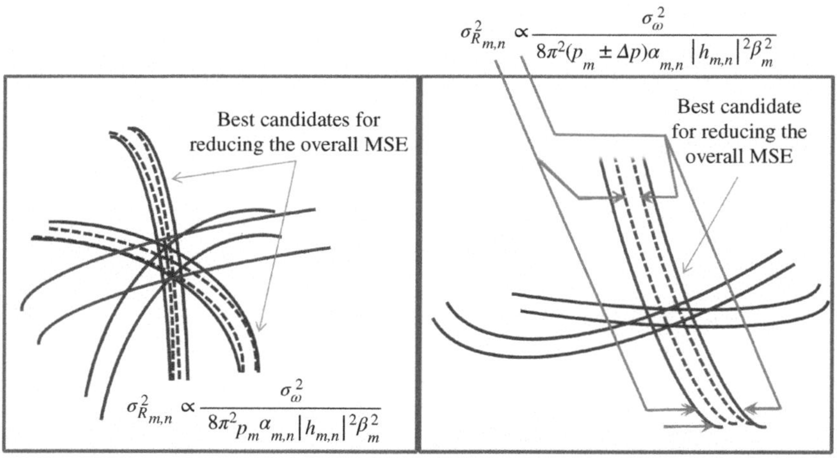

Figure 11.5 Localization MSE with multiple radars.

$R_{m,n}$, includes an estimation error with variance $\sigma^2_{R_{m,n}} \propto \sigma^2_{\tau_{m,n}} = \frac{\sigma^2_w}{8\pi^2 p_m \alpha_{m,n}|h_{m,n}|^2\beta^2_m}$. As such, the ellipse shape has an uncertainty width that is proportional to the measurement variance, and is denoted by $\mu_{m,n}$. This is demonstrated in Figure 11.5 by drawing the elliptical paths that represent all possible target positions that generate a range measurement $R_{m,n} \pm \mu_{m,n}$, i.e., an inner ellipse represents $(R_{m,n} - \mu_{m,n})$, and an outer one represents $(R_{m,n} + \mu_{m,n})$. The width of the uncertainty area, bounded by the inner and outer ellipses, is changing along the contour line. The overall location estimation MSE is determined by the size of the intersection area.

The examples given in Figure 11.5 illustrate the manner in which power allocation increase can be used to improve overall localization accuracy. The MSE may be controlled by increasing the transmit power and/or by choosing transmit-receive pairs that have the target positioned over favorable parts of the ellipse, i.e., locations with lower uncertainty due to geometrical characteristics. In the intersection of multiple ellipses, a more significant effect on the total localization error may be obtained by further narrowing the ones with smaller uncertainty, as the incremental effect of this step is considerably larger than that of reinforcing measurements with higher uncertainty.

11.3 Power Allocation Optimization

The localization accuracy is affected by the system layout, path propagation losses, target RCS properties, noise variance, signal bandwidth, and the radiated power.

These are all incorporated into the elements of the CRLB metric in (11.13) through the vector $\boldsymbol{b}$, vector $\boldsymbol{p}$, and matrix $\boldsymbol{A}$. Some of these are given system characteristics, such as the transmit radar locations, the receive radar locations, the target location (x, y), the target RCS, $\boldsymbol{h}$, the propagation path loss, $\boldsymbol{\alpha}$, and noise variance, σ_w^2. The others are determined by the system operational strategies. These adaptive parameters include the transmit power at each radar, P_m, and the effective signal bandwidth β_m, where the transmit power is constrained by a maximum value $p_{m_{\max}}$, determined by the operational design, and a minimal value $p_{m_{\min}}$, chosen such that it may still be classified as operating in the high SNR region. To address real-life scenarios where system operation may require limiting resource use, this section provides a review of two power optimization schemes:

- *Optimize estimation MSE*: Given a total transmit power budget objective, can operators allocate power among transmit antennas in a manner that achieves better localization accuracy, when compared with uniform power allocation? In this scheme, advantages of using nonuniform power allocation among transmit radars are evaluated.
- *Reduce total power use*: Given an operational localization accuracy target, how should operators allocate power among transmitters such that they reach the localization accuracy threshold while minimizing the total power budget used to attain it?

In order to formulate these power allocation schemes as optimization problems, the CRLB metric expressed in (11.13) is used. The individual transmit antenna power, p_m, $m = 1, \ldots M$, when participating in localization, is constrained by lower and upper limits $p_{m_{\min}} \leq p_m \leq p_{m_{\max}}$, $m = 1, \ldots, M$. Two constraint vectors are defined as $\boldsymbol{p}_{\min} = [p_{1_{\min}}, p_{2_{\min}}, \ldots, p_{M_{\min}}]^{\mathrm{T}}$ and $\boldsymbol{p}_{\max} = [p_{1_{\max}}, p_{2_{\max}}, \ldots, p_{M_{\max}}]^{\mathrm{T}}$.

Notice that for the case of uniform power allocation, the transmit power vector takes the form $\boldsymbol{p}_u = p_u\,[1, 1, \ldots, 1]^{\mathrm{T}}_{M\times 1}$ and thus the trace of the CRLB matrix is

$$\sigma_{x,y}^2\left(\boldsymbol{p}_u\right) = \frac{1}{p_u}\frac{\boldsymbol{b}^{\mathrm{T}}\mathbf{1}}{\mathbf{1}^{\mathrm{T}}\boldsymbol{A}\mathbf{1}}. \tag{11.14}$$

In this case, the total power budget is $P_{\text{total\_u}} = p_u \times M$. Preliminary research into the localization performance of multiple radar systems used uniform power allocation. Here, the purpose of the above two schemes is to improve the use of resources in one of two ways: first, given a total power budget, if allocated uniformly among transmitters, it will result with a vector of $\boldsymbol{p}_u = \frac{P_{\text{total\_u}}}{M}\mathbf{1}^{\mathrm{T}}$ and an associated sum of variances $\sigma_u^2 = \sigma_{x,y}^2\left(\boldsymbol{p}_u\right)$. Is this the best positional accuracy that can be achieved with this power budget? Second, if the needed accuracy threshold amounts to σ_u^2, can less power be used to achieve a similar performance? Next, two schemes are formulated into optimization problems.

11.3.1 Minimize Estimation MSE

The first scenario considers operating under a total power budget constraint. In this case, the total transmit power objective is set to be P_{obj}. The power allocation optimization problem goal is to identify a transmit power allocation vector $\boldsymbol{p}_{\text{MSEo}}$ which uses the given total power budget, i.e., $\mathbf{1}^{\text{T}}\boldsymbol{p}_{\text{MSEo}} = P_{\text{obj}}$, while minimizing localization MSE $\sigma^2_{\text{MSEo}} = \sigma^2_{x,y}\left(\boldsymbol{p}_{\text{MSEo}}\right)$. The sum of localization estimation MSEs, based on the trace of the CRLB matrix, $\text{tr}\left(\boldsymbol{C}_{x,y}\right)$, is used as an objective function. The constraints are to keep all transmit powers within operational range, i.e., $p_{m_{\min}} \leq p_m \leq p_{m_{\max}}, m = 1, \ldots, M$ and keep the total transmit power to $P_{\text{obj}} = \sum_{m=1}^{M} p_m = \mathbf{1}^{\text{T}}\boldsymbol{p}$. This can be formulated into an optimization problem of the form

$$\begin{aligned}
\min_{\boldsymbol{p}} \quad & \text{tr}\left(\boldsymbol{C}_{x,y}\left(\tilde{\boldsymbol{u}}, \boldsymbol{p}\right)\right) = \frac{\boldsymbol{b}_e^{\text{T}}\boldsymbol{p}}{\boldsymbol{p}^{\text{T}}\boldsymbol{A}_e\boldsymbol{p}} \\
\text{s.t.} \quad & \boldsymbol{p} - \boldsymbol{p}_{\max} \preceq \mathbf{0}, \\
& \boldsymbol{p}_{\min} - \boldsymbol{p} \preceq \mathbf{0} \\
& \mathbf{1}^{\text{T}}\boldsymbol{p} - P_{\text{obj}} = 0,
\end{aligned} \tag{11.15}$$

where $\boldsymbol{C}_{x,y}\left(\tilde{\boldsymbol{u}}, \boldsymbol{p}\right)$ is the 2×2 CRLB matrix given in (11.13). The vector $\tilde{\boldsymbol{u}} = \left[\tilde{x}, \tilde{y}, \tilde{\boldsymbol{h}}^{\text{T}}\right]^{\text{T}}$ represents prior estimates of the target location and RCS. When tracking the target, these may be obtained from previous cycles. Alternatively, the search cell center coordinates $\left(x_c, y_c\right)$ may be used as $(\tilde{x}, \tilde{y})$. Vector $\boldsymbol{b}_e = \boldsymbol{b}(\tilde{\boldsymbol{u}})$ and matrix $\boldsymbol{A}_e = \boldsymbol{A}(\tilde{\boldsymbol{u}})$ are calculated for the estimated vector $\tilde{\boldsymbol{u}}$.

The resulting optimization problem in (11.15) is nonconvex due to the objective function being a nonlinear function [21]. One way to deal with this type of problem is to choose an appropriate convex relaxation of the original problem and use the optimal solution to find a local minimum [21–25]. Another way is through the use of Lagrange multipliers and the Karush–Kuhn–Tucker (KKT) conditions and decomposition techniques [26–34]. Both relaxation and domain decomposition methods will be applied to derive potential power allocation algorithms.

11.3.1.1 Relaxation Methods Applied to MLE optimization

In [35], an approximate, convex formulation of the objective problem for (11.15) is proposed, where min $\left\{\text{tr}\left(\boldsymbol{C}_{x,y}\left(\tilde{\boldsymbol{u}}, \boldsymbol{p}\right)\right)\right\}$ is replaced with max $\left\{\text{tr}\left(\boldsymbol{C}^{-1}_{x,y}\left(\tilde{\boldsymbol{u}}, \boldsymbol{p}\right)\right)\right\}$, resulting in the following approximate convex optimization problem

$$\begin{aligned}
\max_{\boldsymbol{p}} \quad & \boldsymbol{b}_e^{\text{T}}\boldsymbol{p} \\
\text{s.t.} \quad & \boldsymbol{p} - \boldsymbol{p}_{\max} \preceq \mathbf{0}, \\
& \boldsymbol{p}_{\min} - \boldsymbol{p} \preceq \mathbf{0}, \\
& \mathbf{1}^{\text{T}}\boldsymbol{p} - P_{\text{obj}} = 0.
\end{aligned} \tag{11.16}$$

The exact solution for this approximate problem can be found using available convex optimization tools, such as CVX [22, 24]. The resulting optimal solution to (11.36) can then be used to find a local minimum. An iterative optimization algorithm is applied to the original nonconvex problem in (11.15), where the initial values are set based on $\boldsymbol{p}^{\star}_{\text{MLE}}$ as follows:

$$\begin{aligned} \boldsymbol{p}_0 &= \boldsymbol{p}^{\star}_{\text{MLE}}, \\ \sigma^2_{x,y}\left(\boldsymbol{p}_0\right) &= \frac{\boldsymbol{b}_e^{\text{T}}\boldsymbol{p}_0}{\boldsymbol{p}_0^{\text{T}}\boldsymbol{A}_e\boldsymbol{p}_0}, \end{aligned} \tag{11.17}$$

and setting initial values for incremental power step $\triangle p_0$, stopping criterion ε, and iteration counter $k = 0$. The search algorithm is implemented as follows:

repeat:

$$\left[\begin{array}{l} k = k + 1 \\ \boldsymbol{p}_k = \arg\min\left\{\sigma^2_{x,y}\left(\boldsymbol{p}^-_{k-1}\right) - \sigma^2_{x,y}\left(\boldsymbol{p}^+_{k-1}\right)\right\} \\ \qquad \text{s.t.} \qquad p_{m_{\text{min}}} \le p_{m_k}; p_{m_{\text{max}}} \ge p_{m_k} \\ \Delta p_k = \Delta p_{k-1}\dfrac{\mathbf{1}^{\text{T}}\boldsymbol{p}_k}{\mathbf{1}^{\text{T}}\boldsymbol{p}_{k-1}} \end{array}\right.$$

while:

$$\sigma^2_{x,y}\left(\boldsymbol{p}_{k-1}\right) - \sigma^2_{x,y}\left(\boldsymbol{p}_k\right) \ge \varepsilon$$

set:

$$\boldsymbol{p}_{\text{MSEo}} = \boldsymbol{p}_k \text{ and } \sigma^2_{\text{MSEo}} = \sigma^2_{x,y}\left(\boldsymbol{p}_{\text{MSEo}}\right).$$

Note that at each iteration step, the power of a single antenna is decreased by Δp_{k-1}, generating the power vector $\boldsymbol{p}^-_{k-1}$. At the same time, power is increased for a different antenna by the same amount, resulting in the power vector $\boldsymbol{p}^+_{k-1}$. The antenna pair with the lowest MSE gap is selected as $\boldsymbol{p}_k$. Thus, the local solution $\boldsymbol{p}_{\text{MSEo}}$ follows the total power allocation constrains $P_{\text{obj}} = \sum_{m=1}^{M} p_{m_{\text{MSEo}}}$ while minimizing the MSE $\sigma^2_{\text{MSEo}} = \sigma^2_{x,y}\left(\boldsymbol{p}_{\text{MSEo}}\right)$ associated with the fixed power budget. The use of relaxation methods here results in a centralized algorithm implementable in a fusion center. Data is collected from all antennas to determine vectors $\boldsymbol{h}$ and $\boldsymbol{\alpha}$ that are used to calculate matrix $\boldsymbol{A}$ and vector $\boldsymbol{b}$.

11.3.1.2 Domain Decomposition Applied to MLE Optimization

An alternative approach is to use domain decomposition methods for the optimization problem in (11.16), by formulating the KKT conditions [26]. To start, the optimization problem in (11.16) is redefined as

$$\begin{aligned}
\min_{z,\boldsymbol{p}} \quad & z \\
\text{s.t.} \quad & \boldsymbol{b}_e^{\mathrm{T}}\boldsymbol{p} - z\boldsymbol{p}^{\mathrm{T}}\boldsymbol{A}_e\boldsymbol{p} \leq 0, \\
& \boldsymbol{p} - \boldsymbol{p}_{\max} \preceq \boldsymbol{0}, \\
& \boldsymbol{p}_{\min} - \boldsymbol{p} \preceq \boldsymbol{0}, \\
& -z \leq 0, \\
& \boldsymbol{1}^{\mathrm{T}}\boldsymbol{p} - P_{\mathrm{obj}} = 0.
\end{aligned} \tag{11.18}$$

The KKT conditions are necessary for a solution to be optimal, provided that some *regularity* conditions are satisfied, i.e., necessary optimality conditions [25, 33] need to be met. If a feasible point satisfies a first-order constraint qualification (CQ1) then it implies that the local minimizer is a KKT point. The most commonly used CQ1 is linear independence constraint qualification (LICQ) [29]. If the gradients are linearly independent, the set of Lagrange multipliers $(\lambda^\star, \vartheta^\star, \boldsymbol{\mu}^\star, \boldsymbol{\nu}^\star)$ is nonempty, bounded, and unique.

Lagrangian Function For this power optimization problem, the appropriate Lagrangian function is

$$\begin{aligned}
\mathcal{L}(z, \boldsymbol{p}, \lambda, \vartheta, \boldsymbol{\mu}, \boldsymbol{\nu}) = & \; z + \lambda\left(\boldsymbol{b}_e^{\mathrm{T}}\boldsymbol{p} - z\boldsymbol{p}^{\mathrm{T}}\boldsymbol{A}_e\boldsymbol{p}\right) + \vartheta\left(\boldsymbol{1}^{\mathrm{T}}\boldsymbol{p} - P_{\mathrm{obj}}\right) \\
& + \boldsymbol{\mu}^{\mathrm{T}}\left(\boldsymbol{p} - \boldsymbol{p}_{\max}\right) + \boldsymbol{\nu}^{\mathrm{T}}\left(\boldsymbol{p}_{\min} - \boldsymbol{p}\right),
\end{aligned} \tag{11.19}$$

and the appropriate KKT conditions are

$$\nabla_z \mathcal{L} = 1 - \lambda^\star \boldsymbol{p}^{\star\mathrm{T}}\boldsymbol{A}_e\boldsymbol{p}^\star = 0, \tag{11.20a}$$

$$\nabla_{\boldsymbol{p}} \mathcal{L} = \lambda^\star\left[\boldsymbol{b}_e - z^\star\left(\boldsymbol{A}_e + \boldsymbol{A}_e^{\mathrm{T}}\right)\boldsymbol{p}^\star\right] + \vartheta^\star\boldsymbol{1} + \boldsymbol{\mu}^\star - \boldsymbol{\nu}^\star = \boldsymbol{0}, \tag{11.20b}$$

$$\lambda^\star\left(\boldsymbol{b}_e^{\mathrm{T}}\boldsymbol{p}^\star - z^\star\boldsymbol{p}^{\star\mathrm{T}}\boldsymbol{A}_e\boldsymbol{p}^\star\right) = 0, \tag{11.20c}$$

$$\mu_m^\star\left(p_m^\star - p_{m_{\max}}\right) = 0, \quad m = 1, \ldots, M, \tag{11.20d}$$

$$\nu_m^\star\left(p_{m_{\min}} - p_m^\star\right) = 0, \quad m = 1, \ldots, M, \tag{11.20e}$$

$$\boldsymbol{1}^{\mathrm{T}}\boldsymbol{p}^\star - P_{\mathrm{obj}} = 0, \tag{11.20f}$$

$$z^\star \geq 0, \lambda^\star \geq 0, \mu_m^\star \geq 0, \nu_m^\star \geq 0. \quad \forall m = 1, \ldots, M. \tag{11.20g}$$

It has been shown in [20] that the LICQ holds for this problem as gradients of the inequality constraints and gradients of the equality constraints are linearly independent. A solution to (11.20a) through (11.20f) may be obtained by choosing $\mu_m^\star = \nu_m^\star = 0, \forall m = 1, \ldots, M$, i.e., ignoring the constrains on power. This results in an elimination of (11.20d) and (11.20e). The Lagrange multipliers $\vartheta^\star$ and $\lambda^\star$ ratio, $\frac{\vartheta^\star}{\lambda^\star}$, can then be expressed as follows (see [20]):

$$\frac{\vartheta^\star}{\lambda^\star} = \sqrt{\frac{\boldsymbol{b}_e^{\mathrm{T}}\left(\boldsymbol{B} - \boldsymbol{B}^{\mathrm{T}}\boldsymbol{A}_e\boldsymbol{B}\right)\boldsymbol{b}_e}{\boldsymbol{1}^{\mathrm{T}}\boldsymbol{B}^{\mathrm{T}}\boldsymbol{A}_e\boldsymbol{B}\boldsymbol{1}}}, \tag{11.21}$$

where $\boldsymbol{B} = \boldsymbol{A}_{ee}^{-1}$ and $\boldsymbol{A}_{ee} = \boldsymbol{A}_e + \boldsymbol{A}_e^{\mathrm{T}}$. The expression for the power assigned to transmitter m, $p_m^{\star}$, results in

$$p_m^{\star} = P_{\mathrm{obj}} \frac{\sum_{m'=1}^{M} \boldsymbol{B}\left[m, m'\right]\left(\boldsymbol{b}_e\left[m'\right] + \frac{\vartheta^{\star}}{\lambda^{\star}}\right)}{\sum_{m=1}^{M}\sum_{m'=1}^{M} \boldsymbol{B}[m, m']\left(\boldsymbol{b}_e[m'] + \frac{\vartheta^{\star}}{\lambda^{\star}}\right)}. \tag{11.22}$$

The solution in (11.22) may be outside the upper and lower bounds on the values of $\boldsymbol{p}$. Thus, an optimal solution to the master optimization problem in (11.18) is obtained by enforcing the global boundary constraints $p_{m_{\min}} \leq p_m^{\star} \leq p_{m_{\max}}$, $\forall m = 1, \dots, M$, on the set of local minimizers. In the following, domain decomposition methods are used to generate a set of $M + 1$ optimization problems used to calculate a set of local minimizers.

Domain Decomposition For the master problem, a set of parallel, nonoverlapping, optimization problems can be derived. The feasible solution sets for the subproblems are then used to derive an optimal power allocation. For the decomposition, the transmit powers are regarded as M vertices, divided into two groups: one includes interior points and the other includes points located on the boundaries.

Definitions:

- A global set of vertices is defined as $\mathcal{P} := \{p_1, p_2, \dots, p_M\}$.
- A set $\mathcal{B}_K$ is defined that includes all possible vectors $\boldsymbol{p}_{2_K} \in \mathbb{R}^{K\times 1}$, $0 \leq K \leq M$. The latter is generated by choosing K different elements out of $\mathcal{P}$, i.e., $\mathcal{B}_K := \left\{\boldsymbol{p}_{2_K} : [p_1, p_2, \dots, p_K]^{\mathrm{T}}, [p_2, p_3, \dots, p_{K+1}]^{\mathrm{T}}, \dots, [p_{M-K+1}, p_{M-K+2}, \dots, p_M]^{\mathrm{T}}\right\}$.
- A set $\mathcal{I}_K$ is defined as a set of complementary vectors $\boldsymbol{p}_{1_K} \in \mathbb{R}^{(M-K)\times 1}$, such that $\mathcal{I}_K := \left\{\boldsymbol{p}_{1_K} : \boldsymbol{p}_{1_K} = \boldsymbol{p}\backslash\boldsymbol{p}_{2_K}\right\}$, where the notation $\left(\boldsymbol{p}\backslash\boldsymbol{p}_{2_K}\right)$ refers to a vector generated by excluding elements of vector $\boldsymbol{p}_{2_K}$ from vector $\boldsymbol{p}$.
- Vectors $\boldsymbol{p}_{1_K}$ in the set $\mathcal{I}_K$ include interior vertices.
- Vectors $\boldsymbol{p}_{2_K}$ in the set $\mathcal{B}_K$ include vertices that lie on the boundaries.
- The vector $\boldsymbol{p}$ is regrouped as $\boldsymbol{p}_K = \left[\boldsymbol{p}_{1_K}^{\mathrm{T}}, \boldsymbol{p}_{2_K}^{\mathrm{T}}\right]^{\mathrm{T}}$, defining a set $\mathcal{U}_K := \left\{\boldsymbol{p}_K : \left[\left(\boldsymbol{p}\backslash\boldsymbol{p}_{2_K}\right)^{\mathrm{T}}, \boldsymbol{p}_{2_K}^{\mathrm{T}}\right]^{\mathrm{T}}\right\}$.
- The matrix $\boldsymbol{A}_e$ and vector $\boldsymbol{b}_e$ are rearranged to conform with the partitioning of the vector $\boldsymbol{p}_K$, resulting in

$$\boldsymbol{A}_{e_K} = \begin{bmatrix} \boldsymbol{A}_{11_{(M-K)\times(M-K)}} & \boldsymbol{A}_{12_{(M-K)\times K}} \\ \boldsymbol{A}_{21_{K\times(M-K)}} & \boldsymbol{A}_{22_{K\times K}} \end{bmatrix}, \tag{11.23}$$

and

$$\boldsymbol{b}_{e_K} = \begin{bmatrix} \boldsymbol{b}_{1_{(M-K)\times 1}} \\ \boldsymbol{b}_{2_{K\times 1}} \end{bmatrix}. \tag{11.24}$$

- The vector $\boldsymbol{p}'_{K_{\text{Bound}}} \in \mathcal{Y}'_K$ is defined, where $\mathcal{Y}'_K := \left\{ \left[p_{1_{\min}}, p_{2_{\min}}, \dots, p_{K_{\min}}\right], \left[p_{1_{\min}}, p_{2_{\min}}, \dots, p_{K_{\max}}\right], \dots, \left[p_{1_{\max}}, p_{2_{\max}}, \dots, p_{K_{\max}}\right] \right\}$ are the possible boundary values assigned to $\boldsymbol{p}_{2_K}$.
- The set of possible values of $\boldsymbol{p}_{K_{\text{Bound}}}$ is $\mathcal{Y}_K := \left\{ \boldsymbol{p}'_{K_{\text{Bound}}} : \mathbf{1}_2^{\text{T}} \boldsymbol{p}'_{K_{\text{Bound}}} \le P_{\mathbf{obj}} \right\}$. The latter condition on the set $\mathcal{Y}_K$ reduces the search to sets that minimize the total transmit power when compared with the simple solution of uniform power allocation.

The master optimization problem in (11.18) can now be decomposed into $M-1$ optimization problems of the form

$$\begin{aligned} \min_{z,\, \boldsymbol{p} \in \mathcal{U}_K} \quad & z \\ \text{s.t.} \quad & \boldsymbol{b}_{e_K}^{\text{T}} \boldsymbol{p}_K - z \boldsymbol{p}_K^{\text{T}} \boldsymbol{A}_{e_K} \boldsymbol{p}_K \le 0, \\ & \mathbf{1}_1^{\text{T}} \boldsymbol{p}_{1_K} + \mathbf{1}_2^{\text{T}} \boldsymbol{p}_{2_K} - P_{\text{obj}} = 0, \\ & \boldsymbol{p}_{2_K} - \boldsymbol{p}_{K_{\text{Bound}}} = \mathbf{0}. \end{aligned} \tag{11.25}$$

The *Lagrangian function* in this case is

$$\begin{aligned} \mathcal{L}(z, \boldsymbol{p}, \lambda, \vartheta, \boldsymbol{\mu}, \boldsymbol{v}) &= z + \lambda \left(\boldsymbol{b}_e^{\text{T}} \boldsymbol{p}_K - z \boldsymbol{p}^{\text{T}} \boldsymbol{A}_e \boldsymbol{p}_K \right) + \vartheta \left(\mathbf{1}^{\text{T}} \boldsymbol{p}_K - P_{\text{obj}} \right) \\ &\quad + \boldsymbol{\mu}^{\text{T}} \left(\mathbf{p}_K - \boldsymbol{p}_{\max} \right) + \boldsymbol{v}^{\text{T}} \left(\boldsymbol{p}_{\min} - \boldsymbol{p}_K \right), \end{aligned} \tag{11.26}$$

and the appropriate KKT conditions for subproblem K are

$$\begin{aligned} &\nabla_z \mathcal{L} = 1 - \lambda^\star \boldsymbol{p}_K^{\text{T}} \boldsymbol{A}_{e_K} \boldsymbol{p}_K^\star = 0, \\ &\nabla_{\boldsymbol{p}_{1_K}} \mathcal{L} = \lambda^\star \left[\boldsymbol{b}_1 - z^\star \left(\boldsymbol{A}_{11} + \boldsymbol{A}_{11}^{\text{T}} \right) \boldsymbol{p}_{1_K}^\star - z^\star \boldsymbol{p}_{K_{\text{Bound}}}^{\text{T}} \left(\boldsymbol{A}_{21} + \boldsymbol{A}_{12}^{\text{T}} \right) \right] + \vartheta^\star \mathbf{1}_1 = \mathbf{0}, \\ &0 = \lambda^\star \left(\boldsymbol{b}_e^T \left[\boldsymbol{p}_{1_K}^{\star\text{T}}, \boldsymbol{p}_{K_{\text{Bound}}} \right]^{\text{T}} - z^\star \left[\boldsymbol{p}_{1_K}^{\star\text{T}}, \boldsymbol{p}_{K_{\text{Bound}}}^{\text{T}} \right] \boldsymbol{A}_{e_K} \left[\boldsymbol{p}_{1_K}^{\star\text{T}}, \boldsymbol{p}_{K_{\text{Bound}}}^{\text{T}} \right]^{\text{T}} \right), \\ &0 = \mathbf{1}_1^{\text{T}} \boldsymbol{p}_{1_K} + \mathbf{1}_2^{\text{T}} \boldsymbol{p}_{2_K} - P_{\text{obj}}, \\ &\mathbf{0} = \boldsymbol{p}_{2_K} - \boldsymbol{p}_{K_{\text{Bound}}}. \end{aligned} \tag{11.27}$$

The condition in (11.27) generates an optimal solution for the allocation of the remaining power to the internal points:

$$\boldsymbol{p}_{1_K}^\star = \frac{1}{z^\star} \left[\left(\boldsymbol{B}_{11} \boldsymbol{b}_1 + \boldsymbol{B}_{12} \boldsymbol{b}_2 \right) + \frac{\vartheta^\star}{\lambda^\star} \left(\boldsymbol{B}_{11} \mathbf{1}_1 + \boldsymbol{B}_{12} \mathbf{1}_2 \right) \right], \tag{11.28}$$

where the conditions in (11.27) are used to obtain

$$z^\star = \frac{\left(\mathbf{1}_1^{\text{T}} \boldsymbol{B}_{11} \boldsymbol{b}_1 + \mathbf{1}_1^{\text{T}} \boldsymbol{B}_{12} \boldsymbol{b}_2 \right)}{P_{\text{obj}} - \boldsymbol{p}_{K_{\text{Bound}}}^{\text{T}} \mathbf{1}_2} + \frac{\left(\mathbf{1}_1^{\text{T}} \boldsymbol{B}_{11} \mathbf{1}_1 + \mathbf{1}_1^{\text{T}} \boldsymbol{B}_{12} \mathbf{1}_2 \right)}{P_{\text{obj}} - \boldsymbol{p}_{K_{\text{Bound}}}^{\text{T}} \mathbf{1}_2} \left(\frac{\vartheta^\star}{\lambda^\star} \right), \tag{11.29}$$

and to generate a solution for the ratio value, $\frac{\vartheta^\star}{\lambda^\star}$, by solving the following quadratic equation:

$$a_{\lambda_{\text{MSEo}}}\left(\frac{\vartheta^\star}{\lambda^\star}\right)^2 + b_{\lambda_{\text{MSEo}}}\left(\frac{\vartheta^\star}{\lambda^\star}\right) + c_{\lambda_{\text{MSEo}}} = 0. \tag{11.30}$$

The expressions for the coefficients $a_{\lambda_{\text{MSEo}}}$, $b_{\lambda_{\text{MSEo}}}$, and $c_{\lambda_{\text{MSEo}}}$ are provided in Appendix 11.A. Evaluating all possible solutions for $\boldsymbol{p}_{K_{\text{Bound}}} \in \mathcal{Y}_K$ for all $\boldsymbol{p}_{2_K} \in \mathcal{B}_K$ results in a set of feasible solutions $\boldsymbol{p}^\star$. The ones that follow the master problem constraints $p_{m_{\min}} \leq p_m^\star \leq p_{m_{\max}}, \forall m = 1, \dots, M$, are added to the feasible set $\mathcal{F}_K'$. Finally an optimal set is found for subproblem K by evaluating

$$\mathcal{F}_K := \left\{ \boldsymbol{p}_K^\star : \arg\min_{\boldsymbol{p}_K^\star \in \mathcal{F}_K'} \left(\frac{\boldsymbol{b}_{e_K}^{\mathrm{T}} \boldsymbol{p}_K^\star}{\boldsymbol{p}_K^{\star\mathrm{T}} \boldsymbol{A}_{e_K} \boldsymbol{p}_K^\star} \right) \right\}. \tag{11.31}$$

The search for potential feasible solutions is improved with the introduction of a modified upper bound $\boldsymbol{p}_{\text{mod}}(K) \prec \boldsymbol{p}_{\max}$, used in the evaluation of $\boldsymbol{p}_K^\star$, $z^\star$, and $\frac{\vartheta^\star}{\lambda^\star}$ in (11.28), (11.29), and (11.30), respectively, and implementation of the following iterative algorithm:

Set initial values:

$$\boldsymbol{p}_{\text{mod}}(K)_0 = 2p_{\boldsymbol{u}}\mathbf{1} \quad (\text{where } 2p_{\boldsymbol{u}}\mathbf{1} < \boldsymbol{p}_{\max})$$

$$\Delta pt = 0.1 p_{\boldsymbol{u}}$$

If

$$\operatorname{tr}\left(\boldsymbol{C}_{x,y}\left(\boldsymbol{p}_K^\star\right)\right)|_{\boldsymbol{p}_{\text{mod}}(K)_i} < \operatorname{tr}\left(\boldsymbol{C}_{x,y}\left(\boldsymbol{p}_K^\star\right)\right)|_{\boldsymbol{p}_{\text{mod}}(K)_{i-1}}$$

then

$$\boldsymbol{p}_{\text{mod}}(K)_i = \boldsymbol{p}_{\text{mod}}(K)_{i-1} + \Delta pt\mathbf{1}$$

otherwise

$$\boldsymbol{p}_{\text{mod}}(K)_i = \boldsymbol{p}_{\text{mod}}(K)_{i-1} - \Delta pt\mathbf{1}.$$

At each iteration, $\boldsymbol{p}_{\text{mod}}(K)$ is updated to follow the direction that minimizes the objective function $\operatorname{tr}\left(\boldsymbol{C}_{x,y}\left(\boldsymbol{p}_K^\star\right)\right)$. The resulting solutions that follow the constraints $p_{m_{\min}} \leq p_m^\star \leq p_{m_{\max}}, \forall m = 1, \dots, M$ are included in the local feasible set $\mathcal{F}$. A unified set of optimal points is created as $\mathcal{F}_{\text{All}} = \mathcal{F}_1 \cup \mathcal{F}_2 \cup \cdots \cup \mathcal{F}_{M+1}$. The set of optimal values for the master problem is then obtained by

$$\boldsymbol{p}_{\text{MSEo}} = \boldsymbol{p}^\star = \arg\min_{\boldsymbol{p} \in \mathcal{F}_{\text{All}}} \left(f_o = \frac{\boldsymbol{b}_e^{\mathrm{T}} \boldsymbol{p}}{\boldsymbol{p}^{\mathrm{T}} \boldsymbol{A}_e \boldsymbol{p}} \right). \tag{11.32}$$

The following process supports parallel processing where each subproblem may be calculated at a different receiver/transmitter or centrally, using parallel computing. The local processing center for the Kth subproblem needs to get the values of the matrix $\boldsymbol{A}_e$ and vectors $\boldsymbol{\alpha}$, $\boldsymbol{h}$, and $\boldsymbol{b}_e$. Information of the local subset of optimal points $\mathcal{F}_K$ is transferred to the central processor. The universal set $\mathcal{F}_{\text{All}}$ is used in the solution of the master problem, as expressed by (11.32).

Decomposition Algorithm Implementation The decomposition method applied to the nonconvex optimization problems supports either centralized or distributed processing. In the latter, the subproblems may be distributed among the transmitters and/or receivers.

The information that needs to be exchanged between the central fusion center and the radars are the values of the matrix $\boldsymbol{A}_e$, vectors $\boldsymbol{b}_e$, $\boldsymbol{p}_{\max}$, and $\boldsymbol{p}_{\min}$. The calculation of the optimal solution is performed $L \times \binom{M}{K}$ times at each of the distributed nodes, where L is the number of iterations using adjustable boundaries value. The resulting local minimum at each radar $\boldsymbol{p}_K^\star = \arg\min_{\boldsymbol{p}_K^\star \in \mathcal{F}_K} \left(\frac{\boldsymbol{b}_{e_K}^{\mathrm{T}} \boldsymbol{p}_K^\star}{\boldsymbol{p}_K^{\star\mathrm{T}} \boldsymbol{A}_{e_K} \boldsymbol{p}_K^\star} \right)$ is then sent to the central fusion center that finds $\boldsymbol{p}_{\text{MSEo}} = \boldsymbol{p}^\star = \arg\min_{\boldsymbol{p}_K^\star} \left(\frac{\boldsymbol{b}_e^{\mathrm{T}} \boldsymbol{p}_K^\star}{\boldsymbol{p}_K^{\star\mathrm{T}} \boldsymbol{A}_e \boldsymbol{p}_K^\star} \right)$.

11.3.1.3 Insights for the MLE Optimization Algorithm

Hereafter, insights into the optimization algorithms are provided:

- The denominator in (11.22) is a normalization factor and the numerator is a weighted sum of all *leveled* values of the vector $\boldsymbol{b}_e$.
- The weights, $\boldsymbol{B}[m, m']$, reflect the overall propagation path conditions.
- The use of the term *channel* in this context refers to the overall performance loss over a given propagation path between transmit radar m, the target, and receive radar n, defined as channel (m, n). The channel conditions incorporate the effects of propagation path fading, target reflectivity, and geometric position of the transmit–receive pair with respect to the target.
- The ratio $\frac{\vartheta^\star}{\lambda^\star}$ defines a *Lagrangian leveling* factor.
- For a given power budget, the Lagrangian leveling factor is set such that the shifted vector $\left(\boldsymbol{b}_e[m'] + \frac{\vartheta^\star}{\lambda^\star}\right)$ reallocates the power budget when multiplied by the weights vectors $\boldsymbol{B}[m, m']$ (see (11.28)).
- The adaptation mechanism supported by the Lagrangian leveling factors shifts the origin of vector $\boldsymbol{b}_e$ elements such that transmitters with better channel conditions, corresponding to higher weights in the relevant row in the matrix $\boldsymbol{B}$ and vector $\boldsymbol{b}_e$, are assigned higher transmit power compared with transmitters with weaker propagation paths.
- The decomposition method applied to the nonconvex optimization problems supports either central or distributed processing. In the latter, the subproblems may be distributed among the transmitters and/or receivers. The information that needs to be exchanged between the central fusion center and the radars

are the values of the matrix $\boldsymbol{A}_e$, vectors $\boldsymbol{b}_e$, $\boldsymbol{p}_{\max}$, and $\boldsymbol{p}_{\min}$. The resulting local minimum at each radar is then sent to all others for final minimization of the global objective function.

As the resulting power allocation strategy favors transmitters with better channel conditions, a natural approach would be to narrow the operation of the system to a minimal subset of transmit and receive radars which deliver the required performance. This idea is developed in [36], where a mathematical framework and techniques for selecting the best radar subset are developed.

11.3.2 Minimize Total Power

For a predetermined target localization MSE threshold, denoted by $\eta_{\max}$, the total radiated power, $\sum_{m=1}^{M} p_m$, may be minimized through appropriate allocation of power among the transmit radars, such that the MSE performance goal is achieved with minimal power requirements. This can be formulated into an optimization problem of the form

$$\begin{aligned} \min_{\boldsymbol{p}} \quad & \mathbf{1}^{\mathrm{T}}\boldsymbol{p} \\ \text{s.t.} \quad & p_m \leq p_{m_{\max}}, \qquad m = 1, \ldots, M, \\ & p_m \geq p_{m_{\min}}, \qquad m = 1, \ldots, M, \\ & \operatorname{tr}\left(\boldsymbol{C}_{x,y}\left(\tilde{\boldsymbol{u}}, \boldsymbol{p}\right)\right) = \eta_{\max}. \end{aligned} \tag{11.33}$$

The optimization problem can be rewritten as

$$\begin{aligned} \min_{\boldsymbol{p}} \quad & \mathbf{1}^{\mathrm{T}}\boldsymbol{p} \\ \text{s.t.} \quad & \boldsymbol{p} - \boldsymbol{p}_{\max} \preceq \mathbf{0}, \\ & \boldsymbol{p}_{\min} - \boldsymbol{p} \preceq \mathbf{0}, \\ & \eta_{\max}\boldsymbol{p}^{\mathrm{T}}\boldsymbol{A}_e\boldsymbol{p} - \boldsymbol{b}_e^{\mathrm{T}}\boldsymbol{p} = 0. \end{aligned} \tag{11.34}$$

The equality constraint will further reduce the transmit power, as any decrease in the power will result in a higher MSE [20]. Thus, for a given solution $\operatorname{tr}\left(\boldsymbol{C}_{x,y}\left(\tilde{\boldsymbol{u}}, \boldsymbol{p}\right)\right) < \eta_{\max}$ with a power allocation vector $\boldsymbol{p}$ there is a power allocation vector $(\boldsymbol{p} - \Delta\boldsymbol{p})$ for which $\operatorname{tr}\left(\boldsymbol{C}_{x,y}\left(\tilde{\boldsymbol{u}}, \boldsymbol{p}\right)\right) < \operatorname{tr}\left(\boldsymbol{C}_{x,y}\left(\tilde{\boldsymbol{u}}, (\boldsymbol{p} - \Delta\boldsymbol{p})\right)\right)$, such that we can find a vector $\operatorname{tr}\left(\boldsymbol{C}_{x,y}\left(\tilde{\boldsymbol{u}}, \boldsymbol{p}\right)\right) < \operatorname{tr}\left(\boldsymbol{C}_{x,y}\left(\tilde{\boldsymbol{u}}, \left(\boldsymbol{p} - \Delta\boldsymbol{p}_{\max}\right)\right)\right) = \eta_{\max}$.

The optimization problem in (11.34) is nonconvex due to the equality constraint [21]. In what follows, both relaxation and decomposition methods are applied to solve the optimization problem defined in (11.34).

11.3.2.1 Relaxation Methods Applied to Total Power Optimization

To solve the optimization problem in (11.34) using relaxation methods, the equality constraint is modified. Since $p_m \neq 0$, $\forall m = 1, 2, \ldots, M$, the constraint is replaced by $\eta_{\max}\boldsymbol{A}_e\boldsymbol{p} - \boldsymbol{b}_e = \boldsymbol{0}$. Minimizing the power vector results in maximizing the MSE, bringing it as close as possible to the threshold value $\eta_{\max}$ [20], i.e., the equality constraint may be replaced by an inequality constraint of the form $\boldsymbol{p}^{\mathrm{T}}\left(\boldsymbol{b}_e - \eta_{\max}\boldsymbol{A}_e\boldsymbol{p}\right) \leq 0$. The following relaxed convex optimization problem can be composed:

$$\begin{aligned} \min_{\boldsymbol{p}} \;\; & \boldsymbol{1}^{\mathrm{T}}\boldsymbol{p} \\ \text{s.t.} \;\; & p_m - p_{m_{\max}} \leq 0, \qquad m = 1, \dots, M, \\ & p_{m_{\min}} - p_m \leq 0, \qquad m = 1, \dots, M, \\ & \boldsymbol{b}_e - \eta_{\max}\boldsymbol{A}_e\boldsymbol{p} \preceq \boldsymbol{0}, \end{aligned} \tag{11.35}$$

and solved using available convex optimization tools, such as CVX [22]. The optimal solution, $\boldsymbol{p}^{\star}_{\mathrm{MLE}}$, is then used as the starting point for a local optimization. An iterative optimization algorithm is applied to the original nonconvex problem (11.34), where the initial values are set based on $\boldsymbol{p}^{\star}_{\mathrm{MLE}}$ as follows:

$$\begin{aligned} \boldsymbol{p}_0 &= \boldsymbol{p}^{\star}_{\mathrm{MLE}}, \\ \sigma^2_{x,y}\left(\boldsymbol{p}_0\right) &= \frac{\boldsymbol{b}_e^{\mathrm{T}}\boldsymbol{p}_0}{\boldsymbol{p}_0^{\mathrm{T}}\boldsymbol{A}_e\boldsymbol{p}_0}, \end{aligned} \tag{11.36}$$

and setting iteration counter $k = 0$. The search algorithm is implemented as follows:

repeat:

$$\left[\begin{aligned} & k = k + 1 \\ & \boldsymbol{p}_k = \arg\min\left\{\sigma^2_{x,y}\left(\boldsymbol{p}_k\right) - \eta_{\max}\right\} \\ & \qquad \text{s.t.} \qquad p_{m_{\min}} \leq p_{m_k}; p_{m_{\max}} \geq p_{m_k} \\ & \triangle p_k = \triangle p_{k-1}\frac{\boldsymbol{1}^{\mathrm{T}}\boldsymbol{p}_k}{\boldsymbol{1}^{\mathrm{T}}\boldsymbol{p}_{k-1}} \end{aligned}\right.$$

while:

$$\eta_{\max} - \sigma^2_{x,y}\left(\boldsymbol{p}_k\right) \geq \epsilon$$

set:

$$\boldsymbol{p}_{_{\mathrm{PWo}}} = \boldsymbol{p}_k \text{ and } \sigma^2_{_{\mathrm{PWo}}} = \sigma^2_{x,y}\left(\boldsymbol{p}_{_{\mathrm{PWo}}}\right).$$

Note that at each iteration step one element of the vector $\boldsymbol{p}$ is decreased by a step value Δp. The one that minimizes the gap to the required threshold is selected as $\boldsymbol{p}_{_{\mathrm{PWo}}}$. The more iterations, the smaller increments of power, Δp, are used, as it gets closer to the MSE threshold. This process is repeated until the change in the MSE from one iteration to the next is smaller than the stopping threshold, ϵ.

11.3.2.2 Domain Decomposition Applied to Total Power Optimization

In this optimization problem, as before, domain decomposition tools are evaluated to find a centralized and decentralized solutions.

Lagrangian Function For the power optimization problem given in (11.34), the Lagrangian function is

$$\begin{aligned}\mathcal{L}(\boldsymbol{p}, \lambda, \boldsymbol{\mu}, \boldsymbol{\nu}) = \mathbf{1}^{\mathrm{T}}\boldsymbol{p} + \lambda\left(\eta_{\max}\boldsymbol{p}^{\mathrm{T}}\boldsymbol{A}_e\boldsymbol{p} - \boldsymbol{b}_e^{\mathrm{T}}\boldsymbol{p}\right)\\ +\boldsymbol{\mu}^{\mathrm{T}}\left(\boldsymbol{p} - \boldsymbol{p}_{\max}\right) + \boldsymbol{\nu}^{\mathrm{T}}\left(\boldsymbol{p}_{\min} - \boldsymbol{p}\right),\end{aligned} \tag{11.37}$$

and the appropriate KKT conditions [26] are

$$\begin{aligned}\nabla_{\boldsymbol{p}}\mathcal{L} = \mathbf{1} + \lambda^{\star}\left[\eta_{\max}\left(\boldsymbol{A}_e + \boldsymbol{A}_e^{\mathrm{T}}\right)\boldsymbol{p}^{\star} - \boldsymbol{b}_e\right] + \boldsymbol{\mu}^{\star} - \boldsymbol{\nu}^{\star} = \mathbf{0},\\ \eta_{\max}\boldsymbol{p}^{\star\mathrm{T}}\boldsymbol{A}_e\boldsymbol{p}^{\star} - \boldsymbol{b}_e^{\mathrm{T}}\boldsymbol{p}^{\star} = 0,\\ \mu_m^{\star}\left(p_m^{\star} - p_{m_{\max}}\right) = 0,\\ \nu_m^{\star}\left(p_{m_{\min}} - p_m^{\star}\right) = 0,\\ \mu_m^{\star} \geq 0,\ \nu_m^{\star} \geq 0.\end{aligned} \tag{11.38}$$

As the LICQs hold, a solution to (11.34) may be obtained by choosing the first and second inequality constraints in (11.38) to be inactive and therefore, $\mu_m^{\star} = \nu_m^{\star} = 0, \forall m = 1, \dots, M$. From the remaining KKT conditions in (11.38), the optimal power allocation vector, $\boldsymbol{p}^{\star}$, may be calculated as

$$\boldsymbol{p}^{\star} = \frac{\boldsymbol{B}}{\eta_{\max}}\left(\boldsymbol{b}_e - \frac{1}{\lambda^{\star}}\mathbf{1}\right), \tag{11.39}$$

and the optimal value of the Lagrange multiplier, $\lambda^{\star}$, may be derived, resulting in

$$\lambda^{\star} = \sqrt{\frac{\mathbf{1}^{\mathrm{T}}\boldsymbol{B}\boldsymbol{A}_e\boldsymbol{B}\mathbf{1}}{\boldsymbol{b}_e^{\mathrm{T}}\left(\boldsymbol{B}\boldsymbol{A}_e\boldsymbol{B} - \boldsymbol{B}\right)\boldsymbol{b}_e}}. \tag{11.40}$$

As before, the solution in (11.39) is obtained while ignoring the upper and lower bounds on the values of $\boldsymbol{p}$. Thus the resulting solution $\boldsymbol{p}_{\mathrm{PWo}} = \boldsymbol{p}^{\star}$ may be outside these boundary constraints. As the objective function is to minimize the total transmit power, a reasonable place to look for feasible solutions is on the boundaries. In the following, domain decomposition methods are used to generate a set of $M + 1$ optimization problems appropriate for this optimization problem.

Domain Decomposition The same process as in Section 11.3.1.2 is applied here and the master optimization problem in (11.34) is decomposed to $M + 1$ optimization problems of the form

$$\begin{aligned}\min_{\boldsymbol{p}_K \in \mathcal{U}_K}\ & \mathbf{1}^{\mathrm{T}}\boldsymbol{p}_K\\ \text{s.t.}\ & \eta_{\max}\boldsymbol{p}_K^{\mathrm{T}}\boldsymbol{A}_{e_K}\boldsymbol{p} - \boldsymbol{b}_{e_K}^{\mathrm{T}}\boldsymbol{p}_K = 0,\end{aligned} \tag{11.41}$$

and the appropriate KKT conditions for this subproblem are

$$\mathbf{1} + \lambda^{\star} \left(\eta_{\max} \boldsymbol{A}_{ee_K} \left[\boldsymbol{p}_K^{\star \mathrm{T}}, \boldsymbol{p}_{K_{\mathrm{Bound}}}^{\mathrm{T}} \right]^{\mathrm{T}} - \boldsymbol{b}_{e_K} \right) = 0, \tag{11.42}$$

and

$$\eta_{\max} \left[\boldsymbol{p}_K^{\star \mathrm{T}}, \boldsymbol{p}_{K_{\mathrm{Bound}}}^{\mathrm{T}} \right] \boldsymbol{A}_{e_K} \left[\boldsymbol{p}_K^{\star \mathrm{T}}, \boldsymbol{p}_{K_{\mathrm{Bound}}}^{\mathrm{T}} \right]^{\mathrm{T}} - \boldsymbol{b}_{e_K}^{\mathrm{T}} \left[\boldsymbol{p}_K^{\star \mathrm{T}}, \boldsymbol{p}_{K_{\mathrm{Bound}}} \right]^{\mathrm{T}} = 0. \tag{11.43}$$

Each problem performs domain boundary search by enforcing $\boldsymbol{p}_{K_{\mathrm{Bound}}}$ on the power allocation vector $\boldsymbol{p}_K$. The resulting optimal power allocation for the internal point vector, $\boldsymbol{p}_K^{\star}$, is of the form

$$\boldsymbol{p}_K^{\star} = \frac{1}{\eta_{\max}} \left[\left(\boldsymbol{B}_{11} \boldsymbol{b}_1 + \boldsymbol{B}_{12} \boldsymbol{b}_2 \right) - \frac{1}{\lambda^{\star}} \left(\boldsymbol{B}_{11} \mathbf{1}_1 + \boldsymbol{B}_{12} \mathbf{1}_2 \right) \right], \tag{11.44}$$

and the optimal Lagrange multiplier value, $\lambda^{\star}$, is obtained by solving the quadratic equation

$$a_{\lambda_{\mathrm{PWo}}} \left(\lambda^{\star} \right)^2 + b_{\lambda_{\mathrm{PWo}}} \lambda^{\star} + c_{\lambda_{\mathrm{PWo}}} = 0, \tag{11.45}$$

where the expressions for the coefficients $a_{\lambda_{\mathrm{PWo}}}$, $b_{\lambda_{\mathrm{PWo}}}$, and $c_{\lambda_{\mathrm{PWo}}}$ are provided in Appendix 11.B.

All possible solutions $\boldsymbol{p}_K^{\star}$ and $\lambda^{\star}$ in (11.44) and (11.45) are evaluated for $\boldsymbol{p}_{K_{\mathrm{Bound}}} \in \mathcal{Y}_K$, and all $\boldsymbol{p}_K \in \mathcal{B}_K$, resulting with a set of feasible solutions $\boldsymbol{p}_K^{\star}$. The ones that follow the constraint $p_{m_{\min}} \leq p_m^{\star} \leq p_{m_{\max}}$, $\forall m = 1, \ldots, M$, are added to a local feasible set $\mathcal{F}_K^{'}$. Finally, an optimal set is found for subproblem K through the evaluation of

$$\mathcal{F}_K := \left\{ \boldsymbol{p}_K^{\star} : \arg \min_{\boldsymbol{p}_K^{\star} \in \mathcal{F}_K^{'}} \left(\mathbf{1}^{\mathrm{T}} \boldsymbol{p}_K^{\star} \right) \right\}. \tag{11.46}$$

The search for potential feasible solutions is performed by introducing modification to the assumed boundary conditions $\boldsymbol{p}_{\max}$ where a modified upper bound $\boldsymbol{p}_{\mathrm{mod}}(K) \prec \boldsymbol{p}_{\max}$ is set and used in the evaluation of $\boldsymbol{p}_K^{\star}$ and $\lambda^{\star}$ in (11.44) and (11.45). The search algorithm follows the iterative algorithm proposed in Section 11.3.1.2. At each iteration, the steps are executed and $\boldsymbol{p}_{\mathrm{mod}}(K)$ is updated to follow the direction that minimizes the objective function $\mathbf{1}^{\mathrm{T}} \boldsymbol{p}_K^{\star}$. The resulting solutions that follow the constraints $p_{m_{\min}} \leq p_m^{\star} \leq p_{m_{\max}}$, $\forall m = 1, \ldots, M$ are included in the local feasible set $\mathcal{F}$.

A universal set of optimal points is created as $\mathcal{F}_{\mathrm{All}} = \mathcal{F}_1 \cup \mathcal{F}_2 \cup \cdots \cup \mathcal{F}_{M+1}$. The set of optimal values for the master problem is then obtained as

$$\boldsymbol{p}_{\mathrm{PWo}} = \boldsymbol{p}^{\star} = \arg \min_{\boldsymbol{p} \in \mathcal{F}_{\mathrm{All}}} \left(f_o = \mathbf{1}^{\mathrm{T}} \boldsymbol{p} \right). \tag{11.47}$$

The decomposition method applied here supports either central or distributed processing, as discussed in Section 11.3.1.2.

11.3.2.3 Insights for the Power Optimization Algorithm

Hereafter, insights into the power optimization algorithms are provided:

- Applying domain decomposition methods and boundary search reveals a Lagrangian leveling mechanism in the form of $\frac{1}{\lambda^\star}\mathbf{1}$ (see (11.39)).
- The Lagrange leveling factors, $\frac{1}{\lambda^\star}$, defines an adaptation of vector $\boldsymbol{b}_e$ values, shifting the origin of its elements such that transmitters with better channel conditions, corresponding to higher weights in the relevant row in the matrix $\boldsymbol{B}$ and vector $\boldsymbol{b}_e$, are assigned higher transmit power compared with transmitters with weaker propagation paths.
- In minimizing the total transmit power, the leveling factor $\lambda^\star$ shifts the origin of the elements of $\boldsymbol{b}_e$, such that some of them become negative. Thus, the weighted sum $\sum_{m'=1}^{M} \boldsymbol{B}\left[m, m'\right]\left(\boldsymbol{b}_e\left[m'\right] - \frac{1}{\lambda^\star}\right)$, in (11.44), is minimized for all transmit radars by allocating more power to stronger propagation paths and minimizing allocation to weaker ones.

A deeper comprehension of the power allocation optimization algorithms, developed in Section 11.3, can be gained from numerical analysis. Some examples are provided in Section 11.4.

11.4 Analysis and Discussion

Multiple propagation paths between the transmit and receive antennas in multiple radar systems exhibit different error characteristics. The expressions in (11.8), (11.9), along with (11.10), (11.11), and (11.12), reveal that localization MSE is reliant on the specific path loss, target reflectivity, effective bandwidth, and transmit power. Power management algorithms, developed in Section 11.3, dilute the error variation through adequate distribution of the transmit power. Numerical analyses offer valuable insight into the factors impacting power allocation. To demonstrate the effect of radar spread and channel conditions on power allocation vector, a 5×5 multiple radar system ($M = 5$ and $N = 5$) is used with three different angular spreads with respect to the target. Radar layouts with different ranges with respect to the target position are chosen. Figure 11.6 demonstrates the topology for the different cases. The target RCS model, used for all cases, is as follows:

$|\boldsymbol{h}_1|^2$ (Tx 1 $\rightarrow$ all receive antennas) $= [0.04, 0.81, 0.72, 0.85, 0.56]^{\mathrm{T}}$,
$|\boldsymbol{h}_2|^2$ (Tx 2 $\rightarrow$ all receive antennas) $= [0.09, 0.9, 0.56, 0.56, 0.36]^{\mathrm{T}}$,
$|\boldsymbol{h}_3|^2$ (Tx 3 $\rightarrow$ all receive antennas) $= [0.01, 0.96, 0.64, 0.25, 0.2]^{\mathrm{T}}$,
$|\boldsymbol{h}_4|^2$ (Tx 4 $\rightarrow$ all receive antennas) $= [0.48, 1, 0.67, 0.64, 0.3]^{\mathrm{T}}$, and
$|\boldsymbol{h}_5|^2$ (Tx 5 $\rightarrow$ all receive antennas) $= [0.15, 0.85, 0.62, 0.5, 0.09]^{\mathrm{T}}$.

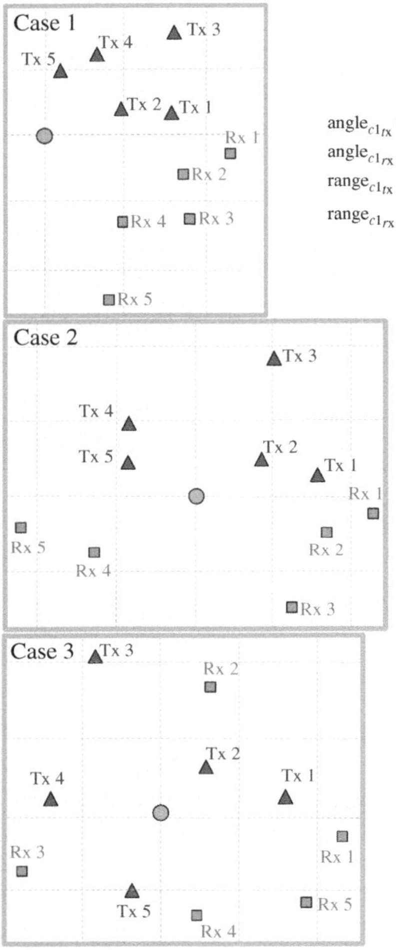

$$\text{angle}_{c1_{tx}} = [10°, 20°, 40°, 60°, 80°]$$
$$\text{angle}_{c1_{rx}} = [355°, 345°, 330°, 310°, 290°]$$
$$\text{range}_{c1_{tx}} = [1, 0.6, 1.35, 0.85, 0.65] \times 10^6$$
$$\text{range}_{c1_{rx}} = [1.5, 1.1, 1.3, 1, 1.4] \times 10^6$$

$$\text{angle}_{c2_{tx}} = [10°, 30°, 60°, 130°, 150°]$$
$$\text{angle}_{c2_{rx}} = [355°, 345°, 310°, 210°, 190°]$$
$$\text{range}_{c2_{tx}} = [1, 0.6, 1.35, 0.85, 0.65] \times 10^6$$
$$\text{range}_{c2_{rx}} = [1.5, 1.1, 1.3, 1, 1.4] \times 10^6$$

$$\text{angle}_{c3_{tx}} = [10°, 50°, 110°, 170°, 250°]$$
$$\text{angle}_{c3_{rx}} = [355°, 70°, 200°, 290°, 330°]$$
$$\text{range}_{c3_{tx}} = [1, 0.6, 1.35, 0.85, 0.65] \times 10^6$$
$$\text{range}_{c3_{rx}} = [1.5, 1.1, 1.3, 1, 1.4] \times 10^6$$

Figure 11.6 Antenna layout for Cases 1, 2, and 3.

Notice that these allocations mimic a scenario with different reflectivity conditions on all transmission paths, where no single transmitter has a significant advantage when compared with the others. With that said, some of the antennas have slight advantages, such as Tx 4 which has stronger channels than the others. It is followed by Tx 2 and Tx 1.

11.4.1 Power Allocation to Minimize MSE

For a given total transmit power budget P_{obj}, a uniform allocation of this power will result in an allocation of $\boldsymbol{p}_u = \left(\frac{P_{\text{obj}}}{5}\right)\mathbf{1}_{5\times1}$. The MSE achievable with this budget is $\sigma^2_{x,y}(\boldsymbol{p}_u) = \frac{5}{P_{\text{obj}}}\frac{\boldsymbol{b}_e^{\mathrm{T}}\mathbf{1}}{\mathbf{1}^{\mathrm{T}}\boldsymbol{A}_e\mathbf{1}}$. Transmit and receive antenna locations are given in terms of range and angular position with respect to the target, as noted in Figure 11.6. Power constraints are set to be equal, where $\boldsymbol{p}_{\min} = 1\times\mathbf{1}$, $\boldsymbol{p}_{\max} = 150\times\mathbf{1}$, for all $m = 1,\dots,M$ (all power units are normalized to a quantity P_0).

In Figures 11.7–11.9, numerical analysis is presented for the radar spreads given in Figure 11.6. For each case, the MSE associated with uniform power allocation, $\sigma^2_{x,y}(\boldsymbol{p}_u)$, is calculated and compared with that resulting from optimal power budget allocation, $\sigma^2_{x,y}\left(\boldsymbol{p}_{\text{MSEo}}\right)$. This power budget is allocated using the algorithms developed in Section 11.3.1.

Simulation results for Case 1 are given in Figure 11.7. They show that the total power budget of $P_{\text{obj}} = 64$ is split between only two of the five transmit radars $p_{2_{\text{MSEo}}} = p_{5_{\text{MSEo}}} = 30.5$. The remaining radars are allocated the minimum power

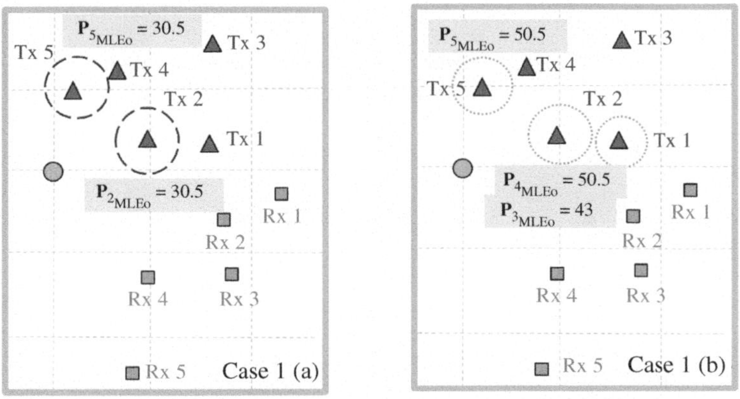

$P_{\text{obj}} = 64$
$\boldsymbol{p}_u = 12.6\times[1,1,1,1,1]$
$\boldsymbol{p}_{\text{MSEo}} = [1, 30.5, 1, 1, 30.5]$
$\sigma^2_{x,y}(\boldsymbol{p}_u) = 8.3855$
$\sigma^2_{x,y}(\boldsymbol{p}_{\text{MSEo}}) = 5.7375$
~32% improvement

$P_{\text{obj}} = 146$
$\boldsymbol{p}_u = 29.2\times[1,1,1,1,1]$
$\boldsymbol{p}_{\text{MSEo}} = [43, 50.5, 1,1, 50.5]$
$\sigma^2_{x,y}(\boldsymbol{p}_u) = 3.672$
$\sigma^2_{x,y}(\boldsymbol{p}_{\text{MSEo}}) = 2.689$
~27% improvement

Figure 11.7 Optimal power allocation for Case 1: (a) for a total power budget of 64 (b) for a total power allocation of 146.

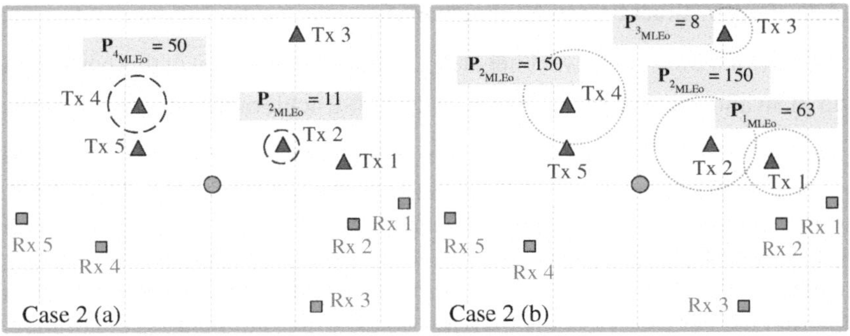

$P_{obj} = 64$
$\boldsymbol{p}_u = 12.6 \times [1,1,1,1,1]$
$\boldsymbol{p}_{MSEo} = [1, 11, 1, 50, 1]$
$\sigma^2_{x,y}(\boldsymbol{p}_u) = 17.238$
$\sigma^2_{x,y}(\boldsymbol{p}_{MSEo}) = 12.525$
~27% improvement

$P_{obj} = 372$
$\boldsymbol{p}_u = 74.4 \times [1,1,1,1,1]$
$\boldsymbol{p}_{MSEo} = [63, 150, 8, 150, 1]$
$\sigma^2_{x,y}(\boldsymbol{p}_u) = 2.986$
$\sigma^2_{x,y}(\boldsymbol{p}_{MSEo}) = 2.518$
~16% improvement

Figure 11.8 Optimal power allocation for Case 2: (a) for a total power budget of 64 (b) for a total power allocation of 372.

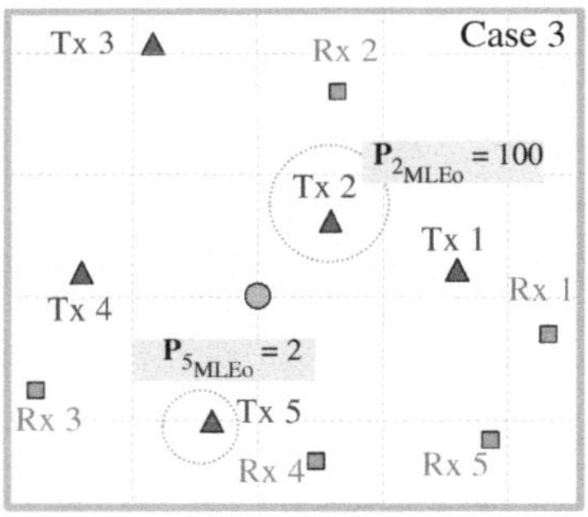

$P_{obj} = 105$
$\boldsymbol{p}_u = 35 \times [1,1,1,1,1]$
$\boldsymbol{p}_{MSEo} = [1, 100, 1, 1, 2]$
$\sigma^2_{x,y}(\boldsymbol{p}_u) = 1.9$
$\sigma^2_{x,y}(\boldsymbol{p}_{MSEo}) = 1.661$

Figure 11.9 Optimal power allocation for Case 2: for a total power budget of 91.

$p_{1_{MSEo}} = p_{3_{MSEo}} = p_{4_{MSEo}} = 1$. The power was allocated to Tx 2 and Tx 5 due to their channel conditions and proximity to the target. When the total power budget is increased from $P_{obj} = 64$ to $P_{obj} = 146$, Tx 1 is added to the first two, representing a tradeoff between channel losses, distance to target, and spatial spread of transmit antennas. In this case, $p_{2_{MSEo}} = p_{5_{MSEo}} = 50.5$ and $p_{1_{MSEo}} = 43$. In both cases, nonuniform power allocation, resulting from the proposed optimization algorithms, offers lower MSE, i.e., $\sigma^2_{x,y}(\boldsymbol{p}_{\text{MSEo}}) \leq \sigma^2_{x,y}(\boldsymbol{p}_u)$. An improvement between 32% and 27% is obtained by allocating the power to the best 2 to 3 radars when compared with equal power allocation among all transmitters.

In Case 2, the same power budget, $P_{\text{obj}} = 64$, is allocated for a different antenna spread. The optimal power allocation results in most power allocated to two antennas $p_{2_{\text{MSEo}}} = 11$ and $p_{4_{\text{MSEo}}} = 50$. The remaining three are allocated minimum power $p_{1_{\text{MSEo}}} = p_{3_{\text{MSEo}}} = p_{5_{\text{MSEo}}} = 1$. When compared with Case 1, we can observe that Tx 5 is replaced here with Tx 4 which has stronger paths to Rx 4 and Rx 5. When power is increased in this case to $P_{\text{obj}} = 372$, the optimal allocation maximizes the power for Tx 2 and Tx 4 and then allocates the majority of the remaining power to Tx 1. As before, this choice extends the spatial diversity of the active transmitters array.

For Case 3, illustrated in Figure 11.9, transmit and receive antennas have a better spatial distribution with respect to the target. This is evident in the higher accuracy it can deliver, where $\sigma^2_{x,y}(\boldsymbol{p}_{\text{MSEo}}) = 1.661$ is lower than any of the previous cases, when even higher total budget was available. This case delivers a moderate MSE reduction of 12.5% when compared with the uniform allocation option. The different MSEs achievable for Case 1, Case 2, and Case 3 are related directly to the location of transmit and receive antennas with respect to the target (as demonstrated in [13]). In Case 3, the advantageous topology results in an interesting allocation, where most power is allocated to a single antenna, in this case, Tx 2 where $p_{2_{\text{MSEo}}} = 100$, $p_{5_{\text{MSEo}}} = 2$, and the remaining three are allocated minimum power $p_{1_{\text{MSEo}}} = p_{3_{\text{MSEo}}} = p_{4_{\text{MSEo}}} = 1$.

In general, power is allocated to a smaller number of antennas while preforming better than when all transmitter operate with equal power. The radars that radiate higher power are those in preferable positions with respect to the receivers and thus generate better multilateration than with others.

11.4.2 Power Allocation to Minimize Total Power

Here, the given MSE threshold is set to $\eta_{\max} = 3\,\text{m}^2$. Power constraints are set to be equal, where $p_{m_{\min}} = 1$ and $p_{m_{\max}} = 100$, for all $m = 1, \dots, M$. The locations for all antennas are the same as in Section 11.4.1 (depicted in Figure 11.6). For this threshold, a uniform power allocation is determined by calculating $P_u\left(\eta_{\max}\right) = \frac{1}{3}\frac{\boldsymbol{b}_e^{\mathrm{T}}\mathbf{1}}{\mathbf{1}^{\mathrm{T}}\boldsymbol{A}_e\mathbf{1}}$. The power allocation vector would be $\boldsymbol{p}_u\left(\eta_{\max}\right) = P_u\left(\eta_{\max}\right)\mathbf{1}_{5\times1}$, with a total transmit power of $P_{\text{total\_u}}\left(\eta_{\max}\right) = 5 \times P_u$. The value of the total transmit power that would have been allocated given a choice of uniform power allocation is calculated, denoted by $P_{\text{total\_u}}\left(\eta_{\max}\right)$. An optimal power allocation vector, $\boldsymbol{p}_{\text{PWo}}$,

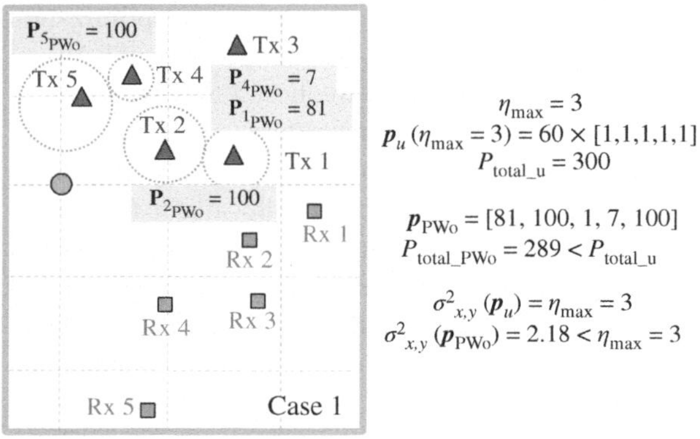

Figure 11.10 Minimizing power, given η_{max} = 3: Antenna layout for Case 1.

is obtained through the use of the domain decomposition algorithm proposed in Section 11.3.2. The resulting minimal total transmit power is $P_{\text{total\_PWo}} = \boldsymbol{p}_{\text{PWo}} \mathbf{1}_{5\times1}$.

In Figure 11.10, numerical analysis is presented for radar spreads in Case 1. The MSE treshhold is set to $\eta_{max} = 3\,\text{m}^2$. The resulting uniform power allocation is $p_{\text{total\_u}}\left(\eta_{max}\right) = 300$. Given this threshold, the optimization algorithm proposed in Section 11.3.2 results in a transmit power vector of $\boldsymbol{p}_{\text{PWo}} = [81, 100, 1, 7, 100]^{\text{T}}$, where $p_{\text{total\_PWo}} = 289$. For this allocation vector, the corresponding MSE is $\sigma^2_{x,y}(\boldsymbol{p}_{\text{PWo}}) = 2.18\,\text{m}^2$, with most of the power allocated to the three transmitters, Tx 1, Tx 2, and Tx 5. This case does not offer significant total power savings when compared with uniform power allocation (about 4% savings), yet it offers significant accuracy improvement (about 27% MSE reduction).

For the same η_{max} threshold, Cases 2 and 3 are evaluated and presented in Figure 11.11. For the layout in Case 2, $P_{\text{total\_}u}\left(\eta_{max}\right) = 322$. Applying the optimization algorithm here results in a power allocation vector of $\boldsymbol{p}_{\text{PWo}} = [1, 60, 1, 100, 1]^{\text{T}}$, where $P_{\text{total\_PWo}} = 163$. For this allocation vector, the corresponding MSE is $\sigma^2_{x,y}(\boldsymbol{p}_{\text{PWo}}) = 3\,\text{m}^2$ with most of the power allocated to the two transmitters 2 and 4. This case does offer significant total power savings when compared with uniform power allocation (about 50% savings) for the same accuracy.

Considering the spread in Case 3, the corresponding uniform allocation will result with $P_{\text{total\_u}}\left(\eta_{max}\right) = 75$. Applying the optimization algorithm here results in a power allocation vector of $\boldsymbol{p}_{\text{PWo}} = [1, 41, 1, 31, 1]^{\text{T}}$, where $P_{\text{total\_PWo}} = 75$. Interestingly, there is no saving in total power in this case, but the same total budget results in a significantly lower MSE $\sigma^2_{x,y}(\boldsymbol{p}_{\text{PWo}}) = 1.78\,\text{m}^2$ offering MSE reduction by a factor of 40%.

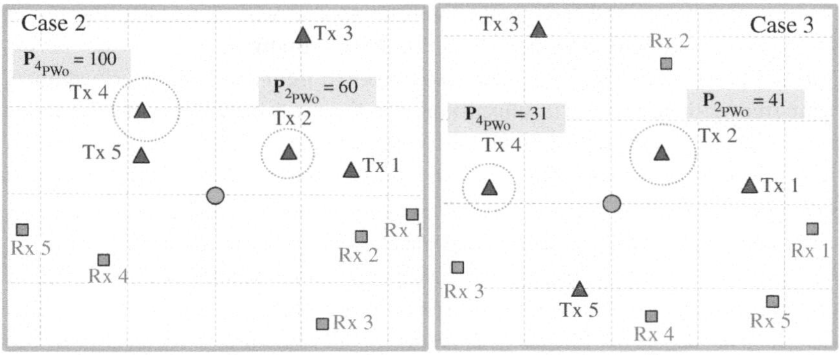

$$\eta_{\max} = 3$$
$$\boldsymbol{p}_u(\eta_{\max} = 3) = 64.4 \times [1,1,1,1,1]$$
$$P_{\text{total\_u}} = 322$$

$$\boldsymbol{p}_{\text{PWo}} = [1, 60, 1, 100, 1]$$
$$P_{\text{total\_PWo}} = 163 < P_{\text{total\_u}}$$

$$\sigma^2_{x,y}(\boldsymbol{p}_u) = \eta_{\max} = 3$$
$$\sigma^2_{x,y}(\boldsymbol{p}_{\text{PWo}}) = 3 = \eta_{\max} = 3$$

$$\eta_{\max} = 3$$
$$\boldsymbol{p}_u(\eta_{\max} = 3) = 15 \times [1,1,1,1,1]$$
$$P_{\text{total\_u}} = 75$$

$$\boldsymbol{p}_{\text{PWo}} = [1, 41, 1, 31, 1]$$
$$P_{\text{total\_PWo}} = 75 = P_{\text{total\_u}}$$

$$\sigma^2_{x,y}(\boldsymbol{p}_u) = \eta_{\max} = 3$$
$$\sigma^2_{x,y}(\boldsymbol{p}_{\text{PWo}}) = 1.78 < \eta_{\max} = 3$$

Figure 11.11 Minimizing power, given $\eta_{\max} = 3$: Antenna layout for Cases 2 and 3.

11.4.3 Discussion

From the body of research on this topic in [20, 35, 37], the following observations can be made:

- When given a total power budget constraint, uniform power allocation will offer inferior accuracy performance when compared with optimal distribution of power, achieved through the proposed power allocation algorithms.
- The allocation of power is determined by factors such as the spatial spread of the radars with respect to the target and their path losses. In most cases, optimal power distribution results in most of the power being allocated to a small subset of transmit antennas that are most advantageous when it comes to path losses and relative locations. This result encourages the use of other resource allocation methods such as subset selections. This idea was explored in detail in [36].
- In the optimization process, higher power is allocated to transmit antennas that are located closer to the target and suffer less path loss. This, combined with higher reflectivity on these paths, makes them better candidates, i.e., energy added to these paths results in larger gain in terms of localization accuracy.
- The angular spread of the transmitters and receivers with respect to the target has a significant impact on the MSE performance, i.e., the higher the angular spread, the lower the MSE. Increasing the transmit power on paths that are

closer to the target and have better angular spread with respect to the target and receivers will have a higher impact on MSE reduction.

- Optimal power allocation will use less power to deliver similar performance, compared with the uniform allocation. For these scenarios, the performance threshold, $\eta_{\max}$, should be defined with a tolerance, i.e., $\overline{\eta}_{\max} = \eta_{\max} + \Delta\eta$, such that $\Delta\eta$ accounts for variations in RCS estimation $\sigma^2_{\tilde{h}_{m,n}}$. The manner in which power is allocated among the various transmitters stays consistent as channels with good propagation properties will have lower estimation errors.
- When tracking multiple targets, the results presented here suggest that similar advantages may be obtained for multiple target scenario. Even for cases in which the effective RCS of the targets with respect to all transmitter–receiver pairs are approximately the same (equivalent to the case of $h_{m,n} \approx h_{\text{const}}, \forall m = 1, \dots, M$ and $\forall n = 1, \dots, N$), power savings may be obtained due to the effects of the locations of the radars with respect to the search area on the localization accuracy.
- The decomposition method applied to the original optimization problems results in a parallel, nonoverlapping, set of $K - 1$ optimization problems that may be solved by distributed processing. The calculation of the optimal solution is performed $L \times \binom{M}{K}$ times at each of the distributed nodes. The original, high dimension, problem is solved with $K - 1$ independent, lower dimension, subproblems. As the computational complexity of these problems grows more than linearly with problem size, substantial savings may be achieved [21, 31, 32].

As the resulting power allocation strategy favors transmitters with better channel conditions, a natural approach would be to narrow the operation of the system to a minimal subset of transmit and receive radars that deliver the required performance. This idea is further developed in [36], where a mathematical framework and techniques for selecting optimal radar subset are developed.

11.5 A Broader Discussion on Resource Allocation

Power allocation schemes, reviewed in this chapter, encompass one facet of resource-aware management potential in multiple radar systems. The concept of large-scale deployment of radars has opened up a plethora of operational opportunities along with complex mission constraints, making smart and efficient use of resources a critical issue.

Antenna power is a resource that, when managed optimally, can conserve energy while maintaining or improving localization accuracy, as discussed in this chapter. Communication considerations are another component that requires attention. Methods for joint optimization of power/localization and bandwidth/communication channel have been evaluated in [38, 39]. Given system constraints on the use of power or localization performance along with bandwidth and communication limits, a CRLB-based problem is formulated and approximate solution is derived.

The operational schemes mentioned so far are based on the notion of full system operation, i.e., all transmit and receive antennas are participating in a given mission. As the radars themselves are a resource, engaging only some of the available antennas, through subset selection, in localization is another way to achieve energy savings, communication efficiency, and reduced computational load. In [36, 40] two antenna selection algorithms are proposed. One minimizes the subset size required to achieve a predetermined localization error threshold and another is optimizing accuracy under subset size constraints. These problems are formulated in a combinatorial optimization framework as a knapsack problem [41, 42], where the goal is to obtain a performance level with the lowest cost, in terms of active system elements. By minimizing the number of operational antennas needed to complete the task, this approach introduces savings in processing load and communication resources, in addition to operational ones.

The application to target tracking has been further explored in [43–46]. These schemes explore adaptive operational modes, among them multiple target tracking via clustering of antennas.

Increasingly, the attention of resource-aware management is drawn to joint resource optimization with tradeoffs among power, communication, topology, and antennas. The fundamentals reviewed in this chapter lay the metrics and mechanisms to address these research topics.

Appendix 11.A Coefficients for Minimize MLE

The values of the coefficients for the quadratic equation given in (11.30) are as follows: the coefficient $a_{\lambda_{\text{MSEo}}}$ is

$$\begin{aligned} a_{\lambda_{\text{MSEo}}} &= \left(\mathbf{1}_1^{\mathrm{T}}\boldsymbol{B}_{11}^{\mathrm{T}} + \mathbf{1}_2^{\mathrm{T}}\boldsymbol{B}_{12}^{\mathrm{T}}\right)\boldsymbol{A}_{11}\left(\boldsymbol{B}_{11}\mathbf{1}_1 + \boldsymbol{B}_{12}\mathbf{1}_2\right) \\ &\quad - u_z\boldsymbol{p}_{K_{\text{Bound}}}^2\left(\boldsymbol{A}_{12} + \boldsymbol{A}_{21}^{\mathrm{T}}\right)\left(\boldsymbol{B}_{11}\mathbf{1}_1 + \boldsymbol{B}_{12}\mathbf{1}_2\right) - u_z^2\boldsymbol{p}_{K_{\text{Bound}}}^{\mathrm{T}}\boldsymbol{A}_{22}\boldsymbol{p}_{K_{\text{Bound}}}, \end{aligned} \tag{11.A.1}$$

the coefficient $b_{\lambda_{\text{MSEo}}}$ is

$$\begin{aligned} b_{\lambda_{\text{MSEo}}} &= \boldsymbol{b}_1^{\mathrm{T}}\left(\boldsymbol{B}_{11}\mathbf{1}_1 + \boldsymbol{B}_{12}\mathbf{1}_2\right) - \left(\mathbf{1}_1^{\mathrm{T}}\boldsymbol{B}_{11}^{\mathrm{T}} + \mathbf{1}_2^{\mathrm{T}}\boldsymbol{B}_{12}^{\mathrm{T}}\right)\left(\boldsymbol{A}_{11} + \boldsymbol{A}_{11}^{\mathrm{T}}\right)\left(\boldsymbol{B}_{11}\boldsymbol{b}_1 + \boldsymbol{B}_{12}\boldsymbol{b}_2\right) \\ &\quad - u_z\boldsymbol{p}_{K_{\text{Bound}}}^{\mathrm{T}}\left(\boldsymbol{A}_{12} + \boldsymbol{A}_{21}^{\mathrm{T}}\right)\left(\boldsymbol{B}_{11}\boldsymbol{b}_1 + \boldsymbol{B}_{12}\boldsymbol{b}_2\right) - 2u_zd_z\boldsymbol{p}_{K_{\text{Bound}}}^{\mathrm{T}}\boldsymbol{A}_{22}\boldsymbol{p}_{K_{\text{Bound}}} \\ &\quad - d_z\boldsymbol{p}_{K_{\text{Bound}}}^{\mathrm{T}}\left(\boldsymbol{A}_{12} + \boldsymbol{A}_{21}^{\mathrm{T}}\right)\left(\boldsymbol{B}_{11}\mathbf{1}_1 + \boldsymbol{B}_{12}\mathbf{1}_2\right) + u_z\boldsymbol{b}_2^{\mathrm{T}}\boldsymbol{p}_{K_{\text{Bound}}}, \end{aligned} \tag{11.A.2}$$

and the coefficient $c_{\lambda_{\text{MSEo}}}$ is

$$\begin{aligned} c_{\lambda_{\text{MSEo}}} &= \boldsymbol{b}_1^{\text{T}} \left(\boldsymbol{B}_{11} \mathbf{1}_1 + \boldsymbol{B}_{12} \mathbf{1}_2 \right) - \left(\boldsymbol{b}_1^{\text{T}} \boldsymbol{B}_{11}^{\text{T}} + \boldsymbol{b}_2^{\text{T}} \boldsymbol{B}_{12}^{\text{T}} \right) \boldsymbol{A}_{11} \left(\boldsymbol{B}_{11} \boldsymbol{b}_1 + \boldsymbol{B}_{12} \boldsymbol{b}_2 \right) \\ &+ d_z \boldsymbol{b}_2^{\text{T}} \boldsymbol{p}_{K_{\text{Bound}}} - d_z \boldsymbol{p}_{K_{\text{Bound}}}^{\text{T}} \left(\boldsymbol{A}_{12} + \boldsymbol{A}_{21}^{\text{T}} \right) \left(\boldsymbol{B}_{11} \boldsymbol{b}_1 + \boldsymbol{B}_{12} \boldsymbol{b}_2 \right) \\ &- d_z^2 \boldsymbol{p}_{K_{\text{Bound}}}^{\text{T}} \boldsymbol{A}_{22} \boldsymbol{p}_{K_{\text{Bound}}}, \end{aligned} \tag{11.A.3}$$

where we define d_z as

$$d_z = \frac{\left(\mathbf{1}_1^{\text{T}} \boldsymbol{B}_{11} \boldsymbol{b}_1 + \mathbf{1}_1^{\text{T}} \boldsymbol{B}_{12} \boldsymbol{b}_2 \right)}{p_{\text{T}} - \boldsymbol{p}_{K_{\text{Bound}}}^{\text{T}} \mathbf{1}_2}, \tag{11.A.4}$$

and u_z as

$$u_z = \frac{\left(\mathbf{1}_1^{\text{T}} \boldsymbol{B}_{11} \mathbf{1}_1 + \mathbf{1}_1^{\text{T}} \boldsymbol{B}_{12} \mathbf{1}_2 \right)}{p_{\text{T}} - \boldsymbol{p}_{K_{\text{Bound}}}^{\text{T}} \mathbf{1}_2}. \tag{11.A.5}$$

Appendix 11.B Coefficients for Minimize Power

The values of the coefficients for the quadratic equation given in (11.45) are as follows: the coefficient a_λ is

$$\begin{aligned} a_{\lambda_{\text{PWo}}} &= \left[\left(\boldsymbol{B}_{11} \boldsymbol{b}_1 + \boldsymbol{B}_{12} \boldsymbol{b}_2 \right)^{\text{T}} \boldsymbol{A}_{11}^{\text{T}} - \boldsymbol{b}_1^{\text{T}} + \eta_{\max} \boldsymbol{p}_{K_{\text{Bound}}}^{\text{T}} \left(\boldsymbol{A}_{12} + \boldsymbol{A}_{21}^{\text{T}} \right) \right] \\ &\times \left(\boldsymbol{B}_{11} \boldsymbol{b}_1 + \boldsymbol{B}_{12} \boldsymbol{b}_2 \right) - \eta_{\max} \boldsymbol{b}_2^{\text{T}} \boldsymbol{p}_{K_{\text{Bound}}} + \eta_{\max}^2 \boldsymbol{p}_{K_{\text{Bound}}}^{\text{T}} \boldsymbol{A}_{22} \boldsymbol{p}_{K_{\text{Bound}}}, \end{aligned} \tag{11.B.1}$$

the coefficient b_λ is

$$\begin{aligned} b_{\lambda_{\text{PWo}}} &= \left[\boldsymbol{b}_1^{\text{T}} - \left(\boldsymbol{B}_{11} \boldsymbol{b}_1 + \boldsymbol{B}_{12} \boldsymbol{b}_2 \right)^{\text{T}} \left(\boldsymbol{A}_{11} + \boldsymbol{A}_{11}^{\text{T}} \right) - \eta_{\max} \boldsymbol{p}_{K_{\text{Bound}}}^{\text{T}} \left(\boldsymbol{A}_{12} + \boldsymbol{A}_{21}^{\text{T}} \right) \right] \\ &\times \left(\boldsymbol{B}_{11} \mathbf{1}_1 + \boldsymbol{B}_{11} \mathbf{1}_2 \right), \end{aligned} \tag{11.B.2}$$

and the coefficient c_λ is

$$c_{\lambda_{\text{PWo}}} = -\left(\boldsymbol{B}_{11} \mathbf{1}_1 + \boldsymbol{B}_{11} \mathbf{1}_2 \right)^{\text{T}} \boldsymbol{A}_{11} \left(\boldsymbol{B}_{11} \mathbf{1}_1 + \boldsymbol{B}_{11} \mathbf{1}_2 \right). \tag{11.B.3}$$

Acknowledgments

The work of Dr. Athina P. Petropulu was supported in part by ARO grant W911NF2320103 and NSF grant ECCS-2320568 and the work of Dr. H. Vincent Poor was supported by NSF grants CNS-2128448 and ECCS-2335876.

References

1 Skolnik, M. (2002). *Introduction to Radar Systems*, 3e. New York: McGraw-Hill Education.

2 Levanon, N. (1988). *Radar Principles*. New York: Wiley.
3 Richards, M.A. (2005). *Fundamentals of Radar Signal Processing*. New York: McGraw-Hill.
4 Zekavat, R. and Buehrer, R.M. (ed.) (2019). *Handbook of Position Location: Theory, Practice, and Advances*. Hoboken, NJ: Wiley-IEEE Press.
5 Lee, H.B. (1975). Accuracy of range-range and range-sum multilateration systems. *IEEE Transactions on Aerospace and Electronic Systems* AES-11: 1346–1361.
6 Weinstein, E. and Weiss, A. (1984). Fundamental limitations in passive time-delay estimation–Part II: wide-band systems. *IEEE Transactions on Acoustics, Speech, and Signal Processing* ASSP-32: 1064–1078.
7 Godrich, H., Haimovich, A.M., and Blum, R.S. (2008). MIMO radar signal processing: concepts and applications of a MIMO radar system with widely separated antennas. In: *MIMO Radar Signal Processing* (ed. J. Li and P. Stoica), 365–410. New York: Wiley.
8 Fishler, E., Haimovich, A., Blum, R.S. et al. (2006). Spatial diversity in radars—models and detection performance. *IEEE Transactions on Signal Processing* 54: 823–838.
9 Yu, Y., Petropulu, A.P., and Poor, H.V. (2010). MIMO radar using compressive sampling. *IEEE Journal of Selected Topics in Signal Processing* 4: 146–163.
10 Li, J. and Stoica, P. (2007). MIMO radar with colocated antennas. *IEEE Signal Processing Magazine* 24: 106–114.
11 Chernyak, V.S. (1998). *Fundamental of Multisite Radar Systems: Multistatic Radars and Multi Radar Systems*. London: CRC Press.
12 Willis, N.J. (2004). *Bistatic Radar*, 2e. Raleigh, NC: SciTech Publishing.
13 Godrich, H., Haimovich, A.M., and Blum, R.S. (2010). Target localization accuracy gain in MIMO radar-based systems. *IEEE Transactions on Information Theory* 56: 2783–2803.
14 Fishler, E., Haimovich, A., Blum, R. et al. (2004). Performance of MIMO radar systems: advantages of angular diversity. *38th Asilomar Conference on Signals, Systems and Computers*, volume 1, 305–309.
15 Bar-Shalom, O. and Weiss, A. (2011). Direct positioning of stationary targets using MIMO radar. *Signal Processing* 91: 2345–2358.
16 Shi, Z., Wang, H., Leung, C.S., and So, H.C. (2020). Robust MIMO radar target localization based on Lagrange programming neural network. *Signal Processing* 174: 107574.

17 Godrich, H., Haimovich, A.M., and Blum, R.S. Target localisation techniques and tools for multiple-input multiple-output radar. *2008 IEEE Radar Conference*, volume 3, 2008. Rome, Italy.

18 He, Q., Blum, R.S., Godrich, H., and Haimovich, A.M. (2010). Target velocity estimation and antenna placement for MIMO radar with widely separated antennas. *IEEE Journal of Selected Topics in Signal Processing* 4: 79–100.

19 Poor, H.V. (1994). *An Introduction to Signal Detection and Estimation*, 2e. New York: Springer.

20 Godrich, H., Petropulu, A., and Poor, H.V. (2011). Power allocation strategies for target localization in distributed multiple-radar architectures. *IEEE Transactions on Signal Processing* 59: 3226–3240.

21 Boyd, S. and Vandenberghe, L. (2004). *Convex Optimization*. Cambridge: Cambridge University Press.

22 Grant, M. and Boyd, S. (2020). CVX: Matlab software for disciplined convex programming. http://cvxr.com/cvx/doc/CVX.pdf (accessed 28 June 2024).

23 Ingber, M.S., Tanski, J.A., and Alsing, P. (2007). A domain decomposition tool for boundary element methods. *Journal of Engineering Analysis with Boundary Elements* 31: 890–896.

24 Blondel, V.D., Boyd, S., and Kimura, H. (2008). *Recent Advances in Learning and Control*. New York: Springer.

25 Bertsekas, D.P. (2003). *Convex Analysis and Optimization*. Belmont, MA: Athena Scientific.

26 Bazaraa, M.S. and Shetty, C.M. (1979). *Nonlinear Programming*. New York: Wiley.

27 Ben-Tal, A., Ghaoui, L.E., and Nemirovski, A. (2009). *Robust Optimization*. Princeton, NJ: Princeton University Press.

28 Borwein, J.M. and Lewis, A.S. (2006). *Convex Analysis and Nonlinear Optimization: Theory and Examples*, 2e. New York: Springer.

29 Gauvin, J. (1977). A necessary and sufficient regularity condition to have bounded multipliers in nonconvex programming. *Journal of Mathematical Programming* 12: 136–138.

30 Reiland, T.W. and Hanson, M.A. (1980). Generalized Kuhn-Tucker conditions and duality for continuous nonlinear programming problems. *Journal of Mathematical Analysis and Applications* 74: 578–598.

31 Chiang, M., Low, S.H., Calderbank, A.R., and Doyle, J.C. (2007). Layering as optimization decomposition: a mathematical theory of network architectures. *Proceedings of the IEEE* 95: 255–312.

32 Bertsekas, D.P. (2016). *Nonlinear Programming*, 3e. Belmont, MA: Athena Scientific.

33 Robinson, S.M. (1976). First order conditions for general nonlinear optimization. *SIAM Journal on Applied Mathematics* 30: 597–607.

34 Hanson, M.A. (1981). On sufficiency of the Kuhn-Tucker conditions. *Journal of Mathematical Analysis and Applications* 80: 545–550.

35 Godrich, H., Petropulu, A., and Poor, H.V. (2010). Power allocation schemes for target localization in widely distributed MIMO radar systems. *Conference om Military Communication*, volume 1, 846–851.

36 Godrich, H., Petropulu, A., and Poor, H.V. (2011). Sensor selection in distributed multiple-radar architectures for localization: a Knapsack problem formulation. *IEEE Transactions on Signal Processing* 60: 247–260.

37 Godrich, H., Petropulu, A.P., and Poor, H.V. (2010). Resource allocation schemes for target localization in distributed multiple radar architectures. *2010 18th European Signal Processing Conference*, volume 1, 1239–1243. Aalborg, Denmark.

38 Garcia, N., Haimovich, A.M., Coulon, M., and Lops, M. (2014). Resource Allocation in MIMO radar with multiple targets for non-coherent localization. *IEEE Transactions on Signal Processing* 62: 2656–2666.

39 Ahmed, A., Zhang, Y.D., and Himed, B. (2019). Distributed dual-function radar-communication MIMO system with optimized resource allocation. *2019 IEEE Radar Conference (RadarConf)*, volume 1. Boston, MA.

40 Godrich, H., Petropulu, A., and Poor, H.V. (2011). A combinatorial optimization framework for subset selection in distributed multiple-radar architectures. *2011 IEEE International Conference on Acoustics, Speech and Signal Processing (ICASSP)*, volume 1, 2796–2799. Prague, Czech Republic.

41 Martello, S. and Toth, P. (1990). *Knapsack Problems: Algorithms and Computer Implementations*. New York: Wiley.

42 Kellerer, H., Pferschy, U., and Pisinger, D. (2004). *Knapsack Problems*. Berlin: Springer-Verlag.

43 Song, X., Zheng, N., and Yan, S. (2018). A joint resource allocation method for multiple targets tracking in distributed MIMO radar systems. *EURASIP Journal on Advances in Signal Processing* 2018: 65.

44 Lu, Y.X., He, Z.S., Zhang, X.J., and Liu, S.L. (2016). Transmit and receive sensors joint selection for MIMO radar tracking based on PCRLB. *2016 IEEE 13th International Conference on Signal Processing (ICSP)*, volume 1, 1551–1555. Chengdu, China.

45 Godrich, H., Tajer, A., and Poor, H.V. (2012). Distributed target tracking in multiple widely separated radar architectures. *2012 IEEE 7th Sensor Array and Multichannel Signal Processing Workshop (SAM)*, volume 1, 153–156. Hoboken, NJ.

46 Shi, C., Ding, L., Wang, F. et al. (2022). Collaborative radar node selection and transmitter resource allocation algorithm for target tracking applications in multiple radar architectures. *Digital Signal Processing* 121: 103318.

12

Information-Theoretic Approach to Fully Adaptive Radar Resource Management

Kristine Bell[1], Chris Kreucher[2], and Muralidhar Rangaswamy[3]

[1] *Metron, Inc., Reston, VA, USA*
[2] *KBR Wyle Services, LLC, Ann Arbor, MI, USA*
[3] *Air Force Research Laboratory, Dayton, OH, USA*

12.1 Introduction

The arrival of agile software-controlled waveform generation and selection in modern radar systems enables new algorithms which select radar transmission parameters to optimize performance. In this chapter, we describe and illustrate via simulation an information-theoretic approach to fully adaptive radar (FAR) resource management (RM) which exploits this new flexibility. FAR, sometimes referred to as "cognitive radar" is a closed-loop process which includes the steps of sensing the environment and responding by selecting and executing the next action(s) for the radar. It takes inspiration from the neurobiological processes of animal cognition to allow radar systems to adapt to their environments [1–5]. Neuropsychologists believe that animal cognition relies on dynamic feedback loops known as perception-action cycles (PACs) to understand and respond to sensory information [6].

Past work in radar resource management (RRM) which has applied perception-action approaches includes [7–18]. While recent research in this area refers to the techniques as "cognitive radar resource management" [15–18], older related work simply uses "sensor management" and/or "resource allocation" [8–14, 19, 20]. Regardless of the name, the algorithms apply a two-step process of perception and action selection. First, the algorithms capture (perceive) the surveillance region using a probabilistic estimate of its state. In the multi-target detection, tracking, and identification context we focus on, this includes estimating probability density functions (PDFs) on target states for each target and probability mass functions (PMFs) on target class for each target. Second, this probabilistic description is used to drive the selection of future sensing actions

Information-Theoretic Radar Signal Processing, First Edition. Edited by Yujie Gu and Yimin D. Zhang.

through an optimization process where the available actions are considered and the action set that maximizes future utility is selected.

The primary challenge of RRM is to balance multiple competing objectives in the presence of constraints on resource usage, such as time and power. In our use case, a single radar system is to perform multi-target track initialization, tracking, and classification. It is given a finite time window and must balance the demands of all the tasks using the available time. This balance is achieved through a single global objective function which calculates the total utility of a set of waveform selections over the time window. This objective function is numerically optimized, subject to the time constraint, to select the future actions the radar is to take.

Developing a single global objective function that captures the goals of the system in a mathematical form amenable to optimization is critical for successful implementation of a fully adaptive radar resource management (FARRM) system. As the number of tasks and actions available for adaptation increases, the complexity of the optimization becomes increasingly more difficult. Broadly speaking, there are two main approaches to defining this objective function: task-driven [18, 21] and information-theoretic [8]. Some work has appeared which has compared these approaches [10, 22].

The task-driven approach first enumerates specific task-level components of the objective function, such as the root mean square error (RMSE) for tracking and correct classification percentage. A user establishes the goal for each of these tasks and defines the utility of being above or below that goal. Then, a multiobjective optimization problem is solved. One method is to define a global task-driven objective function as the weighted sum of the utility of the individual tasks (see [23] for a discussion of multiobjective optimization methods for FAR). The main benefits of this approach are that the objective function is tied to explicit target-level goals and that it directly lays out the relative importance of the various tasks. However, specifying task requirements, costs, and the relative per-target and per-task weightings over the incommensurate task-level measures requires significant domain knowledge [18, 23, 24].

In contrast, the information-theoretic approach employs a global measure of information as its objective function. A number of different measures of information have been studied, including mutual information (MI), entropy, Kullback–Leibler divergence (KLD), and Rényi (alpha) divergence [8, 25–29]. The main benefit of an information-theoretic FARRM objective function is that it implicitly balances the different types of information a radar may acquire – e.g., information about target position and information about target classification are both quantified using a common measure of information [11]. However, an information-theoretic approach does not explicitly optimize task metrics such as RMSE. As such, the information-theoretic objective functions can be difficult for the end-user to understand and connect to specific operational goals [30].

While the information-theoretic approach does allow for separate weightings to give different targets different priorities, and separate weightings to give different types of information different priorities [31], this would require an operator to specify a collection of weights.

The work in this chapter combines and extends our previous work on sensor management [8–14] and FAR [17, 21–23, 32–35]. It is distinguished from our previous efforts in FARRM [22, 35] in that we develop a more comprehensive approach to information-theoretic methods of scheduling, develop a more general and flexible multiple-task tracking and classification model, and provide new simulation examples which use real, collected aircraft trajectories and a tracking and classification scenario where targets enter the surveillance region midway through the vignette.

This chapter proceeds as follows. First, Section 12.2 describes our multiple-task FARRM system model, which is based on a stochastic optimization [36] approach and builds on earlier work in [17, 32]. We describe the two main components of the FARRM PAC – the perceptual and executive processors. The perceptual processor is a recursive Bayesian approach to target state estimation, while the executive processor employs an information-theoretic utility function. The FARRM framework itself is agnostic to the specific objective function and has been illustrated with task-based [17], information-based [22], and even hybrid [35] objective function definitions. In Section 12.3, we develop our information-theoretic utility function, which is based on the MI. In Sections 12.4 and 12.5, we provide the models for the tracking and classification tasks, respectively, including explicit descriptions of the perceptual and executive processors. In Section 12.6, we provide airborne radar simulation results using collected aircraft trajectories where we jointly schedule waveforms for multi-target classification, tracking, and track initialization subject to time constraints. Finally, Section 12.7 presents the conclusions.

12.2 FARRM System Model

The FARRM system model is shown in Figure 12.1. It emulates the PAC through interaction between the perceptual and executive processors. The aim of radar scheduling is to select the radar actions (e.g., parameters of the transmitted waveforms), $\boldsymbol{\theta}_k$, to best estimate a system state $\boldsymbol{x}_k$ to achieve the mission of the radar system. At time k, the perceptual processor receives information about the environment (including targets, clutter, noise, and other environmental parameters) in the form of a measurement $\boldsymbol{z}_k$ and uses the observation to refine its perception of the environment. Here, the perception is captured by the posterior density $f(\boldsymbol{x}_k|\boldsymbol{Z}_k;\boldsymbol{\Theta}_k)$, where $\boldsymbol{Z}_k \doteq \{\boldsymbol{z}_1,\dots,\boldsymbol{z}_k\}$ is the history of measurements up to and including time k and $\boldsymbol{\Theta}_k \doteq \left\{\boldsymbol{\theta}_1,\boldsymbol{\theta}_2,\dots,\boldsymbol{\theta}_k\right\}$ is the history of action vectors.

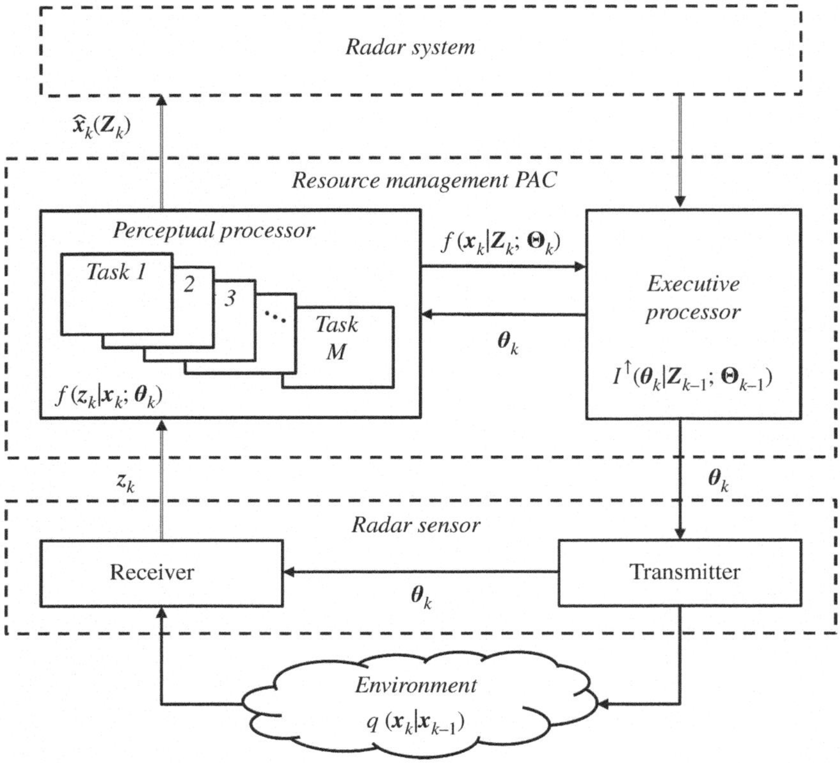

Figure 12.1 FARRM system model.

The executive processor is then responsible for the action selection portion of the PAC, i.e., choosing the next value(s) of $\boldsymbol{\theta}_k$ in response to the latest perception. The new $\boldsymbol{\theta}_k$ is then provided to the radar sensor, which takes the action, forms the new measurement $\boldsymbol{z}_k$, and passes the measurement back to the perceptual processor. This completes one iteration of the PAC.

12.2.1 Multiple-Task PAC

We now develop the explicit mathematical model of the multiple-task PAC. Our model problem is of an agile multi-mode radar which must select radar sensing parameters for a set of M tracking and classification tasks.

Let the resource management frame length be T_F seconds, and let k denote the frame index (time step). We assume there are M dwells in each frame, corresponding to the M tasks. Each of the M dwells may be assigned a different amount of time at each time step. M may change as tasks come and go, for example, when a target enters or exits the surveillance region.

For each of the M task dwells, the radar may elect to transmit nothing (taking up no time and providing no utility) or one of the L waveforms from a library of waveforms. Let $a_l; l = 0, \ldots, L$ denote each of the possible actions (waveforms), where a_0 indicates no waveform is transmitted, and $\mathcal{A} = \{a_0, \ldots, a_L\}$ denotes the set of actions. Each waveform has a time duration, equal to the coherent processing interval (CPI), as well as specific parameters, such as the bandwidth and pulse repetition frequency (PRF), which characterize the waveform and its utility. When a waveform is chosen, it is fixed for the entire task dwell/CPI.

The radar resource parameter vector, or action vector, for the kth frame is defined as the $M \times 1$ vector

$$\boldsymbol{\theta}_k = \left[\theta_{1,k}, \theta_{2,k}, \ldots, \theta_{M,k}\right]^{\mathrm{T}}, \tag{12.1}$$

where $\theta_{m,k}$ is the action chosen from $\mathcal{A}$ for the mth task dwell of the kth frame.

Following the approach in [11, 37], the state vector has the form:

$$\boldsymbol{x}_k = \left[\boldsymbol{x}_{1,k}^{\mathrm{T}}, \boldsymbol{x}_{2,k}^{\mathrm{T}}, \ldots, \boldsymbol{x}_{M,k}^{\mathrm{T}}\right]^{\mathrm{T}}, \tag{12.2}$$

where each $\boldsymbol{x}_{m,k}, m \in \{1, \ldots, M\}$ is a single-task state vector containing components for the particular task of interest. For tracking tasks, we use a tracking state vector which consists of kinematic variables (position, velocity, and acceleration) as well as the received signal-to-noise-ratio (SNR). For classification tasks, the state vector is a scalar classification variable. The target class is a discrete random variable whose value is taken from a discrete, finite set of values, while the tracking state variables are continuous random variables. Therefore, the state of the surveillance region is described using a combination of a PDF for the continuous components and a PMF for the discrete components.

Given the action vector $\boldsymbol{\theta}_k$, the radar sensor transmits the selected waveforms, receives returns, and generates a measurement vector $\boldsymbol{z}_k$, which has the form:

$$\boldsymbol{z}_k = \left[\boldsymbol{z}_{1,k}^{\mathrm{T}}, \boldsymbol{z}_{2,k}^{\mathrm{T}}, \ldots, \boldsymbol{z}_{M,k}^{\mathrm{T}}\right]^{\mathrm{T}}, \tag{12.3}$$

where $\boldsymbol{z}_{m,k}$ is the measurement vector for the mth task during the kth frame. It may be a vector, a scalar, or empty.

The system state $\boldsymbol{x}_k$ evolves according to a state transition model, which we assume to be a first-order Markov model with initial target state PDF/PMF $q(\boldsymbol{x}_0)$ and a state transition PDF/PMF $q(\boldsymbol{x}_k|\boldsymbol{x}_{k-1})$. The measurement model is characterized by the conditional PDF/PMF, or likelihood function, $f(\boldsymbol{z}_k|\boldsymbol{x}_k; \boldsymbol{\theta}_k)$, which is a function of the system state vector $\boldsymbol{x}_k$ and the action vector $\boldsymbol{\theta}_k$.

The first element of the perception-action cycle is the perceptual processor. The goal of the perceptual processor is to combine known sensing actions and received measurements with the models of system state evolution and measurement statistics to construct a posterior PDF/PMF $f(\boldsymbol{x}_k|\boldsymbol{Z}_k;\boldsymbol{\Theta}_k)$ on the system state and a corresponding system state estimate $\hat{\boldsymbol{x}}_k(\boldsymbol{Z}_k)$.

For our first-order Markov motion model, the posterior PDF/PMF of $\boldsymbol{x}_k$ given $\boldsymbol{Z}_k$ can be obtained using the Bayes–Markov recursion:

$$\begin{aligned} f^{+}\left(\boldsymbol{x}_0\right) &= q\left(\boldsymbol{x}_0\right) \\ f^{-}\left(\boldsymbol{x}_k\right) \doteq f\left(\boldsymbol{x}_k|\boldsymbol{Z}_{k-1};\boldsymbol{\Theta}_k\right) &= \int q(\boldsymbol{x}_k|\boldsymbol{x}_{k-1})f^{+}(\boldsymbol{x}_{k-1})d\boldsymbol{x}_{k-1} \\ f^{-}\left(\boldsymbol{z}_k\right) \doteq f(\boldsymbol{z}_k|\boldsymbol{Z}_{k-1};\boldsymbol{\Theta}_k) &= \int f(\boldsymbol{z}_k|\boldsymbol{x}_k;\theta_k)f^{-}(\boldsymbol{x}_k)d\boldsymbol{x}_k \\ f^{+}(\boldsymbol{x}_k) \doteq f\left(\boldsymbol{x}_k|\boldsymbol{Z}_k;\boldsymbol{\Theta}_k\right) &= \frac{f(\boldsymbol{z}_k|\boldsymbol{x}_k;\theta_k)f^{-}(\boldsymbol{x}_k)}{f^{-}\left(\boldsymbol{z}_k\right)}, \end{aligned} \tag{12.4}$$

where $f^{-}(\boldsymbol{x}_k)$ is the predicted PDF/PMF obtained from the motion update step and $f^{+}(\boldsymbol{x}_k)$ is the posterior PDF/PMF obtained from the information update step.

The state estimation performance is characterized by the posterior Bayes risk, which is the expected value of a perceptual processor error function, $\epsilon\left(\hat{\boldsymbol{x}}(\boldsymbol{Z}_k),\boldsymbol{x}_k\right)$, with respect to the posterior PDF/PMF,

$$R^{+}(\boldsymbol{Z}_k;\boldsymbol{\Theta}_k) = \mathbb{E}_k^{+}\left\{\epsilon\left(\hat{\boldsymbol{x}}(\boldsymbol{Z}_k),\boldsymbol{x}_k\right)\right\}, \tag{12.5}$$

where $\mathbb{E}_k^{+}\{\cdot\}$ denotes expectation with respect to $f^{+}(\boldsymbol{x}_k)$.[1] The state estimate is found by minimizing the posterior Bayes risk:

$$\hat{\boldsymbol{x}}_k(\boldsymbol{Z}_k) = \underset{\hat{\boldsymbol{x}}(\boldsymbol{Z}_k)}{\arg\min}\ R^{+}(\boldsymbol{Z}_k;\boldsymbol{\Theta}_k). \tag{12.6}$$

The second element of the perception-action cycle is the executive processor. The function of the executive processor is to select future sensing actions to optimize the performance of the state estimator that will include both the future and past observations. For single time-step scheduling, we define the joint conditional PDF/PMF of $\boldsymbol{x}_k$ and $\boldsymbol{z}_k$ conditioned on $\boldsymbol{Z}_{k-1}$ as:

$$f^{\uparrow}(\boldsymbol{x}_k,\boldsymbol{z}_k) \doteq f(\boldsymbol{x}_k,\boldsymbol{z}_k|\boldsymbol{Z}_{k-1};\boldsymbol{\Theta}_k) = f(\boldsymbol{z}_k|\boldsymbol{x}_k;\theta_k)f\left(\boldsymbol{x}_k|\boldsymbol{Z}_{k-1};\boldsymbol{\Theta}_k\right). \tag{12.7}$$

From (12.4), the joint conditional PDF can also be expressed as:

$$f^{\uparrow}(\boldsymbol{x}_k,\boldsymbol{z}_k) = f^{+}(\boldsymbol{x}_k)f^{-}\left(\boldsymbol{z}_k\right). \tag{12.8}$$

If we take the expectation of the error function with respect to the joint conditional PDF, we obtain the predicted conditional (PC)–Bayes risk as

$$R^{\uparrow}(\theta_k|\boldsymbol{Z}_{k-1};\boldsymbol{\Theta}_{k-1}) = \mathbb{E}_k^{\uparrow}\{\epsilon(\hat{\boldsymbol{x}}(\boldsymbol{Z}_k),\boldsymbol{x}_k)\}, \tag{12.9}$$

---

1 The most commonly used error functions are the squared error, $\left(\hat{\boldsymbol{x}}(\boldsymbol{Z}_k)-\boldsymbol{x}_k\right)^2$, for continuous random variables, and the indicator function, $\mathbf{1}_{\boldsymbol{x}_k}\left(\hat{\boldsymbol{x}}(\boldsymbol{Z}_k)\right)$, for discrete random variables. The resulting Bayes risks are the mean square error and probability of correct decision, respectively.

where $\mathbb{E}_k^{\uparrow}\{\cdot\}$ denotes expectation with respect to $f^{\uparrow}(\boldsymbol{x}_k, \boldsymbol{z}_k)$. Using (12.5) and (12.8), the PC–Bayes risk can also be expressed as the expectation of the posterior Bayes risk with respect to $f^{-}\left(\boldsymbol{z}_k\right)$, i.e.,

$$R^{\uparrow}(\theta_k|\boldsymbol{Z}_{k-1};\boldsymbol{\Theta}_{k-1}) = \mathbb{E}_{\boldsymbol{z}_k}^{-}\{R^{+}(\boldsymbol{Z}_k;\boldsymbol{\Theta}_k)\}, \tag{12.10}$$

where $\mathbb{E}_{\boldsymbol{z}_k}^{-}\{\cdot\}$ denotes expectation with respect to $f^{-}\left(\boldsymbol{z}_k\right)$.

Ideally, the executive processor would choose the next action vector to minimize the PC–Bayes risk. However, the PC–Bayes risk does not generally have a closed-form analytical expression. As a substitute, we use an information-theoretic measure of utility that is analytically tractable and reflects the quality of the target state estimate through the amount of information it contains, which we denote as $I^{\uparrow}(\theta_k|\boldsymbol{Z}_{k-1};\boldsymbol{\Theta}_{k-1})$. The executive processor optimization problem is then to select the next action vector to maximize the information,

$$\theta_k = \arg\max_{\theta} \; I^{\uparrow}(\theta|\boldsymbol{Z}_{k-1};\boldsymbol{\Theta}_{k-1}). \tag{12.11}$$

12.2.2 Independent Tasks

We assume that the task transition models are independent across tasks and that the measurements in each dwell are independent. Under these assumptions, the prior and transition PDF/PMFs and the likelihood function factor into a product of individual task terms. They are expressed as:

$$\begin{aligned}
q(\boldsymbol{x}_0) &= \prod_{m=1}^{M} q(\boldsymbol{x}_{m,0}) \\
q(\boldsymbol{x}_k|\boldsymbol{x}_{k-1}) &= \prod_{m=1}^{M} q(\boldsymbol{x}_{m,k}|\boldsymbol{x}_{m,k-1}) \\
f(\boldsymbol{z}_k|\boldsymbol{x}_k;\theta_k) &= \prod_{m=1}^{M} f(\boldsymbol{z}_{m,k}|\boldsymbol{x}_{m,k};\theta_{m,k}).
\end{aligned} \tag{12.12}$$

With this factorization, the joint Bayes–Markov recursion in (12.4) decouples into M individual recursions, and the posterior PDF/PMF as well as the joint conditional PDF/PMF have the factored forms:

$$\begin{aligned}
f\left(\boldsymbol{x}_k|\boldsymbol{Z}_k;\boldsymbol{\Theta}_k\right) &= \prod_{m=1}^{M} f(\boldsymbol{x}_{m,k}|\boldsymbol{Z}_{m,k};\boldsymbol{\Theta}_{m,k}) \\
f(\boldsymbol{x}_k,\boldsymbol{z}_k|\boldsymbol{Z}_{k-1};\boldsymbol{\Theta}_k) &= \prod_{m=1}^{M} f(\boldsymbol{x}_{m,k},\boldsymbol{z}_{m,k}|\boldsymbol{Z}_{m,k-1};\boldsymbol{\Theta}_{m,k}),
\end{aligned} \tag{12.13}$$

where $\boldsymbol{Z}_{m,k} \doteq \{\boldsymbol{z}_{m,1},\dots,\boldsymbol{z}_{m,k}\}$ and $\boldsymbol{\Theta}_{m,k} \doteq \left\{\theta_{m,1},\theta_{m,2},\dots,\theta_{m,k}\right\}$ are the history of measurements and actions, respectively, for the mth task.

We further assume that the error function $\epsilon\left(\hat{\boldsymbol{x}}(\boldsymbol{Z}_k), \boldsymbol{x}_k\right)$ is the sum of task error functions,[2]

$$\epsilon\left(\hat{\boldsymbol{x}}(\boldsymbol{Z}_k), \boldsymbol{x}_k\right) = \sum_{m=1}^{M} \epsilon\left(\hat{\boldsymbol{x}}_m(\boldsymbol{Z}_k), \boldsymbol{x}_{m,k}\right). \tag{12.14}$$

Then the Bayes risk is the sum of task Bayes risks,

$$R^{+}(\boldsymbol{Z}_k; \boldsymbol{\Theta}_k) = \sum_{m=1}^{M} R^{+}(\boldsymbol{Z}_{m,k}; \boldsymbol{\Theta}_{m,k}), \tag{12.15}$$

and the task state estimates can be obtained separately by minimizing the task Bayes risk:

$$\hat{\boldsymbol{x}}_{m,k}(\boldsymbol{Z}_{m,k}) = \underset{\hat{\boldsymbol{x}}_m(\boldsymbol{Z}_{m,k})}{\arg\min}\; R^{+}(\boldsymbol{Z}_{m,k}; \boldsymbol{\Theta}_{m,k}). \tag{12.16}$$

The PC–Bayes risk is also the sum of task PC–Bayes risks,

$$R^{\uparrow}(\theta_k | \boldsymbol{Z}_{k-1}; \boldsymbol{\Theta}_{k-1}) = \sum_{m=1}^{M} R^{\uparrow}(\theta_{m,k} | \boldsymbol{Z}_{m,k-1}; \boldsymbol{\Theta}_{m,k-1}), \tag{12.17}$$

and the information-theoretic utility function (if chosen as described in Section 12.3) is the sum of task information functions,

$$I^{\uparrow}(\theta_k | \boldsymbol{Z}_{k-1}; \boldsymbol{\Theta}_{k-1}) = \sum_{m=1}^{M} I_m^{\uparrow}(\theta_{m,k} | \boldsymbol{Z}_{m,k-1}; \boldsymbol{\Theta}_{m,k-1}). \tag{12.18}$$

The executive processor still optimizes the joint information-theoretic utility function to determine the next set of actions. It should be emphasized that the optimization is in a global sense and the next action may not be the optimal solution for a particular radar task in isolation.

12.3 Information-Theoretic Utility Function

We now develop the executive processor information-theoretic utility function.

12.3.1 Definitions

We begin by defining some basic continuous and discrete random variables and their PDFs and PMFs and then provide a review of the definitions of various information-theoretic quantities and some relationships between them. The material is taken from [25] and [26].

2 The task error functions may be weighted to account for different priorities among the tasks.

Let ξ denote a discrete random variable that takes on one of N_ξ values in the set $C = \{\xi_1, \xi_2, \dots, \xi_{N_\xi}\}$ and let $f(\xi)$ denote its PMF. We denote the components of $f(\xi)$ as $f_i(\xi) = P(\xi = \xi_i); i = 1, \dots, N_\xi$. Let d denote a discrete random variable that takes on one of N_d values in the set $\mathcal{D} = \{d_1, d_2, \dots, d_{N_d}\}$ and let $f(d)$ denote its PMF. Let $f(\xi|d)$ denote the conditional PMF of ξ given d. We denote the components of $f(\xi|d)$ as $f_{i|j}(\xi|d) = P(\xi = \xi_i|d = d_j); i = 1, \dots, N_\xi; j = 1, \dots, N_d$.

The KLD between two PMFs for the discrete random variable ξ is defined as:

$$D(f(\xi)||g(\xi)) \doteq \sum_{i=1}^{N_\xi} f_i(\xi) \ln \frac{f_i(\xi)}{g_i(\xi)}. \tag{12.19}$$

For a discrete random variable, the entropy characterizes the level of uncertainty or information. The entropy of ξ is defined as:

$$H_\xi \doteq -\sum_{i=1}^{N_\xi} f_i(\xi) \ln f_i(\xi). \tag{12.20}$$

The entropy has the property $0 \le H_\xi \le \ln(N_\xi)$. The entropy is low when the PMF is concentrated on one of the classes and high when the PMF is distributed across the classes. The maximum value is obtained when all classes have the same probability. The conditional entropy of ξ given d is defined as:

$$H_{\xi|d} \doteq -\sum_{j=1}^{N_d} f_j(d) \sum_{i=1}^{N_\xi} f_{i|j}(\xi|d) \ln f_{i|j}(\xi|d). \tag{12.21}$$

The MI between ξ and d is defined as:

$$I_{\xi d} \doteq \mathbb{E}_d\left\{D(f(\xi|d)||f(\xi))\right\} = \sum_{j=1}^{N_d} f_j(d) \left\{ \sum_{i=1}^{N_\xi} f_{i|j}(\xi|d) \ln \frac{f_{i|j}(\xi|d)}{f_i(\xi)} \right\}, \tag{12.22}$$

thus the MI is the expected KLD between the conditional PMF of ξ given d and the marginal PMF of ξ, where the expectation is with respect to d. There are many relationships between these quantities, including:

$$H_{\xi|d} = H_\xi - I_{\xi d}. \tag{12.23}$$

Let $\boldsymbol{x}$ and $\boldsymbol{z}$ denote vectors of continuous random variables with PDFs $f(\boldsymbol{x})$ and $f(\boldsymbol{z})$, respectively. Let $f(\boldsymbol{x}|\boldsymbol{z})$ denote the conditional PDF of $\boldsymbol{x}$ given $\boldsymbol{z}$. The KLD between two PDFs for the continuous random variable $\boldsymbol{x}$ is defined as:

$$D(f(\boldsymbol{x})||g(\boldsymbol{x})) \doteq \int f(\boldsymbol{x}) \ln \frac{f(\boldsymbol{x})}{g(\boldsymbol{x})} d\boldsymbol{x}. \tag{12.24}$$

For continuous random variables, the entropy of $\boldsymbol{x}$ is defined as:[3]

$$H_{\boldsymbol{x}} \doteq -\int f(\boldsymbol{x}) \ln f(\boldsymbol{x}) d\boldsymbol{x}, \tag{12.25}$$

3 For continuous random variables, the standard terminology for the definitions in (12.25) and (12.26) are the differential entropy and conditional differential entropy, and the standard

and the conditional entropy of $\boldsymbol{x}$ given $\boldsymbol{z}$ is defined as:

$$H_{\boldsymbol{x}|\boldsymbol{z}} \doteq -\int f(\boldsymbol{z})\left\{\int f(\boldsymbol{x}|\boldsymbol{z})\ln f(\boldsymbol{x}|\boldsymbol{z})d\boldsymbol{x}\right\}d\boldsymbol{z}. \tag{12.26}$$

The MI between $\boldsymbol{x}$ and $\boldsymbol{z}$ is defined as:

$$I_{\boldsymbol{xz}} \doteq \mathbb{E}_{\boldsymbol{z}}\left\{D\left(f(\boldsymbol{x}|\boldsymbol{z})||f(\boldsymbol{x})\right)\right\} = \int f(\boldsymbol{z})\left\{\int f(\boldsymbol{x}|\boldsymbol{z})\ln\frac{f(\boldsymbol{x}|\boldsymbol{z})}{f(\boldsymbol{x})}d\boldsymbol{x}\right\}d\boldsymbol{z}, \tag{12.27}$$

and we have the following relationship:

$$H_{\boldsymbol{x}|\boldsymbol{z}} = H_{\boldsymbol{x}} - I_{\boldsymbol{xz}}. \tag{12.28}$$

12.3.2 Objective Function

In Section 12.2.1, we showed that the expected task performance is characterized by the PC–Bayes risk, which is the expected value of the posterior–Bayes risk over the predicted PDF of the next measurement $f^{-}\left(\boldsymbol{z}_k\right)$. Since the PC–Bayes risk is difficult to compute analytically, we seek an information-theoretic measure of utility that is easier to analyze and results in an executive processor optimization problem whose solution is the same as, or close to, the solution that would be obtained if the PC–Bayes risk was used.

The foundation of our approach is the observation that the entropy of the posterior distribution is a suitable information-theoretic surrogate for the posterior Bayes risk. It is low when the posterior distribution contains a lot of information about the state and good state estimates with low Bayes risk can be obtained. If we take the expected value of the posterior entropy with respect to $f^{-}\left(\boldsymbol{z}_k\right)$, we obtain the corresponding surrogate for the PC–Bayes risk, which turns out to be the conditional entropy, defined in (12.21) and (12.26).

The entropy of the predicted and posterior PDF/PMFs is defined as:

$$\begin{aligned} H_k^{-}\left(\boldsymbol{Z}_{k-1};\boldsymbol{\Theta}_{k-1}\right) &= -\int f^{-}(\boldsymbol{x}_k)\ln f^{-}(\boldsymbol{x}_k)d\boldsymbol{x}_k \\ H_k^{+}\left(\boldsymbol{z}_k|\boldsymbol{Z}_{k-1};\theta_k,\boldsymbol{\Theta}_{k-1}\right) &= -\int f^{+}(\boldsymbol{x}_k)\ln f^{+}(\boldsymbol{x}_k)d\boldsymbol{x}_k, \end{aligned} \tag{12.29}$$

and the conditional entropy is

$$H^{\dagger}\left(\theta_k|\boldsymbol{Z}_{k-1};\boldsymbol{\Theta}_{k-1}\right) = -\int f^{-}(\boldsymbol{z}_k)H_k^{+}\left(\boldsymbol{z}_k|\boldsymbol{Z}_{k-1};\theta_k,\boldsymbol{\Theta}_{k-1}\right)d\boldsymbol{z}_k. \tag{12.30}$$

Using the property in (12.28), the conditional entropy can be written as

$$H^{\dagger}\left(\theta_k|\boldsymbol{Z}_{k-1};\boldsymbol{\Theta}_{k-1}\right) = H_k^{-} - I_{\boldsymbol{xz}}(\theta_k|\boldsymbol{Z}_{k-1};\boldsymbol{\Theta}_{k-1}). \tag{12.31}$$

---

notations are $h_{\boldsymbol{x}}$ and $h_{\boldsymbol{x}|\boldsymbol{z}}$, respectively. However, since we are using a combination of continuous and discrete random variables, we follow the convention in [26] and use the same terminology (entropy) and notation ($H_{\boldsymbol{x}}$) for both continuous and discrete random variables.

where

$$I_{xz}(\theta_k|\boldsymbol{Z}_{k-1};\boldsymbol{\Theta}_{k-1}) \doteq \mathbb{E}_{\boldsymbol{z}_k}^{-}\left\{\int f^{+}(\boldsymbol{x}_k)\ln\frac{f^{+}(\boldsymbol{x}_k)}{f^{-}(\boldsymbol{x}_k)}d\boldsymbol{x}_k\right\} \tag{12.32}$$

is the MI (or expected KLD) between the target state and the next measurement, defined in (12.22) and (12.27).

Since H_k^- in (12.31) is not a function of the action vector $\boldsymbol{\theta}_k$, the MI is an equivalent information-theoretic utility function to the conditional entropy, which can be *maximized* to obtain the next action vector.

The MI is a measure of the expected *gain* in information [8, 10, 13, 16, 29]. For our independent task model, the global MI decomposes into the sum of the task MIs as:

$$I_{xz}(\theta_k|\boldsymbol{Z}_{k-1};\boldsymbol{\Theta}_{k-1}) = \sum_{m=1}^{M} I_{xz;m}(\theta_{m,k}|\boldsymbol{Z}_{m,k-1};\boldsymbol{\Theta}_{m,k-1}). \tag{12.33}$$

To summarize, our information-theoretic method for selecting the next action vector is to find the action vector that maximizes the MI,

$$\boldsymbol{\theta}_k = \arg\max_{\theta} I_{xz}(\theta|\boldsymbol{Z}_{k-1};\boldsymbol{\Theta}_{k-1}). \tag{12.34}$$

12.4 Tracking Task

For a tracking task, the tracking state vector consists of kinematic variables (position, velocity, and possibly acceleration) and the received SNR. For three-dimensional (3D) tracking, the state vector has the form:

$$\boldsymbol{x}_{m,k} = \left[x_{m,k}, \dot{x}_{m,k}, \ddot{x}_{m,k}, y_{m,k}, \dot{y}_{m,k}, \ddot{y}_{m,k}, h_{m,k}, \dot{h}_{m,k}, \ddot{h}_{m,k}, s_{m,k}\right]^{\mathrm{T}}. \tag{12.35}$$

We assume an initial tracking state distribution that is multivariate Gaussian with mean $\boldsymbol{\mu}_{m,0}$ and covariance matrix $\boldsymbol{\Sigma}_{m,0}$, which we denote as $q(\boldsymbol{x}_{m,0}) = \mathcal{N}\left(\boldsymbol{x}_{m,0}; \boldsymbol{\mu}_{m,0}, \boldsymbol{\Sigma}_{m,0}\right)$. We assume a linear motion model of the form:

$$\boldsymbol{x}_{m,k} = \boldsymbol{F}_m \boldsymbol{x}_{m,k-1} + \boldsymbol{e}_{m,k}, \tag{12.36}$$

where $\boldsymbol{F}_m$ is the state transition matrix and $\boldsymbol{e}_{m,k}$ is zero-mean additive white Gaussian noise (AWGN) with covariance matrix $\boldsymbol{Q}_m$. The transition PDF is then $q(\boldsymbol{x}_{m,k}|\boldsymbol{x}_{m,k-1}) = \mathcal{N}\left(\boldsymbol{x}_{m,k}; \boldsymbol{F}_m \boldsymbol{x}_{m,k-1}, \boldsymbol{Q}_m\right)$.

We assume that when new targets enter the surveillance region, they are detected by a separate surveillance process and provided to the system with an initial tracking state distribution. Targets may arrive at the onset of the simulation or part way through.

We assume measurements are received with detection probability $P_D\left(\boldsymbol{x}_{m,k}; \theta_{m,k}\right)$ and false alarm probability P_F. The detection probability is determined by the detection threshold, which is set to achieve a desired P_F, and the received SNR, which is a function of the sensor parameters and target state. When measurements are received, we assume they follow a nonlinear, AWGN measurement model of the form:

$$\boldsymbol{z}_{m,k} = \boldsymbol{h}_{m,k}\left(\boldsymbol{x}_{m,k}\right) + \boldsymbol{n}_{m,k}, \tag{12.37}$$

where $\boldsymbol{h}_{m,k}\left(\boldsymbol{x}_{m,k}\right)$ is a nonlinear transformation from the target state space to the radar measurement space and $\boldsymbol{n}_{m,k}$ is the measurement error, which is modeled as a zero-mean Gaussian random vector with covariance matrix $\boldsymbol{R}_{m,k}\left(\theta_{m,k}\right)$. The single target likelihood function is then

$$f(\boldsymbol{z}_{m,k}|\boldsymbol{x}_{m,k};\theta_{m,k}) = P_D\left(\boldsymbol{x}_{m,k};\theta_{m,k}\right)\mathcal{N}\left(\boldsymbol{z}_{m,k};\boldsymbol{h}_{m,k}\left(\boldsymbol{x}_{m,k}\right),\boldsymbol{R}_{m,k}\left(\theta_{m,k}\right)\right). \tag{12.38}$$

In the perceptual processor, we implement the Bayes–Markov recursion expressions in (12.4) using the extended Kalman filter (EKF). It requires evaluating the Jacobian matrix, which is defined as:

$$\tilde{\boldsymbol{H}}_{m,k}(\boldsymbol{x}) = \left[\nabla_{\boldsymbol{x}}\boldsymbol{h}_{m,k}^{\mathrm{T}}(\boldsymbol{x})\right]^{\mathrm{T}}. \tag{12.39}$$

The EKF recursion is as follows:

Initialization:

$$\begin{aligned}\boldsymbol{\mu}_{m,0}^{+} &= \boldsymbol{\mu}_{m,0}\\ \boldsymbol{P}_{m,0}^{+} &= \boldsymbol{\Sigma}_{m,0}.\end{aligned} \tag{12.40}$$

Motion update:

$$\begin{aligned}\boldsymbol{\mu}_{m,k}^{-} &= \boldsymbol{F}_m\boldsymbol{\mu}_{m,k-1}^{+}\\ \boldsymbol{P}_{m,k}^{-} &= \boldsymbol{F}_m\boldsymbol{P}_{m,k-1}^{+}\boldsymbol{F}_m^{\mathrm{T}} + \boldsymbol{Q}_m.\end{aligned} \tag{12.41}$$

Information update, when a measurement is received:

$$\begin{aligned}\boldsymbol{H}_{m,k} &= \tilde{\boldsymbol{H}}_{m,k}(\boldsymbol{\mu}_{m,k}^{-})\\ \boldsymbol{K}_{m,k} &= \boldsymbol{P}_{m,k}^{-}\boldsymbol{H}_{m,k}^{\mathrm{T}}\left[\boldsymbol{H}_{m,k}\boldsymbol{P}_{m,k}^{-}\boldsymbol{H}_{m,k}^{\mathrm{T}} + \boldsymbol{R}_{m,k}(\theta_{m,k})\right]^{-1}\\ \boldsymbol{\mu}_{m,k}^{+} &= \boldsymbol{\mu}_{m,k}^{-} + \boldsymbol{K}_{m,k}\left[\boldsymbol{z}_{m,k} - \boldsymbol{h}_{m,k}(\boldsymbol{\mu}_{m,k}^{-})\right]\\ \boldsymbol{P}_{m,k}^{+} &= \boldsymbol{P}_{m,k}^{-} - \boldsymbol{K}_{m,k}\boldsymbol{H}_{m,k}\boldsymbol{P}_{m,k}^{-}.\end{aligned} \tag{12.42}$$

Information update, when no measurement is received:

$$\begin{aligned}\boldsymbol{\mu}_{m,k}^{+} &= \boldsymbol{\mu}_{m,k}^{-}\\ \boldsymbol{P}_{m,k}^{+} &= \boldsymbol{P}_{m,k}^{-}.\end{aligned} \tag{12.43}$$

The posterior–Bayes risk for the tracking state vector is the trace of the posterior mean square error (MSE) matrix. The solution to (12.16) is the mean of the posterior PDF:

$$\hat{\boldsymbol{x}}_{m,k}(\boldsymbol{Z}_{m,k}) = \mathbb{E}_k^+ \left\{\boldsymbol{x}_{m,k}\right\} = \boldsymbol{\mu}_{m,k}^+. \tag{12.44}$$

In the executive processor, the MI has the form:

$$\begin{aligned} &I_{\boldsymbol{xz};m}(\theta_{m,k}|\boldsymbol{Z}_{m,k-1};\boldsymbol{\Theta}_{m,k-1}) \\ &= \frac{P_D\left(\boldsymbol{x}_{m,k};\theta_{m,k}\right)}{2}\left(\ln|\boldsymbol{P}_{m,k}^-| + \ln\left|\left[\boldsymbol{P}_{m,k}^-\right]^{-1} + \boldsymbol{H}_{m,k}^{\mathrm{T}}\boldsymbol{R}_{m,k}(\theta_{m,k})^{-1}\boldsymbol{H}_{m,k}\right|\right), \end{aligned} \tag{12.45}$$

where $|\cdot|$ denotes the determinant of a matrix.

The probability of detection $P_D\left(\boldsymbol{x}_{m,k};\theta_{m,k}\right)$ and the estimation covariance matrix $\boldsymbol{R}_{m,k}(\theta_{m,k})$ depend on the characteristics of the radar waveform and generally have closed-form analytical expressions. The probability of detection also depends on the current state of the target $\boldsymbol{x}_{m,k}$. To evaluate the MI, we substitute the probability of detection evaluated at the predicted state estimate, $P_D\left(\boldsymbol{\mu}_{m,k}^-;\theta_{m,k}\right)$. The remaining terms in the MI are available from the EKF recursion, so they are straightforward to evaluate.

12.5 Classification Task

For a classification task, the state vector is a scalar classification variable, $x_{m,k}$, which takes on one of a discrete set of N_c values in the set C,

$$x_{m,k} \in C = \left\{1, 2, \ldots, N_c\right\}. \tag{12.46}$$

The prior PMF $q(x_{m,0})$ is represented by the $N_c \times 1$ vector $\boldsymbol{q}_m$, which consists of the N_c probabilities

$$\left[\boldsymbol{q}_m\right]_i = P(x_{m,0} = i); \qquad i = 1, \ldots, N_c. \tag{12.47}$$

The transition model $q(x_{m,k}|x_{m,k-1})$ is represented by the $N_c \times N_c$ transition matrix $\boldsymbol{\Upsilon}_m$, where

$$\left[\boldsymbol{\Upsilon}_m\right]_{ij} = P(x_{m,k} = i|x_{m,k-1} = j); \qquad i,j = 1, \ldots, N_c. \tag{12.48}$$

Depending on the application, switching between classes may or may not be possible. For example, if the class represents a behavior of the target, then it is possible to change classes; however, if the class represents a fixed property of the target such as the type of vehicle or aircraft, then the class cannot change, and $\boldsymbol{\Upsilon}_m$ is equal to the identity matrix.

We assume that when new targets enter the surveillance region, they are detected by a separate surveillance process and provided to the system with an initial classification state distribution.

We assume that the radar data is processed to produce a discrete-valued measurement of target class, i.e., $z_{m,k} \in C = \{1, 2, \ldots, N_c\}$ and the likelihood function $f(z_{m,k}|x_{m,k}; \theta_{m,k})$ is represented by the $N_c \times N_c$ likelihood matrix $\boldsymbol{L}_m(\theta_{m,k})$, where

$$\left[\boldsymbol{L}_m\right]_{ij}(\theta_{m,k}) = P(z_{m,k} = i|x_{m,k} = j; \theta_{m,k}); \qquad i,j = 1, \ldots, N_c. \tag{12.49}$$

In the perceptual processor, the Bayes–Markov recursion can be computed exactly. Let $f_i^-(x_{m,k}) \doteq P(x_{m,k} = i|\boldsymbol{Z}_{m,k-1}; \boldsymbol{\Theta}_{m,k-1})$ denote the predicted PMF and $f_i^+(x_{m,k}) \doteq P(x_{m,k} = i|\boldsymbol{Z}_{m,k}; \boldsymbol{\Theta}_{m,k})$ denote the posterior PMF. The recursion is as follows:

$$\begin{aligned}
f_i^+(x_{m,0}) &= \left[\boldsymbol{q}_m\right]_i; \qquad i = 1, \ldots, N_c \\
f_i^-(x_{m,k}) &= \sum_{j=1}^{N_c} \left[\boldsymbol{\Upsilon}_m\right]_{ij} f_j^+(x_{m,k-1}); \qquad i = 1, \ldots, N_c \\
f^-(z_{m,k}) &= \sum_{j=1}^{N_c} \left[\boldsymbol{L}_m\right]_{z_{m,k},j}(\theta_{m,k}) f_j^-(x_{m,k}) \\
f_i^+(x_{m,k}) &= \frac{\left[\boldsymbol{L}_m\right]_{z_{m,k},i}(\theta_{m,k}) f_i^-(x_{m,k})}{f^-(z_{m,k})}; \quad i = 1, \ldots, N_c.
\end{aligned} \tag{12.50}$$

The posterior–Bayes risk for the classification state vector is the posterior probability of incorrect classification. The solution to (12.16) is the maximum of the posterior PMF:

$$\hat{x}_{m,k}(\boldsymbol{Z}_{m,k}) = \arg\max_{i \in C} f_i^+\left(x_{m,k}\right). \tag{12.51}$$

In the executive processor, to compute the classification MI, we first define the prior probability of receiving classification measurement $z_{m,k}$ using the notation

$$f_j^-(z_{m,k}) \doteq P(z_{m,k} = j|\boldsymbol{Z}_{m,k-1}; \boldsymbol{\Theta}_{m,k}) \tag{12.52}$$

and the posterior probability of class $x_{m,k}$ after receiving observation $z_{m,k}$ using the notation

$$f_{i|j}^+(x_{m,k}|z_{m,k}) \doteq P(x_{m,k} = i|z_{m,k} = j, \boldsymbol{Z}_{m,k-1}; \boldsymbol{\Theta}_{m,k}). \tag{12.53}$$

These quantities are computed for every $z_{m,k}$ using the expressions in (12.50). Then the MI is computed directly from the definition in (12.22):

$$\begin{aligned}
&I_{xz;m}(\theta_{m,k}|\boldsymbol{Z}_{m,k-1}; \boldsymbol{\Theta}_{m,k-1}) \\
&\quad = \sum_{j=1}^{N_c} f_j^-(z_{m,k}) \left\{ \sum_{i=1}^{N_c} f_{i|j}^+(x_{m,k}|z_{m,k}) \ln \frac{f_{i|j}^+(x_{m,k}|z_{m,k})}{f_j^-(z_{m,k})} \right\}.
\end{aligned} \tag{12.54}$$

12.6 Simulation Example

This section demonstrates our information-theoretic approach to FARRM in a multitarget tracking, classification, and track initialization simulation.

12.6.1 Simulation Scenario

Our model problem consists of an agile multi-mode radar platform and three airborne targets as illustrated in Figure 12.2. The radar platform is charged with performing tracking and classification of the targets in the surveillance region, which is indicated by the box. Targets 1 and 3 are always in the surveillance region, while Target 2 enters about one-third of the way into the 250-second simulation.

The radar platform flies at a constant heading with constant velocity of 200 m/s. The three targets are taken from the TrajAir [38] data collection, which is a set of recorded trajectories over the Pittsburgh–Butler Regional Airport in Pennsylvania. Since the TrajAir targets are general aviation aircraft operating over a narrow region, we spatially shifted and scaled the recorded flight data to construct speeds and locations consistent with maneuvering targets. Target 1 has a range between 40 and 60 km to the radar platform. Target 2 has a range between 50 km (at entry into the surveillance region) and 35 km to the radar platform (at the end of the simulation). Target 3 has a relatively constant range of approximately 45 km to the radar platform. All targets vary their speeds between 100 and 300 m/s during the 250-second simulation in accordance with the (scaled) recorded data. The simulation has four tasks: track the three targets and identify (classify) Target 1.

12.6.2 Sensor and Task Models

The tracking state vector $\boldsymbol{x}_{m,k}$ consists of the three-dimensional position, velocity, acceleration, and SNR, as specified in (12.35). We define the SNR in decibels

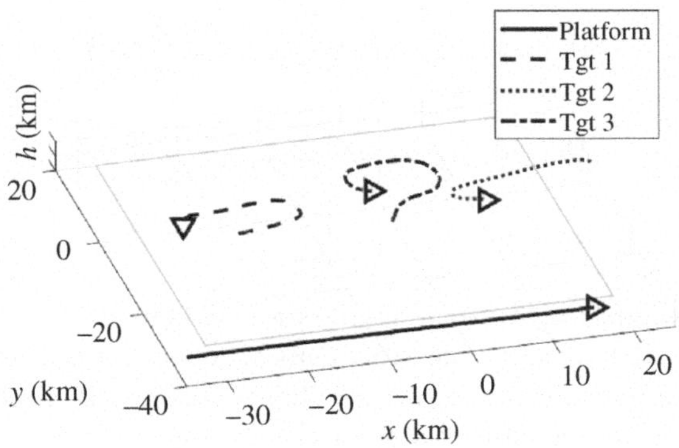

Figure 12.2 Our simulation has an airborne radar platform and three airborne targets. Target trajectories are derived from the TrajAir [38] collected aircraft dataset. The surveillance region is indicated by the solid box. Source: Adapted from [38].

as $s_{m,k} = 10\log_{10}\zeta_{m,k}$, where $\zeta_{m,k}$ is the SNR in linear scale. The classification state variable $x_{m,k}$ is assumed to be one of $N_c = 5$ classes. For tracking, we use the Singer model [39] for target motion as the tracking transition model and for classification, we assume a transition matrix with diagonal entries $\left[\boldsymbol{\Upsilon}_m\right]_{ii} = 0.95$ and off-diagonal entries $\left[\boldsymbol{\Upsilon}_m\right]_{ij} = 0.0125$. Although the target class should not change, providing some probability for transition forces the radar to make periodic classification measurements and prevents the state estimate from getting stuck in an incorrect class due to an incorrect measurement.

We assume the radar transmits a waveform and receives returns through an antenna with a fixed azimuth beamwidth ($\Delta\phi$) and elevation beamwidth ($\Delta\theta$). The transmitted waveform is characterized by its center frequency (f_c), pulse bandwidth (B_p), PRF (f_p), and number of pulses (N_p). We model the tracking measurement process as providing detections which are estimates of target range (R), range-rate ($\dot{R}$), azimuth angle (ϕ), elevation angle (θ), and SNR in decibels ($s = 10\log_{10}\zeta$). For this simulation, we fix f_c, $\Delta\phi$, and $\Delta\theta$ and assume that the radar can select B_p, f_p, and N_p from a collection of options. We also assume that the detection threshold and the false alarm rate P_F are fixed. Let $B_{p;m,k}, f_{p;m,k}$, and $N_{p;m,k}$ denote the parameters of the selected waveform for the mth tracking task. The detection probability is given by [40]:

$$P_D(\zeta_{m,k};\theta_{m,k}) = Q_{MAR}\left(\sqrt{2N_{p;m,k}\zeta_{m,k}}, \sqrt{-2\ln P_F}\right), \tag{12.55}$$

where $Q_{MAR}(a,b)$ is the Marcum Q-function.

The estimation covariance matrix is a diagonal matrix whose components are [41, 42]:

$$\begin{aligned}
\left[\boldsymbol{R}_{m,k}(\theta_{m,k})\right]_R &= \left[2N_{p;m,k}\zeta_{m,k}\left(\frac{2}{c}\right)^2\left(3B_{p;m,k}\right)^2\right]^{-1} \\
\left[\boldsymbol{R}_{m,k}(\theta_{m,k})\right]_{\dot{R}} &= \left[2N_{p;m,k}\zeta_{m,k}\left(\frac{4\pi f_c}{c}\right)^2\left(\frac{1}{12B_{p;m,k}^2} + \frac{(N_{p;m,k}^2 - 1)}{12f_{p;m,k}^2}\right)\right]^{-1} \\
\left[\boldsymbol{R}_{m,k}(\theta_{m,k})\right]_\phi &= \left[2N_{p;m,k}\zeta_{m,k}\left(\frac{1.782\pi}{\Delta\phi}\right)^2\right]^{-1} \\
\left[\boldsymbol{R}_{m,k}(\theta_{m,k})\right]_\theta &= \left[2N_{p;m,k}\zeta_{m,k}\left(\frac{1.782\pi}{\Delta\theta}\right)^2\right]^{-1} \\
\left[\boldsymbol{R}_{m,k}(\theta_{m,k})\right]_s &= \left(\frac{10}{\ln(10)}\right)^2,
\end{aligned} \tag{12.56}$$

where $c = 3 \times 10^8$ m/s is the speed of light.

Classification measurements are characterized similarly. The radar is assumed to transmit and receive a CPI, process the data through a target recognition algorithm, and return a discrete classification call. We characterize the performance of a classification waveform by the probability that the discrete classification call is correct, denoted by p_{dc}. Let $p_{dc;m,k}$ denote the value corresponding to the selected waveform for the mth task. The classification likelihood matrix has the form:

$$[\boldsymbol{L}_m]_{ij}(\theta_{m,k}) = \begin{cases} p_{dc;m,k} & i = j \\ \dfrac{1 - p_{dc;m,k}}{N_c - 1} & i \neq j. \end{cases} \tag{12.57}$$

12.6.3 Waveforms

We assume that the radar resource management frame length is the same as the track update interval, which is $T_F = 100$ ms. The 100-ms frame is used to perform both surveillance (which detects new targets) and service of existing targets by performing tracking and classification tasks. Our focus in this study is on scheduling the agile radar to best perform the latter of those two functions, tracking and classification of known targets.

As such, we assume that 90 ms of the 100-ms frame time is reserved for performing the surveillance function, leaving 10 ms to perform tracking and classification of the known targets. The surveillance function is assumed to be performed using some standard method (e.g., periodic scan with detection revisits and coarse initialization) and provides coarsely initialized targets with a large covariance which then becomes part of the tracking and classification tasks.

The resource management algorithm is free to perform a measurement for each task during the tracking and classification portion of the timeline or elect to measure any subset of the tasks as long as the total measurement time is less than or equal to the 10-ms time budget. The waveforms available for tracking and classification, their parameters, and dwell times are given in Table 12.1. Also included is the "no observation" waveform #0. This allows the scheduler to choose not to measure some of the tasks in the current frame, thereby leaving extra time for other tasks. The fixed tracking waveform parameters are $f_c = 3$ GHz, $\Delta\phi = 2°$, $\Delta\theta = 6°$, and $P_F = 10^{-6}$.

The set of waveforms allows the scheduler to perform tradeoff between receiving information on different tasks as well as receiving different types of information. For example, it may select a longer CPI (yielding higher P_D or p_{dc}) for one task at the expense of a shorter CPI (yielding lower P_D or p_{dc}) or no CPI for other tasks.

12.6.4 Waveform Optimization

The predicted utility of a sensing action is scored using the information-theoretic utility function in (12.33). The objective is to maximize the total information

Table 12.1 Our airborne radar scheduler selects between classification and tracking waveforms.

Waveform		Parameter(s)		Dwell Time
0		N/A		0.0
#	B_p (MHz)	f_p (kHz)	N_p	T (ms)
1,2,3	2.5, 5, 7.5	20	25	1.25
4,5,6	2.5, 5, 7.5	10	25	2.50
7,8,9	2.5, 5, 7.5	10	50	5.00
#		p_{dc}		T (ms)
10		0.6		1.0
11		0.75		2.5

When selecting a tracking waveform, the scheduler chooses bandwidth, pulse count, and PRF. When selecting classification waveform, the scheduler chooses the classification accuracy.

gain over the four tasks subject to the timing constraint. This is accomplished by selecting an information-optimal set of waveforms to transmit for the collection of tasks at each time step.

The numerical waveform optimization proceeds as follows. We first compute the information-theoretic utility of each waveform for each task. The computational complexity is $\mathcal{O}(LM)$, where L is the number of waveforms and M is the number of tasks. Next, we record the maximum utility waveform from among all waveforms having the same CPI for each task. This reduces the number of permutations under consideration by discarding any equal-CPI waveforms that have lower utility than another choice for the same task. As such, we do not explicitly enumerate the L^M combinations of waveforms that could be transmitted together. Finally, we apply a discrete double auction parameter selection (DDAPS) [43, 44] approach to find the set of M waveforms that maximize the total MI subject to the timing constraint.

12.6.5 Simulation Results

Simulation example results for a single trial are shown in Figures 12.3 and 12.4. Figure 12.3 shows the coarsest level description of results, indicating how much time (of the 10-ms budget) is allocated to each task at each of the time steps. Since each time step is 100 ms, there are 2500 time steps in the 250-second simulation.

Since Target 2 does not enter the surveillance region until $t = 100$ s, no resources are allocated to it until that point. At the beginning of the simulation ($t = 0$ s to $t = 5$ s), the scheduler elects to split its time between tracking dwells on Targets 1 and 3 and classification dwells on Target 1. Once Target 1 is identified

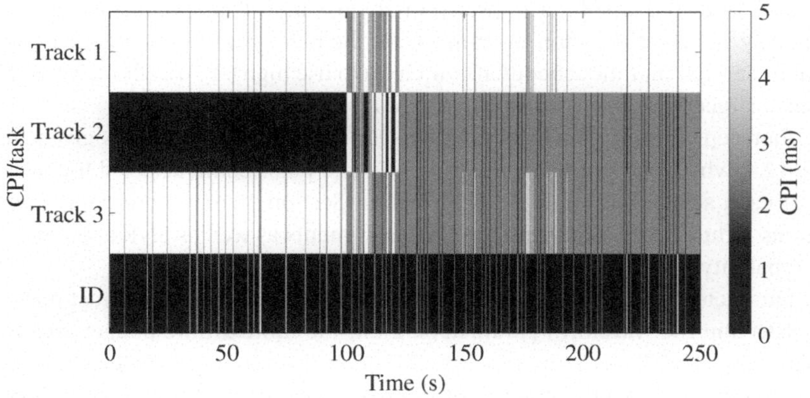

Figure 12.3 The time (out of 10 ms) the scheduler gives to each task for a single simulation. At the beginning, the scheduler performs some classification dwells and then rarely thereafter. Track dwells are initially divided between Tracks 1 and 3. Once Target 2 enters, it is preferentially measured. Later in the simulation, the scheduler gives extra resources to Track 1 due to its distance.

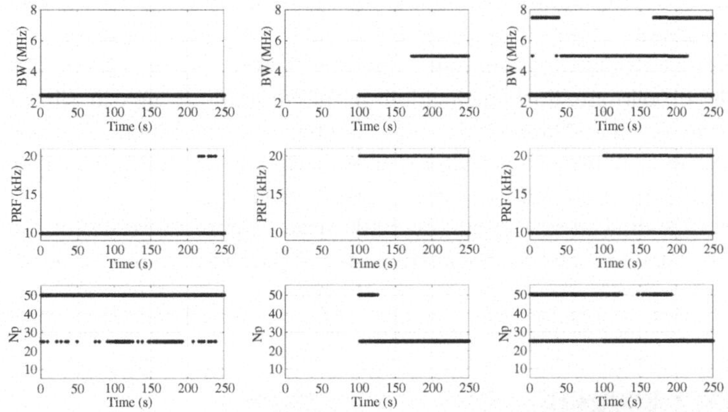

Figure 12.4 The selected waveforms for tracking Target 1 (left), Target 2 (middle), and Target 3 (right) for a single simulation. The scheduler varies the bandwidth, PRF, and pulse count to trade between detectability and range/range-rate resolution subject to the total time constraint.

(at approximately $t = 5$ s), the scheduler divides its time between tracking dwells on Targets 2 and 3, only occasionally revisiting with a classification dwell on Target 1 since our model includes some time-induced class confusion for the reasons described in Section 12.6.2. When Target 2 enters at $t = 100$ s, the surveillance process provides a very coarse estimate of target state. This large track covariance leads the scheduler to preferentially use high-CPI dwells on Target 2 temporarily. Shortly thereafter, the range from the radar platform to Target 1 has increased enough that the scheduler now elects to use high CPI dwells on Track 1 to receive larger P_D at the expense of reduced P_D on Tracks 2 and 3.

A richer depiction of the scheduling choices for the three targets is given in Figure 12.4, which shows the bandwidth, PRF, and pulse count selected for each track at each scheduling time. As described in Section 12.6.2, these parameter selections influence the statistics of the measurements as well as SNR and detection probability.

The parameter selections correspond to the tradeoffs the scheduler must make when selecting the waveform parameters. First, the limited time budget means that the scheduler elects to use a longer CPI on one task to improve detectability that necessarily reduces the amount of time available for other tasks. Additionally, the scheduler may choose to get improved range resolution through higher bandwidth but that leads to reduced detection probability due to the correspondingly larger noise bandwidth entering the radar. This can be compensated for by increasing CPI at the expense of other targets. We see from the figure that the scheduler opts to choose different bandwidths, PRFs, and CPIs (via selection of PRF and pulse count) throughout the simulation as the received detections and track estimate warrant.

We now turn to a Monte Carlo analysis of the simulation. Figures 12.5 to 12.9 show the result of 100 Monte Carlo realizations of the simulation scenario. All trials share the same sensor and target trajectories, but the measurement realizations are drawn randomly each time. This leads to variations in the scheduling decisions and the MI, RMSE, and probability of correct class for each trial.

First, Figure 12.5 shows the average (over 100 trials) amount of time (of the 10 ms) that is allocated to each task at each time step in the simulation. This result

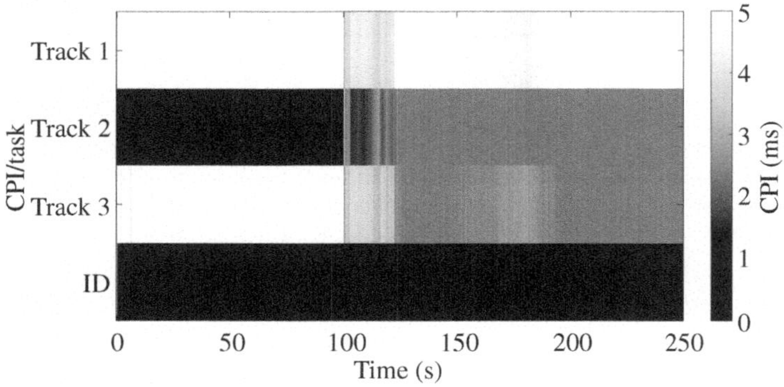

Figure 12.5 The average (over 100 trials) amount of time (out of 10 ms) the scheduler dedicates to each task during the simulation.

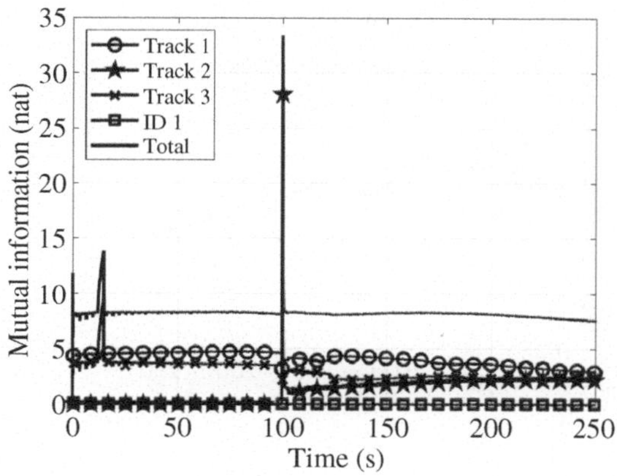

Figure 12.6 The average (over 100 trials) of the MI for each task and the total MI.

is similar to that of the corresponding single simulation figure, showing how the scheduler elects to move its attention between the four tasks as the situation evolves. The prominent features include dividing attention equally between Tracks 1 and 3 early; focusing (briefly) on Track 2 at $t = 100\,\text{s}$; and giving extra time to Track 1 near the end when its range to the platform is largest.

Next, Figure 12.6 illustrates the average MI for each of the four tasks (three tracking task and one classification task) for the waveforms selected as well as the total MI as a function of time. The MI for each waveform is computed as described in (12.46), where the information flow that the observation is expected to provide is weighted by the detection probability. The set of waveforms to transmit at a particular time is chosen to maximize the total MI subject to the total time constraint. Therefore, while the total MI represents the maximum total MI possible given the time constraints, the MI of the individual tasks is not necessarily the largest MI that any waveform could produce on that task.

The total MI is approximately constant over the simulation. Notable exceptions happen at $t \approx 15$ s, where Track 3 performs a maneuver and at $t = 100$ s, when Track 2 enters the region and has a large covariance provided by the detection engine. An additional feature this analysis reveals is that the MI of Track 1 declines near the end of the simulation as its range increases and detection probability is correspondingly decreased.

A more detailed look at the MI and its optimization is given in Figure 12.7. For each of the four tasks, we compare the MI (again, averaged over 100 trials) in the four-task simulation with the MI in a simulation that has only a single task. Broadly speaking, we find that there is a small loss in MI when all four tasks are present, resulting from the time constraint which prevents the scheduler from selecting the best possible waveform for each particular task at all times. A second

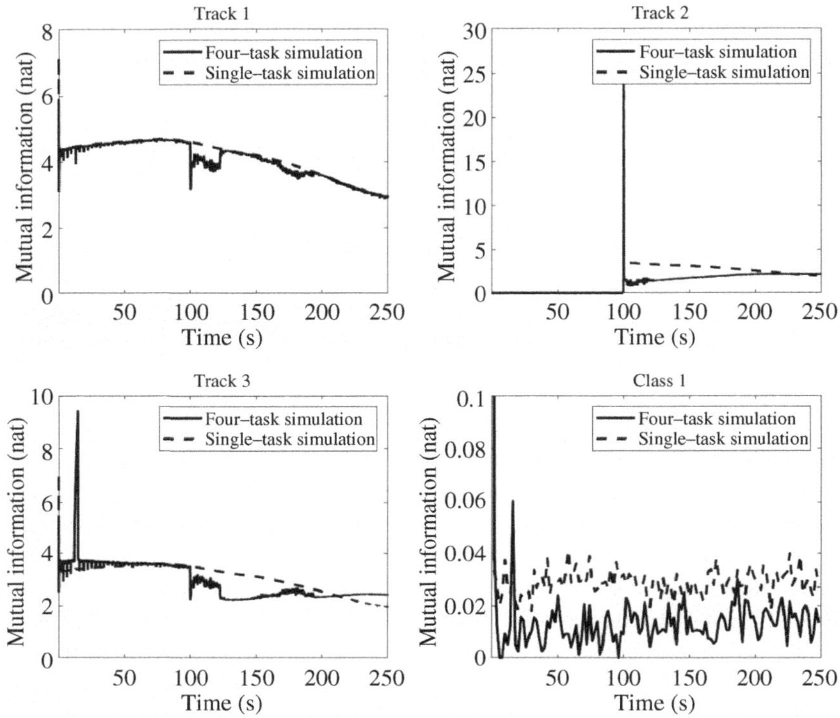

Figure 12.7 A comparison of the average (over 100 trials) MI in a four-task simulation with the MI in a simulation that has only a single task.

feature of the comparison is that occasionally the four-task approach has more MI than the single-task MI (e.g., Track 3 at $t = 15\,\mathrm{s}$). This is because the MI is a measure of the information *gain* the measurement provides. In the single-task case, the scheduler is able to use the full CPI to follow Track 3, meaning that it is well followed even as it maneuvers. In contrast, when the time is to be divided among many tasks, Track 3 is not as well followed and is temporarily out of track as it maneuvers. This leads to a large expected information flow from making an observation immediately after the maneuver.

Figure 12.8 shows the average position and velocity RMSE as a function of time for the four-task simulation. Track 1 and Track 3 are detected at onset, while Track 2 is detected after $t = 100\,\mathrm{s}$. Track 2 is initialized with a large covariance and has a short period of larger RMSE while the track is refined. After that period, the RMSEs of both position and velocity are similar for the three targets until the end of the simulation where Target 1's distance makes it significantly more difficult to detect.

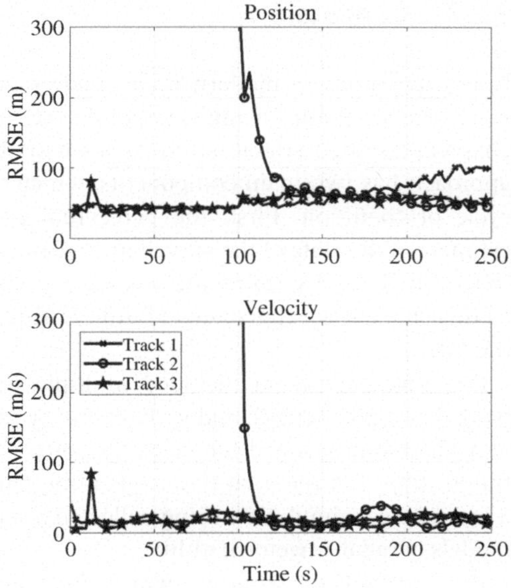

Figure 12.8 The average (over 100 trials) position and velocity RMSE tracking error for the three targets as a function of time.

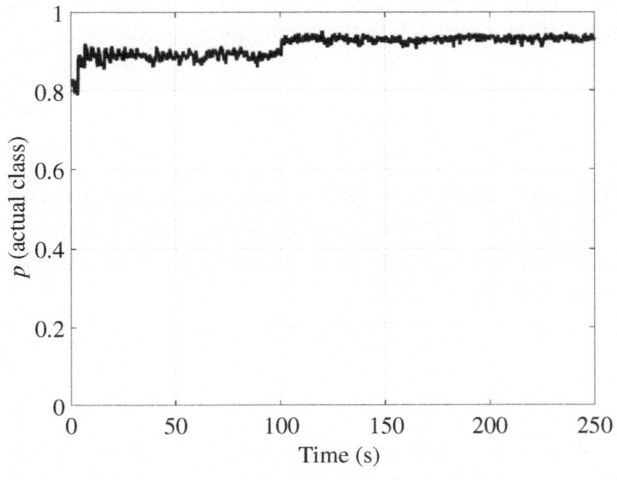

Figure 12.9 The average (over 100 trials) posterior probability in the correct class for the identification task.

Finally, Figure 12.9 shows the posterior probability of correct class for Target 1 averaged over 100 trials. As mentioned earlier, our classification state transition model diffuses the class probability over time, requiring occasional revisits with a classification waveform to update the probability. We find empirically that the system maintains a high probability in the correct class.

12.7 Conclusion

New capabilities in modern multi-mode radars enable techniques to optimize waveform selection for multi-target detection, tracking, and classification. This chapter presented a rigorous model-based mathematical approach to FARRM. The approach has two main components, which are based on the perception-action cycle of cognition. First, the perceptual processor constructs a probabilistic estimate of the state of the surveillance region. Next, the executive processor uses this estimate to determine the sensing actions expected to yield the most utility. In this chapter, we characterized utility using an information-theoretic measure, the MI.

Our model problem included multiple tracking and classification tasks, a bank of variable-length, and variable-utility waveforms with a fixed time budget. Our technique selected the collection of waveforms to transmit during a radar frame subject to the total time constraint. Our techniques were illustrated with a multi-target simulation using collected aircraft trajectories and first-principles models of radar waveform utility.

Recent trends in cognitive radar include using data-driven and learning-based methods in place of or in conjunction with model-based methods. Future work might include a comparison of our approach to these emerging techniques.

Acknowledgment

This material is based upon work supported by the Air Force Research Laboratory (AFRL) under Contract No. FA8649-20-P-0940. Any opinions, findings, and conclusions or recommendations expressed in this material are those of the authors and do not necessarily reflect the views of AFRL.

References

1 Haykin, S. (2006). Cognitive radar: a way of the future. *IEEE Signal Processing Magazine* 23(1): 30–40.

2 Haykin, S. (2012). *Cognitive Dynamic Systems: Perception-Action Cycle, Radar and Radio*. Cambridge University Press.

3 Guerci, J.R. (2010). *Cognitive Radar: The Knowledge-Aided Fully Adaptive Approach*. Artech House.

4 Greco, M., Gini, F., Stinco, P., and Bell, K. (2018). Cognitive radars: on the road to reality: progress thus far and possibilities for the future. *IEEE Signal Processing Magazine* 35(4): 112–125.

5 Gurbuz, S., Griffiths, H., Charlish, A. et al. (2019). An overview of cognitive radar: past, present, and future. *IEEE Aerospace and Electronic Systems Magazine* 34(12): 6–18.

6 Fuster, J.M. (2010). *Cortex and Mind: Unifying Cognition*. Oxford University Press.

7 Hernandez, M., Kirubarajan, T., and Bar-Shalom, Y. (2004). Multisensor resource deployment using posterior Cramér-Rao bounds. *IEEE Transactions on Aerospace and Electronic Systems* 40(2): 399–416.

8 Kreucher, C., Kastella, K., and Hero, A. (2003). Information-based sensor management for multitarget tracking. *Proceedings of the SPIE* 5204: 480–489.

9 Kreucher, C., Kastella, K., and Hero, A. (2003). A Bayesian method for integrated multitarget tracking and sensor management. *Proceedings of the 6th International Conference on Information Fusion*, 704–711.

10 Kreucher, C., Hero, A., and Kastella, K. (2005). A comparison of task driven and information driven sensor management for target tracking. *Proceedings of the 44th IEEE Conference on Decision and Control*, 4004–4009.

11 Kreucher, C., Kastella, K., and Hero, A. (2005). Sensor management using an active sensing approach. *Signal Processing* 85(3): 607–624.

12 Kreucher, C., Blatt, D., Hero, A., and Kastella, K. (2006). Adaptive multi-modality sensor scheduling for detection and tracking of smart targets. *Digital Signal Processing* 16(5): 546–567.

13 Kreucher, C., Kastella, K., and Hero, A. (2005). Multi-platform information-based sensor management. *Proceedings of the SPIE* 5820: 141–151.

14 Kreucher, C., Hero, A., Kastella, K., and Morelande, M. (2007). An information-based approach to sensor management in large dynamic networks. *Proceedings of the IEEE* 95(5): 978–999.

15 Chavali, P. and Nehorai, A. (2012). Scheduling and power allocation in a cognitive radar network for multiple-target tracking. *IEEE Transactions on Signal Processing* 60(2): 715–729.

16 Romero, R. and Goodman, N. (2013). Cognitive radar network: cooperative adaptive beamsteering for integrated search-and-track application. *IEEE Transactions on Aerospace and Electronic Systems* 49(2): 915–931.

17 Bell, K., Baker, C., Smith, G. et al. (2015). Cognitive radar framework for target detection and tracking. *IEEE Journal of Selected Topics in Signal Processing* 9(8): 1427–1439.

18 Charlish, A. and Hoffmann, F. (2017). Cognitive radar management. In: *Novel Radar Techniques and Applications Volume 2: Waveform Diversity and Cognitive Radar, and Target Tracking and Data Fusion*, 157–193. Institution of Engineering and Technology.

19 Schmaedeke, W. (1993). Information-based sensor management. *Proceedings of the SPIE* 1955: 156–164.

20 Manyika, J. and Durrant-Whyte, H.F. (1992). Information-theoretic approach to management in decentralized data fusion. *Proceedings of the SPIE* 1828: 202–213.

21 Bell, K., Smith, G.E., Mitchell, A.E., and Rangaswamy, M. (2018). Multiple task fully adaptive radar. *Proceedings of the 52nd Asilomar Conference on Signals, Systems, and Computers*, 1344–1348.

22 Bell, K., Kreucher, C., and Rangaswamy, M. (2021). An evaluation of task and information driven approaches for radar resource allocation. *Proceedings of 2021 IEEE Radar Conference.*

23 Mitchell, A.E., Smith, G.E., Bell, K.L. et al. (2018). Cost function design for the fully adaptive radar framework. *IET Radar, Sonar and Navigation* 12(12): 1380–1389.

24 Nadjiasngar, R. and Charlish, A. (2015). Quality of service resource management for a radar network. *Proceedings of 2015 IEEE Radar Conference*, 344–349.

25 Cover, T. and Thomas, J. (1991). *Elements of Information Theory*. NY: Wiley.

26 Kay, S.M. (2020). *Information Theoretic Signal Processing and its Applications.* Sachuest Point Publishers.

27 Liese, F. and Vajda, I. (2006). On divergences and informations in statistics and information theory. *IEEE Transactions on Information Theory* 52(10): 4394–4412.

28 Aughenbaugh, J.M. and LaCour, B.R. (2008). Metric selection for information theoretic sensor management. *Proceedings of the 11th International Conference on Information Fusion.*

29 Yang, C., Kadar, I., Blasch, E., and Bakich, M. (2011). Comparison of information theoretic divergences for sensor management. *Proceedings of the SPIE* 8050: 199–208.

30 Castañón, D.A., Mahler, R., Hintz, K.J. et al. (2006). Issues in resource management with applications to real-world problems. *Proceedings of the SPIE* 6235: 604–658.

31 Kreucher, C. (2005). An information-based approach to sensor resource allocation. PhD thesis. University of Michigan.

32 Mitchell, A.E., Smith, G.E., Bell, K.L. et al. (2018). Hierarchical fully adaptive radar. *IET Radar, Sonar and Navigation* 12(12): 1371–1379.

33 Smith, G.E., Cammenga, Z., Mitchell, A.E. et al. (2016). Experiments with cognitive radar. *IEEE Aerospace and Electronic Systems Magazine* 31(12): 34–46.

34 Bell, K., Smith, G., Mitchell, A., and Rangaswamy, M. (2019). Fully adaptive radar for target classification. *Proceedings of 2019 IEEE Radar Conference.*

35 Bell, K., Kreucher, C., Brandewie, A., and Johnson, J. (2023). Fully adaptive radar resource allocation for tracking and classification. In: *Next Generation Cognitive Radar*, Chapter 9, Radar, Electromagnetics and Signal Processing

Technologies Series (ed. K.V. Mishra, B.M. Shankar, and M. Rangaswamy), 277–312. IET Press.

36 Charlish, A., Bell, K., and Kreucher, C. (2020). Implementing perception-action cycles using stochastic optimization. *Proceedings of 2020 IEEE Radar Conference.*

37 Kreucher, C., Kastella, K., and Hero, A. (2005). Multitarget tracking using the joint multitarget probability density. *IEEE Transactions on Aerospace and Electronic Systems* 41(4): 1396–1414.

38 Patrikar, J., Moon, B., Oh, J., and Scherer, S. (2022). Predicting like a pilot: dataset and method to predict socially-aware aircraft trajectories in non-towered terminal airspace. *Proceedings of 2022 International Conference on Robotics and Automation (ICRA)*, 2525–2531.

39 Singer, R.A. (1970). Estimating optimal tracking filter performance for manned maneuvering targets. *IEEE Transactions on Aerospace and Electronic Systems* 6(4): 473–483.

40 Richards, M. (2005). *Fundamentals of Radar Signal Processing*. McGraw-Hill.

41 Van Trees, H.L. (2002). *Optimum Array Processing*. Wiley.

42 Dogandzic, A. and Nehorai, A. (2001). Cramér-Rao bounds for estimating range, velocity, and direction with an active array. *IEEE Transactions on Signal Processing* 49(6): 1122–1137.

43 Smith, V.L. (1962). An experimental study of competitive market behavior. *Journal of Political Economy* 70(2): 111–137.

44 Friedman, D. (1993). The double auction market institution: a survey. In: *The Double Auction Market: Institutions, Theories and Evidence*, Chapter 1 (ed. D. Friedman and J. Rust), 3–25. Perseus Publishing.

13

Information-Theoretic Limits of Integrated Sensing and Communications

Yifeng Xiong[1], Fuwang Dong[2], and Fan Liu[2]

[1] School of Information and Communication Engineering, Beijing University of Posts and Telecommunications, Beijing, China
[2] Department of Electronic and Electrical Engineering, Southern University of Science and Technology, Shenzhen, China

13.1 Introduction

The integrated sensing and communication (ISAC) technology has recently gained recognition from the international telecommunication union (ITU) as a key component within the 6G landscape, encompassing diverse usage scenarios [1, 2]. Through the utilization of unified hardware platforms and shared radio waveforms, this integration facilitates resource-efficient collaboration among sensing and communications (S&C) subsystems, supporting a range of emerging applications, notably vehicular-to-everything (V2X) networking, extended reality (XR), and digital twins.

ISAC system was originally designed to extend the capabilities of existing infrastructures, resulting in the development of two primary paradigms, namely, *sensing-centric* and *communication-centric*. The main idea of sensing-centric schemes is to convey communication information through the additional degree of freedom (DoF) of radar systems, such as through radar sidelobes and the waveform–antenna mapping of multiple-input multiple-output (MIMO) radars [3]. On the other hand, representative communication-centric schemes adopt the orthogonal frequency-division multiplexing (OFDM) waveforms to sense the target of interest [4]. However, the above-mentioned schemes do not specifically aim for maximal S&C performance enhancements. To further exploit the benefit of ISAC (termed as the "integration gain" [5]), recent advances have given rise to *joint design* schemes [6], which aim to conceive innovative joint signaling strategies from the ground up, effectively enabling both S&C tasks

Information-Theoretic Radar Signal Processing, First Edition. Edited by Yujie Gu and Yimin D. Zhang.

simultaneously. Despite these efforts, the fundamental limits analysis based on information theory and the resulting performance tradeoff between S&C have remained long-standing challenges within the research community [7]. Such information theory-based ISAC research direction may provide vital guidelines for practical system design and prompt a deeper understanding of the integration mechanism.

When examining a conventional separated S&C system, it becomes evident that their fundamental performance limitations stem primarily from resource constraints. For instance, the well-known capacity formula for the scalar Gaussian channel underscores that communication performance, specifically the achievable rate, hinges upon the availability of power and bandwidth resources. In stark contrast, the performance boundaries of ISAC systems diverge significantly. Essentially, designing an ISAC system presents itself as a multiobjective optimization problem. Unlike the separated S&C systems, the performance in both S&C domains mutually constrains each other within ISAC systems besides the resource budget. Intuitively, S&C tasks serve distinct objectives, thereby giving rise to varying "preference discrepancies" in system configurations. With only rare exceptions, it is improbable that S&C systems can simultaneously attain their optimal performance within the ISAC framework, implying the existence of a fundamental tradeoff.

For decades, S&C systems have been recognized as an intriguing pairing of information theory principles, intricately intertwined with each other [8]. Broadly, the tradeoff inherent in S&C can be comprehended through the perspective of a *signal-and-system* duality. To illustrate this, consider a straightforward linear Gaussian model

$$\boldsymbol{Y} = \boldsymbol{H}\boldsymbol{X} + \boldsymbol{Z}, \tag{13.1}$$

where $\boldsymbol{X}$, $\boldsymbol{Y}$, $\boldsymbol{H}$, and $\boldsymbol{Z}$ represent the transmitted ISAC signal, the received signal, the channel, and the noise, respectively. The communication task is to decode the message encoded in $\boldsymbol{X}$, whereas in sensing systems, the goal is to extract information from the channel $\boldsymbol{H}$. Inspired by this, we may regard $\boldsymbol{H}$ as the "transmitted signal" from the environment in the sensing system, while considering $\boldsymbol{X}$ as the passed-through "channel." This conceptual analogy suggests that communication and sensing tasks can be viewed as two distinct "virtual users" operating within a single multiple access channel framework, inevitably entailing a performance tradeoff between these two users. However, such a rough portrayal of the S&C tradeoff cannot meet the requirements for ISAC system designers. The pioneering mutual information and minimum mean square error (I-MMSE) equation illustrates the connection between estimation and information theories [9]. Furthermore, the authors of [10] have evaluated the inner bounds on the performance of a radar and communications coexistence system by introducing a novel radar estimation rate. Moreover, the authors of [11, 12] consider a scenario where the transmitter (Tx) communicates with a user through a memoryless state-dependent channel, simultaneously estimating the state from generalized feedback. The capacity-distortion-cost tradeoff of this channel is characterized to illustrate the optimal achievable rate for reliable communication while adhering to a preset state estimation distortion. Nevertheless, the fundamental limits of conceiving a joint signal for a general scenario where the Tx senses arbitrary targets rather than the specified communication state still remain widely open.

In this chapter, we will comprehensively synthesize current knowledge pertaining to various facets of the S&C tradeoff, employing the perspective of information theory [12–16]. Specifically, in Section 13.2, we shall commence with the basic capacity-distortion (C-D) theory including the general framework and how to efficiently compute the C-D boundary. In Section 13.3, we further illustrate the C-D theory by studying a typical parameter estimation task for sensing. We will show that the S&C tradeoff is a twofold tradeoff, namely, the resource limitation (subspace tradeoff (ST)) and different preferences on the signal patterns (deterministic-random tradeoff (DRT)). In Section 13.4, we will discuss the ST and DRT in target detection problems. Finally, we conclude this chapter as well as highlight several open challenges in Section 13.5. It should be highlighted that, throughout this chapter, a, $\mathbf{\mathsf{a}}$, and $\mathbf{\mathsf{A}}$ represent random variables (scalars), random vectors, and random matrices, respectively. The corresponding deterministic quantities are denoted by a, $\boldsymbol{a}$, and $\boldsymbol{A}$, respectively.

13.2 Capacity-Distortion Theory

The C-D theory was originally introduced in the realm of rate-distortion theory for lossy data compression [17]. Just as its name implies, the C-D theory explores the tradeoff between communication capacity and sensing distortion, which is a powerful tool for building a general analytical framework for S&C tradeoff in ISAC system [11–14]. Specifically, distortion encompasses a broad range of functions with the expression of $d(\mathsf{\eta}, \hat{\mathsf{\eta}})$. These functions operate on the true value of a specific quantity (referred to as the sensing parameter in sensing problems) $\mathsf{\eta}$ and its corresponding estimate $\hat{\mathsf{\eta}}$.

As for the ISAC waveform design, the waveform used for sensing is a random signal since it will communicate messages simultaneously. In this case, the expectation of the distortion is a reasonable performance metric to provide a holistic assessment of sensing performance over an extended duration. Let us take the typical estimation and detection tasks as examples.

- *Estimation task*: The widely adopted distortion function is the squared Euclidean distance, represented as $d(\boldsymbol{\eta}, \hat{\boldsymbol{\eta}}) = ||\boldsymbol{\eta} - \hat{\boldsymbol{\eta}}||^2$, whose expectation yields the mean square error (MSE).
- *Detection task*: In binary detection problems (e.g., target presence or absence), the objective is to determine the value of a binary variable $\eta \in \{0, 1\}$. A valid distortion function is the Hamming distance $d(\eta, \hat{\eta}) = \eta \oplus \hat{\eta}$, where $\oplus$ denotes the modulo-2 sum operator. Then, the associated expectation can be expressed by

$$\begin{aligned}\mathbb{E}\{\eta \oplus \hat{\eta}\} &= (1 \oplus 1)\Pr\{\hat{\eta} = 1|\eta = 1\} + (0 \oplus 0)\Pr\{\hat{\eta} = 0|\eta = 0\} \\ &\quad + (1 \oplus 0)\Pr\{\hat{\eta} = 1|\eta = 0\} + (0 \oplus 1)\Pr\{\hat{\eta} = 0|\eta = 1\} \\ &= 1 - P_D + P_{\mathrm{FA}}, \end{aligned} \tag{13.2}$$

where $\mathbb{E}\{\cdot\}$ denotes the expectation operator. $P_D \triangleq 1 - \Pr\{\hat{\eta} = 0|\eta = 1\}$ and $P_{\mathrm{FA}} \triangleq \Pr\{\hat{\eta} = 1|\eta = 0\}$ are the detection and false alarm probabilities, respectively. It is noteworthy that constant false alarm rate (CFAR) detectors, which adhere to the Neyman–Pearson criterion with a fixed P_{FA} [18], equate the minimization of the Hamming distance with the maximization of the detection probability P_D.

13.2.1 General Framework

For sensing tasks, it is important to note that the design of the channel input $\mathbf{X}_n$ for the nth channel use must adhere to causality principles. Consequently, it is not feasible to base this design on real-time channel output $\mathbf{Y}_n$ or the parameter value $\boldsymbol{\eta}_n$. Instead, one must rely exclusively on a *delayed feedback* signal denoted as $\mathbf{Z}_{n-1}$, whose value may rely on the channel output $\mathbf{Y}_n$, or the echoes originating from the target.

Let us consider such a memoryless ISAC channel model termed as state-dependent memoryless channel with delayed feedback (SDMC-DF), as shown in Figure 13.1. Consider a sequence of $(2^{NR}, R)$ codes $\{\mathbf{X}^N | N \in \mathbb{N}\}$ encoding the message $\mathsf{W} \in \{1, \ldots, 2^{NR}\}$, where N and R are the length of the coding block and the coding rate, respectively. Then, the effect of an SDMC-DF is given as [11]

$$\begin{aligned} &p_{\mathsf{W},\mathbf{Y}^N,\mathbf{Z}^N,\mathbf{X}^N,\boldsymbol{\eta}^N}(W, \boldsymbol{Y}, \boldsymbol{Z}, \boldsymbol{X}, \boldsymbol{\eta}) = p_{\mathsf{W}}(W) \\ &\quad\times \prod_{n=1}^{N} p_{\boldsymbol{\eta}}(\boldsymbol{\eta}_n) p_{\mathbf{X}|\mathsf{W},\mathbf{Z}}(\boldsymbol{X}_n | \mathsf{W}, \boldsymbol{Z}_{n-1}) p_{\mathbf{Y},\mathbf{Z}|\mathbf{X},\boldsymbol{\eta}}(\boldsymbol{Y}_n, \boldsymbol{Z}_n | \boldsymbol{X}_n, \boldsymbol{\eta}_n), \end{aligned} \tag{13.3}$$

and the per-block resource budgets and distortion metrics for such channels can be expressed by

$$\begin{aligned}\mathbb{E}\left\{b\left(\mathbf{X}^N\right)\right\} &= \frac{1}{N}\sum_{n=1}^{N}\mathbb{E}\{b(\mathbf{X}_n)\},\\ \mathbb{E}\left\{d\left(\boldsymbol{\eta}^N,\hat{\boldsymbol{\eta}}^N\right)\right\} &= \frac{1}{N}\sum_{n=1}^{N}\mathbb{E}\{d(\boldsymbol{\eta}_n,\hat{\boldsymbol{\eta}}_n)\},\end{aligned} \tag{13.4}$$

where $b(\cdot)$ denotes the instantaneous cost of a single codeword. Moreover, the values of the sensing parameters are allowed to vary across time in general, denoted by $\boldsymbol{\eta}_n$ at the nth channel use, constituting a parameter sequence $\boldsymbol{\eta}^N$. Thus, by defining the block error probability as $P_e^{(N)} \triangleq \Pr\{\hat{\mathsf{W}} \neq \mathsf{W}\}$ with length N, the achievability conditions can be concretized as

$$\limsup_{N\to\infty} \frac{1}{N}\sum_{n=1}^{N}\mathbb{E}\{d(\boldsymbol{\eta}_n,\hat{\boldsymbol{\eta}}_n)\} \leqslant D, \tag{13.5a}$$

$$\limsup_{N\to\infty} \frac{1}{N}\sum_{n=1}^{N}\mathbb{E}\{b(\mathbf{X}_n)\} \leqslant B, \tag{13.5b}$$

$$\limsup_{N\to\infty} P_e^{(N)} = 0, \tag{13.5c}$$

where D and B represent the expected distortion and resource budget, respectively. Note that when we temporarily drop the sensing distortion constraint (13.5a), the capacity aligns with the classical result of "capacity-with-cost" [19]

$$C_B = \max_{p_{\mathbf{X}}(\boldsymbol{X})\in\mathcal{P}_B} I(\mathbf{X};\mathbf{Y}), \tag{13.6}$$

where

$$\mathcal{P}_B = \{p_{\mathbf{X}}(\boldsymbol{X})|\mathbb{E}\{b(\mathbf{X})\} \leqslant B\}.$$

The result (13.6) suggests that the capacity subjected to cost budget constraint can be elegantly expressed in a single-letter form. This representation would significantly streamline subsequent analytical procedures. In pursuit of establishing a single-letter representation for the C-D function, we demonstrate that the optimal sensing distortion can be attained through the utilization of a straightforward letter-wise minimum expected posterior distortion estimator [12] with the form of $\left[\hat{\boldsymbol{\eta}}^\star(\mathbf{X}_1,\mathbf{Z}_1),\ldots,\hat{\boldsymbol{\eta}}^\star(\mathbf{X}_N,\mathbf{Z}_N)\right]$, where

$$\hat{\boldsymbol{\eta}}^\star(\boldsymbol{X},\boldsymbol{Z}) = \arg\min_{\boldsymbol{\eta}'} \int d(\boldsymbol{\eta},\boldsymbol{\eta}')p_{\boldsymbol{\eta}|\mathbf{X},\mathbf{Z}}(\boldsymbol{\eta}|\boldsymbol{X},\boldsymbol{Z})d\boldsymbol{\eta}, \tag{13.7}$$

and $p_{\boldsymbol{\eta}|\mathbf{X},\mathbf{Z}}(\boldsymbol{\eta}|\boldsymbol{X},\boldsymbol{Z})$ is the *a posteriori* distribution of $\boldsymbol{\eta}$ given the transmitted codeword $\mathbf{X}$ (known at the Tx) and the delayed feedback $\mathbf{Z}$. We can learn from this result that past feedback does not enhance the estimation of the current channel

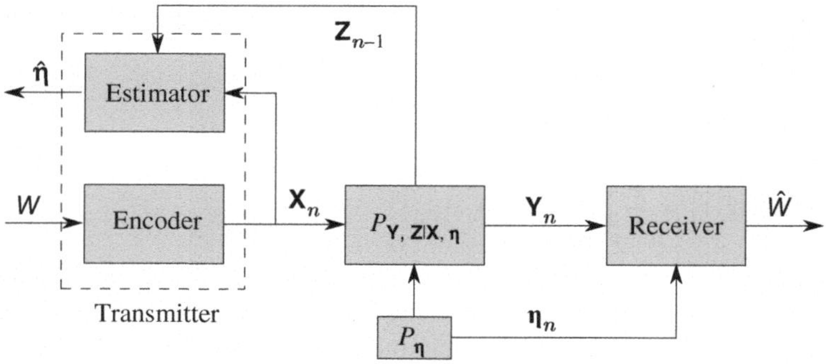

Figure 13.1 The structure of an SDMC-DF.

state in memoryless channels. It can be observed that the expected distortion also admits a single-letter representation

$$c(\boldsymbol{X}) = \mathbb{E}\left\{d\left(\boldsymbol{\eta}, \hat{\boldsymbol{\eta}}^{\star}(\mathbf{X}, \mathbf{Z})\right) | \mathbf{X} = \boldsymbol{X}\right\}. \tag{13.8}$$

Given this perspective, the C-D function can be regarded as a capacity that incurs two distinct costs [12]

$$C_B(D) = \sup_{p_{\mathbf{X}}(\boldsymbol{X}) \in \mathcal{P}_B \cap \mathcal{P}_D} I(\mathbf{X}; \mathbf{Y}|\boldsymbol{\eta}), \tag{13.9a}$$

$$\mathcal{P}_B = \{p_{\mathbf{X}}(\boldsymbol{X}) | \mathbb{E}\{b(\mathbf{X})\} \leqslant B\}, \tag{13.9b}$$

$$\mathcal{P}_D = \{p_{\mathbf{X}}(\boldsymbol{X}) | \mathbb{E}\{c(\mathbf{X})\} \leqslant D\}. \tag{13.9c}$$

The result (13.9) indicates an intriguing perspective, wherein the anticipated sensing distortion may alternatively be conceptualized as a form of "sensing-induced communication cost." This conceptual framework broadens the applicability of the C-D theory to a wider range of scenarios, even those where the performance of sensing cannot be adequately characterized by a conventional distortion function. Consider, for instance, the widely employed Cramér-Rao bound (CRB) as a performance metric for estimation problems. The CRB exhibits independence from any specific estimator $\hat{\boldsymbol{\eta}}$ and, in certain cases, it does not rely on the true value of the parameter $\boldsymbol{\eta}$. Consequently, the CRB cannot be classified as a distortion function. Nevertheless, it exhibits an association with the transmitted codeword $\mathbf{X}$ (as we will elucidate later), allowing for the generalized application of the C-D theory.

On the other hand, involving diverse channel models is another way of generalizing the C-D methodology. In practice, the sensing channel and the communication channel often exhibit distinctions, where the capacity-distortion approach can be applied with minor adjustments: replacing (13.9a) with

$$C_B(D) = \sup_{p_{\mathsf{X}}(\boldsymbol{X}) \in \mathcal{P}_B \cap \mathcal{P}_D} I(\mathbf{X}; \mathbf{Y}_c | \boldsymbol{\eta}). \tag{13.10}$$

13.2.2 Computing the C-D Boundary

The result (13.9) stands as a comprehensive representation of the C-D region from a purely theoretical perspective, as opposed to the operational definition (13.5). However, it is imperative to bear in mind that our initial aspiration for the C-D theory was to understand the nature of the S&C tradeoff, hopefully beyond the ST. Regrettably, digging the essence of ST itself from (13.9) proves to be an arduous task, let alone delving deeper into insight.

In this subsection, we provide a viable approach involving the explicit computation of C-D functions across representative scenarios in pursuit of insightful revelations. Notably, the renowned Blahut–Arimoto (B-A) algorithm, initially conceived for the computation of unconstrained capacity by Blahut [20], can be adapted to compute C-D functions with appropriate modifications. To elaborate further, the core of the original B-A algorithm, which addresses the maximization problem $C = \max_{p_{\mathsf{X}}(\boldsymbol{X})} I(\mathbf{X}; \mathbf{Y})$ given an initial trial distribution $r(\boldsymbol{X})$ for $p_{\mathsf{X}}(\boldsymbol{X})$ is embodied by a fixed-point iteration that iteratively executes the following two steps:

1. Update the trial distribution $q(\boldsymbol{X}|\boldsymbol{Y})$ for the *a posteriori* distribution $p_{\mathsf{X}|\mathsf{Y}}(\boldsymbol{X}|\boldsymbol{Y})$ according to

$$q(\boldsymbol{X}|\boldsymbol{Y}) \leftarrow \frac{p_{\mathsf{X}}(\boldsymbol{X}) p_{\mathsf{Y}|\mathsf{X}}(\boldsymbol{Y}|\boldsymbol{X})}{\int r(\boldsymbol{X}) p_{\mathsf{Y}|\mathsf{X}}(\boldsymbol{Y}|\boldsymbol{X}) d\boldsymbol{X}}; \tag{13.11}$$

2. Update $r(\boldsymbol{X})$ by

$$r(\boldsymbol{X}) \leftarrow \frac{e^{\int p_{\mathsf{Y}|\mathsf{X}}(\boldsymbol{Y}|\boldsymbol{X}) \log q(\boldsymbol{X}|\boldsymbol{Y}) d\boldsymbol{Y}}}{\int e^{\int p_{\mathsf{Y}|\mathsf{X}}(\boldsymbol{Y}|\boldsymbol{X}) \log q(\boldsymbol{X}|\boldsymbol{Y}) d\boldsymbol{Y}} d\mathbf{X}}, \tag{13.12}$$

which is equivalent to solving the following constrained differential entropy maximization problem:

$$\begin{aligned} r(\boldsymbol{X}) \leftarrow \arg\max_{p(\boldsymbol{X})} & -\int p(\boldsymbol{X}) \log p(\boldsymbol{X}) d\boldsymbol{X}, \\ \text{s.t.} \quad & p(\boldsymbol{X}) \geqslant 0 \ \forall \boldsymbol{X}, \ \int p(\boldsymbol{X}) d\boldsymbol{X} = 1, \\ & \mathbb{E}_{p(\boldsymbol{X})} \left\{ \int p_{\mathsf{Y}|\mathsf{X}}(\boldsymbol{Y}|\boldsymbol{X}) \log q(\boldsymbol{X}|\boldsymbol{Y}) d\boldsymbol{Y} \right\} \geqslant t, \end{aligned} \tag{13.13}$$

for some constant t.

In our scheme, the objective function is changed to the conditional mutual information $I(\mathbf{X}; \mathsf{Y}|\eta)$. Thus, formula (13.11) will be replaced by

$$q(\boldsymbol{X}|\boldsymbol{Y},\boldsymbol{\eta}) \leftarrow \frac{p_{\mathsf{X}}(\boldsymbol{X})p_{\mathsf{Y}|\mathsf{X},\eta}(\boldsymbol{Y}|\boldsymbol{X},\boldsymbol{\eta})}{\int p_{\mathsf{X}}(\boldsymbol{X})p_{\mathsf{Y}|\mathsf{X},\eta}(\boldsymbol{Y}|\boldsymbol{X},\boldsymbol{\eta})d\boldsymbol{X}}. \tag{13.14}$$

Additionally, since two extra constraints (13.9b) and (13.9c) are now enforced, (13.12) should be modified by

$$r(\boldsymbol{X}) \leftarrow \frac{e^{\int [p_{\eta}(\boldsymbol{\eta})p_{\mathsf{Y}|\mathsf{X},\eta}(\boldsymbol{Y}|\boldsymbol{X},\boldsymbol{\eta})\log q(\boldsymbol{X}|\boldsymbol{Y},\boldsymbol{\eta})-\lambda b(\boldsymbol{X})-\mu c(\boldsymbol{X})]d\boldsymbol{\eta}d\boldsymbol{Y}}}{\int e^{\int [p_{\eta}(\boldsymbol{\eta})p_{\mathsf{Y}|\mathsf{X},\eta}(\boldsymbol{Y}|\boldsymbol{X},\boldsymbol{\eta})\log q(\boldsymbol{X}|\boldsymbol{Y},\boldsymbol{\eta})-\lambda b(\boldsymbol{X})-\mu c(\boldsymbol{X})]d\boldsymbol{\eta}d\boldsymbol{Y}}d\boldsymbol{X}}, \tag{13.15}$$

where λ and μ are the Lagrangian multipliers corresponding to the constraints (13.9b) and (13.9c), respectively.

In the preceding discussions, it exhibits a vague sense that the imposition of sensing distortion requirements on the ISAC signal $\mathbf{X}$ has the effect of diminishing its randomness. These constraints, by design, introduce limitations to the entropy maximization problem. However, this observation does not constitute a rigorously defined statement, since the entropy maximization serves as an intermediary step within the context of the B-A algorithm. In Section 13.2.3, we will provide a case study of the single-input single-output (SISO) Gaussian channel model to validate our intuition.

13.2.3 Case Study: Real-Valued SISO Gaussian Channel

The real-valued SISO Gaussian channel model with Rayleigh fading can be written as

$$\mathsf{Y}_n = \eta_n \mathsf{X}_n + \mathsf{N}_n, \tag{13.16}$$

with X_n being the ISAC waveform which meets the power budget

$$\limsup_{N\to\infty} \frac{1}{N}\sum_{n=1}^{N} \mathbb{E}\{|\mathsf{X}_n|^2\} \leqslant P_T = 10\,\text{dB}. \tag{13.17}$$

η_n and N_n are mutually independent zero-mean Gaussian variables with unit variance. The sensing-optimal estimator is the letter-wise minimum mean square error (MMSE) estimator given by $\hat{\eta}_{n,\text{MMSE}} = \mathbb{E}\{\eta_n|\mathsf{X}_n, \mathsf{Y}_n\}$, assuming the use of a quadratic sensing distortion $d(\eta, \hat{\eta}) = (\eta - \hat{\eta})^2$. Moreover, we consider the perfect feedback scenario, i.e., $\mathsf{Z}_n = \mathsf{Y}_n$. Thus, the MSE can be computed as follows:

$$\frac{1}{N}\sum_{n=1}^{N} \mathbb{E}\{(\eta - \hat{\eta}_{n,\text{MMSE}})^2\} = \mathbb{E}\left\{\frac{1}{1+|\mathsf{X}|^2}\right\}, \tag{13.18}$$

where X is the single-letter representation of the channel input following the probability distribution $p_{\mathsf{X}}(X)$.

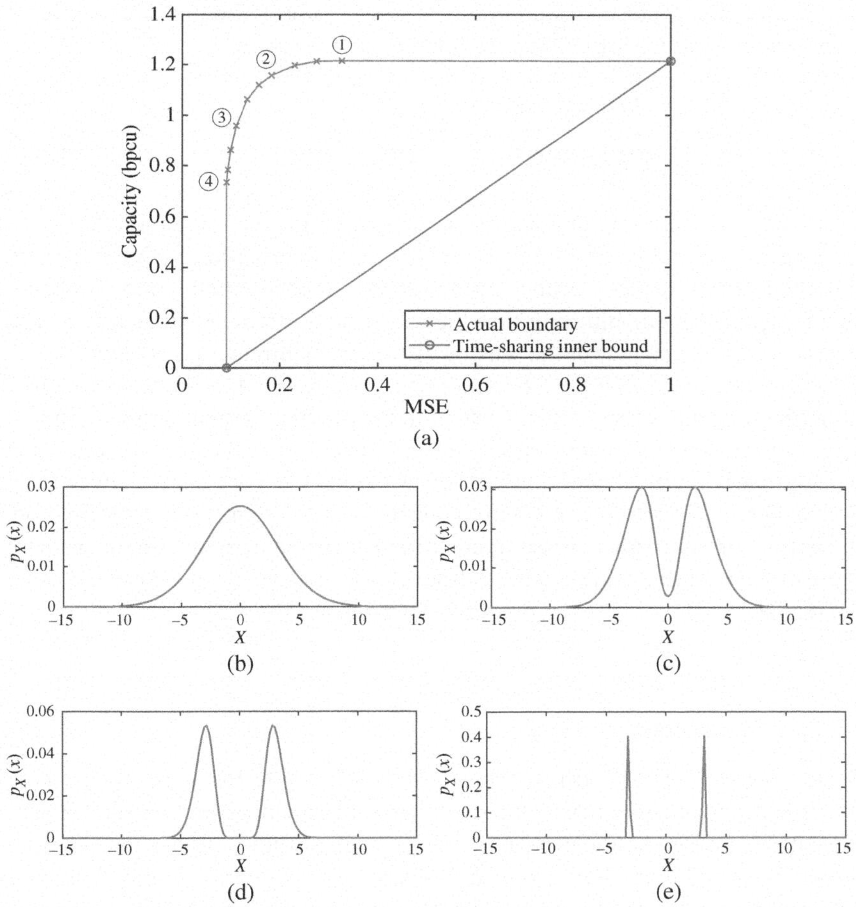

Figure 13.2 The C-D boundary of the real-valued SISO Gaussian channel scenario with $P_T = 10$ dB, as well as the Pareto-optimal input distributions $p_X(X)$ along the boundary. (a) The C-D boundary, (b) $p_X(X)$ at ①, (c) $p_X(X)$ at ②, (d) $p_X(X)$ at ③, and (e) $p_X(X)$ at ④.

In accordance with the previous assumptions, one can numerically determine the C-D boundary by employing the adapted B-A algorithm, as illustrated in Figure 13.2. At the communication-optimal point ①, the input distribution is Gaussian, yielding the following results:

$$C(D) = \frac{1}{2}\mathbb{E}\{\log_2(1 + P_T|\eta|^2)\} \approx 1.213 \text{ bits},$$
$$\mathbb{E}\{c(\mathsf{X})\} = \mathbb{E}\left\{\frac{1}{1+|\mathsf{X}|^2}\right\} \approx 0.327. \tag{13.19}$$

By contrast, the input distribution at the sensing-optimal point ④ tends to be binary phase-shift keying (BPSK) modulation, and we have

$$C(D) \approx 0.733 \text{ bits},$$
$$\mathbb{E}\{c(\mathsf{X})\} = \frac{1}{1+P_T} \approx 0.091. \tag{13.20}$$

As we traverse the C-D boundary from the communication-optimal point ① to the sensing-optimal point ④, the associated input distribution $p_{\mathsf{X}}(X)$ undergoes a gradual shift from a Gaussian distribution toward BPSK modulation. This transition aligns with our intuition that the ISAC signal becomes progressively less random as the stringency of the sensing distortion requirement increases.

It should be noted that the ST cannot be clearly justified in the SISO scenario. Consequently, the observed S&C tradeoff in the previously mentioned example is primarily a result of the differing signal preferences between communication and sensing tasks. Intuitively, communication tasks tend to favor signals characterized by a higher degree of randomness to enable the transmission of more information. In contrast, sensing tasks prefer signals with a degree of determinism to facilitate the distinction of received signals (such as echoes) associated with varying sensing parameters. This tradeoff is commonly referred to as DRT [15].

13.2.4 Discussions

As we showed in this section, the C-D theory, while being highly general, lacks illustrative qualities. Moreover, the numerical computation of the capacity-distortion boundary using the B-A algorithm can become computationally prohibitive. For instance, consider the scenario where there are K samples per dimension. Thus, the total number of samples N_{MC} required by the widely adopted Monte Carlo integration method is on the order of $N_{\text{MC}} = \mathcal{O}\left(K^{N_{\mathsf{Y}}+N_{\mathsf{X}}+N_{\boldsymbol{\eta}}}\right)$, where N_{Y}, N_{X}, and $N_{\boldsymbol{\eta}}$ are the dimensionalities of $\mathbf{Y}$, $\mathbf{X}$, and $\boldsymbol{\eta}$, respectively. This exponential growth in sample complexity can readily exceed the computational capacity of most devices, even when dealing with small-scale MIMO systems. Consequently, there is a need for less general yet more illustrative approaches, which may involve the utilization of specific sensing performance metrics with favorable properties.

In Sections 13.3 and 13.4, we endeavor to identify distortion functions that are amenable to rigorous analysis while also possessing the requisite generality to reflect the characteristics of ST and DRT. These distortion functions, or alternatively, nondistortion sensing-induced communication costs as implied by (13.9), serve as key elements in our investigation.

13.3 Parameter Estimation

Numerous performance metrics exist for parameter estimation, including both distortion and nondistortion metrics. While the MSE serves as a straightforward metric, its computation depends on the possibly complicated MMSE estimator. By contrast, CRB, a lower bound for all unbiased estimators, usually facilitates more tractable analytical expressions. Furthermore, the CRB is attainable in the regime of high signal-to-noise-ratio (SNR) [18]. In this section, we shift our focus from the general distortion measure $d(\boldsymbol{\eta}, \hat{\boldsymbol{\eta}})$ to the CRB of target parameter estimation and investigate the interplay between CRB and communication rate as a special case of the C-D tradeoff.

13.3.1 CRB-Rate Region

The general received signal models for communication and sensing can be expressed by

$$\mathbf{Y}_c = \mathbf{H}_c\mathbf{X} + \mathbf{Z}_c, \tag{13.21a}$$

$$\mathbf{Y}_s = \mathbf{H}_s(\boldsymbol{\eta})\mathbf{X} + \mathbf{Z}_s, \tag{13.21b}$$

where the sensing channel $\mathbf{H}_s \in \mathbb{C}^{N_s \times M}$ is a deterministic function (possibly non-linear) of the random parameter $\boldsymbol{\eta} \in \mathbb{R}^K$. A typical example of an angular MIMO radar channel can be found in [21]. Under the assumption of a memoryless channel, both the sensing parameter $\boldsymbol{\eta}$ and communication channel $\mathbf{H}_c \in \mathbb{C}^{N_c \times M}$ vary independent and identically distributed (i.i.d.) every T symbols. $\mathbf{X} \in \mathbb{C}^{M \times T}$, $\mathbf{Z}_c$, and $\mathbf{Z}_s$ represent the transmitted ISAC signal matrix, the Gaussian noise with zero mean, and variances of σ_c^2 and σ_s^2, respectively. Here, M, N_s, N_c, and K denote the numbers of transmitting antennas, sensing receiving antennas, communication receiving antennas, and the parameters of interest, respectively. To avoid deviation from the core idea, we assume that the ISAC Tx has perfect knowledge on $\mathbf{H}_c$.

The application scenario of the separate channel models in (13.21) is illustrated in Figure 13.3. The ISAC Tx communicates with a single communication Rx while sensing one or more targets simultaneously by transmitting a dual-functional signal $\mathbf{X}$. The sensing Rx can be situated in one of two configurations, namely, the monostatic sensing model and the bistatic sensing model (linked to the Tx via a wired connection). In both scenarios, the Tx and sensing Rx have the perfect knowledge of the ISAC signal $\mathbf{X}$, whereas X is unknown to the communication Rx since it contains useful information intended for the communication user. Consequently, we regard the transmitted ISAC signal $\mathbf{X} \sim p_{\mathbf{X}}(X)$ as a random

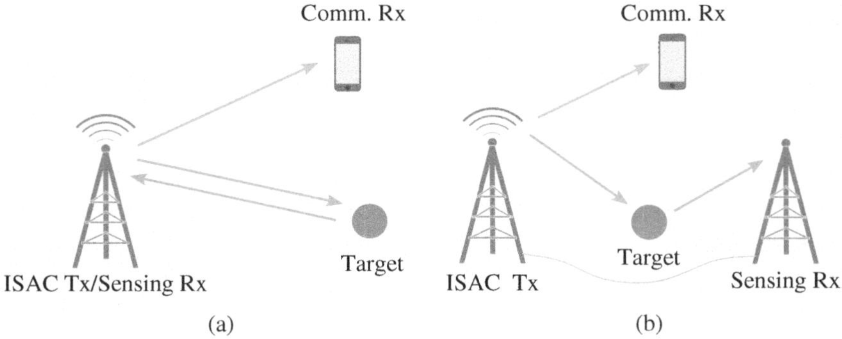

Figure 13.3 The ISAC scenarios described in (13.21), where the dual-functional waveform $\mathbf{X}$ is known to both the ISAC Tx and sensing receiver (Rx). (a) Monostatic sensing and (b) bistatic sensing.

matrix, with its realization perfectly known at both the ISAC Tx and sensing Rx, but unknown at the communication Rx.

The aforementioned ISAC signal framework does not impose any new challenge on the communication performance metric which can be directly evaluated by the mutual information $I(\mathbf{X}; \mathbf{Y}_c|\mathbf{H}_c)$. However, due to the randomness of $\mathbf{X}$, the sensing CRB becomes a function of $\mathbf{X}$, which is different from the commonly used CRB within the deterministic signal frameworks. One potential strategy involves treating $\mathbf{X}$ as a potential nuisance parameter. This could entail either incorporating it as an integral component of the elusive sensing parameter or integrating it out from the likelihood function that characterizes the observations. This approach yields conventional expressions for the hybrid and marginal CRB, respectively. However, neither of these methods fully captures a crucial aspect of the ISAC system: the fact that the random signal $\mathbf{X}$ is explicitly known to the receiving Rx, evidently leading to a loose bound in general. Consequently, we adopt a Miller–Chang type CRB, which entails the computation of the CRB for a specific realization of $\mathbf{X}$, followed by taking the expectation over the random variable $\mathbf{X}$ [22]. Thus, the MSE for any weakly unbiased estimator is lower bounded by the Miller–Chang Bayesian CRB in the form of

$$\mathrm{MSE}_{\boldsymbol{\eta}}(\hat{\boldsymbol{\eta}}) \geqslant \mathbb{E}\left\{\mathrm{tr}(\mathbf{J}_{\boldsymbol{\eta}|\mathbf{X}}^{-1})\right\}, \tag{13.22}$$

where $\mathrm{tr}(\cdot)$ is the matrix trace operator, and the expectation is taken with respect to $\mathbf{X}$. The Bayesian–Fisher information matrix (BFIM) of $\boldsymbol{\eta}$ can be expressed by

$$\begin{aligned}\mathbf{J}_{\boldsymbol{\eta}|\mathbf{X}} &= \mathbb{E}\left\{\left.\frac{\partial \log p_{\mathbf{Y}_s|\mathbf{X},\boldsymbol{\eta}}(\boldsymbol{Y}_s|\boldsymbol{X},\boldsymbol{\eta})}{\partial \boldsymbol{\eta}} \frac{\partial \log p_{\mathbf{Y}_s|\mathbf{X},\boldsymbol{\eta}}(\boldsymbol{Y}_s|\boldsymbol{X},\boldsymbol{\eta})}{\partial \boldsymbol{\eta}^{\mathrm{T}}}\right| \mathbf{X}\right\} \\ &+ \mathbb{E}\left\{\frac{\partial \log p_{\boldsymbol{\eta}}(\boldsymbol{\eta})}{\partial \boldsymbol{\eta}} \frac{\partial \log p_{\boldsymbol{\eta}}(\boldsymbol{\eta})}{\partial \boldsymbol{\eta}^{\mathrm{T}}}\right\}. \end{aligned} \tag{13.23}$$

Next, we will show that the BFIM can be further transformed as an affine map of the sample covariance matrix $\mathbf{R_X} = T^{-1}\mathbf{XX}^{\mathrm{H}}$ [15], where $(\cdot)^{\mathrm{H}}$ denotes the Hermitian transpose operator.

The sensing model (13.21) may be rewritten as

$$\mathrm{vec}(\mathbf{Y}_s^{\mathrm{T}}) = (\boldsymbol{I}_{N_s} \otimes \mathbf{X}^{\mathrm{T}})\mathrm{vec}(\mathbf{H}_s^{\mathrm{T}}) + \mathrm{vec}(\mathbf{Z}_s^{\mathrm{T}}), \tag{13.24}$$

where $\mathrm{vec}(\cdot)$ and $\otimes$ denote the vectorization operator and Kronecker product, respectively. By defining an extended parameter vector

$$\boldsymbol{\theta} = \begin{bmatrix} (\boldsymbol{I}_{N_s} \otimes \mathbf{X}^{\mathrm{T}})\mathrm{vec}(\mathbf{H}_s^{\mathrm{T}}) \\ (\boldsymbol{I}_{N_s} \otimes \mathbf{X}^{\mathrm{H}})\mathrm{vec}(\mathbf{H}_s^{\mathrm{H}}) \end{bmatrix},$$

the BFIM of $\boldsymbol{\theta}$ can be expressed as

$$\boldsymbol{J}_{\boldsymbol{\theta}} = \sigma_s^{-2}\boldsymbol{I}_{2N_sT}. \tag{13.25}$$

Thus, the BFIM of $\mathbf{h}_s = [\mathrm{vec}\,(\mathbf{H}_s^{\mathrm{T}})^{\mathrm{T}}, \mathrm{vec}\,(\mathbf{H}_s^{\mathrm{T}})^{\mathrm{H}}]^{\mathrm{T}}$ can be written as

$$\mathbf{J}_{\mathbf{h}_s|\mathbf{X}} = \mathbb{E}_{\mathbf{h}_s}\left\{\left(\frac{\partial\boldsymbol{\theta}}{\partial\mathbf{h}_s}\right)^{*}\boldsymbol{J}_{\boldsymbol{\theta}}\left(\frac{\partial\boldsymbol{\theta}}{\partial\mathbf{h}_s}\right)^{\mathrm{T}}\right\} = \frac{T}{\sigma_s^2}\begin{bmatrix} \boldsymbol{I}_{N_s} \otimes \mathbf{R}_{\mathbf{X}}^{\mathrm{T}} & \boldsymbol{0} \\ \boldsymbol{0} & \boldsymbol{I}_{N_s} \otimes \mathbf{R}_{\mathbf{X}} \end{bmatrix}, \tag{13.26}$$

where $(\cdot)^*$ denotes the complex conjugate operators. Let us denote $\mathbf{F} = \frac{\partial\mathbf{h}_s^*}{\partial\boldsymbol{\eta}} = [\mathbf{F}_1, \mathbf{F}_2, \ldots, \mathbf{F}_{2N_s}] \in \mathbb{C}^{K\times 2N_sT}$, then we have

$$\begin{aligned}\mathbf{J}_{\boldsymbol{\eta}|\mathbf{X}} &= \frac{T}{\sigma_s^2}\mathbb{E}_{\boldsymbol{\eta}}\left\{\mathbf{F}\begin{bmatrix} \boldsymbol{I}_{N_s} \otimes \mathbf{R}_{\mathbf{X}}^{\mathrm{T}} & \boldsymbol{0} \\ \boldsymbol{0} & \boldsymbol{I}_{N_s} \otimes \mathbf{R}_{\mathbf{X}} \end{bmatrix}\mathbf{F}^{\mathrm{H}}\right\} + \boldsymbol{J}_{\mathrm{P}} \\ &= \frac{T}{\sigma_s^2}\mathbb{E}_{\boldsymbol{\eta}}\left\{\sum_{i=1}^{N_s}\mathbf{F}_i\mathbf{R}_{\mathbf{X}}^{\mathrm{T}}\mathbf{F}_i^{\mathrm{H}} + \mathbf{F}_{N_s+i}\mathbf{R}_{\mathbf{X}}\mathbf{F}_{N_s+i}^{\mathrm{H}}\right\} + \boldsymbol{J}_{\mathrm{P}},\end{aligned} \tag{13.27}$$

where $\boldsymbol{J}_{\mathrm{P}}$ is defined by the second term in (13.23), relying on the prior distribution $p_{\boldsymbol{\eta}}(\boldsymbol{\eta})$. Let us define a pair of linear super-operators $\boldsymbol{\Phi}_1(\cdot)$ and $\boldsymbol{\Phi}_2(\cdot)$, characterized by

$$\boldsymbol{\Phi}_1(\boldsymbol{A}) = \sum_{i=1}^{N_s}\mathbf{F}_i\boldsymbol{A}\mathbf{F}_i^{\mathrm{H}}, \quad \boldsymbol{\Phi}_2(\boldsymbol{A}) = \sum_{i=1}^{N_s}\mathbf{F}_{N_s+i}\boldsymbol{A}\mathbf{F}_{N_s+i}^{\mathrm{H}}. \tag{13.28}$$

Note that the following matrix

$$\boldsymbol{\Psi}_1 = \sum_{i=1}^{M}\sum_{j=1}^{M}\boldsymbol{E}_{ij} \otimes \boldsymbol{\Phi}_1(\boldsymbol{E}_{ij}), \tag{13.29}$$

is a faithful representation of $\boldsymbol{\Phi}_1$, where $\boldsymbol{E}_{ij}$ is an $M \times M$ matrix whose (i, j)th entry equals to 1, while all other entries are zero. This follows the Choi representation based on Choi–Jamiolkowski isomorphism [15]. Then we have

$$\boldsymbol{\Psi}_1 = \sum_{i=1}^{N_s}\left(\mathrm{vec}(\mathbf{F}_i)\right)\left(\mathrm{vec}(\mathbf{F}_i)\right)^{\mathrm{H}}. \tag{13.30}$$

By defining $\widetilde{\boldsymbol{\Psi}}_1 = \mathbb{E}\{\boldsymbol{\Psi}_1\}$, the eigendecomposition of $\widetilde{\boldsymbol{\Psi}}_1$ can be expressed by

$$\widetilde{\boldsymbol{\Psi}}_1 = \sum_{i=1}^{r_1} \left(\sqrt{\lambda_i}\boldsymbol{u}_i\right)\left(\sqrt{\lambda_i}\boldsymbol{u}_i\right)^{\mathrm{H}}, \tag{13.31}$$

where λ_i and $\boldsymbol{u}_i$ are the ith nonzero eigenvalue and the corresponding eigenvector, respectively. r_1 is the number of the nonzero eigenvalues. Note that there exist a linear one-to-one correspondence between every realization of $\boldsymbol{\Phi}_1$ and its Choi representation of $\boldsymbol{\Psi}_1$, we have that $\widetilde{\boldsymbol{\Psi}}_1 = \mathbb{E}\{\boldsymbol{\Psi}_1\}$ is also the Choi representation of $\mathbb{E}\{\boldsymbol{\Phi}_1(\boldsymbol{A})\}$, and thus

$$\mathbb{E}\{\boldsymbol{\Phi}_1(\boldsymbol{A})\} = \sum_{i=1}^{r_1} \widetilde{\boldsymbol{F}}_i\boldsymbol{A}\widetilde{\boldsymbol{F}}_i^{\mathrm{H}} \tag{13.32}$$

with r_1 being the rank of $\widetilde{\boldsymbol{\Psi}}_1$ and $\widetilde{\boldsymbol{F}}_i = \sqrt{\lambda_i}\mathrm{mat}(\boldsymbol{u}_i)$, where $\mathrm{mat}_{K\times M}(\boldsymbol{u})$ represents the $K \times M$ dimensional matrix satisfying $\mathrm{vec}(\mathrm{mat}_{K\times M}(\boldsymbol{u})) = \boldsymbol{u}$. Similar arguments also apply to $\boldsymbol{\Phi}_2(\cdot)$, hence we have

$$\mathbb{E}\{\boldsymbol{\Phi}_2(\boldsymbol{A})\} = \sum_{i=1}^{r_2} \widetilde{\boldsymbol{G}}_i\boldsymbol{A}\widetilde{\boldsymbol{G}}_i^{\mathrm{H}}, \tag{13.33}$$

where $\widetilde{\boldsymbol{G}}_i = \sqrt{\overline{\lambda}_i}\mathrm{mat}(\overline{\boldsymbol{u}}_i)$. Here, r_2, $\overline{\lambda}_i$, and $\overline{\boldsymbol{u}}_i$ have the similar definitions in (13.31). Consequently, we have

$$\mathbf{J}_{\eta|\mathbf{X}} = \frac{T}{\sigma_s^2}\boldsymbol{\Phi}(\mathbf{R}_{\mathbf{X}}) = \sum_{i=1}^{r_1} \widetilde{\boldsymbol{F}}_i\mathbf{R}_{\mathbf{X}}^{\mathrm{T}}\widetilde{\boldsymbol{F}}_i^{\mathrm{H}} + \sum_{j=1}^{r_2} \widetilde{\boldsymbol{G}}_j\mathbf{R}_{\mathbf{X}}\widetilde{\boldsymbol{G}}_j^{\mathrm{H}} + \frac{\sigma_s^2}{T}\boldsymbol{J}_{\mathrm{P}}. \tag{13.34}$$

It can be observed from (13.22) that the Miller–Chang CRB can be interpreted as an analogous concept of "expected sensing distortion," although it does not serve as a real distortion metric. This interpretation allows us to regard the CRB as a cost attributed to sensing that impacts the allocation of signaling resources. Even in scenarios involving high-dimensional systems, valuable insights regarding the tradeoff between the CRB and the achievable data rate can still be gleaned through the exploitation of the affine structure inherent in the BFIM, as elaborated upon in the Sections 13.3.2–13.3.4.

13.3.2 CRB-Rate Tradeoff

To clarify the CRB-rate tradeoff, we construct the corresponding Pareto optimization problem as follows

$$\min_{p_{\mathbf{X}}(\boldsymbol{X})} \ \alpha\mathbb{E}\left\{\mathrm{tr}([\boldsymbol{\Phi}(\mathbf{R}_{\mathbf{X}})]^{-1})\right\} - (1-\alpha)I(\mathbf{X};\mathbf{Y}_c|\mathbf{H}_c), \tag{13.35a}$$

$$\text{s.t.} \quad \mathbb{E}\left\{\mathrm{tr}(\mathbf{R}_{\mathbf{X}})\right\} = P_{\mathrm{T}}, \tag{13.35b}$$

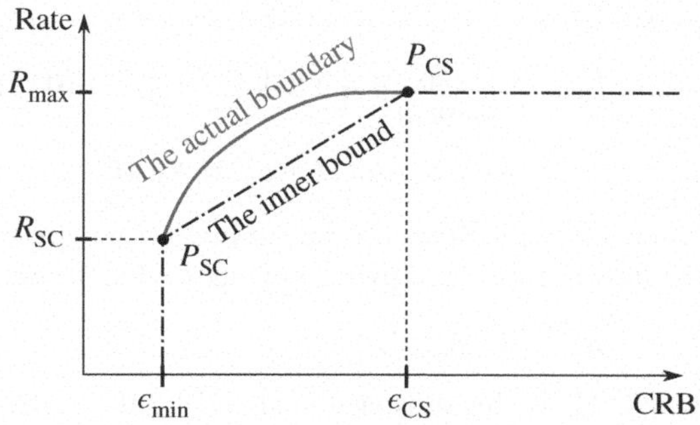

Figure 13.4 CRB-rate region, Pareto boundary, and time-sharing inner bound.

where weighting factor $\alpha \in [0, 1]$ is introduced to adjust the priority of the S&C performance. It should be highlighted that problem (13.35) can be equivalently reformulated by a capacity maximization problem subjected to two cost constraints when we move the CRB from objective (13.35a) to (13.35b).

Naturally, the complete representation of the Pareto boundary within the CRB-rate region would impose prohibitively high computational burdens. This arises from the necessity of numerically searching for the optimal probability density function $p_{\mathbf{X}}(\boldsymbol{X})$ by using the modified B-A algorithm provided previously. Instead of fully depicting the Pareto boundary, we further investigate two special corner points over the Pareto frontier which are remarked in Figure 13.4 to reveal fundamental insights into the CRB-rate tradeoff. These two points include

- P_{CS}: the optimal CRB under the constraint of maximizing communication capacity,
- P_{SC}: the optimal achievable rate constrained by the minimum CRB.

Obviously, all the points located on the line segment connected to these two points are achievable by a time-sharing strategy. Thus, we mainly focus on the characteristics of the S&C performance at the two points.

13.3.3 P_{CS} Performance Characterization

It is widely acknowledged that the Gaussian distribution is the sole capacity-achieving input distribution subjected to an average power constraint [23]. This case corresponds to problem (13.35) with only power constraint while setting $\alpha = 0$. Furthermore, note that each column of $\mathbf{X}$ follows the distribution of $\mathcal{CN}(\mathbf{0}, \widetilde{\boldsymbol{R}}_{\mathbf{X}}^{CS})$, where the statistical covariance matrix $\widetilde{\boldsymbol{R}}_{\mathbf{X}}^{CS}$ can be calculated by maximizing the achievable rate as follows:

$$
\begin{aligned}
R_{\max} &= \max_{\widetilde{\boldsymbol{R}} \succcurlyeq \mathbf{0}, \widetilde{\boldsymbol{R}} = \widetilde{\boldsymbol{R}}^{\mathrm{H}}} \mathbb{E}\left\{ \log\det\{\boldsymbol{I} + \sigma_c^{-2}\mathbf{H}_c\widetilde{\boldsymbol{R}}\mathbf{H}_c^{\mathrm{H}}\}\right\} \text{ s.t. } \mathrm{tr}(\widetilde{\boldsymbol{R}}) \leqslant P_{\mathrm{T}}, \\
&= \mathbb{E}\left\{ \log\det\{\boldsymbol{I} + \sigma_c^{-2}\mathbf{H}_c\widetilde{\boldsymbol{R}}_{\mathbf{X}}^{\mathrm{CS}}\mathbf{H}_c^{\mathrm{H}}\}\right\},
\end{aligned} \tag{13.36}
$$

where $\det\{\cdot\}$ denotes the determinant of a matrix. The optimal solution of (13.36) exhibits the special structure of eigenvalue decomposition as follows:

$$
\widetilde{\boldsymbol{R}}_{\mathbf{X}}^{\mathrm{CS}} = \boldsymbol{U}_c\boldsymbol{\Lambda}_c\boldsymbol{U}_c^{\mathrm{H}}, \tag{13.37}
$$

where $\boldsymbol{U}_c$ is the subspace spanned by the right singular vectors of $\mathbf{H}_c$, and $\boldsymbol{\Lambda}_c$ is a diagonal matrix whose diagonal elements are obtained by the classical water-filling method. Consequently, the optimal ISAC signal at P_{CS} can be readily identified with the following structure

$$
\mathbf{X}^{\mathrm{CS}} = \boldsymbol{U}_c\boldsymbol{\Lambda}_c^{\frac{1}{2}}\mathbf{D}, \tag{13.38}
$$

where the entries of $\mathbf{D}$ are i.i.d. with $\mathcal{CN}(0,1)$.

At point P_{CS}, the utilization of a Gaussian random signal prompts an intriguing inquiry regarding its impact on sensing performance. Note that the CRB relies on the sample covariance matrix $\mathbf{R}_{\mathbf{X}}$ which follows a complex Wishart distribution, rather than the statistical covariance matrix $\widetilde{\boldsymbol{R}}_{\mathbf{X}} = \mathbb{E}\{\mathbf{R}_{\mathbf{X}}\}$ as shown from (13.22) to (13.34). Let us denote the CRB at P_{CS} as ϵ_{CS}, we have

$$
\begin{aligned}
\epsilon_{\mathrm{CS}} &\triangleq \frac{T}{\sigma_s^2}\mathbb{E}\left\{\mathrm{tr}\left(\left[\boldsymbol{\Phi}\left(\mathbf{R}_{\mathbf{X}}\right)\right]^{-1}\right)\right\} \geqslant \frac{T}{\sigma_s^2}\,\mathrm{tr}\left(\left(\boldsymbol{\Phi}\left[\mathbb{E}\left\{\mathbf{R}_{\mathbf{X}}\right\}\right]\right)^{-1}\right) \\
&= \frac{T}{\sigma_s^2}\,\mathrm{tr}\left(\left[\boldsymbol{\Phi}\left(\widetilde{\boldsymbol{R}}_{\mathbf{X}}^{\mathrm{CS}}\right)\right]^{-1}\right),
\end{aligned} \tag{13.39}
$$

which holds for arbitrarily distributed $\mathbf{R}_{\mathbf{X}}$ based on Jensen's inequality, due to the convexity of the function $\mathrm{tr}\left(\left[\boldsymbol{\Phi}\left(\mathbf{R}_{\mathbf{X}}\right)\right]^{-1}\right)$ in terms of $\mathbf{R}_{\mathbf{X}}$. Formula (13.39) implies a sensing performance loss brought by the Wishart-distributed $\mathbf{R}_{\mathbf{X}}$. It is because the Jensen lower bound is attained when $\mathbf{R}_{\mathbf{X}} = \mathbb{E}\{\mathbf{R}_{\mathbf{X}}\} = \widetilde{\boldsymbol{R}}_{\mathbf{X}}^{\mathrm{CS}}$, which holds for a Wishart matrix only when $T \to \infty$.

Furthermore, the work of [15] has provided a more explicit upper bound of ϵ_{CS} as

$$
\mathbb{E}\left\{\mathrm{tr}\left(\left[\boldsymbol{\Phi}\left(\mathbf{R}_{\mathbf{X}}\right)\right]^{-1}\right)\right\} \leqslant \frac{T\cdot\mathrm{Tr}\,\boldsymbol{\Phi}\left(\widetilde{\boldsymbol{R}}_{\mathbf{X}}^{\mathrm{CS}}\right)^{-1}}{T-\min\left\{K, \mathrm{rank}\left(\widetilde{\boldsymbol{R}}_{\mathbf{X}}^{\mathrm{CS}}\right)\right\}}, \tag{13.40}
$$

which clearly indicates that the maximum sensing performance loss at P_{CS} is jointly determined by the number of sensing parameters K and the rank[1] of $\widetilde{\boldsymbol{R}}_{\mathbf{X}}^{\mathrm{CS}}$.

1 In high-SNR regime we have $\mathrm{rank}\left(\widetilde{\boldsymbol{R}}_{\mathbf{X}}^{\mathrm{CS}}\right) = \mathrm{rank}(\mathbf{H}_c)$.

For the ideal scenarios with $T \to \infty$, there is no sensing performance loss since the upper bound converges to its lower counterpart.

13.3.4 P_{SC} Performance Characterization

Compared to the point P_{CS}, the achieving strategy for the other corner point P_{SC} is difficult to obtain in general. Let us denote $\epsilon_{\min}$ as the minimum achievable CRB. According to Jensen's inequality,

$$\begin{aligned}\frac{T}{\sigma_s^2}\mathbb{E}\left\{\mathrm{tr}\left(\left[\boldsymbol{\Phi}\left(\mathbf{R}_{\mathbf{X}}\right)\right]^{-1}\right)\right\} &\geqslant \frac{T}{\sigma_s^2}\,\mathrm{tr}\left(\left[\boldsymbol{\Phi}\left(\mathbb{E}\left\{\mathbf{R}_{\mathbf{X}}\right\}\right)\right]^{-1}\right)\\ &= \frac{T}{\sigma_s^2}\,\mathrm{tr}\left(\left[\boldsymbol{\Phi}\left(\widetilde{\boldsymbol{R}}_{\mathbf{X}}^{\mathrm{SC}}\right)\right]^{-1}\right) \triangleq \epsilon_{\min} \end{aligned} \tag{13.41}$$

holds for any complex semidefinite matrix $\mathbf{R}_{\mathbf{X}}$. Here, $\widetilde{\boldsymbol{R}}_{\mathbf{X}}^{\mathrm{SC}}$ represents the optimal solution of the following deterministic CRB minimization problem

$$\widetilde{\boldsymbol{R}}_{\mathbf{X}}^{\mathrm{SC}} = \underset{\widetilde{\boldsymbol{R}}\succeq 0,\ \widetilde{\boldsymbol{R}}=\widetilde{\boldsymbol{R}}^{\mathrm{H}}}{\arg\min}\ \mathrm{tr}\left(\left[\boldsymbol{\Phi}\left(\widetilde{\boldsymbol{R}}\right)\right]^{-1}\right) \quad s.t.\ \mathrm{tr}\left(\widetilde{\boldsymbol{R}}\right) \leqslant P_{\mathrm{T}}. \tag{13.42}$$

The optimal solution of the convex semidefinite programming (SDP) problem (13.42) is not necessarily unique since it is not strictly convex. All the solutions of an SDP belong to the subspace spanned by the maximum-rank solution. Hence, they may be parameterized as

$$\widetilde{\boldsymbol{R}}_{\mathbf{X}}^{\mathrm{SC}} = \boldsymbol{U}_s\boldsymbol{\Lambda}_s\boldsymbol{U}_s^{\mathrm{H}}, \tag{13.43}$$

where $\boldsymbol{U}_s$ consists of the eigenvectors of the maximum-rank sensing-optimal solution corresponding to the nonzero eigenvalues, and $\boldsymbol{\Lambda}_s$ is a positive semidefinite Hermitian matrix.

The equality in (13.41) holds if and only if

$$\mathbf{R}_{\mathbf{X}} = \mathbb{E}\{\mathbf{R}_{\mathbf{X}}\} = \widetilde{\boldsymbol{R}}_{\mathbf{X}}^{\mathrm{SC}}, \tag{13.44}$$

which implies that the global minimum of the CRB $\epsilon_{\min}$ can be achieved when the sample covariance matrix $\mathbf{R}_{\mathbf{X}}$ becomes a deterministic matrix. Meanwhile, this prompts the question of whether any communication DoF remains in the ISAC signal at P_{SC}. A random signal $\mathbf{X}$ can also have the deterministic covariance matrix with the following structure

$$\mathbf{X}^{\mathrm{SC}} = \sqrt{T}\left(\widetilde{\boldsymbol{R}}_{\mathbf{X}}^{\mathrm{SC}}\right)^{\frac{1}{2}}\mathbf{Q} = \sqrt{T}\boldsymbol{U}_s\boldsymbol{\Lambda}_s^{\frac{1}{2}}\mathbf{Q}, \tag{13.45}$$

where $\mathbf{Q} \in \mathbb{C}^{\mathsf{M}_{\mathrm{SC}}\times T}$ is a random semi-unitary matrix such that $\mathbf{Q}\mathbf{Q}^{\mathrm{H}} = \boldsymbol{I}$ and $\mathsf{M}_{\mathrm{SC}} = \mathrm{rank}\left(\widetilde{\boldsymbol{R}}_{\mathbf{X}}^{\mathrm{SC}}\right)$. Since $\left(\widetilde{\boldsymbol{R}}_{\mathbf{X}}^{\mathrm{SC}}\right)^{\frac{1}{2}}$ is deterministic, the communication DoFs at P_{SC} are only contributed by the randomness of $\mathbf{Q}$.

Given the fixed structure in (13.45), maximizing the achievable rate at P_{SC} is equivalently transformed into finding optimal distribution $p_{\mathbf{Q}}(\boldsymbol{Q})$ over the set of all $\mathsf{M}_{SC} \times T$ semi-unitary matrices. In other words, we aim to find semi-unitary matrices over the Stiefel manifold $\mathcal{S}(T, \mathsf{M}_{SC})$ to maximize the mutual information $I(\mathbf{Q}; \mathbf{Y}_c|\mathbf{H}_c)$. In the regime of high-SNR, the task at hand is equivalent to tackling the sphere packing problem within the Stiefel manifold. Here, the optimal probability density function $p_{\mathbf{X}}(\boldsymbol{X})$ takes the form of a uniform distribution, which subsequently yields the ensuing asymptotic result [15]

$$R_{SC} = \mathbb{E}\left\{\left(1 - \frac{\mathsf{M}_{SC}}{2T}\right) \log\det\{\sigma_c^{-2}\mathbf{H}_c\widetilde{\boldsymbol{R}}_{\mathbf{X}}^{SC}\mathbf{H}_c^{H}\} + \mathsf{c}_0\right\} + O(\sigma_c^2), \tag{13.46}$$

where $O(\sigma_c^2)$ represents the residual term with the same order of σ_c^2, and

$$\mathsf{c}_0 = \frac{\mathsf{M}_{SC}}{T}\left[\left(T - \frac{\mathsf{M}_{SC}}{2}\right)\log\frac{T}{e} - \log\Gamma(T) + \log(2\sqrt{\pi})\right], \tag{13.47}$$

converges to zero as $T \to \infty$ with $\Gamma(\cdot)$ being a Gamma function. It implies that the communication DoFs are lossless with $T \to \infty$. This is due to the fact that as T increases, even a Gaussian matrix will exhibit rows that are asymptotically orthogonal, rendering it asymptotically equivalent to a semi-unitary matrix.

13.3.5 Discussions

The above findings clearly illustrate the impact of both ST and DRT in an ISAC system. When we examine the leftmost segments of the matrices $\mathbf{X}^{CS}$ and $\mathbf{X}^{SC}$, it becomes evident that the ISAC signals optimized for communication and sensing should align with these respective orthogonal bases denoted as $\mathbf{U}_c$ and $\mathbf{U}_s$, respectively. In essence, ST entails the transfer of signal power from the subspace spanned by span($\boldsymbol{U}_c$) to the one spanned by span($\boldsymbol{U}_s$). Additionally, when we compare the rightmost segments of $\mathbf{X}^{CS}$ and $\mathbf{X}^{SC}$, specifically $\mathbf{D}$ and $\mathbf{Q}$, we observe that signals optimized for communication and sensing adopt Gaussian and semi-unitary codebooks, respectively, further emphasizing the role of DRT. As the ISAC system traverses the Pareto frontier from P_{CS} to P_{SC}, the distribution $p_{\mathbf{X}}(\boldsymbol{X})$ gradually transitions from Gaussian to a less random distribution, eventually becoming a uniform distribution over semi-unitary matrices. In this context, the DRT depicted in Figure 13.2 represents a one-dimensional special case, where the semi-unitary matrix reduces to constant-modulus signals in their scalar form, namely, the BPSK modulation as shown in Figure 13.2(e).

Beyond the tradeoff in input distributions, the DRT also manifests in the achievable DoFs for communication and sensing at the two extreme points. At P_{CS}, the communication subsystem evidently achieves the full DoF of the Gaussian channel, denoted as M_{CS}, equivalent to the rank of $\widetilde{\boldsymbol{R}}_{\mathbf{X}}^{CS}$. However, it can be observed from (13.40) that the equivalent number of individual observations of the target reduces due to the utilization of Gaussian signaling, which is up to $\min\{K, \mathsf{M}_{\text{CS}}\}$. Conversely, at P_{SC}, the sensing subsystem attains the full DoFs of T thanks to the deterministic signal sample covariance matrix. In contrast, the semi-unitary signaling incurs a communication DoF loss of $\frac{\mathsf{M}_{\text{SC}}^2}{2T}$, as indicated by (13.46).

13.4 Target Detection

13.4.1 Subspace Tradeoff: Error Exponent

The ISAC system exhibits obvious resource competition relationships between the S&C subsystems. Motivated by this, most of the existing literature on ISAC system design focuses on resource allocation problems such as beamforming [6] and subcarrier power allocation [24]. In this subsection, we further discuss this kind of relationship by taking the target detection task as an example.

13.4.1.1 Rate-Exponent Region

In the detection task, the state parameter of a target takes the value on set $\eta \in \{0, 1\}$, representing the absence or presence of the target. During the nth channel use, the considered ISAC system model can be expressed by

$$\mathbf{Y}_{c,n} = \boldsymbol{H}_c\mathbf{X}_n + \mathbf{Z}_{c,n}, \tag{13.48a}$$

$$\mathbf{Y}_{s,n} = \eta\boldsymbol{H}_s\mathbf{X}_n + \mathbf{Z}_{s,n}, \tag{13.48B}$$

where $\mathbf{Y}_{c,n}$, $\mathbf{Y}_{s,n}$, $\boldsymbol{H}_c$, and $\boldsymbol{H}_s$ are the received signal by communication Rx, the echo signal, the communication channel, and the sensing channel, respectively. Here, the noise matrices $\mathbf{Z}_{s,n}$ and $\mathbf{Z}_{c,n}$ follow complex Gaussian distribution. We adopt the classical achievable rate as the communication metric and the error exponent[2] as the sensing metric. The error exponent can be defined by [26]

$$E = \lim_{N\to\infty} \frac{1}{N} \log \frac{1}{\delta_N}, \tag{13.49}$$

where

$$\delta_N = \max_{w\in[M_N]} \max_{s\in\{0,1\}} \Pr\left\{\hat{W}\left(\mathbf{Y}_s^N, \mathbf{X}^N(\mathsf{W})\right) \neq \eta | \eta = s, \mathsf{W} = w\right\}$$

is the maximum detection error probability over all codewords and target states with $\mathbf{Y}_s^N = \{\mathbf{Y}_{s,1}, \mathbf{Y}_{s,2}, \dots, \mathbf{Y}_{s,N}\}$, $\mathbf{X}^N(W) = \{\mathbf{X}_1, \mathbf{X}_2, \dots, \mathbf{X}_N\}$, and $\mathsf{W} \in \{M_N\}$ is the encoded message. Then, we define the rate-exponent region by the set of

2 The error exponent is related to a class of more frequently used sensing performance metrics, i.e., statistical divergences, via the large deviation theory [25].

all achievable (R, E). By denoting $\widetilde{\boldsymbol{R}}_{\mathbf{X}} = \mathbb{E}\left\{\frac{1}{N}\sum_{n=1}^{N}\mathbf{X}_n\mathbf{X}_n^{\mathrm{H}}\right\}$ as the statistical covariance matrix of the transmitted ISAC signal, the rate-exponent region can be characterized by [26]

$$R \leqslant \log\det\left\{\boldsymbol{I} + \sigma_c^{-2}\boldsymbol{H}_c\widetilde{\boldsymbol{R}}_{\mathbf{X}}\boldsymbol{H}_c^{\mathrm{H}}\right\}, \tag{13.50a}$$

$$E \leqslant \frac{1}{4}\,\mathrm{tr}\left(\sigma_s^{-2}\boldsymbol{H}_s\widetilde{\boldsymbol{R}}_{\mathbf{X}}\boldsymbol{H}_s^{\mathrm{H}}\right). \tag{13.50b}$$

It can be observed from (13.50) that the S&C performance is coupled through the statistical covariance matrix $\widetilde{\boldsymbol{R}}_{\mathbf{X}}$, which satisfies a power budget constraint $\mathrm{tr}(\widetilde{\boldsymbol{R}}_{\mathbf{X}}) \leqslant P_T$. Thus, the structures of $\widetilde{\boldsymbol{R}}_{\mathbf{X}}$ for optimizing S&C performance can be respectively expressed by

- *Communication-optimal*: The eigenvectors of $\widetilde{\boldsymbol{R}}_{\mathbf{X}}$ match those of communication channel $\boldsymbol{H}_c$, and the corresponding eigenvalues are determined by water-filling strategy [25].
- *Sensing-optimal*: $\widetilde{\boldsymbol{R}}_{\mathbf{X}}$ would be aligned with the dominant eigenvector (possibly having multiplicity larger than 1, though) of sensing channel $\boldsymbol{H}_s$.

Apparently, the S&C subsystems cannot achieve their optimal performance simultaneously with a high probability, hence exhibiting an S&C tradeoff. Additionally, (13.50) provides an explicit characterization of the proximity between the communication user and the sensing target, which can be quantified by assessing the discrepancy between the "communication subspace" $\mathrm{span}(\boldsymbol{H}_c)$ and the "sensing subspace" represented by the dominant eigenvector of $\boldsymbol{H}_s$, which reflects again the ST tradeoff.

13.4.1.2 Case Study

In this subsection, we take a practical example to further illustrate the ST tradeoff. Let us consider a base station (BS) communicating to a single-antenna user while sensing a point-like target. The uniform linear arrays are equipped at both transmitting and receiving sides with the same number of antennas $N_s = M = 10$ for sensing. Thus, the communication and sensing channels can be expressed by

$$\boldsymbol{h}_c = \alpha_c\boldsymbol{a}^{\mathrm{H}}(\theta_c), \quad \boldsymbol{H}_s = \alpha_s\boldsymbol{a}(\theta_s)\boldsymbol{a}^{\mathrm{H}}(\theta_s), \tag{13.51}$$

where α_c and α_s denote the scalar channel coefficients, θ_c and θ_s denote the bearing angles of the user and the target relative to the BS, respectively. The steering vector of uniform linear array is given by

$$\boldsymbol{a}(\theta) = [1,\ e^{\jmath\pi\sin(\theta)},\ e^{\jmath 2\pi\sin(\theta)},\ \ldots,\ e^{\jmath\pi(N_s-1)\sin(\theta)}]^{\mathrm{T}}, \tag{13.52}$$

with $\jmath$ being the imaginary unit.

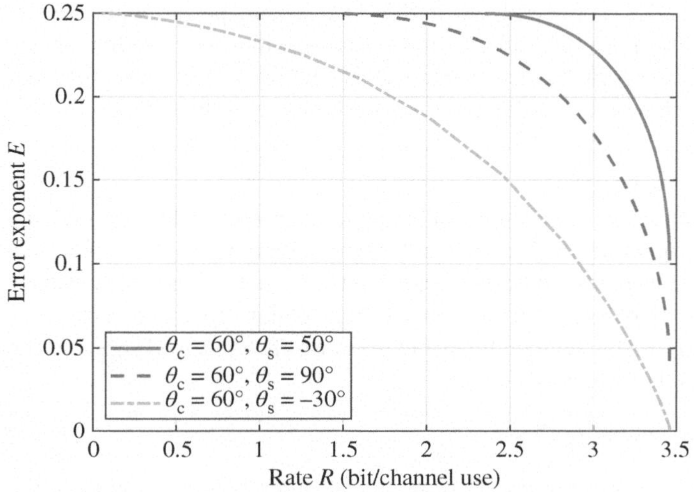

Figure 13.5 The rate-exponent region of the single-antenna user, point-like target scenario, with S&C SNRs equal to 10 and 0 dB, respectively.

In this scenario, the rate-exponent region (13.50) can be recast by

$$R(\tilde{\lambda}) \leqslant \log(1 + P_T \sigma_c^{-2} |\alpha_c|^2 |\mathbf{r}^{\mathrm{H}}(\tilde{\lambda}) \boldsymbol{a}(\theta_c)|^2), \tag{13.53a}$$

$$E(\tilde{\lambda}) \leqslant \frac{1}{4} \operatorname{tr} \left(P_T \sigma_s^{-2} |\alpha_s|^2 |\mathbf{r}^{\mathrm{H}}(\tilde{\lambda}) \boldsymbol{a}(\theta_s)|^2 \right), \tag{13.53b}$$

$$\mathbf{r}(\tilde{\lambda}) = \underset{\boldsymbol{v}:\|\boldsymbol{v}\|=1}{\operatorname{argmax}}\ (1 - \tilde{\lambda}) |\boldsymbol{v}^{\mathrm{H}} \boldsymbol{a}(\theta_c)|^2 + \tilde{\lambda} |\boldsymbol{v}^{\mathrm{H}} \boldsymbol{a}(\theta_s)|^2, \ \tilde{\lambda} \in [0, 1], \tag{13.53c}$$

where $\tilde{\lambda}$ is the weighting factor. Moreover, the communication and sensing SNRs may be expressed as

$$\mathrm{SNR}_c = P_T \|\boldsymbol{H}_c\|_F^2 \sigma_c^{-2}, \ \ \mathrm{SNR}_s = P_T \|\boldsymbol{H}_s\|_F^2 \sigma_s^{-2}, \tag{13.54}$$

where $\| \cdot \|_F$ denotes Frobenius norm of a matrix.

In Figure 13.5, we show the rate-exponent for several fixed angle pairs (θ_c, θ_s). Thanks to the vector channel, the closeness between the S&C subspaces can be clearly characterized by the angle between $\boldsymbol{a}(\theta_c)$ and $\boldsymbol{a}(\theta_s)$. Figure 13.5 illustrates that the tradeoff between S&C becomes increasingly pronounced as the angular separation between the target and the user widens.

13.4.2 Deterministic-Random Tradeoff: Time-Sharing Strategy

The error exponent can only depict the subspace tradeoff since it is only defined in the regime of infinitely large T. In fact, target detection problems also have their own DRT, but the DRT is only visible with the aid of sensing performance metrics defined over a finite T. In this subsection, we shall see that the detection probability is indeed such a performance metric, but its DRT works in a rather different mechanism from that of the CRB in parameter estimation problems.

13.4.2.1 System Model

Let us consider the simplified example of binary target presence detection relying on MIMO radars, in which the transmit and receive antenna arrays are colocated. Assuming that the region of interest is located in the far field of the array, we may use a single angle of arrival (which is also the angle of departure due to the colocated assumption) to characterize the array response. The received signal at the radar Rx can thus be written as

$$\mathbf{Y}_s = \eta \cdot \mathsf{A}e^{j\theta}\boldsymbol{a}(\varphi)\boldsymbol{a}^{\mathrm{H}}(\varphi)\mathbf{X} + \mathbf{Z}_s, \tag{13.55}$$

where $\eta \in \{0, 1\}$ indicates the presence/absence of the target, φ is the direction in which the target presence is to be tested, $\boldsymbol{a}(\varphi) \in \mathbb{C}^{N_s}$ denotes the corresponding array steering vector, $\mathsf{A}e^{j\theta}$ is the unknown complex scattering coefficient of the target, and $\mathbf{X} \in \mathbb{C}^{N_s \times T}$ denotes the transmitted ISAC signal, while $\mathbf{Z}_s \in \mathbb{C}^{N_s \times T}$ denotes the noise. Adding the communication link to the model, we have

$$\mathbf{Y}_c = \mathbf{H}_c\mathbf{X} + \mathbf{Z}_c, \tag{13.56}$$

where $\mathbf{H}_c \in \mathbb{C}^{N_c \times N_s}$ and $\mathbf{Z}_c \in \mathbb{C}^{N_c \times T}$ denote the communication channel and the noise, respectively.

Next, let us derive the probability of detection and the false alarm rate when the optimal detector is employed. First note that the likelihood ratio may be expressed as

$$L_{\mathbf{Y}_s}(\boldsymbol{Y}_s) \propto \mathbb{E}_{\mathsf{A}}\left\{ e^{-\frac{\mathsf{A}^2}{N_0}\sum_{t=1}^{T}\|\boldsymbol{a}(\varphi)\boldsymbol{a}^{\mathrm{H}}(\varphi)\mathbf{x}(t)\|^2} \times I_0\left(\frac{2\mathsf{A}}{N_0}\left|\sum_{t=1}^{T}\mathbf{y}_s^{\mathrm{H}}(t)\boldsymbol{a}(\varphi)\boldsymbol{a}^{\mathrm{H}}(\varphi)\mathbf{x}(t)\right|\right)\right\}, \tag{13.57}$$

where $I_0(\cdot)$ denotes the zeroth-order modified Bessel function. The optimal detector is thus the likelihood ratio detector having the test statistic of

$$\mathsf{T} = \left|\sum_{t=1}^{T}\mathbf{y}_s^{\mathrm{H}}(t)\boldsymbol{a}(\varphi)\boldsymbol{a}^{\mathrm{H}}(\varphi)\mathbf{x}(t)\right|^2 \underset{\mathcal{H}_0}{\overset{\mathcal{H}_1}{\gtrless}} \xi, \tag{13.58}$$

where $\mathcal{H}_0$ and $\mathcal{H}_1$ represent the absence and presence of the target, respectively.

Assume now that the amplitude of the target reflection coefficient A is Rayleigh distributed, leading to $\mathsf{A}e^{j\theta} \sim \mathcal{CN}(0, \overline{\mathsf{A}^2})$. The conditional expectation of the test statistic can now be expressed as

$$\mathbb{E}\left\{\mathsf{Z}|\mathcal{H}_1\right\} = \overline{\mathsf{A}^2}T^2\mathrm{tr}^2\left(\left(\boldsymbol{a}(\varphi)\boldsymbol{a}^{\mathrm{H}}(\varphi)\right)^2\mathbf{R}_{\mathbf{X}}\right) + T\mathrm{tr}\left(\left(\boldsymbol{a}(\varphi)\boldsymbol{a}^{\mathrm{H}}(\varphi)\right)^2\mathbf{R}_{\mathbf{X}}\right), \tag{13.59a}$$

$$\mathbb{E}\left\{\mathsf{Z}|\mathcal{H}_0\right\} = TN_0\ \mathrm{tr}\left(\left(\boldsymbol{a}(\varphi)\boldsymbol{a}^{\mathrm{H}}(\varphi)\right)^2\mathbf{R}_{\mathbf{X}}\right). \tag{13.59b}$$

The instantaneous SNR of a single realization of $\mathbf{X}$ takes the form of

$$\mathrm{SNR} = \frac{\overline{\mathsf{A}^2}T}{N_0}\ \mathrm{tr}\left(\left(\boldsymbol{a}(\varphi)\boldsymbol{a}^{\mathrm{H}}(\varphi)\right)^2\mathbf{R}_{\mathbf{X}}\right) = \rho. \tag{13.60}$$

We may then obtain the false alarm rate and detection probability as

$$P_{\mathrm{FA}} = e^{-\frac{\lambda}{TN_0\ \mathrm{tr}\left(\left(\boldsymbol{a}(\varphi)\boldsymbol{a}^{\mathrm{H}}(\varphi)\right)^2\mathbf{R}_{\mathbf{X}}\right)}}, \tag{13.61a}$$

$$P_D = e^{-\frac{\lambda}{\overline{\mathsf{A}^2}T^2\mathrm{tr}^2\left(\left(\boldsymbol{a}(\varphi)\boldsymbol{a}^{\mathrm{H}}(\varphi)\right)^2\mathbf{R}_{\mathbf{X}}\right)+TN_0\ \mathrm{tr}\left(\left(\boldsymbol{a}(\varphi)\boldsymbol{a}^{\mathrm{H}}(\varphi)\right)^2\mathbf{R}_{\mathbf{X}}\right)}}, \tag{13.61b}$$

respectively. This further implies that P_D may be represented in terms of P_{FA} as follows:

$$P_D = P_{\mathrm{FA}}^{\frac{1}{1+\rho}} = P_{\mathrm{FA}}^{\frac{1}{1+\frac{\overline{\mathsf{A}^2}T}{N_0}\ \mathrm{tr}\left(\left(\boldsymbol{a}(\varphi)\boldsymbol{a}^{\mathrm{H}}(\varphi)\right)^2\mathbf{R}_{\mathbf{X}}\right)}}. \tag{13.62}$$

Note that the likelihood ratio detector (13.58) belongs to the class of CFAR detector, hence the false alarm rate P_{FA} can be viewed as a fixed input parameter.

13.4.2.2 Sensing-Optimal $p_{\mathbf{X}}(\boldsymbol{X})$

Next, let us consider the sensing-optimal signal distribution. The problem of designing such a distribution can be formulated as

$$\max_{p_{\mathbf{X}}(\boldsymbol{X})}\ \mathbb{E}\left\{P_D\right\},\ \text{s.t. } \mathbb{E}\left\{\mathrm{tr}(\mathbf{R}_{\mathbf{X}})\right\} \leqslant P_T. \tag{13.63}$$

We first note that both the objective function and the constraint are related to $\mathbf{X}$ only via $\mathbf{R}_{\mathbf{X}}$. Therefore, without loss of generality, we may simply design $p_{\mathbf{R}_{\mathbf{X}}}(\boldsymbol{R}_{\mathbf{X}})$ rather than $p_{\mathbf{X}}(\boldsymbol{X})$. Next, note that P_D is a function of $\boldsymbol{A}\mathbf{R}_{\mathbf{X}}$, with $\boldsymbol{A} = (\boldsymbol{a}(\varphi)\boldsymbol{a}^{\mathrm{H}}(\varphi))^2$. We may thus denote P_D as $P_D(\boldsymbol{A}\mathbf{R}_{\mathbf{X}})$. Expanding the expectations in (13.63) as integrals, the design problem now reads

$$\max_{p_{\mathbf{R}_{\mathbf{X}}}(\boldsymbol{R}_{\mathbf{X}})}\ \int P_D(\mathrm{tr}(\boldsymbol{A}\boldsymbol{R}_{\mathbf{X}}))p_{\mathbf{R}_{\mathbf{X}}}(\boldsymbol{R}_{\mathbf{X}})d\boldsymbol{R}_{\mathbf{X}}, \tag{13.64a}$$

$$\text{s.t. } p_{\mathbf{R}_{\mathbf{X}}}(\boldsymbol{R}_{\mathbf{X}}) \geqslant 0\ \forall\boldsymbol{R}_{\mathbf{X}},\ \int \mathrm{tr}(\boldsymbol{R}_{\mathbf{X}})p_{\mathbf{R}_{\mathbf{X}}}(\boldsymbol{R}_{\mathbf{X}})d\boldsymbol{R}_{\mathbf{X}} = P_T, \tag{13.64b}$$

$$\int p_{\mathbf{R}_{\mathbf{X}}}(\boldsymbol{R}_{\mathbf{X}})d\boldsymbol{R}_{\mathbf{X}} = 1, p_{\mathbf{R}_{\mathbf{X}}}(\boldsymbol{R}_{\mathbf{X}})\left(1 - \mathbb{I}\{\boldsymbol{R}_{\mathbf{X}} \succcurlyeq \mathbf{0}\}\right) = 0\ \forall\boldsymbol{R}_{\mathbf{X}}. \tag{13.64c}$$

To avoid cluttered notations, we denote $p_{\mathbf{R_X}}(\boldsymbol{R}_{\mathbf{X}})$ as $p(\boldsymbol{R}_{\mathbf{X}})$ hereafter. Using the Karush–Kuhn–Tucker (KKT) conditions, we obtain the following conditions for the optimal $p(\boldsymbol{R}_{\mathbf{X}})$:

$$\int \mathrm{tr}(\boldsymbol{R}_{\mathbf{X}})p(\boldsymbol{R}_{\mathbf{X}})d\boldsymbol{R}_{\mathbf{X}} = P_T, \tag{13.65a}$$

$$P_D(\mathrm{tr}(\boldsymbol{A}\boldsymbol{R}_{\mathbf{X}})) \leqslant \lambda_1 + \lambda_2\ \mathrm{tr}(\boldsymbol{R}_{\mathbf{X}}),\ \forall \boldsymbol{R}_{\mathbf{X}} \succcurlyeq \mathbf{0}, \tag{13.65b}$$

$$\mathrm{supp}(p(\boldsymbol{R}_{\mathbf{X}})) = \{\boldsymbol{R}_{\mathbf{X}} | P_D(\mathrm{tr}(\boldsymbol{A}\boldsymbol{R}_{\mathbf{X}})) = \lambda_1 + \lambda_2\ \mathrm{tr}(\boldsymbol{R}_{\mathbf{X}})\}. \tag{13.65c}$$

The result (13.65b) further implies

$$\max_{\boldsymbol{R}_{\mathbf{X}} \in \mathbb{S}_+} \ [P_D(\mathrm{tr}(\boldsymbol{A}\boldsymbol{R}_{\mathbf{X}})) - \lambda_2\ \mathrm{tr}(\boldsymbol{R}_{\mathbf{X}})] \leqslant \lambda_1, \tag{13.66}$$

where $\mathbb{S}_+$ denotes the positive semidefinite cone $\{\boldsymbol{R}_{\mathbf{X}} | \boldsymbol{R}_{\mathbf{X}} \succcurlyeq \mathbf{0},\ \boldsymbol{R}_{\mathbf{X}} = \boldsymbol{R}_{\mathbf{X}}^{\mathrm{H}}\}$. The probability density function $p(\boldsymbol{R}_{\mathbf{X}})$ is nonzero only when the equality in (13.66) is achieved, for an appropriate pair of (λ_1, λ_2) satisfying (13.65a) and (13.65b). Geometrically, each pair of (λ_1, λ_2) determines a tangent line of $P_D(\cdot)$, as portrayed in Figure 13.6.

Although it is not yet straightforward how such λ_1 and λ_2 can be determined, we first observe that they necessarily satisfy $\lambda_1 \geqslant P_D(0),\ \lambda_2 \geqslant 0$, otherwise the condition (13.65b) would always be violated by certain choices of $\boldsymbol{R}_{\mathbf{X}}$. Furthermore, we can also see that $\lambda_2 \neq 0$ for any finite P_T. Thus, we may obtain the following relaxed condition for the support of $\boldsymbol{R}_{\mathbf{X}}$: $\mathrm{supp}(p(\boldsymbol{R}_{\mathbf{X}})) \subseteq \bigcup_{\lambda_2 > 0} S_{\mathrm{opt}}(\lambda_2)$, where

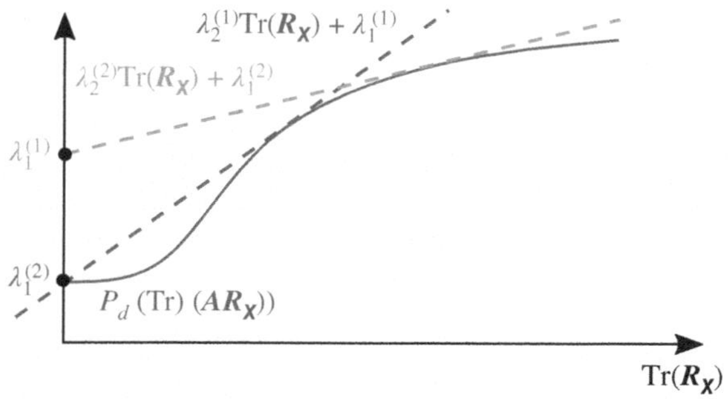

Figure 13.6 Geometrical illustration of (13.66).

$\mathcal{S}_{\text{opt}}(\lambda_2) = \{\boldsymbol{R}_{\mathbf{X}} | \boldsymbol{R}_{\mathbf{X}} = \text{argmax}_{\boldsymbol{R}_{\mathbf{X}} \in \mathbb{S}_+} P_D(\text{tr}(\boldsymbol{A}\boldsymbol{R}_{\mathbf{X}})) - \lambda_2 \, \text{tr}(\boldsymbol{R}_{\mathbf{X}})\}$. Let us consider the decomposition of

$$\max_{\boldsymbol{R}_{\mathbf{X}} \in \mathbb{S}_+} [P_D(\text{tr}(\boldsymbol{A}\boldsymbol{R}_{\mathbf{X}})) - \lambda_2 \text{tr}(\boldsymbol{R}_{\mathbf{X}})] = \max_{C \geqslant 0} \left\{ \left[\max_{\substack{\boldsymbol{R}_{\mathbf{X}} \in \mathbb{S}_+ \\ \text{tr}(\boldsymbol{R}_{\mathbf{X}}) = C}} P_D(\text{tr}(\boldsymbol{A}\boldsymbol{R}_{\mathbf{X}})) \right] - \lambda_2 \tilde{C} \right\}. \tag{13.67}$$

Note that the maximum is achieved at a finite $\tilde{C}$ when $\lambda_2 > 0$. By observing that P_D is a monotonically increasing function of its argument, we have

$$\max_{\substack{\boldsymbol{R}_{\mathbf{X}} \in \mathbb{S}_+ \\ \text{tr}(\boldsymbol{R}_{\mathbf{X}}) = \tilde{C}}} P_D(\text{tr}(\boldsymbol{A}\boldsymbol{R}_{\mathbf{X}})) = P_D \left\{ \max_{\substack{\boldsymbol{R}_{\mathbf{X}} \in \mathbb{S}_+ \\ \text{tr}(\boldsymbol{R}_{\mathbf{X}}) = \tilde{C}}} \text{tr}(\boldsymbol{A}\boldsymbol{R}_{\mathbf{X}}) \right\} = P_D \left(\lambda_{\max}(\boldsymbol{A}) \cdot \tilde{C} \right). \tag{13.68}$$

Since $\lambda_{\max}(\boldsymbol{A})$ has multiplicity 1, the optimal $\boldsymbol{R}_{\mathbf{X}}$ is given by $\boldsymbol{R}_{\mathbf{X}} = \tilde{C}\boldsymbol{v}\boldsymbol{v}^{\text{H}}$ with $\boldsymbol{v}$ being the eigenvector associated with $\lambda_{\max}(\boldsymbol{A})$. We now arrive at

$$\max_{\tilde{C} \geqslant 0} P_D(\lambda_{\max}(\boldsymbol{A}) \cdot \tilde{C}) - \lambda_2 \tilde{C} \leqslant \lambda_1. \tag{13.69}$$

As for the set $\mathcal{S}_{\text{opt}}(\lambda_2)$, we also have $\mathcal{S}_{\text{opt}}(\lambda_2) = \{\boldsymbol{R}_{\mathbf{X}} | \boldsymbol{R}_{\mathbf{X}} = g(\lambda_2)\boldsymbol{v}\boldsymbol{v}^{\text{H}}\}$, where $g(\lambda_2) = \text{argmax}_{\tilde{C} \geqslant 0} P_D(\lambda_{\max}(\boldsymbol{A}) \cdot \tilde{C}) - \lambda_2 \tilde{C}$, implying that

$$\text{supp}(p(\boldsymbol{R}_{\mathbf{X}})) \subseteq \{\boldsymbol{R}_{\mathbf{X}} | \boldsymbol{R}_{\mathbf{X}} = a\boldsymbol{v}\boldsymbol{v}^{\text{H}}, \ a \geqslant 0\}. \tag{13.70}$$

In light of (13.70), (13.65c) may be rewritten as

$$\text{supp}(p(\boldsymbol{R}_{\mathbf{X}})) = \{a\boldsymbol{v}\boldsymbol{v}^{\text{H}} | a \in \mathcal{A}(\lambda_1, \lambda_2)\}, \tag{13.71a}$$

$$\mathcal{A}(\lambda_1, \lambda_2) = \{a | P_D(a\lambda_{\max}(\boldsymbol{A})) = \lambda_1 + \lambda_2 a\}, \tag{13.71b}$$

where λ_1 and λ_2 can be determined using the following conditions

$$P_D(b\lambda_{\max}(\boldsymbol{A})) \leqslant \lambda_1 + \lambda_2 b \ \forall b \geqslant 0, \tag{13.72a}$$

$$\text{Card}(\mathcal{A}(\lambda_1, \lambda_2)) > 0, \tag{13.72b}$$

$$\min_{a \in \mathcal{A}(\lambda_1, \lambda_2)} a \leqslant P_T \leqslant \max_{a \in \mathcal{A}(\lambda_1, \lambda_2)} a, \tag{13.72c}$$

where $\text{Card}(\cdot)$ denote the cardinality of a set. Finally, we arrive at the following result:

Proposition 13.1 *The optimal probability distribution is given by*

$$p(\boldsymbol{R}_X) = \begin{cases} \frac{P^\star - P_T}{P^\star}\delta(\boldsymbol{R}_\mathbf{X}) + \frac{P_T}{P^\star}\delta(\boldsymbol{R}_\mathbf{X} - P^\star \boldsymbol{v}\boldsymbol{v}^\mathrm{H}), & P_T \leqslant P^\star; \\ \delta(\boldsymbol{R}_\mathbf{X} - P_T \boldsymbol{v}\boldsymbol{v}^\mathrm{H}), & P_T > P^\star. \end{cases} \tag{13.73}$$

where $P^\star$ is the solution to $P'_D(x)x = P_D(x) - P_D(0)$.

13.4.2.3 Discussions and Numerical Results

Proposition 13.1 implies that the sensing-optimal transmission strategy is an ON-OFF keying-like strategy: Transmitting signals have sample covariance matrix of $P^\star \boldsymbol{v}\boldsymbol{v}^\mathrm{H}$ with probability $P_T/P^\star$, while keeping silence with probability $(P^\star - P_T)/P^\star$. This is in stark contrast to the optimal strategy in the parameter estimation tasks with CRB being the performance metric, for which $\mathbf{R}_\mathbf{X}$ is unique. The reason why this happens is that the CRB is convex in $\mathbf{R}_\mathbf{X}$, while the detection probability P_D is not convex in $\mathbf{R}_\mathbf{X}$, as can be observed from (13.62). Although P_D does not have a constant convexity/concavity, the DRT persists, which follows from the sparsity of set $\mathcal{A}(\lambda_1, \lambda_2)$ corresponding to the set of tangent lines of $P_D(\cdot)$ (defined in (13.71b)).

The aforementioned phenomenon has a vivid information-theoretical interpretation. To elaborate, let us consider the deterministic counterpart of (13.63):

$$\max_{p(\boldsymbol{R}_\mathbf{X})} \ P_D(\boldsymbol{A}\boldsymbol{R}_\mathbf{X}), \ \text{s.t. } \operatorname{tr}(\boldsymbol{R}_\mathbf{X}) \leqslant P_T. \tag{13.74}$$

Note that (13.74) corresponds to the multiobjective optimization problem of

$$\max_{p(\boldsymbol{R}_\mathbf{X})} \ \left(P_D(\boldsymbol{A}\boldsymbol{R}_\mathbf{X}), -\operatorname{tr}(\boldsymbol{R}_\mathbf{X})\right). \tag{13.75}$$

The solutions to this problem form the Pareto boundary.

Now we may see that the solutions to (13.63) in fact correspond to the result of employing the **time-sharing** strategy to the solutions to (13.74). Geometrically speaking, the optimal P_D-SNR boundary is the convex envelope of the Pareto boundary of (13.74). To see this more clearly, let us consider a tangible example. For simplicity, we assume that $\frac{\overline{\mathsf{A}}^2 T}{N_0}\lambda_{\max}(\boldsymbol{A}) = 1$. Considering the scenario of $P_{\mathrm{FA}} = 10^{-5}$, we may plot the corresponding P_D in Figure 13.7, along with the corresponding convex envelope obtained by the ON–OFF keying time-sharing strategy. In this case, we have $P^\star \approx 9.407$. When $P_T < P^\star$, we see that the sensing-optimal distribution $p_\rho(\rho)$ takes the two-point form. By contrast, when $P_T \geqslant P^\star$, the sensing-optimal distribution is a Dirac delta distribution peaked at P_T. These numerical results clearly corroborate our analysis.

Finally, it is worth noting that the line of reasoning in this subsection can be generalized to generic sensing performance metrics and tasks beyond target detection. Interested readers may refer to [27].

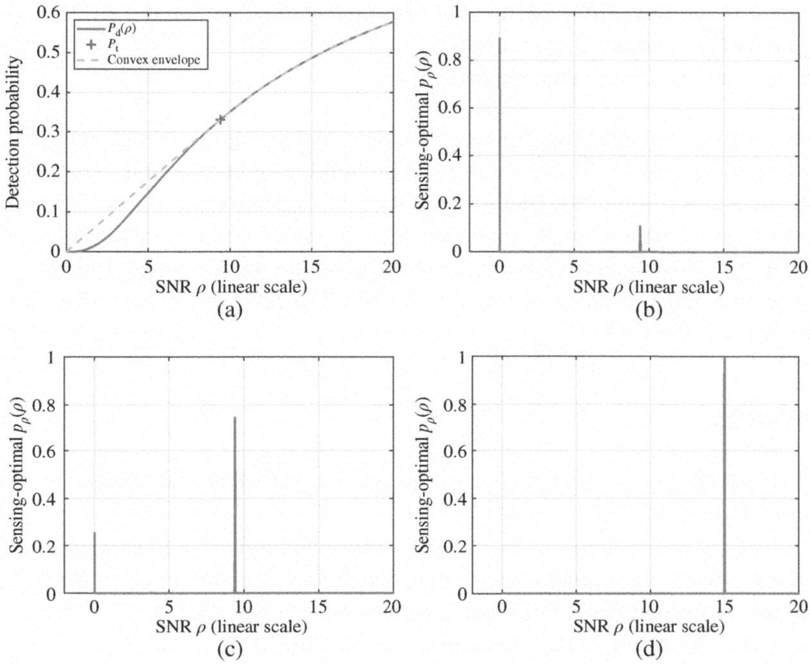

Figure 13.7 Numerical results of the detection problem with $\frac{\overline{A^2}T}{N_0}\lambda_{\max}(\boldsymbol{A}) = 1$ and $P_{\mathrm{FA}} = 10^{-5}$. (a) $P_D(\rho)$ and its convex envelope, (b) $p_\rho(\rho)$ at $P_T = 1 < P^\star$, (c) $p_\rho(\rho)$ at $P_T = 7 < P^\star$, (d) $p_\rho(\rho)$ at $P_T = 15 > P^\star$.

13.5 Conclusions

This chapter provided the recent research progress of the information-theoretical limits for ISAC systems. We began by introducing the general framework of C-D theory, followed by a comprehensive analysis of both target estimation and detection scenarios. Specifically, we revealed a twofold tradeoff in ISAC systems, namely, the ST and DRT. Beyond this fundamental tradeoff, numerous critical issues remain unresolved within the domain of S&C tradeoffs. Several of these inquiries include the following:

1. How does the DRT manifest itself under generic sensing performance metrics?
2. How do we characterize the S&C tradeoff when the channels are not memoryless?

3. How do we design ISAC systems that are capable of achieving the entire C-D boundary (not just the corner points)?
4. Can we unify S&C performance metrics?

These formidable inquiries underscore the extensive groundwork required before the full potential of ISAC can be comprehended and harnessed. Nevertheless, the decomposition of ST-DRT, often referred to as the "projector metaphor," is poised to serve as a valuable meta-intuition in forthcoming investigations of ISAC systems. This metaphor elucidates that the fundamental tradeoff in ISAC is discerned through variations in preferences between S&C tasks, encompassing both the ST and the DRT.

References

1 ITU-R WP5D (2023). Draft New Recommendation ITU-R M. [IMTFRAMEWORK FOR 2030 AND BEYOND].

2 Liu, F., Cui, Y., Masouros, C. et al. (2022). Integrated sensing and communications: towards dual-functional wireless networks for 6G and beyond. *IEEE Journal on Selected Areas in Communications* 40(6): 1728–1767.

3 Hassanien, A., Amin, M.G., Aboutanios, E., and Himed, B. (2019). Dual-function radar communication systems: a solution to the spectrum congestion problem. *IEEE Signal Processing Magazine* 36(5): 115–126.

4 Sturm, C. and Wiesbeck, W. (2011). Waveform design and signal processing aspects for fusion of wireless communications and radar sensing. *Proceedings of the IEEE* 99(7): 1236–1259.

5 Cui, Y., Liu, F., Jing, X., and Mu, J. (2021). Integrating sensing and communications for ubiquitous IoT: applications, trends, and challenges. *IEEE Network* 35(5): 158–167.

6 Liu, F., Liu, Y.F., Li, A. et al. (2022). Cramér-Rao bound optimization for joint radar-communication beamforming. *IEEE Transactions on Signal Processing* 70: 240–253.

7 Liu, A., Huang, Z., Li, M. et al. (2022). A survey on fundamental limits of integrated sensing and communication. *IEEE Communications Surveys and Tutorials* 24(2): 994–1034.

8 Bliss, D.W. (2014). Cooperative radar and communications signaling: the estimation and information theory odd couple. *Proceedings of 2014 IEEE Radar Conference*, 0050–0055. Covington, KY, USA.

9 Guo, D., Shamai, S., and Verdu, S. (2005). Mutual information and minimum mean-square error in Gaussian channels. *IEEE Transactions on Information Theory* 51(4): 1261–1282.

10 Chiriyath, A.R., Paul, B., Jacyna, G.M., and Bliss, D.W. (2016). Inner bounds on performance of radar and communications co-existence. *IEEE Transactions on Signal Processing* 64(2): 464–474.

11 Kobayashi, M., Caire, G., and Kramer, G. (2018). Joint state sensing and communication: optimal tradeoff for a memoryless case. *2018 IEEE International Symposium on Information Theory (ISIT)*, 111–115.

12 Ahmadipour, M., Kobayashi, M., Wigger, M., and Caire, G. (2024). An information-theoretic approach to joint sensing and communication. *IEEE Transactions on Information Theory* 70(2): 1124–1146.

13 Kobayashi, M., Hamad, H., Kramer, G., and Caire, G. (2019). Joint state sensing and communication over memoryless multiple access channels. *Proceedings of 2019 IEEE International Symposium on Information Theory (ISIT)*, 270–274. Paris, France.

14 Ahmadipour, M., Wigger, M., and Kobayashi, M. (2021). Joint sensing and communication over memoryless broadcast channels. *Proceedings of 2020 IEEE Information Theory Workshop (ITW)*, 1–5. Riva del Garda, Italy.

15 Xiong, Y., Liu, F., Cui, Y. et al. (2023). On the fundamental tradeoff of integrated sensing and communications under Gaussian channels. *IEEE Transactions on Information Theory* 69(9): 5723–5751.

16 Li, X., Zhang, D., Lv, Q. et al. (2017). IndoTrack: device-free indoor human tracking with commodity Wi-Fi. *Proceedings of the ACM on Interactive, Mobile, Wearable and Ubiquitous Technologies* 1(3): 1–22.

17 Berger, T. (1971). *Rate Distortion Theory: A Mathematical Basis for Data Compression*. Prentice-Hall.

18 Kay, S.M. (1998). *Fundamentals of Statistical Signal Processing*. Englewood Cliffs, NJ: Prentice Hall.

19 El Gamal, A. and Kim, Y.H. (2011). *Network Information Theory*, 1e. Cambridge University Press.

20 Blahut, R. (1972). Computation of channel capacity and rate-distortion functions. *IEEE Transactions on Information Theory* 18(4): 460–473.

21 Li, J. and Stoica, P. (2007). MIMO radar with colocated antennas. *IEEE Signal Processing Magazine* 24(5): 106–114.

22 Miller, R. and Chang, C. (1978). A modified Cramér-Rao bound and its applications (Corresp.). *IEEE Transactions on Information Theory* 24(3): 398–400.

23 Polyanskiy, Y. and Wu, Y. (2023). *Information Theory: From Coding to Learning*, 1e. Cambridge University Press.

24 Bică, M. and Koivunen, V. (2019). Radar waveform optimization for target parameter estimation in cooperative radar-communications systems. *IEEE Transactions on Aerospace and Electronic Systems* 55(5): 2314–2326.

25 Cover, T.M. and Thomas, J.A. (2006). *Elements of Information Theory, 2e*. Hoboken, NJ: Wiley-Interscience.

26 Joudeh, H. and Willems, F.M.J. (2022). Joint communication and binary state detection. *IEEE Journal on Selected Areas in Information Theory* 3(1): 113–124.

27 Xiong, Y., Liu, F., and Lops, M. (2023). Generalized deterministic-random tradeoff in integrated sensing and communications: the sensing-optimal operating point. *2024 IEEE International Conference on Acoustics, Speech and Signal Processing (ICASSP)*, 8481–8485. Seoul, Republic of Korea.

14

Ziv–Zakai Bound for Multisource DOA Estimation

Zongyu Zhang[1], Zhiguo Shi[1], and Arye Nehorai[2]

[1] *College of Information Science and Electronic Engineering, Zhejiang University, Hangzhou, Zhejiang, China*
[2] *Department of Electrical and Systems Engineering, Washington University in St. Louis, St. Louis, MO, USA*

14.1 Introduction

As a fundamental problem in array signal processing, direction-of-arrival (DOA) estimation has been widely applied in radar, sonar, navigation, and wireless communications [1] where multisource estimation is a more general problem compared with single-source estimation. In the past few decades, the major research on this topic has focused on the algorithm design to achieve higher estimation accuracy, better angular resolution, etc. Some popular techniques include multiple signal classification (MUSIC) [2–4], estimation of signal parameters via rotational invariance techniques (ESPRIT) [5–7], sparse reconstruction-based algorithms [8–10], and machine learning-based algorithms [11–13].

To evaluate the estimation accuracy, mean square error (MSE) is a well-accepted metric, which, however, does not have a closed-form expression for its exact minimum value in DOA estimation. Hence, information-theoretic MSE lower bounds become commonly used powerful tools for performance evaluation (see [14] and the references therein). As the most widely used bound for unbiased estimators, Cramér–Rao bound (CRB) [15] can be easily obtained from the inverse of a Fisher information matrix. More specifically, a closed-form expression of the CRB for DOA estimation is derived in [16, 17], where both conditional and unconditional received signal models are considered. In [18, 19], the authors further derive the coarray CRB to account for virtual coarrays in underdetermined estimation scenarios. However, since Fisher information only captures small estimation errors, CRB becomes a local bound that only provides a tight bound asymptotically (e.g., at a high signal-to-noise-ratio (SNR)). In the non-asymptotic region, on

Information-Theoretic Radar Signal Processing, First Edition. Edited by Yujie Gu and Yimin D. Zhang.

the other hand, the MSE depends not only on the small estimation error that is lower bounded by the CRB but also on the large estimation error affected by the *a priori* distribution of the parameters. Thus, the classic CRB is a non-Bayesian bound that does not effectively account for the *a priori* probability density of the parameters. In the past few decades, numerous Bayesian bounds that utilize the *a priori* information of the parameters have been developed to overcome the locality of the classic CRB. Among them, covariance inequality-based bound [20] is a typical class of Bayesian bounds, which include Bayesian CRB (BCRB) [1], Weiss–Weinstein bound (WWB) [21], Reuven–Messer bound (RMB) [22], Bayesian–Abel bound (RAB) [23], and Bobrovsky–Zakai bound (BZB) [24]. Although these bounds introduce different covariance inequalities to pursue global tight expressions, there is always a tradeoff between the global tightness and the computational complexity (see higher-order derivative, nonconvex optimization over free variables, and empirical testing point selection) [14].

Different from the covariance inequality-based bounds, Ziv–Zakai bound (ZZB) utilizes both the *a priori* information of the parameters and the Fisher information from the observation data to provide a global tight bound. By exploiting the relationship between the minimum probability of error of a binary hypothesis testing problem and the MSE lower bound, the ZZB can provide a global tight bound without the tradeoff required by the covariance inequality-based bounds. ZZB is first proposed in [25–27], where a scalar time delay estimation problem is considered as an example. Then, ZZB has been considered in many specific parameter estimation problems over the past few decades [28–37]. Specifically, a single-source DOA estimation problem using linear arrays is considered in [28, 32]. In [30], the authors extend the ZZB from the estimation of scalar parameters to that of vector parameters and provide an explicit ZZB expression for single-source two-dimensional DOA estimation in [29]. As a typical vector parameter estimation problem, the ZZB derivation in [29] also provides a promising framework for multisource DOA estimation. However, generalizing the framework for multisource DOA estimation remains challenging. It requires solving high-dimensional integration and constrained optimization problems involved in the existing ZZB framework, and generalizing the Sherman–Morrison formula and the matrix determinant lemma in [29] for a single-source case to a multisource case. Moreover, under the Bayesian estimation framework, it is reasonable to assume a uniform distribution on the parameters to be estimated (i.e., DOAs) when there is no *a priori* information available. In this regard, it still remains challenging to assume a uniform *a priori* distribution on DOA θ itself to derive the ZZB instead of assuming a uniform distribution on $\sin\theta$ as in [29].[1]

1 Uniform assumption on $\sin\theta$ implies that the source signal prefers to impinge on the array from the boresight. With the DOA deviating from the boresight, the probability of the DOA gradually decreases.

There are three main challenges preventing the derivation of ZZB for multisource DOA estimation:

- An explicit ZZB expression is essential because the computational burden for numeric solutions becomes too heavy to be acceptable when the number of parameters is high.
- In multisource DOA estimation, an ordering process for both the estimated and the true DOAs must be adopted before MSE calculation in order to eliminate the permutation ambiguity. However, such an ordering process also brings potential dependence among multisource DOAs and implicitly changes the MSE convergence in the *a priori* performance region. Thus, the ZZB derived based on the natural assumption of independent DOAs becomes invalid for evaluating the MSE in the non-asymptotic region.
- Different from a single-source case, the mutual coherence among multiple sources is generally unavoidable due to multipath propagation (see [38–40] and the references therein). In this case, the effect of mutual coherence on the ZZB is still unclear.

To solve these challenges, we elaborate the explicit ZZB expression for hybrid coherent/uncorrelated multisource DOA estimation [41] in this chapter. Following the basic idea of the ZZB for vector parameters estimation, we generalize the ZZB developed for single-source DOA estimation to multisource DOA estimation by adopting the Woodbury matrix identity and Sylvester's determinant theorem, where each of the multisource DOA is assumed to independently follow a uniform *a priori* distribution. Clearly, such a natural assumption does not consider the effect of the ordering process in multisource DOA estimation, such that the resulting ZZB is invalid for evaluating the estimation performance in the non-asymptotic region. Hence, we analyze the effect of the ordering process on the MSE convergence in the *a priori* performance region by utilizing the order statistics, such that the derived ZZB remains to be globally tight for multisource DOA estimation. The obtained ZZB is an explicit expression with respect to the number of sources, the number of snapshots, the number of array sensors, the *a priori* distribution of DOAs, the SNR of the sources, and the coherence coefficients. Moreover, we provide a unified expression of the ZZB for both overdetermined and underdetermined DOA estimation scenarios and reveal how the number of sources and the *a priori* range of parameters affect the MSE convergence in the *a priori* performance region. We perform extensive simulations to show that the ZZB outperforms the commonly used CRBs in terms of the global tightness and the threshold SNR of asymptotic region prediction.

The main contributions of this chapter are listed as follows:

- We provide an explicit ZZB for hybrid coherent/uncorrelated multisource DOA estimation.

Table 14.1 List of notations.

Notation	Description	Notation	Description
$\mathbb{R}$	Set of real numbers	$\mathbf{0}_M$	M-dimensional all-zero vector
$\mathbb{C}$	Set of complex numbers	$\mathbf{1}_M$	M-dimensional all-one vector
$[\cdot]^{\mathrm{T}}$	Transpose	$\boldsymbol{I}_M$	$M \times M$-dimensional identity matrix
$[\cdot]^{\mathrm{H}}$	Hermitian transpose	$\mathbb{E}\{\cdot\}$	Statistical expectation
$\mathrm{Tr}\{\cdot\}$	Trace of a matrix	$\mathrm{diag}\,[\cdot]$	Diagonal matrix
$\lvert\cdot\rvert$	Determinant of a matrix	$\mathrm{block\,diag}\,[\cdot]$	Block diagonal matrix
$\mathrm{Pr}\,(\cdot)$	Probability	$\lVert\cdot\rVert_2$	ℓ_2 norm
$\mathrm{vec}(\cdot)$	Vectorization	$\otimes$	Kronecker product
$K!$	Factorial of positive integer K		

- We incorporate the order statistics to analyze how the ordering process affects the ZZB in the non-asymptotic region.
- The derived ZZB is a unified expression for both overdetermined and underdetermined DOA estimation.
- The derived ZZB reveals the relationship between MSE convergence performance and the number of sources in the *a priori* performance region.

The rest of this chapter is organized as follows. In Section 14.2, we introduce preliminaries on the basic array signal model, the definition of MSE in DOA estimation, and the basic concepts of CRB and ZZB. Then, in Section 14.3, we provide the detailed derivation of the ZZB and formulate the ZZB as an explicit expression with respect to multiple sources considering the effect of the ordering process. We perform extensive simulations in Section 14.4 to demonstrate the effectiveness of the derived ZZB. Finally, we make our conclusions and point out future directions in Section 14.5. The notations used in this chapter are summarized in Table 14.1.

14.2 Preliminaries

In this section, we introduce the array signal model, the definitions of MSE in multisource DOA estimation, and the basic concepts of CRB and ZZB.

14.2.1 Array Signal Model

We assume a linear array with M sensors located at $\{d_1, d_2, \dots, d_M\}$, where K far-field narrow-band signals impinges on the array from directions

$\boldsymbol{\theta} = [\theta_1, \theta_2, \dots, \theta_K]^{\mathrm{T}}$. Therein, the first L $(1 \leq L \leq K)$ sources are mutually coherent, and the other $K - L$ sources are uncorrelated and independent of the first L sources. Thus, the array observation data become

$$\begin{aligned} \boldsymbol{x}(t) &= \sum_{k=1}^{L} \boldsymbol{a}(\theta_k)\beta_k s_1(t) + \sum_{k=L+1}^{K} \boldsymbol{a}(\theta_k)s_k(t) + \boldsymbol{n}(t) \\ &= \boldsymbol{A}(\boldsymbol{\theta})\boldsymbol{s}(t) + \boldsymbol{n}(t), \forall t = 1, 2, \dots, T, \end{aligned} \tag{14.1}$$

where $\boldsymbol{A}(\boldsymbol{\theta}) = [\boldsymbol{a}(\theta_1), \boldsymbol{a}(\theta_2), \dots, \boldsymbol{a}(\theta_K)] \in \mathbb{C}^{M\times K}$ is the steering matrix with steering vector corresponding to the kth source as

$$\boldsymbol{a}(\theta_k) = \left[e^{-j\frac{2\pi}{\lambda}d_1 \sin\theta_k}, \dots, e^{-j\frac{2\pi}{\lambda}d_M \sin\theta_k}\right]^{\mathrm{T}}, \tag{14.2}$$

$\boldsymbol{s}(t) = [\boldsymbol{\beta}^{\mathrm{T}} s_1(t), s_{L+1}(t), \dots, s_K(t)]^{\mathrm{T}} \in \mathbb{C}^K$ describes the K signal waveforms, $\boldsymbol{\beta} = [\beta_1, \dots, \beta_L]^{\mathrm{T}} \in \mathbb{C}^L$ with β_l denoting the coherence coefficient of the lth $(1 \leq l \leq L)$ coherent signal with respect to the first signal $s_1(t)$, and $\boldsymbol{n}(t) \sim \mathcal{CN}(\boldsymbol{0}_M, \sigma_n^2 \boldsymbol{I}_M)$ is the complex zero-mean additive white Gaussian noise with power σ_n^2. The noise is independent of all the signals. Further, λ in (14.2) represents the wavelength, and T is the number of snapshots.

Without loss of generality, we make the following three fundamental assumptions:

Assumption 14.1 (Unconditional Model) All the uncorrelated signals $s_k(t)$, $k = 1, L+1, \dots, K$ are obtained from stationary zero-mean complex Gaussian stochastic processes.

Assumption 14.2 All the DOAs are distinct (say, $\theta_\imath \neq \theta_\jmath\ \forall \imath \neq \jmath$).

Assumption 14.3 Each DOA follows *a priori* uniform distribution $\theta_k \sim \mathcal{U}[\vartheta_{\min}, \vartheta_{\max}], \forall k = 1, 2, \dots, K$.

According to Assumption 14.1, the theoretical covariance matrix of the array observation data $\boldsymbol{x}(t)$ is

$$\boldsymbol{R}_{\boldsymbol{x}|\boldsymbol{\theta}} = \mathbb{E}\left\{\boldsymbol{x}(t)\boldsymbol{x}^{\mathrm{H}}(t)\right\} = \boldsymbol{A}(\boldsymbol{\theta})\boldsymbol{\Sigma}\boldsymbol{A}^{\mathrm{H}}(\boldsymbol{\theta}) + \sigma_n^2\boldsymbol{I}_M, \tag{14.3}$$

where

$$\boldsymbol{\Sigma} = \mathbb{E}\left\{\boldsymbol{s}(t)\boldsymbol{s}^{\mathrm{H}}(t)\right\} = \text{block diag}\left[\boldsymbol{\Sigma}_{co}, \boldsymbol{\Sigma}_{un}\right] \tag{14.4}$$

with $\boldsymbol{\Sigma}_{co} = \boldsymbol{\beta}\boldsymbol{\beta}^{\mathrm{H}}\sigma_1^2$ denoting the covariance matrix of the coherent sources and $\boldsymbol{\Sigma}_{un} = \text{diag}\left[\sigma_{L+1}^2, \dots, \sigma_K^2\right]$ representing the power of the uncorrelated sources. Obviously, the diagonal elements in $\boldsymbol{\Sigma}$ represents the source power. For example, $|\beta_k|^2\sigma_1^2$ $(k = 1, \dots, L)$ denotes the power of the kth coherent source that depends on the coherence coefficient β_k, while σ_k^2 $(k = L+1, \dots, K)$ is the power of the kth uncorrelated source.

14.2.2 MSE Definition in DOA Estimation

The MSE for multisource DOA estimation is defined as

$$\mathrm{MSE} = \frac{1}{K}\sum_{k=1}^{K}\mathbb{E}\left\{(\hat{\theta}_k - \theta_k)^2\right\} = \frac{1}{K}\,\mathrm{Tr}\{\boldsymbol{R}_\epsilon\}, \tag{14.5}$$

where $\hat{\theta}_k$ is the estimate of θ_k, and

$$\boldsymbol{R}_\epsilon = \mathbb{E}\left\{\epsilon\epsilon^{\mathrm{T}}\right\} = \mathbb{E}\left\{(\hat{\boldsymbol{\theta}} - \boldsymbol{\theta})(\hat{\boldsymbol{\theta}} - \boldsymbol{\theta})^{\mathrm{T}}\right\} \tag{14.6}$$

is the error correlation matrix. It is observed in (14.5) that the lower bound on the MSE can be obtained by extracting the diagonal elements of the lower bound of the error correlation matrix. Moreover, the definition of MSE in (14.5) implies that there is not a specific DOA that is more important than others during the computation of the MSE. This fact leads to the following assumption:

Assumption 14.4 All DOAs share an equal weight in the computation of the MSE.

14.2.3 Basic Concept of CRB

The CRB derived from the inverse of a Fisher information is the most commonly used lower bound in evaluating the accuracy of DOA estimation. According to the zero-mean Gaussian unconditional model in Assumption 14.1, the $\imath\jmath$th entry of the Fisher information matrix $\boldsymbol{J}$ with respect to $\boldsymbol{\theta}$ is given by [1]

$$\boldsymbol{J}_{\imath\jmath} = T\mathrm{Tr}\left\{\frac{\partial \boldsymbol{R}_{\boldsymbol{x}|\boldsymbol{\theta}}}{\partial \theta_\imath}\boldsymbol{R}_{\boldsymbol{x}|\boldsymbol{\theta}}^{-1}\frac{\partial \boldsymbol{R}_{\boldsymbol{x}|\boldsymbol{\theta}}}{\partial \theta_\jmath}\boldsymbol{R}_{\boldsymbol{x}|\boldsymbol{\theta}}^{-1}\right\}, \tag{14.7}$$

based on which, the CRB for DOA estimation is explicitly given in [16, 17]. However, the Fisher information matrix suffers from rank deficiency problem in underdetermined DOA estimation problems. To overcome this problem, Wang and Nehorai [18] and Liu and Vaidyanathan [19] derive the coarray Fisher information matrix as

$$\tilde{\boldsymbol{J}}_{\imath\jmath} = T\left[\mathrm{vec}\left(\frac{\partial \boldsymbol{R}_{\boldsymbol{x}|\boldsymbol{\theta}}}{\partial \theta_\imath}\right)\right]^{\mathrm{H}}\left(\boldsymbol{R}_{\boldsymbol{x}|\boldsymbol{\theta}}^{\mathrm{T}} \otimes \boldsymbol{R}_{\boldsymbol{x}|\boldsymbol{\theta}}\right)^{-1}\mathrm{vec}\left(\frac{\partial \boldsymbol{R}_{\boldsymbol{x}|\boldsymbol{\theta}}}{\partial \theta_\jmath}\right). \tag{14.8}$$

As the Bayesian version of CRB, the BCRB considers the Fisher information derived from the *a priori* probability density of the parameters. However, under the *a priori* uniform distribution given in Assumption 14.3, there is no *a priori* information of $\boldsymbol{\theta}$ contributing to the Fisher information [1]. In this case, the BCRB degrades to the classic CRB.

14.2.4 Basic Concept of ZZB

The ZZB for a vector-parameter estimation problem is given in [30, Eq. (32)] as[2]

$$\boldsymbol{w}^{\mathrm{T}}\boldsymbol{R}_{\epsilon}\boldsymbol{w} \geq \frac{1}{2}\int_{0}^{\infty} \max_{\boldsymbol{\delta}:\boldsymbol{w}^{\mathrm{T}}\boldsymbol{\delta}=h} \left[\int_{\mathbb{R}^K} \left(f_{\theta}(\boldsymbol{\varphi}) + f_{\theta}(\boldsymbol{\varphi}+\boldsymbol{\delta})\right) P_{\min}(\boldsymbol{\varphi}, \boldsymbol{\varphi}+\boldsymbol{\delta})d\boldsymbol{\varphi}\right] hdh, \quad (14.9)$$

where $\boldsymbol{w} \in \mathbb{R}^K$ is the weight vector with unit norm $\|\boldsymbol{w}\|_2 = 1$, $f_{\theta}(\boldsymbol{\varphi})$ represents the *a priori* probability density function (PDF) of $\boldsymbol{\theta}$ given $\boldsymbol{\theta} = \boldsymbol{\varphi}$, and $P_{\min}(\boldsymbol{\varphi}, \boldsymbol{\varphi}+\boldsymbol{\delta})$ is the minimum probability of error of the binary hypothesis testing problem

$$\begin{aligned} &\mathcal{H}_0 : \boldsymbol{\theta} = \boldsymbol{\varphi}; \qquad \boldsymbol{x} \sim f(\boldsymbol{x}|\boldsymbol{\varphi}), \\ &\mathcal{H}_1 : \boldsymbol{\theta} = \boldsymbol{\varphi}+\boldsymbol{\delta}; \quad \boldsymbol{x} \sim f(\boldsymbol{x}|\boldsymbol{\varphi}+\boldsymbol{\delta}), \end{aligned} \quad (14.10)$$

with

$$\begin{aligned} &\Pr(\mathcal{H}_0) = \frac{f_{\theta}(\boldsymbol{\varphi})}{f_{\theta}(\boldsymbol{\varphi}) + f_{\theta}(\boldsymbol{\varphi}+\boldsymbol{\delta})}, \\ &\Pr(\mathcal{H}_1) = 1 - \Pr(\mathcal{H}_0). \end{aligned} \quad (14.11)$$

Here, $f(\boldsymbol{x}|\boldsymbol{\varphi})$ and $f(\boldsymbol{x}|\boldsymbol{\varphi}+\boldsymbol{\delta})$ are PDFs of $\boldsymbol{x}$ given $\boldsymbol{\theta} = \boldsymbol{\varphi}$ and $\boldsymbol{\theta} = \boldsymbol{\varphi}+\boldsymbol{\delta}$, respectively. As the ZZB is a function of the weight vector, we adopt the following weights

$$\boldsymbol{w} = \frac{1}{\sqrt{K}}\mathbf{1}_K, \quad (14.12)$$

for multisource DOA estimation according to Assumption 14.4. Thus, (14.9) accordingly becomes

$$\frac{1}{K}\mathbf{1}_K^{\mathrm{T}}\boldsymbol{R}_{\epsilon}\mathbf{1}_K \geq \frac{1}{2}\int_{0}^{\infty} \max_{\boldsymbol{\delta}:\mathbf{1}_K^{\mathrm{T}}\boldsymbol{\delta}=\sqrt{K}h} \left[\int_{\mathbb{R}^K} \left(f_{\theta}(\boldsymbol{\varphi}) + f_{\theta}(\boldsymbol{\varphi}+\boldsymbol{\delta})\right) P_{\min}(\boldsymbol{\varphi}, \boldsymbol{\varphi}+\boldsymbol{\delta})d\boldsymbol{\varphi}\right] hdh. \quad (14.13)$$

Specifically, for a single-source case, vector $\boldsymbol{\theta}$ degrades to a scalar θ and $\delta = h$. As a result, (14.13) becomes

$$\mathbb{E}\left\{\left(\hat{\theta}-\theta\right)^2\right\} \geq \frac{1}{2}\int_{0}^{\infty}\int_{-\infty}^{+\infty} \left(f_{\theta}(\varphi) + f_{\theta}(\varphi+h)\right) P_{\min}(\varphi, \varphi+h)d\varphi hdh, \quad (14.14)$$

which was considered in [29].

14.3 ZZB Derivation for Multisource Estimation

In this section, we provide the detailed ZZB derivation for multisource DOA estimation. Naturally, the DOAs of the sources are independently distributed

2 The valley filling operation in [[30], Eq. (32)] is ignored here.

in the space. We first generalize the ZZB for single-source estimation [29] to multisource estimation in Section 14.3.1. During the MSE calculation for multisource DOA estimation, the ordering process for both the estimated and the true DOAs must be taken into account, thus introducing implicit extra dependence among the DOAs. The ordering process makes the generalized ZZB depicted in Section 14.3.1 invalid outside the asymptotic region. Considering the effect of the ordering process on the ZZB, we then adopt the order statistics in Section 14.3.2 to make the ZZB tight for the performance evaluation of multisource DOA estimation from the *a priori* performance region to the asymptotic region.

14.3.1 Generalized ZZB for Multisource Estimation

The Sherman–Morrison formula and the matrix determinant lemma involved in [29] make the ZZB derivation only adoptable for single-source DOA estimation. Based on the hybrid coherent/uncorrelated multisource model, we adopt the Woodbury matrix identity and Sylvester's determinant theorem to generalize the ZZB from single-source estimation to multisource estimation, following the naturally independent multisource DOA assumption. According to (14.9), deriving the minimum probability of error $P_{\min}$ is the essential step for ZZB derivation. Hence, we first derive explicit expression of $P_{\min}$ in Section 14.3.1.1 and, based on which, we further solve the integration and the optimization problem in Section 14.3.1.2.

14.3.1.1 Minimum Probability of Error

For the array observation data matrix $\boldsymbol{X} = [\boldsymbol{x}(1), \dots, \boldsymbol{x}(T)] \in \mathbb{C}^{M\times T}$, $P_{\min}(\boldsymbol{\varphi}, \boldsymbol{\varphi} + \boldsymbol{\delta})$ is lower bounded by $P(\boldsymbol{\delta})$ as [[42], p. 79]

$$P_{\min}(\boldsymbol{\varphi}, \boldsymbol{\varphi} + \boldsymbol{\delta}) \geq P(\boldsymbol{\delta}) = \exp\left\{ \mu(p; \boldsymbol{\delta}) + \frac{1}{8}\frac{\partial^2 \mu(p; \boldsymbol{\delta})}{\partial p^2} \right\} Q\left(\frac{1}{2}\sqrt{\frac{\partial^2 \mu(p; \boldsymbol{\delta})}{\partial p^2}} \right)\Bigg|_{p=\frac{1}{2}}, \tag{14.15}$$

where

$$Q(z) = \int_z^{\infty} \frac{1}{\sqrt{2\pi}} e^{-\frac{v^2}{2}} dv, \tag{14.16}$$

is the tail distribution function of the standard normal distribution, and

$$\begin{aligned}\mu(p; \boldsymbol{\delta}) &= \ln \int f(\boldsymbol{X}|\boldsymbol{\varphi} + \boldsymbol{\delta})^p\, f(\boldsymbol{X}|\boldsymbol{\varphi})^{1-p} d\boldsymbol{X} \\ &= T\left[p \ln |\boldsymbol{R}_{\boldsymbol{x}|\boldsymbol{\varphi}}| + (1-p)\ln |\boldsymbol{R}_{\boldsymbol{x}|\boldsymbol{\varphi}+\boldsymbol{\delta}}| - \ln |p\boldsymbol{R}_{\boldsymbol{x}|\boldsymbol{\varphi}} + (1-p)\boldsymbol{R}_{\boldsymbol{x}|\boldsymbol{\varphi}+\boldsymbol{\delta}}| \right],\end{aligned} \tag{14.17}$$

is the semi-invariant moment generating function for the T-snapshot observation data matrix $\boldsymbol{X}$ [42, p. 67]. Here, $\boldsymbol{R}_{\boldsymbol{x}|\boldsymbol{\varphi}}$ and $\boldsymbol{R}_{\boldsymbol{x}|\boldsymbol{\varphi}+\boldsymbol{\delta}}$ are the covariance matrices of $\boldsymbol{x}$ given $\boldsymbol{\theta} = \boldsymbol{\varphi}$ and $\boldsymbol{\theta} = \boldsymbol{\varphi} + \boldsymbol{\delta}$, respectively. Given $p = \frac{1}{2}$, (14.17) and its second derivative with respect to p respectively become

$$\mu(p;\boldsymbol{\delta})|_{p=\frac{1}{2}} = T\left[\frac{\ln\left(|\boldsymbol{R}_{\boldsymbol{x}|\boldsymbol{\varphi}}||\boldsymbol{R}_{\boldsymbol{x}|\boldsymbol{\varphi}+\boldsymbol{\delta}}|\right)}{2} - \ln\left|\frac{\boldsymbol{R}_{+}}{2}\right|\right], \tag{14.18}$$

and

$$\left.\frac{\partial^2\mu(p;\boldsymbol{\delta})}{\partial p^2}\right|_{p=\frac{1}{2}} = 4T\mathrm{Tr}\left\{\left(\boldsymbol{R}_{+}^{-1}\boldsymbol{R}_{-}\right)^2\right\}, \tag{14.19}$$

where $\boldsymbol{R}_{-} = \boldsymbol{R}_{\boldsymbol{x}|\boldsymbol{\varphi}} - \boldsymbol{R}_{\boldsymbol{x}|\boldsymbol{\varphi}+\boldsymbol{\delta}}$ and $\boldsymbol{R}_{+} = \boldsymbol{R}_{\boldsymbol{x}|\boldsymbol{\varphi}} + \boldsymbol{R}_{\boldsymbol{x}|\boldsymbol{\varphi}+\boldsymbol{\delta}}$.

Then, $\mu(p;\boldsymbol{\delta})|_{p=\frac{1}{2}}$ and $\frac{\partial^2\mu(p;\boldsymbol{\delta})}{\partial p^2}|_{p=\frac{1}{2}}$ can be respectively approximated as

$$\left.\mu(p;\boldsymbol{\delta})\right|_{p=\frac{1}{2}} \approx -\frac{1}{8}\boldsymbol{\delta}^{\mathrm{T}}\boldsymbol{J}\boldsymbol{\delta}, \tag{14.20}$$

and

$$\left.\frac{\partial^2\mu(p;\boldsymbol{\delta})}{\partial p^2}\right|_{p=\frac{1}{2}} \approx \boldsymbol{\delta}^{\mathrm{T}}\boldsymbol{J}\boldsymbol{\delta}, \tag{14.21}$$

by adopting the second-order Taylor series expansion around $\boldsymbol{\delta} = \boldsymbol{0}_K$. Hence, $P(\boldsymbol{\delta})$ is approximated as

$$P_S(\boldsymbol{\delta}) = Q\left(\frac{1}{2}\sqrt{\boldsymbol{\delta}^{\mathrm{T}}\boldsymbol{J}\boldsymbol{\delta}}\right), \tag{14.22}$$

by substituting (14.20) and (14.21) into (14.15).

Although the result is consistent with [29], the derivation of $P_S(\boldsymbol{\delta})$ here is appropriate for a multisource case. Specifically, Bell et al. [29] solves the determinants in $\mu(p;\boldsymbol{\delta})|_{p=\frac{1}{2}}$ using the matrix determinant lemma, from which $P_S(\boldsymbol{\delta})$ is further derived. Obviously, the matrix determinant lemma only works for a single-source case, such that $P_S(\boldsymbol{\delta})$ derived there is only appropriate for a single source. In contrast, we preserve the matrix determinants and perform the Taylor expansion directly on $\mu(p;\boldsymbol{\delta})|_{p=\frac{1}{2}}$ and $\frac{\partial^2\mu(p;\boldsymbol{\delta})}{\partial p^2}|_{p=\frac{1}{2}}$. Hence, the derivation of $P_S(\boldsymbol{\delta})$ does not depend on the number of sources, and it is appropriate for multisource cases.

However, $P(\boldsymbol{\delta}) \approx P_S(\boldsymbol{\delta})$ only holds when $\boldsymbol{\delta}$ is located in a small region around $\boldsymbol{0}_K$. For large $\boldsymbol{\delta}$ outside the region, we use the following derivation to provide a good approximation for $P(\boldsymbol{\delta})$.

According to Sylvester's determinant theorem

$$|\boldsymbol{I}_M + \boldsymbol{U}\boldsymbol{V}| = |\boldsymbol{I}_K + \boldsymbol{V}\boldsymbol{U}|, \tag{14.23}$$

for $\boldsymbol{U} \in \mathbb{C}^{M\times K}$ and $\boldsymbol{V} \in \mathbb{C}^{K\times M}$, the determinants in (14.18) respectively become

$$|\boldsymbol{R}_{\boldsymbol{x}|\boldsymbol{\varphi}}| = \sigma_n^{2M}\left|\boldsymbol{I}_K + \frac{1}{\sigma_n^2}\boldsymbol{A}^{\mathrm{H}}(\boldsymbol{\varphi})\boldsymbol{A}(\boldsymbol{\varphi})\boldsymbol{\Sigma}\right|, \tag{14.24}$$

$$|\boldsymbol{R}_{\boldsymbol{x}|\boldsymbol{\varphi}+\boldsymbol{\delta}}| = \sigma_n^{2M} \left| \boldsymbol{I}_K + \frac{1}{\sigma_n^2} \boldsymbol{A}^{\mathrm{H}}(\boldsymbol{\varphi}+\boldsymbol{\delta})\boldsymbol{A}(\boldsymbol{\varphi}+\boldsymbol{\delta})\boldsymbol{\Sigma} \right|, \tag{14.25}$$

and

$$\left| \frac{1}{2}\boldsymbol{R}_+ \right| = \sigma_n^{2M} \left| \boldsymbol{I}_{2K} + \frac{1}{2\sigma_n^2} \boldsymbol{B}^{\mathrm{H}}\boldsymbol{B}\boldsymbol{D} \right|, \tag{14.26}$$

where

$$\boldsymbol{B} = [\boldsymbol{A}(\boldsymbol{\varphi}), \boldsymbol{A}(\boldsymbol{\varphi}+\boldsymbol{\delta})], \tag{14.27}$$

and

$$\boldsymbol{D} = \text{block diag}\,[\boldsymbol{\Sigma}_{co}, \boldsymbol{\Sigma}_{un}, \boldsymbol{\Sigma}_{co}, \boldsymbol{\Sigma}_{un}]. \tag{14.28}$$

Similar to the discussion in [29], $\boldsymbol{A}^{\mathrm{H}}(\boldsymbol{\varphi})\boldsymbol{A}(\boldsymbol{\varphi})$, $\boldsymbol{A}^{\mathrm{H}}(\boldsymbol{\varphi}+\boldsymbol{\delta})\boldsymbol{A}(\boldsymbol{\varphi}+\boldsymbol{\delta})$, and $\boldsymbol{B}^{\mathrm{H}}\boldsymbol{B}$ are related to the array response, and the minima of $P(\boldsymbol{\delta})$ occurs at the null point of the array response as

$$\left| \boldsymbol{a}^{\mathrm{H}}(\theta)\boldsymbol{a}(\theta+\delta) \right| = 0, \tag{14.29}$$

for a single-source case. It implies that there is a lower bound for $P(\boldsymbol{\delta})$ when the steering vector $\boldsymbol{a}(\theta+\delta)$ is orthogonal to $\boldsymbol{a}(\theta)$. Hence, for a multisource case, we generalize such approximation (14.29) to

$$\boldsymbol{A}^{\mathrm{H}}(\boldsymbol{\theta})\boldsymbol{A}(\boldsymbol{\theta}) \approx M\boldsymbol{I}_K, \tag{14.30}$$

i.e., the steering vectors are approximately orthogonal to each other. Then, we have

$$\boldsymbol{B}^{\mathrm{H}}\boldsymbol{B} \approx M\boldsymbol{I}_{2K}, \tag{14.31}$$

and (14.18) is approximated as

$$\begin{aligned} \mu(p;\boldsymbol{\delta})|_{p=\frac{1}{2}} &\approx T \ln \frac{\left| \boldsymbol{I}_K + \frac{M}{\sigma_n^2}\boldsymbol{\Sigma} \right|}{\left| \boldsymbol{I}_{2K} + \frac{M}{2\sigma_n^2}\boldsymbol{D} \right|} \\ &= T \ln \frac{4(1+M\|\boldsymbol{\beta}\|_2^2\eta_1)}{(2+M\|\boldsymbol{\beta}\|_2^2\eta_1)^2} + T \sum_{k=L+1}^{K} \ln \frac{4(1+M\eta_k)}{(2+M\eta_k)^2}, \end{aligned} \tag{14.32}$$

where $\eta_k = \frac{\sigma_k^2}{\sigma_n^2}$ denotes the SNR of the kth source.

Now, we provide a good approximation for (14.19). With the approximation (14.31), the inverse of $\boldsymbol{R}_+$ becomes

$$\boldsymbol{R}_+^{-1} \approx \frac{1}{2\sigma_n^2}\boldsymbol{I}_M - \frac{1}{4\sigma_n^4}\boldsymbol{B}\boldsymbol{D}\boldsymbol{Y}\boldsymbol{B}^{\mathrm{H}}, \tag{14.33}$$

by adopting the Woodbury matrix identity, where

$$\boldsymbol{Y} = \left(\boldsymbol{I}_{2K} + \frac{M}{2\sigma_n^2}\boldsymbol{D}\right)^{-1} = \text{block diag}\,[\boldsymbol{Y}_{co}, \boldsymbol{Y}_{un}, \boldsymbol{Y}_{co}, \boldsymbol{Y}_{un}], \tag{14.34}$$

with

$$\boldsymbol{Y}_{co} = \left(\boldsymbol{I}_L + \frac{M}{2\sigma_n^2}\boldsymbol{\Sigma}_{co}\right)^{-1} = \boldsymbol{I}_L - \frac{M\eta_1}{2 + M\|\boldsymbol{\beta}\|_2^2\eta_1}\boldsymbol{\beta}\boldsymbol{\beta}^{\mathrm{H}} \tag{14.35}$$

using Sherman–Morrison formula, and

$$\boldsymbol{Y}_{un} = \left(\boldsymbol{I}_{K-L} + \frac{M}{2\sigma_n^2}\boldsymbol{\Sigma}_{un}\right)^{-1} = \text{diag}\left[\left(1 + \frac{M\eta_{L+1}}{2}\right)^{-1}, \ldots, \left(1 + \frac{M\eta_K}{2}\right)^{-1}\right]. \tag{14.36}$$

Accordingly, $(\boldsymbol{R}_+^{-1}\boldsymbol{R}_-)^2$ in (14.19) can be approximated as

$$(\boldsymbol{R}_+^{-1}\boldsymbol{R}_-)^2 \approx \frac{M}{4\sigma_n^4}\left(\boldsymbol{B}\boldsymbol{D}^2\boldsymbol{B}^{\mathrm{H}} - \frac{M}{\sigma_n^2}\boldsymbol{B}\boldsymbol{D}^3\boldsymbol{Y}\boldsymbol{B}^{\mathrm{H}} + \frac{M^2}{4\sigma_n^4}\boldsymbol{B}\boldsymbol{D}^4\boldsymbol{Y}^2\boldsymbol{B}^{\mathrm{H}}\right). \tag{14.37}$$

Then, (14.19) is approximated as

$$\left.\frac{\partial^2\mu(p;\boldsymbol{\delta})}{\partial p^2}\right|_{p=\frac{1}{2}} \approx \frac{TM^2}{\sigma_n^4}\,\mathrm{Tr}\left\{\left(\boldsymbol{D}^2 - \frac{M}{\sigma_n^2}\boldsymbol{D}^3\boldsymbol{Y} + \frac{M^2}{4\sigma_n^4}\boldsymbol{D}^4\boldsymbol{Y}^2\right)\right\}, \tag{14.38}$$

where

$$\mathrm{Tr}\{\boldsymbol{D}^2\} = 2\|\boldsymbol{\beta}\|_2^4\sigma_1^4 + \sum_{k=L+1}^{K} 2\sigma_k^4, \tag{14.39}$$

$$\mathrm{Tr}\{\boldsymbol{D}^3\boldsymbol{Y}\} = \frac{4\|\boldsymbol{\beta}\|_2^6\sigma_1^6}{2 + M\|\boldsymbol{\beta}\|_2^2\eta_1} + \sum_{k=L+1}^{K}\frac{4\sigma_k^6}{2 + M\eta_k}. \tag{14.40}$$

and

$$\mathrm{Tr}\{\boldsymbol{D}^4\boldsymbol{Y}^2\} = \frac{8\|\boldsymbol{\beta}\|_2^8\sigma_1^8}{\left(2 + M\|\boldsymbol{\beta}\|_2^2\eta_1\right)^2} + \sum_{k=L+1}^{K}\frac{8\sigma_k^8}{(2 + M\eta_l)^2}. \tag{14.41}$$

As such, (14.19) becomes

$$\left.\frac{\partial^2\mu(p;\boldsymbol{\delta})}{\partial p^2}\right|_{p=\frac{1}{2}} \approx 8T\left[\left(\frac{M\|\boldsymbol{\beta}\|_2^2\eta_1}{2 + M\|\boldsymbol{\beta}\|_2^2\eta_1}\right)^2 + \sum_{k=L+1}^{K}\left(\frac{M\eta_k}{2 + M\eta_k}\right)^2\right]. \tag{14.42}$$

Accordingly, $P(\boldsymbol{\delta})$ for large $\boldsymbol{\delta}$ is approximated as

$$P_L = \exp\left\{T\left[\ln\frac{4(1+M\|\boldsymbol{\beta}\|_2^2\eta_1)}{(2+M\|\boldsymbol{\beta}\|_2^2\eta_1)^2} + \left(\frac{M\|\boldsymbol{\beta}\|_2^2\eta_1}{2+M\|\boldsymbol{\beta}\|_2^2\eta_1}\right)^2\right]\right.$$
$$\left. + T\sum_{k=L+1}^{K}\left[\ln\frac{4(1+M\eta_k)}{(2+M\eta_k)^2} + \left(\frac{M\eta_k}{2+M\eta_k}\right)^2\right]\right\} \tag{14.43}$$
$$\times Q\left(\sqrt{2T\left[\left(\frac{M\|\boldsymbol{\beta}\|_2^2\eta_1}{2+M\|\boldsymbol{\beta}\|_2^2\eta_1}\right)^2 + \sum_{k=L+1}^{K}\left(\frac{M\eta_k}{2+M\eta_k}\right)^2\right]}\right),$$

by substituting (14.32) and (14.42) into (14.15).

Here, we adopt the Woodbury matrix identity and Sylvester's determinant theorem to derive P_L for multisource scenario. Moreover, the derived P_L is approximately[3] same as that in [29] when $K = 1$.

Recalling (14.22) and (14.43), $P(\boldsymbol{\delta})$ in (14.15) is approximated as

$$P(\boldsymbol{\delta}) \approx \begin{cases} P_S(\boldsymbol{\delta}), & \boldsymbol{\delta} \in \Delta \\ P_L, & \boldsymbol{\delta} \notin \Delta \end{cases}, \tag{14.44}$$

where Δ denotes the region

$$\Delta = \left\{\boldsymbol{\delta} : \boldsymbol{\delta}^{\mathrm{T}}\boldsymbol{J}\boldsymbol{\delta} \le 8T\left[\left(\frac{M\|\boldsymbol{\beta}\|_2^2\eta_1}{2+M\|\boldsymbol{\beta}\|_2^2\eta_1}\right)^2 + \sum_{k=L+1}^{K}\left(\frac{M\eta_k}{2+M\eta_k}\right)^2\right]\right\}. \tag{14.45}$$

Obviously, the boundary of this region occurs at $\boldsymbol{\delta}$ that satisfies $P_S(\boldsymbol{\delta}) = P_L$. However, the boundary cannot be solved in an explicit way. Because $Q(z)$ is a decreasing function with respect to z, the region Δ with respect to $\boldsymbol{\delta}$ can be given by [29]

$$\frac{\sqrt{\boldsymbol{\delta}^{\mathrm{T}}\boldsymbol{J}\boldsymbol{\delta}}}{2} \le \sqrt{2T\left[\left(\frac{M\|\boldsymbol{\beta}\|_2^2\eta_1}{2+M\|\boldsymbol{\beta}\|_2^2\eta_1}\right)^2 + \sum_{k=L+1}^{K}\left(\frac{M\eta_k}{2+M\eta_k}\right)^2\right]}, \tag{14.46}$$

which leads to (14.45).

---

3 Although the derivation of P_L is consistent with [29, Eq. (C. 13)], [29, Eq. (C. 13)] introduces an extra approximation. Hence, we use the expression "approximately" here.

14.3.1.2 Integration and Optimization Solution

Although the signals can be coherent, the DOAs are assumed to be independent, based on which the K-dimensional integration in (14.13) becomes

$$\begin{aligned}\frac{1}{2}\int_{\mathbb{R}^K}\left(f_\theta(\boldsymbol{\varphi})+f_\theta(\boldsymbol{\varphi}+\boldsymbol{\delta})\right)P_{\min}(\boldsymbol{\varphi},\boldsymbol{\varphi}+\boldsymbol{\delta})d\boldsymbol{\varphi} &\geq \frac{P(\boldsymbol{\delta})}{\zeta^K}\int_\Phi d\boldsymbol{\varphi}\\ &= \frac{P(\boldsymbol{\delta})}{\zeta^K}\prod_{k=1}^K(\zeta-\delta_k),\end{aligned}\tag{14.47}$$

where

$$\Phi=\left\{\boldsymbol{\varphi}|\varphi_k\in\left[\vartheta_{\min},\vartheta_{\max}-\delta_k\right],k=1,2,\dots,K\right\},\tag{14.48}$$

is the K-dimensional integration region, $\zeta=\vartheta_{\max}-\vartheta_{\min}$ denotes the range of DOAs. By substituting (14.44) into (14.47), we have

$$\frac{P(\boldsymbol{\delta})}{\zeta^K}\prod_{k=1}^K(\zeta-\delta_k)\approx\frac{P_L}{\zeta^K}\prod_{k=1}^K(\zeta-\delta_k)+\begin{cases}P_S(\boldsymbol{\delta})-P_L & \boldsymbol{\delta}\in\Delta\\ 0 & \boldsymbol{\delta}\notin\Delta\end{cases},\tag{14.49}$$

such that the optimization problem in (14.13) can be separated as three terms

$$\begin{aligned}&\max_{\boldsymbol{\delta}:\mathbf{1}_K^{\mathrm{T}}\boldsymbol{\delta}=\sqrt{K}h}\frac{P(\boldsymbol{\delta})}{\zeta^K}\prod_{k=1}^K(\zeta-\delta_k)\\ &\approx\max_{\boldsymbol{\delta}:\mathbf{1}_K^{\mathrm{T}}\boldsymbol{\delta}=\sqrt{K}h}\frac{P_L}{\zeta^K}\prod_{k=1}^K(\zeta-\delta_k)+\max_{\boldsymbol{\delta}\in\Delta:\mathbf{1}_K^{\mathrm{T}}\boldsymbol{\delta}=\sqrt{K}h}P_S(\boldsymbol{\delta})-P_L.\end{aligned}\tag{14.50}$$

According to the arithmetic–geometric mean inequality, it is easy to prove that the first term in (14.50) becomes

$$\max_{\boldsymbol{\delta}:\mathbf{1}_K^{\mathrm{T}}\boldsymbol{\delta}=\sqrt{K}h}\frac{P_L}{\zeta^K}\prod_{k=1}^K(\zeta-\delta_k)=P_L\left(1-\frac{h}{\sqrt{K}\zeta}\right)^K.\tag{14.51}$$

Similar to the discussion in [29], the second term is

$$\max_{\boldsymbol{\delta}\in\Delta:\mathbf{1}_K^{\mathrm{T}}\boldsymbol{\delta}=\sqrt{K}h}P_S(\boldsymbol{\delta})=Q\left(\frac{\sqrt{K}h}{2\sqrt{\mathbf{1}_K^{\mathrm{T}}\boldsymbol{J}^{-1}\mathbf{1}_K}}\right),\tag{14.52}$$

at the point

$$\boldsymbol{\delta}=\sqrt{K}h\frac{\boldsymbol{J}^{-1}\mathbf{1}_K}{\mathbf{1}_K^{\mathrm{T}}\boldsymbol{J}^{-1}\mathbf{1}_K}.\tag{14.53}$$

Thus, (14.50) is further written as

$$\max_{\boldsymbol{\delta}:\mathbf{1}_K^{\mathrm{T}}\boldsymbol{\delta}=\sqrt{K}h}\frac{P(\boldsymbol{\delta})}{\zeta^K}\prod_{k=1}^K(\zeta-\delta_k)\approx P_L\left(1-\frac{h}{\sqrt{K}\zeta}\right)^K+\begin{cases}Q\left(\frac{\sqrt{K}h}{2\sqrt{\mathbf{1}^{\mathrm{T}}\boldsymbol{J}^{-1}\mathbf{1}}}\right)-P_L & 0\leq h\leq\tilde{h},\\ 0 & h>\tilde{h},\end{cases}\tag{14.54}$$

where

$$\tilde{h} \approx \min\left[\sqrt{\frac{8T\mathbf{1}_K^{\mathrm{T}}\boldsymbol{J}^{-1}\mathbf{1}_K}{K}\left[\left(\frac{M\|\boldsymbol{\beta}\|_2^2\eta_1}{2+M\|\boldsymbol{\beta}\|_2^2\eta_1}\right)^2+\sum_{k=L+1}^{K}\left(\frac{M\eta_k}{2+M\eta_k}\right)^2\right]},\sqrt{K}\zeta\right], \tag{14.55}$$

is the threshold given by (14.46) and constraint $0 \leq \tilde{h} \leq \sqrt{K}\zeta$.

After solving the optimization problem and the integration with respect to h, the generalized ZZB for lower bounding the error correlation matrix is derived as

$$\frac{1}{K}\mathbf{1}_K^{\mathrm{T}}\boldsymbol{R}_\epsilon\mathbf{1}_K \geq \frac{12P_L\mathbf{1}_K^{\mathrm{T}}\boldsymbol{R}_\theta\mathbf{1}_K}{(K+1)(K+2)} + \Gamma_{\frac{3}{2}}(\tilde{u})\frac{\mathbf{1}_K^{\mathrm{T}}\boldsymbol{J}^{-1}\mathbf{1}_K}{K}, \tag{14.56}$$

where

$$\boldsymbol{R}_\theta = \frac{\zeta^2}{12}\boldsymbol{I}_K, \tag{14.57}$$

is the *a priori* covariance matrix of $\boldsymbol{\theta}$,

$$\Gamma_{\frac{3}{2}}(\tilde{u}) = \frac{1}{\Gamma\left(\frac{3}{2}\right)}\int_0^{\tilde{u}} e^{-\xi}\xi^{\frac{1}{2}}\,d\xi, \tag{14.58}$$

is the normalized incomplete Gamma function with

$$\tilde{u} = \frac{K\tilde{h}^2}{8\mathbf{1}_K^{\mathrm{T}}\boldsymbol{J}^{-1}\mathbf{1}_K}, \tag{14.59}$$

and $\Gamma(\frac{3}{2}) = \frac{\sqrt{\pi}}{2}$.

As discussed in Section 14.2.2, the lower bound on the MSE can be obtained by extracting the diagonal elements from the error correlation matrix, such that the generalized ZZB for the MSE of mutlisource DOAs estimation becomes

$$\mathrm{MSE} \geq \frac{12P_L\,\mathrm{Tr}\{\boldsymbol{R}_\theta\}}{(K+1)(K+2)} + \Gamma_{\frac{3}{2}}(\tilde{u})\frac{\mathrm{Tr}\{\boldsymbol{J}^{-1}\}}{K}. \tag{14.60}$$

It can be observed that only the first term contains the *a priori* information of $\boldsymbol{\theta}$ (see $\boldsymbol{R}_\theta$). By contrast, the second term is determined by the Fisher information matrix.

14.3.2 ZZB for Estimation with Ordering Process

The generalized ZZB we derived in Section 14.3.1 is invalid in the non-asymptotic region, because it does not consider the effect of the essential ordering process for both the estimated and the true DOAs on the MSE calculation. In other words, since we cannot exactly know that each estimated DOA belongs to which true

DOA, we have to introduce a certain criterion to realize a pair match between an estimated DOA and a true DOA to avoid the permutation ambiguity. The following example illustrates this.

Example 14.1 Two signals impinge on the array from DOAs 30° and 45°, while the estimates are obtained as 29° and 44°. Here, there is two potential root MSE (RMSE) results obtained due to the permutation ambiguity. The first one is $\sqrt{\frac{(30^\circ-29^\circ)^2+(45^\circ-44^\circ)^2}{2}} = 1.00^\circ$ whereas the other is $\sqrt{\frac{(30^\circ-44^\circ)^2+(45^\circ-29^\circ)^2}{2}} \approx 15.03^\circ$. Obviously, in order to obtain accurate MSE result, we have to introduce a certain criterion to avoid the permutation ambiguity.

In one-dimensional DOA estimation, only small estimation error occurs in the asymptotic region, because the estimates are located close to the true DOA counterparts. Hence, the ordering process matches each estimate to the respective true DOA. In the non-asymptotic region, on the other hand, the large estimation errors start to occur, which are distributed over the whole parameter space. In such case, the ordering process still tries to match an estimated DOA to the closest one of the true DOAs. When the estimate belongs to another true DOA rather than the one being matched, the resulting MSE involves the implicit dependence among different DOAs, which is actually introduced by the ordering process.

Specifically, a large estimation error dominates in the *a priori* performance region, such that each θ_k would be randomly estimated via the *a priori* distribution $\theta_k \sim \mathcal{U}[\vartheta_{\min}, \vartheta_{\max}]$. Accordingly, the MSE of each θ_k is determined by the *a priori* variance $\sigma^2_{\theta_k}$. On the contrary, after the ordering process, the MSE actually becomes

$$\text{MSE} = \frac{1}{K}\sum_{k=1}^{K}\mathbb{E}\left\{\left(\hat{\theta}_{(k)} - \theta_{(k)}\right)^2\right\}, \tag{14.61}$$

where $\hat{\theta}_{(k)}$ and $\theta_{(k)}$ are the kth smallest order statistic in $\hat{\boldsymbol{\theta}}$ and $\boldsymbol{\theta}$. Here, the subscript (k) indicates the kth order statistic. It is clear that $\theta_{(k)}$ tends to be randomly estimated in the interval affected by $\theta_{(k-1)}$ and $\theta_{(k+1)}$[4] after the ordering process rather than its natural interval $[\vartheta_{\min}, \vartheta_{\max}]$. Thus, the MSE convergence of $\theta_{(k)}$ is determined by the *a priori* variance $\sigma^2_{\theta_{(k)}}$. It is easy to understand that as K increases, the *a priori* variance $\sigma^2_{\theta_{(k)}}$ becomes smaller than the original *a priori* variance $\sigma^2_{\theta_k}$, leading to a lower MSE in the *a priori* performance region.

Hence, the effect of the ordering process on the MSE calculation is described by the following ratio

$$\kappa = \frac{\text{Tr}\{\boldsymbol{R}_{(\theta)}\}}{\text{Tr}\{\boldsymbol{R}_{\theta}\}}, \tag{14.62}$$

---

4 The interval is affected by $\vartheta_{\min}$ and $\theta_{(2)}$ for $\theta_{(1)}$, and $\theta_{(K-1)}$ and $\vartheta_{\max}$ for $\theta_{(K)}$, respectively.

where

$$\mathrm{Tr}\{\boldsymbol{R}_{(\theta)}\} = \sum_{k=1}^{K} \sigma^2_{\theta_{(k)}}, \tag{14.63}$$

is the trace of the *a priori* covariance matrix of the order statistics $\{\theta_{(1)}, \ldots, \theta_{(K)}\}$.

The *a priori* PDF of $\theta_{(k)}$ is given by [43, p. 229]

$$f(\theta_{(k)}) = \frac{K!\left(\frac{\theta_{(k)}}{\zeta}\right)^{k-1}\left(1 - \frac{\theta_{(k)}}{\zeta}\right)^{K-k}}{\zeta(k-1)!(K-k)!}, \tag{14.64}$$

such that the *a priori* variance of $\theta_{(k)}$ is

$$\sigma^2_{\theta_{(k)}} = \int_0^{\zeta} \left(\theta_{(k)} - \mathbb{E}\left\{\theta_{(k)}\right\}\right)^2 f(\theta_{(k)}) d\theta_{(k)} = \frac{\zeta^2(K+1-k)k}{(K+1)^2(K+2)}, \tag{14.65}$$

where

$$\mathbb{E}\left\{\theta_{(k)}\right\} = \int_0^{\zeta} \theta_{(k)} f(\theta_{(k)}) d\theta_{(k)} = \frac{k\zeta}{K+1}, \tag{14.66}$$

is the expectation of the variable $\theta_{(k)}$. Accordingly, (14.63) becomes

$$\mathrm{Tr}\{\boldsymbol{R}_{(\theta)}\} = \frac{K\zeta^2}{6(K+1)}, \tag{14.67}$$

and the ratio κ in (14.62) becomes

$$\kappa = \frac{2}{K+1}, \tag{14.68}$$

which is a function of the number of sources K. It is worth noting that we have $\kappa = 1$ when there is only a single source, which indicates that there is no need to perform ordering process. Furthermore, with the number of sources increasing, the difference between the MSE convergence in the *a priori* performance region before and after the ordering process becomes larger because the ratio κ becomes smaller.

According to above analysis, we incorporate the ratio κ (14.68) into the first term of the generalized ZZB (14.60), such that the resulting ZZB

$$\mathrm{MSE} \geq 2P_L \frac{K\zeta^2}{(K+1)^2(K+2)} + \Gamma_{\frac{3}{2}}(\tilde{u}) \frac{\mathrm{Tr}\{\boldsymbol{J}^{-1}\}}{K}. \tag{14.69}$$

is globally valid for evaluating multisource DOA estimation with the ordering process.

Now, it is observed that the derived ZZB is a linear combination between the so-called *a priori* bound (APB) and the CRB with the linear combination coefficient $2P_L$ and $\Gamma_{\frac{3}{2}}(\tilde{u})$, respectively. Specifically, the ZZB converges to the APB in the *a priori* performance region as

$$\lim_{\forall \eta_k \to 0} \mathrm{MSE} \geq \frac{K\zeta^2}{(K+1)^2(K+2)}, \tag{14.70}$$

since

$$\lim_{\forall \eta_k \to 0} 2P_L = 1, \tag{14.71}$$

and

$$\lim_{\forall \eta_k \to 0} \Gamma_{\frac{3}{2}}(\tilde{u}) = 0. \tag{14.72}$$

In the asymptotic region, the ZZB converges to the CRB as

$$\lim_{\forall \eta_k \to +\infty} \text{MSE} \geq \frac{\text{Tr}\{\boldsymbol{J}^{-1}\}}{K}, \tag{14.73}$$

since

$$\lim_{\forall \eta_k \to +\infty} 2P_L = 0, \tag{14.74}$$

and

$$\lim_{\forall \eta_k \to +\infty} \Gamma_{\frac{3}{2}}(\tilde{u}) = 1. \tag{14.75}$$

The APB (14.70) is a function of only the *a priori* range of DOAs and the number of sources, which indicates that the estimation performance cannot be improved by adopting more array sensors or more snapshots once the estimation falls into the *a priori* performance region. However, more array sensors or more snapshots can enable the estimation to leave the *a priori* performance region at a lower SNR, according to the expression of $2P_L$ (14.43).

When the sources are uncorrelated, P_L (14.43) and $\tilde{u}$ (14.59) respectively degrade to

$$P_L = \exp\left\{T\sum_{k=1}^{K}\left[\ln\frac{4(1+M\eta_k)}{(2+M\eta_k)^2} + \left(\frac{M\eta_k}{2+M\eta_k}\right)^2\right]\right\} Q\left(\sqrt{2T\sum_{k=1}^{K}\left(\frac{M\eta_k}{2+M\eta_k}\right)^2}\right), \tag{14.76}$$

and

$$\tilde{u} \approx \min\left[T\sum_{k=1}^{K}\left(\frac{M\eta_k}{2+M\eta_k}\right)^2, \frac{K^2\zeta^2}{8\mathbf{1}_K^{\mathrm{T}}\boldsymbol{J}^{-1}\mathbf{1}_K}\right]. \tag{14.77}$$

For single-source DOA estimation, the ZZB is simplified to

$$\begin{aligned}\mathbb{E}\left\{(\hat{\theta}-\theta)^2\right\} \geq\ & 2\exp\left\{T\left[\ln\frac{4(1+M\eta)}{(2+M\eta)^2} + \left(\frac{M\eta}{2+M\eta}\right)^2\right]\right\} \\ & \times Q\left(\sqrt{2T}\frac{M\eta}{2+M\eta}\right)\sigma_\theta^2 + \Gamma_{\frac{3}{2}}(\tilde{u})J^{-1},\end{aligned} \tag{14.78}$$

where $\sigma_\theta^2 = \frac{\zeta^2}{12}$ is the *a priori* variance of θ, and J is the scalar Fisher information. Here, the ZZB (14.78) degrades to [29, Eq. (C.30)] if we ignore the implicit condition $\tilde{h} < \sqrt{K}\zeta$ and the extra approximation in [29, Eq. (C.14)].

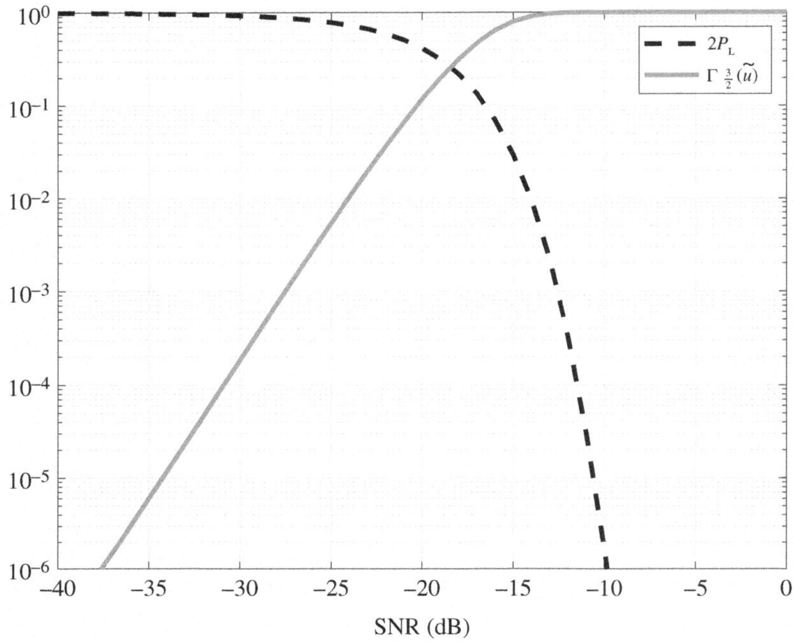

Figure 14.1 Combination coefficients versus the SNR for single source.

Since the derived ZZB is a linear combination between the APB and the CRB, we investigate the effect of SNR on the linear combination coefficients in Figure 14.1, where we consider a simple single-source case. The DOA follows *a priori* distribution $\mathcal{U}[-90°, 90°]$. We adopt a uniform linear array (ULA) with $M = 20$ sensors with half-wavelength inter-element spacing and use $T = 40$ snapshots. For each SNR point, we perform 10,000 Monte Carlo trials. We can see that the coefficient $2P_L$ decreases from 1 to 0 as the SNR increases. On the other hand, the coefficient $\Gamma_{\frac{3}{2}}(\tilde{u})$ behaves with an opposite trend. Obviously, the APB and CRB dominate the ZZB respectively in the *a priori* performance region and the asymptotic region.

Since both combination coefficients $2P_L$ and $\Gamma_{\frac{3}{2}}(\tilde{u})$ are functions of the coherence coefficient $\boldsymbol{\beta}$ under coherent multisource case, we further investigate the effect of the coherence coefficients on the combination coefficients in Figure 14.2. We assume that there are two coherent sources with the coherence coefficient $[1, \beta_2]$, such that the second source has the waveform $s_2 = \beta_2 s_1$ with the power $\sigma_2^2 = |\beta_2|^2\sigma_1^2$. For better illustration, we consider the corresponding uncorrelated source case as reference. Different from the coherent source case, we fix the source power σ_1^2 and $\sigma_2^2 = |\beta_2|^2\sigma_1^2$ while use the independent waveform s_1 and s_2. It is observed that there is a larger difference between the coherent and uncorrelated combination coefficients with $|\beta_2|$ increasing. This is because the norm $\|\boldsymbol{\beta}\|_2$ in

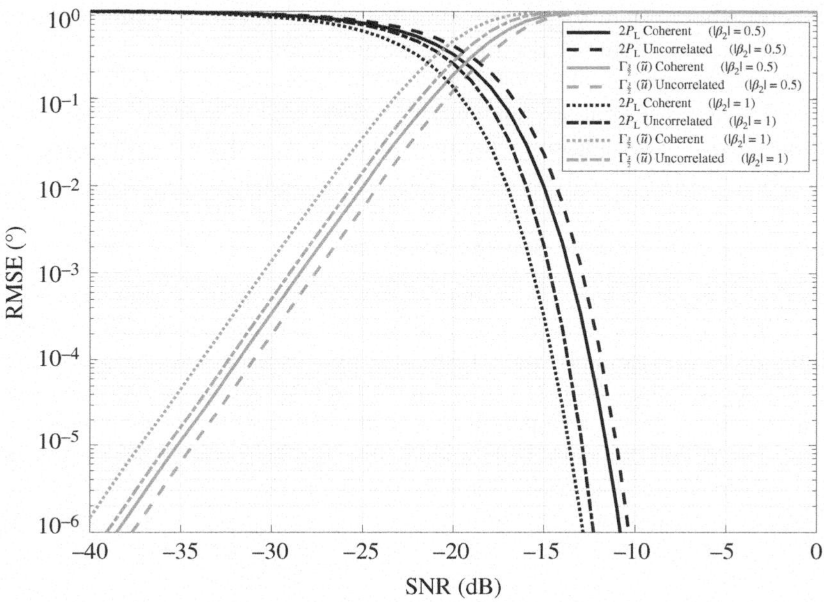

Figure 14.2 Combination coefficients versus the SNR for two sources with different coherence coefficients.

(14.43) and (14.55) is larger, such that the combination coefficients $2P_L$ and $\Gamma_{\frac{3}{2}}(\tilde{u})$ become more distinguishable in the coherent case. For larger $|\beta_2|$, $2P_L$ leaves from 1 and $\Gamma_{\frac{3}{2}}(\tilde{u})$ reaches to 1 at a lower SNR. Thus, the ZZB will leave the *a priori* performance region and reach the asymptotic region at a lower SNR. Moreover, compared with the uncorrelated sources case, $2P_L$ and $\Gamma_{\frac{3}{2}}(\tilde{u})$ for coherent sources respectively decrease from 1 and reach 1 earlier, such that the coherent ZZB will leave from the APB and reach the CRB at a lower SNR than the uncorrelated ZZB.

14.4 Simulation Results

In this section, we evaluate the derived ZZB for multisource DOA estimation, where both the CRB and the APB are plotted for reference. As a Bayesian bound, the ZZB requires a known *a priori* distribution for the parameter, such that we assume each DOA to follow a uniform distribution $\mathcal{U}[-60°, 60°]$ in our simulations. In all simulations, we run $\mathcal{L}$ =10,000 Monte Carlo trials for each SNR point to evaluate the bounds and the RMSE

$$\text{RMSE} = \sqrt{\frac{1}{\mathcal{L}K}\sum_{\ell=1}^{\mathcal{L}}\sum_{k=1}^{K}\left(\hat{\theta}_{\ell,(k)} - \theta_{\ell,(k)}\right)^2} \tag{14.79}$$

of DOA estimation. Here, $\hat{\theta}_{\ell,(k)}$ is the estimate of $\theta_{\ell,(k)}$, the kth DOA after ordering process in the ℓth Monte Carlo trial. We assume a minimum interval of 10° in each $\boldsymbol{\theta}$ in multisource case to guarantee resolvable sources, i.e., all the K DOAs in each Monte Carlo trial are randomly generated with at least 10° separation. Without otherwise specification, in our simulations, we adopt $T = 40$ snapshots and a ULA with $M = 20$ half-wavelength inter-element spacing sensors.[5]

14.4.1 Example 1: Single-Source Estimation

In the first example, we consider a single-source DOA estimation where the existing ZZB in [29] is only valid for this case. We compare the derived ZZB (14.78) with the ZZB in [29] and plot the MUSIC algorithm as a reference. It is observed from Figure 14.3 that, compared with the ZZB in [29], the derived ZZB reaches the asymptotic region at a slightly higher SNR because of the extra approximation introduced in [29, Eq. (C.14)]. Similar trend can be found in the numerical bound shown in [29, Figure 4]. Besides, both the derived ZZB and the ZZB in [29] converge to the same APB and CRB in the *a priori* performance region and

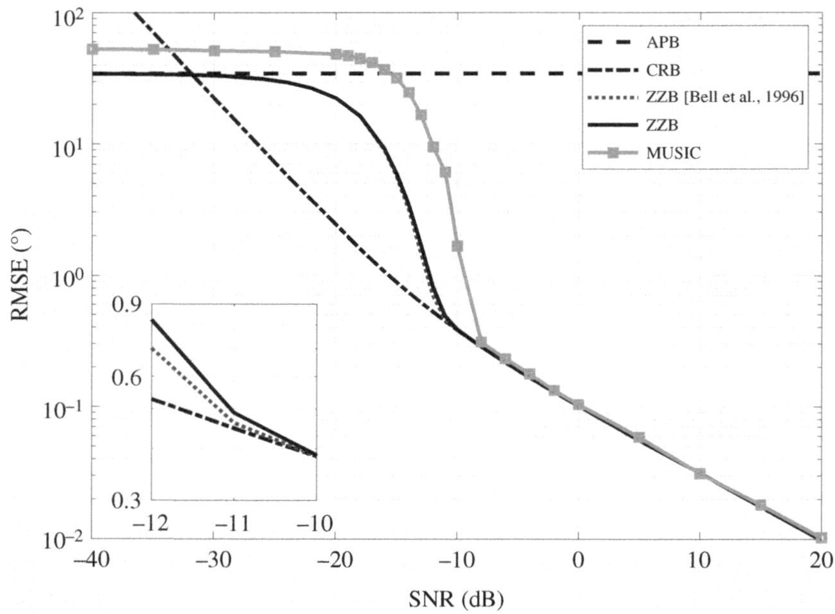

Figure 14.3 Comparison of bounds for single-source case ($K = 1$).

5 While the ULA is used here as an example in the simulations, other linear array configurations are also applicable.

the asymptotic region, respectively. Compared with the CRB, the ZZB is tighter outside the asymptotic region, and successfully predicts the threshold SNR of the asymptotic region for MUSIC algorithm. In contrast, the CRB does not provide valid bounds outside the asymptotic region and exceeds the MSE of the MUSIC algorithm in the *a priori* performance region.

14.4.2 Example 2: Uncorrelated Multisource Estimation

In the second example, we consider an uncorrelated multisource scenario. We assume that all K sources have the same SNRs. From Figure 14.4 it is observed that the generalized ZZB and the ZZB enter the asymptotic region at the same SNR and then reach the CRB in the asymptotic region. However, in the transition region, the generalized ZZB increases faster than the ZZB with the SNR decreasing, and the generalized ZZB converges to a higher level exceeding the MSE of the MUSIC algorithm. Hence, the generalized ZZB does not provide a valid bound in the non-asymptotic region. Similar to the single-source case, the CRB is still loose in the transition region and becomes invalid in the *a priori* performance region. On the contrary, the ZZB provides a global tight bound in both the asymptotic region and the non-asymptotic region, because the ZZB accounts for the effect

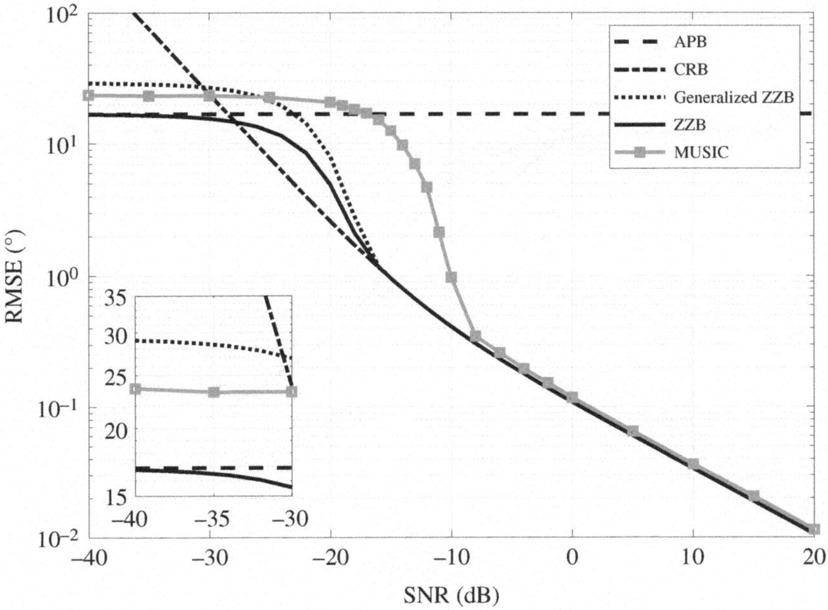

Figure 14.4 Comparison of bounds for multisource case ($K = 5$).

of ordering process in multisource DOA estimation. Hence, the ZZB is the only global tight bound for multisource DOA estimation. Besides, compared with the single-source case in Figure 14.3, the asymptotic region threshold in Figure 14.4 appears at a lower SNR. The reason is that, with K increasing, the coefficients $2P_L$ and $\Gamma_{\frac{3}{2}}(\tilde{u})$ more rapidly reach 0 and 1, respectively.

14.4.3 Example 3: The Effect of Number of Snapshots

In the third example, we investigate the effect of the number of snapshots T on the ZZB, where other simulation settings are kept same as Example 2. In Figure 14.5, regardless of the number of snapshots, the ZZB converges to the same APB in the *a priori* performance region because the APB does not depend on the number of snapshots. Even so, it is preferred to adopt more snapshots to make the ZZB leave the *a priori* performance region at a lower SNR. On the other hand, the ZZB reaches the CRB at a lower SNR with T increasing, bringing a narrower transition region. It can be observed that, when the number of snapshots doubles, the threshold SNR of the asymptotic region decreases by about 3 dB, thus enabling us to predict the threshold SNR of the ZZB for different numbers of snapshots.

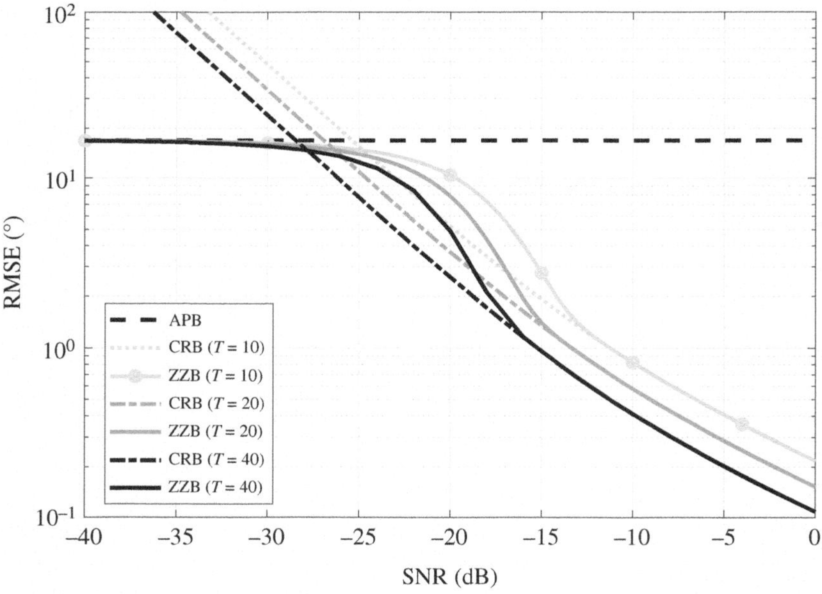

Figure 14.5 Effect of the number of snapshots on ZZB.

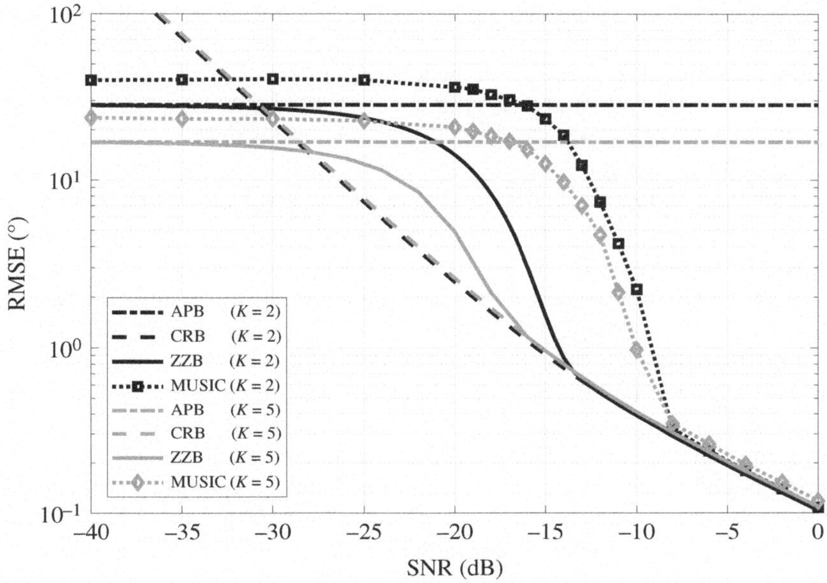

Figure 14.6 Effect of the number of sources on ZZB.

14.4.4 Example 4: The Effect of Number of Sources

In the fourth example, we evaluate the effect of the number of sources K on the ZZB, and other simulation settings also follow those in Example 2. It is observed from Figure 14.6 that the ZZB for different numbers of sources K always lower bounds the MSE of the corresponding MUSIC algorithm and converges to the corresponding APB (14.70) and the CRB in the *a priori* performance region and the asymptotic region. The ZZB reaches the corresponding CRB at a lower SNR for larger number of sources K because $\Gamma_{\frac{3}{2}}(\tilde{u})$ increases to 1 more rapidly with a larger K. Also, the ZZB converges a lower APB and a higher CRB in the *a priori* performance region and the asymptotic region, respectively. Obviously, different from the ZZB that always converges to the APB and provides global tight bound, the CRB becomes invalid in the non-asymptotic region.

14.4.5 Example 5: Hybrid Coherent/Uncorrelated Multisource Estimation

In the fifth example, we investigate the ZZB for the hybrid coherent/uncorrelated multisource DOA estimation. $K = 5$ sources are assumed. The first $L = 3$ sources are mutually coherent with the coefficient $\boldsymbol{\beta} = [1, 0.9e^{j\phi_1}, 0.8e^{j\phi_2}]$, and the last $K - L = 2$ sources are uncorrelated. Both ϕ_1 and ϕ_2 are random phases with a uniform distribution $\mathcal{U}[-\pi, \pi]$. For better demonstration, we also plot corresponding uncorrelated bounds with the same sources power as a reference, i.e., the SNRs are $\eta_2 = |\beta_2|^2\eta_1$, $\eta_3 = |\beta_3|^2\eta_1$, $\eta_4 = \eta_5 = \eta_1$, respectively.

In Figure 14.7, both the coherent ZZB and uncorrelated ZZB converge to the same APB in the *a priori* performance region, because the coherence coefficient does not affect the APB. In the asymptotic region, coherent ZZB and uncorrelated ZZB respectively reach the corresponding CRBs. Here, compared with the uncorrelated ZZB, the coherent ZZB is lower in the transition region. The reason is that, as shown in Figure 14.2, the coefficient $2P_L$ for coherent ZZB varies more rapidly than that for the uncorrelated ZZB. Thus, the coherent ZZB leaves from the APB more rapidly. Actually, compared with the uncorrelated sources, the coherence among the sources brings more determinacy in the estimation, making the coherent model being closer to the deterministic model. Thus, the coherent bound is lower than the uncorrelated bound theoretically. It is noted that the estimators usually require extra steps to eliminate the effect of coherence and realize effective estimation, during which the estimation accuracy may be sacrificed to some extent. However, the coherent ZZB implies that the coherence among sources may be utilized to realize better estimation accuracy.

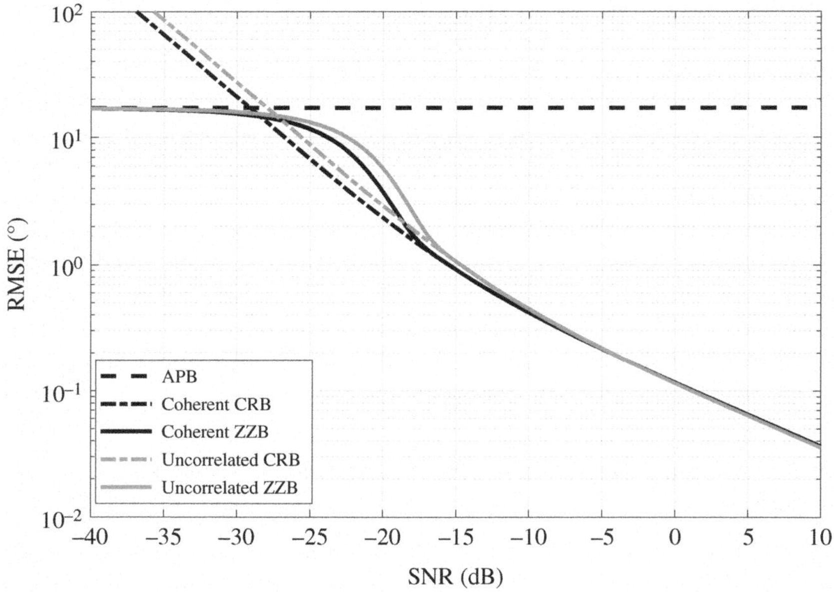

Figure 14.7 Effect of the mutual coherence on ZZB.

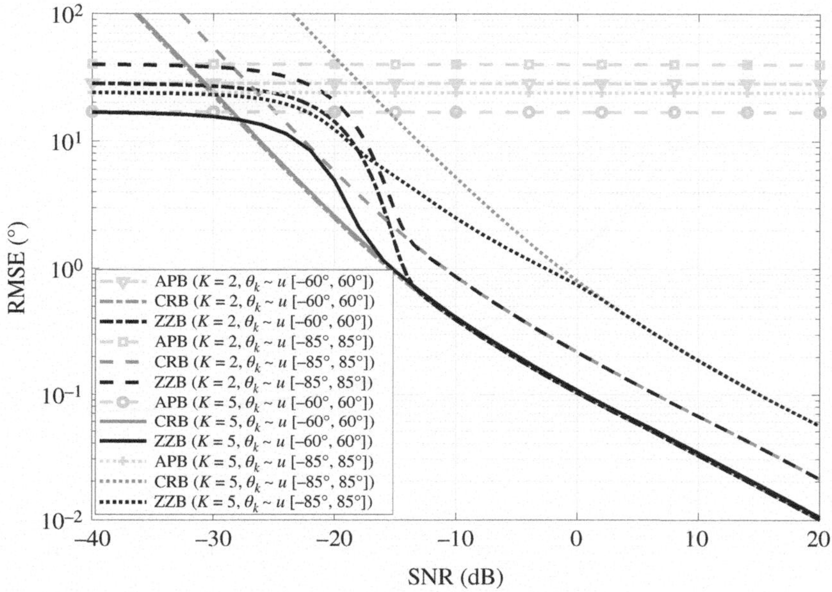

Figure 14.8 Effect of the *a priori* distribution of DOAs on ZZB.

14.4.6 Example 6: The Effect of *A Priori* Parameter Range

In the sixth example, we investigate the effect of the *a priori* distribution of DOAs on the ZZB for different numbers of sources. Since the ULA does not perform well when the directions are close to the array endfires, we consider the *a priori* uniform distribution with two different ranges $\mathcal{U}[-85°, 85°]$ and $\mathcal{U}[-60°, 60°]$ in Figure 14.8. Obviously, when the number of sources is fixed, the wider the *a priori* parameter range, the larger ZZB. On the other hand, the difference between the ZZBs with two difference *a priori* parameter ranges becomes larger in the asymptotic region with the number of sources increasing. Obviously, the effective array aperture is reduced for the range outside of $[-60°, 60°]$. When there are multiple impinging sources, the estimation accuracy becomes worse.

14.4.7 Example 7: Coprime Array Underdetermined Estimation

The ZZB is applicable for underdetermined DOA estimation with virtual coarray by incorporating the coarray Fisher information matrix into (14.69). Here, we consider a (3, 5) pair coprime linear array with a total number of $M = 2 \times 3 + 5 - 1 = 10$ physical sensors [44, 45] to estimate $K = 11$ sources following *a priori* distribution $\mathcal{U}[-60°, 60°]$. In each Monte Carlo trial, the DOAs are sampled with a least

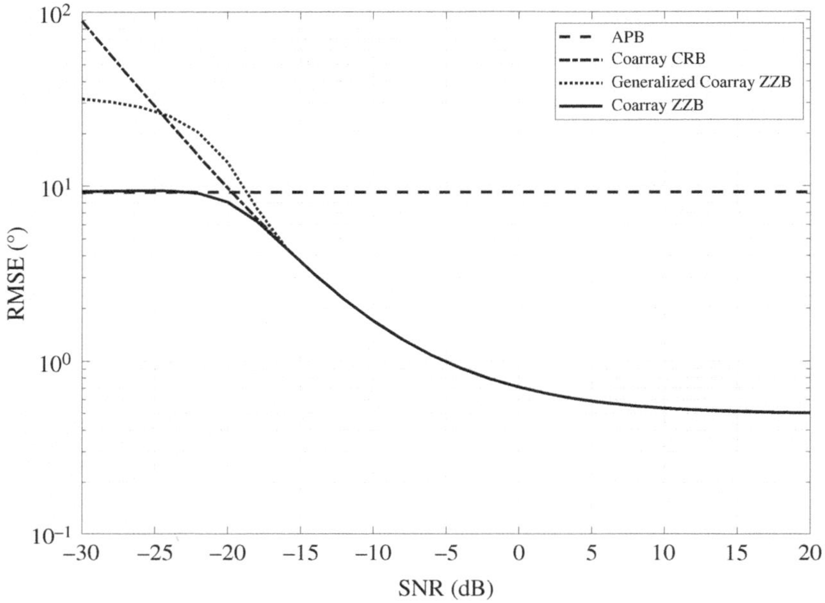

Figure 14.9 ZZB for underdetermined DOA estimation.

separation 5° from each other. Also, we plot the coarray CRB and the generalized coarray ZZB in Figure 14.9 for references. We assume there is no coherence among different sources, and each source has the same source power. It is observed that, in the asymptotic region, the coarray ZZB reaches the coarray CRB and becomes flatten out with the SNR increasing. Another observation is, different from coarray CRB, the coarray ZZB converges to the APB in the *a priori* performance region such that the ZZB provides a unified global bound for both overdetermined DOA estimation and underdetermined DOA estimation.

14.5 Conclusions and Future Directions

In this chapter, we provided an explicit ZZB expression for multisource DOA estimation with a hybrid coherent/uncorrelated unconditional model. After generalizing the ZZB derivation for single-source estimation to multisource estimation, we analyzed the effect of the ordering process between the estimates and true DOAs during MSE calculation on the MSE convergence in the non-asymptotic region, from which the order statistics are incorporated into ZZB derivation and makes the derived ZZB global tight for multisource DOA estimation. The derived ZZB revealed how the number of sources affects the MSE convergence and revealed that the MSE convergence in the *a priori* performance region does not depend on the number of snapshots and the array configuration. Extensive simulations were performed to investigate the advantages of the ZZB over existing CRB over a wide range of SNRs.

It is noted that the derived ZZB in this chapter considers one-dimensional DOA estimation problems, where the ordering process is a well-accepted criterion to realize the good match between the estimates and the true DOAs. However, for two-dimensional multisource DOA estimation problem, it is preferred to adopt the two-dimensional minimum Euclidean distance criterion rather than the ordering process along either azimuth or elevation, which brings much more complicated effect on the MSE convergence in the *a priori* performance region and cannot be described via the order statistics. We believe there remain challenges for ZZB derivation in two-dimensional multisource DOA estimation.

Acknowledgment

The work of Zongyu Zhang and Zhiguo Shi was supported by the National Natural Science Foundation of China under Grant U21A20456.

References

1 Van Trees, H.L. (2002). *Detection, Estimation, and Modulation Theory, Part IV: Optimum Array Processing*. Wiley.
2 Schmidt, R.O. (1986). Multiple emitter location and signal parameter estimation. *IEEE Transactions on Antennas and Propagation* 34(3): 276–280.
3 Rao, B.D. and Hari, K.V.S. (1989). Performance analysis of root-MUSIC. *IEEE Transactions on Acoustics, Speech, and Signal Processing* 37(12): 1939–1949.
4 Liu, C.-L. and Vaidyanathan, P.P. (2015). Remarks on the spatial smoothing step in coarray MUSIC. *IEEE Signal Processing Letters* 22(9): 1438–1442.
5 Roy, R. and Kailath, T. (1989). ESPRIT-estimation of signal parameters via rotational invariance techniques. *IEEE Transactions on Acoustics, Speech, and Signal Processing* 37(7): 984–995.
6 Han, F.-M. and Zhang, X.-D. (2005). An ESPRIT-like algorithm for coherent DOA estimation. *IEEE Antennas and Wireless Propagation Letters* 4: 443–446.
7 Chen, F.-J., Kwong, S., and Kok, C.-W. (2010). ESPRIT-like two-dimensional DOA estimation for coherent signals. *IEEE Transactions on Aerospace and Electronic Systems* 46(3): 1477–1484.

8 Malioutov, D., Çetin, M., and Willsky, A.S. (2005). A sparse signal reconstruction perspective for source localization with sensor arrays. *IEEE Transactions on Signal Processing* 53(8): 3010–3022.

9 Zhou, C., Gu, Y., Fan, X. et al. (2018). Direction-of-arrival estimation for coprime array via virtual array interpolation. *IEEE Transactions on Signal Processing* 66(22): 5956–5971.

10 Zhou, C., Gu, Y., Shi, Z., and Zhang, Y.D. (2018). Off-grid direction-of-arrival estimation using coprime array interpolation. *IEEE Signal Processing Letters* 25(11): 1710–1714.

11 Elbir, A.M. (2020). DeepMUSIC: multiple signal classification via deep learning. *IEEE Sensors Letters* 4(4): 1–4.

12 Barthelme, A. and Utschick, W. (2021). A machine learning approach to DoA estimation and model order selection for antenna arrays with subarray sampling. *IEEE Transactions on Signal Processing* 69: 3075–3087.

13 Papageorgiou, G.K., Sellathurai, M., and Eldar, Y.C. (2021). Deep networks for direction-of-arrival estimation in low SNR. *IEEE Transactions on Signal Processing* 69: 3714–3729.

14 Van Trees, H.L. and Bell, K.L. (2007). *Bayesian Bounds for Parameter Estimation and Nonlinear Filtering/Tracking*. New York: Wiley.

15 Cramér, H. (1946). *Mathematical Methods of Statistics*. Princeton University Press.

16 Stoica, P. and Nehorai, A. (1989). MUSIC, maximum likelihood and Cramér-Rao bound. *IEEE Transactions on Acoustics, Speech, and Signal Processing* 37(5): 720–741.

17 Stoica, P. and Nehorai, A. (1990). Performance study of conditional and unconditional direction-of-arrival estimation. *IEEE Transactions on Acoustics, Speech, and Signal Processing* 38(10): 1783–1795.

18 Wang, M. and Nehorai, A. (2017). Coarrays, MUSIC, and the Cramér-Rao bound. *IEEE Transactions on Signal Processing* 65(4): 933–946.

19 Liu, C.-L. and Vaidyanathan, P.P. (2017). Cramér-Rao bounds for coprime and other sparse arrays, which find more sources than sensors. *Digital Signal Processing* 61: 43–61.

20 Renaux, A., Forster, P., Larzabal, P. et al. (2008). A fresh look at the Bayesian bounds of the Weiss-Weinstein family. *IEEE Transactions on Signal Processing* 56(11): 5334–5352.

21 Weiss, A.J. and Weinstein, E. (1985). A lower bound on the mean-square error in random parameter estimation. *IEEE Transactions on Information Theory* 31(5): 680–682.

22 Reuven, I. and Messer, H. (1997). A Barankin-type lower bound on the estimation error of a hybrid parameter vector. *IEEE Transactions on Information Theory* 43(3): 1084–1093.

23 Renaux, A., Forster, P., Larzabal, P., and Richmond, C.D. (2006). The Bayesian Abel bound on the mean square error. *Proceedings of IEEE International Conference on Acoustics, Speech, and Signal Processing*, volume 3, III. Toulouse, France.

24 Bobrovsky, B.Z. and Zakai, M. (1976). A lower bound on the estimation error for certain diffusion processes. *IEEE Transactions on Information Theory* 22(1): 45–52.

25 Ziv, J. and Zakai, M. (1969). Some lower bounds on signal parameter estimation. *IEEE Transactions on Information Theory* 15(3): 386–391.

26 Seidman, L.P. (1970). Performance limitations and error calculations for parameter estimation. *Proceedings of the IEEE* 58(5): 644–652.

27 Chazan, D., Zakai, M., and Ziv, J. (1975). Improved lower bounds on signal parameter estimation. *IEEE Transactions on Information Theory* 21(1): 90–93.

28 Gu, X. and Wong, M. (1991). A modified Ziv-Zakai lower bound and its application in array processing. *Proceedings of International Conference on Acoustics, Speech, and Signal Processing*, volume 2, 1477–1480. Toronto, ON, Canada.

29 Bell, K.L., Ephraim, Y., and Van Trees, H.L. (1996). Explicit Ziv-Zakai lower bound for bearing estimation. *IEEE Transactions on Signal Processing* 44(11): 2810–2824.

30 Bell, K.L., Steinberg, Y., Ephraim, Y., and Van Trees, H.L. (1997). Extended Ziv-Zakai lower bound for vector parameter estimation. *IEEE Transactions on Information Theory* 43(2): 624–637.

31 Sadler, B.M., Liu, N., and Xu, Z. (2010). Ziv-Zakai bounds on time delay estimation in unknown convolutive random channels. *IEEE Transactions on Signal Processing* 58(5): 2729–2745.

32 Khan, D. and Bell, K.L. (2013). Explicit Ziv-Zakai bound for analysis of DOA estimation performance of sparse linear arrays. *Signal Processing* 93(12): 3449–3458.

33 Decarli, N. and Dardari, D. (2014). Ziv-Zakai bound for time delay estimation of unknown deterministic signals. *Proceedings of IEEE International Conference on Acoustics, Speech, and Signal Processing*, 4673–4677. Florence, Italy.

34 Chiriac, V.M., He, Q., Haimovich, A.M., and Blum, R.S. (2015). Ziv-Zakai bound for joint parameter estimation in MIMO radar systems. *IEEE Transactions on Signal Processing* 63(18): 4956–4968.

35 Mishra, K.V. and Eldar, Y.C. (2017). Performance of time delay estimation in a cognitive radar. *Proceedings of IEEE International Conference on Acoustics, Speech, and Signal Processing*, 3141–3145. New Orleans, LA.

36 Laas, T. and Xu, W. (2021). On the Ziv-Zakai bound for time difference of arrival estimation in CP-OFDM systems. *Proceedings of IEEE Wireless Communications and Networking Conference (WCNC)*, 1–5. Nanjing, China.

37 Zhang, Z., Shi, Z., Zhou, C. et al. (2022). Ziv-Zakai bound for compressive time delay estimation. *IEEE Transactions on Signal Processing* 70: 4006–4019.

38 Yang, Z., Stoica, P., and Tang, J. (2019). Source resolvability of spatial-smoothing-based subspace methods: a Hadamard product perspective. *IEEE Transactions on Signal Processing* 67(10): 2543–2553.

39 Zheng, Z., Huang, Y., Wang, W.-Q., and So, H.C. (2020). Direction-of-arrival estimation of coherent signals via coprime array interpolation. *IEEE Signal Processing Letters* 27: 585–589.

40 Zheng, H., Zhou, C., Shi, Z., and Gu, Y. (2022). Structured tensor reconstruction for coherent DOA estimation. *IEEE Signal Processing Letters* 29: 1634–1638.

41 Zhang, Z., Shi, Z., and Gu, Y. (2023). Ziv-Zakai bound for DOAs estimation. *IEEE Transactions on Signal Processing* 71: 136–149.

42 Van Trees, H.L. (2001). *Detection, Estimation, and Modulation Theory, Part III: Radar-Sonar Signal Processing and Gaussian Signal in Noise*. Wiley.

43 Casella, G. and Berger, R.L. (2002). *Statistical Inference*. Thomson Learning.

44 Pal, P. and Vaidyanathan, P.P. (2011). Coprime sampling and the MUSIC algorithm. *Proceedings of IEEE Digital Signal Processing Workshop / IEEE Signal Processing Education Workshop (DSP/SPE)*, 289–294. Sedona, AZ.

45 Zhang, Z., Zhou, C., Gu, Y. et al. (2019). An IDFT approach for coprime array direction-of-arrival estimation. *Digital Signal Processing* 94: 45–55.

Index

d

e

m

n

q

t

Wiley The manufacturer's authorized representative according to the EU General Product Safety Regulation is Wiley-VCH GmbH, Boschstr. 12, 69469 Weinheim, Germany, e-mail: Product_Safety@wiley.com.

Printed and bound by CPI Group (UK) Ltd, Croydon, CR0 4YY
19/08/2026
02201088-0002